INTERNATIONAL JOURNAL OF APPLIED ELECTROMAGNETICS AND MECHANICS

INTERNATIONAL JOURNAL OF

APPLIED ELECTROMAGNETICS AND MECHANICS

Editor-in-Chief
T. Takagi

Regional Editor
G.A. Maugin

Volume 13, 2001/2002

IOS Press
Ohmsha

Amsterdam • Washington, DC • Tokyo

Printed in the Netherlands

CONTENTS

International Journal of Applied Electromagnetics and Mechanics 13 (2001/2002) 1–2
IOS Press

Preface

The present and forthcoming volumes of the *International Journal of Applied Electromagnetics and Mechanics* contain the reviewed papers of the ISEM01.

The tenth International Symposium on Applied Electromagnetics and Mechanics, ISEM-Tokyo, was held at Toshi Center Hotel in Tokyo from May 13th through 16th in 2001. It was organized by the Japan Society of Applied Electromagnetics and Mechanics and the University of Tokyo. The ISEM is now a well known international forum on applied electromagneics. It started in 1988 in Japan and since then was held in Japan for five times and once in Korea, the United Kingdom, Germany and Italy. This tenth ISEM-Tokyo, back to Japan again, was planned to a tribute to Professor Kenzo Miya to honour his achievements as ISEM's creator and its first chairperson.

The symposium was composed of fourteen organized sessions and it covers the following topics in the area of applied electromagnetics and mechanics as; Strategies for new research and their realization, Advanced dynamics and control, Advanced magnetic engineering, Advanced magnetic fusion nuclear research, Analysis and control of magnetic field, Applied superconductivity, Beam science and technology, Biomagnetics and medical engineering, Electromagnetic functional materials and adaptive systems, Electromagnetic nondestructive evaluation, Electromagnetic sensors and actuators, Inverse problems and new application, NDE and Magnetic detection of material degradation and damage evaluation, and Virtual reality and smart visualization.

During the symposium 334 papers were presented including three plenary lectures, 36 invited lectures, three panel presentations and 292 poster presentations. Three plenary lectures are: " New trend of science and technology in the 21th century" by Prof. Akito Arima, a member of the House of Councilors, "The next step forward in magnetic fusion research: ITER" by Dr. Robert Aymar of ITER Garching Joint Work Site, "An effort for construction of universum in technology" by Prof. Kenzo Miya of Keio University. These three lectures stimulated us to move to a new research direction in the 21st century. More than 370 participants registered from 25 countries and had a fruitful discussion in the lounge and at the banquet as well as in the session.

Papers were submitted to ISEM-Tokyo in two steps. The proceedings book, which was published as "Applied Electromagnetcis and Mechanics" in JSAEM Studies in Applied Electromagnetics and Mechanics 9, and was distributed at the symposium, comprised more than 350 papers which were reviewed for the first step two-page papers. In the second step 245 extended full papers were submitted at the symposium site and were reviewed at least two referees. Finally 221 papers were accepted for the publication in these special issues of the International Journal of Applied Electromagnetics and Mechanics.

We would like to express our sincere gratitude to the principal sponsors, namely The Commemorative Association for the Japan World Exposition, Tokyo Convention & Visitors Bureau, The Kajima Foundation, The Asahi Glass Foundation, The Ministry of Education, Culture, Sports, Science and Technology, and Atomic Energy Society of Japan. We are now confident that the ISEM-Tokyo was successful due to their generous contribution and that the success will be followed by the ISEM-Versailles to be chaired by Prof. D. Lesselier, LSS and Prof. A. Razek, LGEP in May 2003 in France.

Finally we should like to express our sincere thanks to session organizers and referees those have contributed to the publication of the ISEM-Tokyo Proceedings and these special issue papers, especially to the chairpersons of Editorial Board, Prof. K. Nagaya, Gunma University and Prof. I. Nishiguchi, Kanagawa Institute of Technology. We also sincerely appreciate Ms. Junko Saito of ISEM-Tokyo Secretariat for her extra effort in the editing process.

Toshiyuki Takagi
Academic Chairperson
Tohoku University, Japan

Mitsuru Uesaka
Conference Chairperson
The University of Tokyo, Japan

International Journal of Applied Electromagnetics and Mechanics 13 (2001/2002) 3–11
IOS Press

Neural networks for environmental data processing: a research project of relevant national interest for the Italian Government[1]

Pietro Burrascano
Project's scientific co-ordinator Department Ingegneria Industriale – Univ. Perugia; Via G. Duranti 1/A-4 I-06125 Perugia, Italy
E-mail: burrasc@unipg.it; Fax: +39 0744 492925

Abstract. A group of Italian researchers in the field of neural processing systems proposed to the Italian government a project in the field of Neural Networks for environmental data processing. The project has been funded, and started to be operative at the beginning of 2001. A major objective of the project is that of focusing the potential of the research group in the field of neural modelling to the problem of environmental management in order to produce, at the end of the project period, powerful prediction tools in environmental applications. Air and electromagnetic pollution together with water resources management are the specific environmental problems to which the predictive neural models will be particularised and applied.

1. Introduction

The awareness of the ever-increasing effects of human civilisation on the environmental equilibrium is growing up to a noticeable interest, either in the scientific community and in the public opinion, about the topics concerning the sanitary, economical and social impact of environmental problems. Topics as chemical, electromagnetic, and acoustic pollution and about the re-equilibrium of the available environmental resources are nowadays carried to the attention of everyone. The importance of the implied phenomena has moved several initiatives, also concerning the definition of appropriate limiting laws. The involved physical phenomena are rather complex, and this limits the availability of reliable prediction instruments useful for both monitoring in an efficient way the present situation as well as for predicting the effects produced by some modifications oriented to healing or to building-up new installation.

It is thus very important to have available forecasting models. The availability of such instruments, which also take into account the simultaneous presence of several pollution agents, is of great importance and usefulness. In particular it can orient short-term as well as long-term interventions, as stopping the vehicular traffic, resizing electromagnetic sources, planning of territorial exploitation and localisation of new settlings.

Such a kind of forecasting activity implies to understand the interaction between different agents, and to model these dependencies as well as to build-up prototypes able to simulate the behaviour of complex

[1]This work has been financially supported by the italian government (MURST Projects).

systems with the aim of performing forecasting and estimation. Neural networks proved in several applications to be an effective instrument for non-linear multiple-variable interaction modelling. It has been indeed proved by many practical and theoretical studies that artificial neural networks result to be particularly well suited to complex non-linear function approximation. This characteristic is maintained even when only some values of these function, corresponding to some combinations of the independent variable, are known; such ability generally termed generalisation property, refers to the capability of approximating the mathematical law that explains the observed data in the best possible way.

The target of the present research project is the employment of neural techniques for defining models able to link a set of variables, used as gauge of the phenomenon under analysis, to a specific kind of environmental effect. This result makes available forecasting instruments to be employed for decision support.

In the following a brief description is given of the project organisation; then some sketches of a few among early results obtained in the different fields will be summarised.

2. Structure of the project and main research topics

The research project "Neural Networks for environmental data processing" co-ordinates the activities of researchers from seven different university laboratories: the universities of Ancona, Bari, Catania, Cagliari, Perugia, Reggio Calabria, Roma "La Sapienza" are involved in the project. The different research units develop specific tasks defined within three main streams: forecasting of gaseous and electromagnetic pollutant and optimal management of hydrological resources. A task of the overall project is, moreover, the development of hardware modules for real-time operations.

A fundamental aspect of the project is its practical usefulness: consequently a relevant aspect of the project is to make available measured data and share them among the different units. The quality of the developed neural forecaster will be thus tested on common real world data. The latter aspect was considered of relevance in defining the project, and connections with existing projects were pursued.

About gaseous pollutant prediction, the main data source comes from the measurements taken by means of monitoring stations located in particularly critical zones on the territory. The connection with the SMAURN project (Urban Environment Monitoring System With Neural Networks – In Italian "Sistema di Monitoraggio dell'Ambiente Urbano di Villa S.G. con l'utilizzo delle Reti Neurali") is particularly important, because it allows for the availability of multivariate environmental data; this connection is granted from the Reggio Calabria unit.

It is well known that one among the main causes of air pollution comes from the vehicular traffic: it has thus been considered very important to insert expertise granting the correlation of transportation demand to the availability of the transportation network. Forecasting models allow for predicting the levels of vehicular traffic on the basis of the transportation demand, and thus to infer the levels of produced pollutant. Such contribution is granted from the units of Roma "La Sapienza" and Reggio Calabria.

A strong relationship with other national projects has been defined also for the parts of the project concerning the electromagnetic pollutant. Two pre existing local projects contribute to the present one: the ANERT project, through the Catania unit, and the similar one developed by Perugia unit in collaboration with Regione Umbria: both of them are finalised to the construction of databases about the electromagnetic sources currently active on the territory. The first target is to employ such databases to obtain local maps of field levels; conversely, such interaction can give to the management of such projects a useful information about the characteristics of the sources needed for their better characterisation and fully detailed field map formation.

The unit of Cagliari will find useful support from the units of Bari, Perugia and Roma "La Sapienza" about the development of multivariate data processing algorithms applied to forecasting for optimal management of surface water fluxes and to underground water resources monitoring. This working group proposes the use of neuro-fuzzy algorithms, non-linear forecasting techniques and feature selection methods for data complexity reduction. The same group will operate to ultrasonic techniques for revealing the presence of biomass in the ground (in connection to the CNR-IMAE institute of Sassari).

Expertise about the hardware implementation has been included in the project, in order to implement some of the proposed algorithms: particular emphasis will be given to monitoring of gaseous pollutants, whose rapid temporal variation requires real-time signal processing facilities. The units of Ancona and Reggio Calabria will usefully co-operate by sharing their experience on this field.

The results obtained during the present year of activity will be presented in a workshop to be held in Perugia in January 2002. Information will be available at the address [1].

3. Early results

In the following a schematic description of some of the early results obtained in the different fields by the research units will be briefly described. Far from the idea of giving a sufficient technical description of the single aspects of the work in progress, the following part of this contribution is intended to give an idea of the specific topics considered in the project.

– Prediction of ELF electromagnetic environmental fields: *Catania and Perugia research units*

The problem addressed is to predict the ELF magnetic field distribution in the neighbourhood of electric plants, from a few measures taken along the walls and the soil surrounding the apparatus. For this purpose the Laplacian behaviour of each component Bk ($k = x, y, z$) of the stationary magnetic fields inside current-free regions is exploited. In this case the modulus of the ELF stationary magnetic field inside a finite volume Ω enclosed by a closed surface $\partial\Omega$ can be evaluated from the three magnetic field components on the boundary surface, provided that the relevant Green's function G for the considered domain is known:

$$\nabla^2 B_k = 0 \quad B_{k \text{ inside } \Omega} = \int_{\partial\Omega} G B_k d\Omega \quad k = x, y, z \tag{1}$$

The above allows for a direct computation of the field inside the volume starting from a number of measurements on the boundary. The Green function for a generic volume shape is generally not analytically known, except for few simple shapes: a Multilayer Perceptron (MLP) has been used to identify and approximate the appropriate Green function for such domains. This has permitted ELF magnetic field values to be predicted inside the volume based solely on input data consisting of a few experimental measures on the volume surfaces.

Figure 1 gives some results on synthetic and real data. It shows the results coming out from testing performed on synthetic data at 216 internal points for which a very satisfactory absolute MSE of 6.7 10^{-5} was obtained. The curves depicting simulated and predicted values almost completely overlap in the figure.

By using the trained MLP network several experiments for a case study are currently carried out, concerning the prediction of the ELF magnctic field intensity outside an electrical local substation. A complete ELF magnetic field map has been created from a limited number of measured data along the walls and the soil of the substation. The intensity of the predicted magnetic field component Bx taken

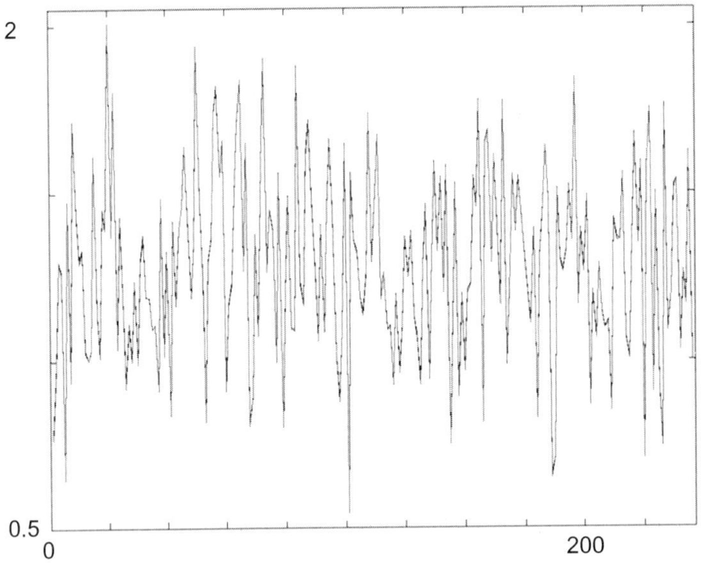

Fig. 1. prediction in 216 internal points -synthetic data; MSE 6.7 10^{-5}.

at a distance of 30 cm from the front wall of the electrical substation has been found to be in close agreement with the measured values

– O_3 and acoustic noise time series forecasting by a Twofold Neural Predictor: *Roma research unit*

Environmental data sequences are often characterised by a chaotic behaviour eventually corrupted by noise. The chaotic character of a signal is well measured by its embedding dimension D_s: it can be evaluated by means of the Spatio-Temporal Entropy (STE), which takes on values close to 50% in the case of chaotic signals of low dimensionality and tends to 100% in the case of noise.

Chaotic signals are predictable: let $S(n)$ be the signal of interest and $e(n)$ the prediction error $e(n) = S(n) - S'(n)$. If $S(n)$ is chaotic, then $e(n)$ will be more chaotic than $S(n)$, i.e., it will be characterised by an embedding dimension D_s larger than that of $S(n)$. From the above it can be supposed that there exists the possibility of improving the performance of a given predictor P by simply applying it also to the prediction error $e(n)$. The overall prediction accuracy can be thus improved by combining the prediction of the input sequence with that of its prediction error. A twofold predictor can be used in this regard. The tests sequences taken into account for the experiments are obtained by measuring the level of O_3 $[\mu g]$ and of acoustic noise $[dB]$ in a given location of the downtown of Rome (Italy) at a sampling rate of 5 minutes. The neural networks have been trained on the first 2000 samples of $S(n)$ and $e(n)$, while the predictor performances have been tested on the successive 600 samples. The prediction results of both experiments are reported in Table 2.

Sequence of O_3 values: the sequence $S(n)$ to be predicted is characterised by a chaotic behaviour ($STE = 58\%$; small value of the embedding dimension $D_s = 5$). The predictor of $S(n)$ produces error sequences with a high presence of noise (STE values close to 75% and limited accuracy of the predictor of $e(n)$). It is trained on an error training set characterised by $D_e = 7$.

The absolute performance of the predictor of $S(n)$ would be already satisfactory: however we obtain a relative improvement of 9.02% by using the twofold predictor. In fact, the $\mathrm{NMSE}_s = 1.33 \cdot 10^{-1}$ is reduced up to $\mathrm{NMSE}_t = 1.21 \cdot 10^{-1}$.

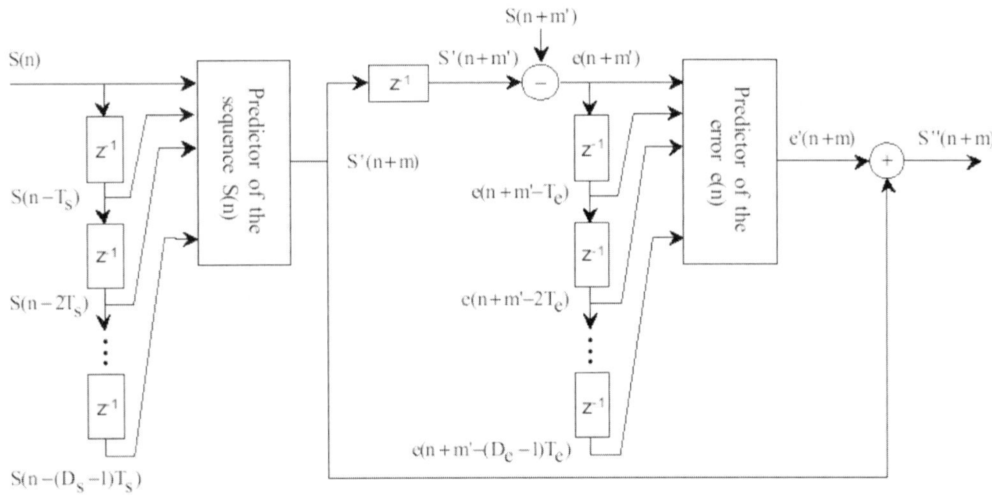

Fig. 2. Scheme of the proposed twofold predictor. The final improved predicted value is S”$(n + m)$.

Table 1
Prediction results on the sequence of O_3 and Acoustic noise values

Prediction task		O_3 values	Acoustic noise
Prediction of $S(n)$:	NMSE_s	$1.33 \cdot 10^{-1}$	$2.46 \cdot 10^{-1}$
Prediction of $e(n)$:	NMSE_e	$7.85 \cdot 10^{-1}$	$7.35 \cdot 10^{-1}$
Twofold prediction of $S(n)$:	NMSE_t	$1.21 \cdot 10^{-1}$	$1.82 \cdot 10^{-1}$

Sequence of acoustic noise values: the sequence manifest a partially chaotic behaviour, also corrupted by noise (high value of Ds and by STE $= 65\%$). The predictor of $S(n)$ produces error sequences characterised by STE values of about 70%, -prevalently noisy- with $D_e = 12$. The noise evidently present in $S(n)$ limits its prediction accuracy. Nevertheless, the prediction of $e(n)$ is sufficient to ensure an improvement of 26.01% in the twofold operation, since we pass from $\mathrm{NMSE}_s = 2.46 \cdot 10^{-1}$ to $\mathrm{NMSE}_t = 1.82 \cdot 10^{-1}$.

– Blind Separation for e.m. source localization and assessment: *Perugia research unit*

If the electromagnetic field measured in the environment and due to different emitting antennas exceeds the allowed levels, it is necessary to reduce the radiation from each emitter according to how much it contributes to the overall level. The Italian law, for instance, defines a detailed procedure to perform such reduction. In order to apply this procedure it is necessary to know what emitting antennas significantly contribute and what is the amount of their contribution; this, in turn, requires to know their localisation and emitting characteristics. Procedures must be defined in order to detect, localise and electromagnetically characterise such sources by means of field measurements.

The aim of blind source separation (BSS) is to separate out unobservable or latent source signals when mixed signals only are observed, with little or no knowledge about the source statistics nor their temporal characteristics: a mixture of independent source signals is supposed to be observed $m(t) = M^H s(t)$ where M is a constant full-rank m-by-m mixing matrix and $s(t)$ is the vector-stream containing the m source signals to be separated; the superscript ‘H’ denotes Hermitian transpose. For separating out the linearly mixed independent sources, we use a neural network with m inputs and m outputs, described by the relationships $o(t) = W^H m(t)$ where m is the network input vector, o denotes the output vector and

W is the weight-matrix. As the mixing model is linear, a linear separating structure is effective; thus the output $o(t)$ is taken as an estimate of the true source stream $s(t)$. Also, the matrix W may be retained as an estimate of the true mixing matrix M.

In the present application of Blind Source Separation we suppose the sources to be antennas, and the receivers to be electromagnetic probes. We further suppose the propagation to be cylindrical and both the sources and sensors radiation diagram to be hysotropic with good faith. In this case, each source T_i is described by a pair of coordinates (x_i^T, y_i^T) and each receiver R_i is described (x_i^R, y_i^R), $i = 1 \ldots m$. Then, the $(r; c)$th entry of the propagation matrix M is described by the phasor:

$$M_{rc} = \frac{1}{d_{rc}} \exp[-j\gamma d_{rc}]\ d_{rc} = \sqrt{(x_r^T - x_c^R)^2 + (y_r^T - y_c^R)^2} \tag{2}$$

where gamma denotes a constant of propagation and a two-dimensional space has been considered for notation simplicity. In the present context we need an algorithm that: (i) exhibits good estimation precision regardless of the computational burden, because the signal processing operations need not necessarily to be performed on-line, and (ii) allows complex-valued signals handling. For these reasons, we chose to employ the JADE (Joint Approximate Diagonalization of Eigen-matrices) package by Cardoso. It is worth remarking that the JADE algorithm estimates the source signals and normalize them to have unit power, so that the true source power hides in the rows of the estimated inverse mixing operator.

Let us suppose the inverse approximated mixing operator has been estimated through JADE: We now wish to design an algorithm for recovering the relevant information of the sources, namely their position and the power of emitted field. In our measurement model, we have many sources and only one probe, which we may move in known locations to take meaningful measures. The matrix containing the moduli of the inverse propagation operator writes:

$$\begin{bmatrix} A_1/d_{11}\ A_1/d_{12} \cdots \\ A_2/d_{21}\ A_2/d_{22} \cdots \\ \vdots \end{bmatrix}$$

where both the signal powers Ai and the distances d_{rc} are unknown. The dependence on the A_i may be easily eliminated by normalising each row about its first element, which gives rise to the knowledge of the ratios $D_{rc} = d_c r / d_{cc}$. If $m \geqslant 3$, we can easily locate each source by simple triangularisation. In our experiment we supposed $m = 4$, and the closed form expression of 1st source location is, for instance:

$$y_1^T = \frac{(8 - 3D_{4232})Y}{(4 - 2D_{4232})Y}; x_1^T \sqrt{\frac{3Y^3 - 2Y y_1^T}{4 - 2D_{4232)Y}}}; D_{4232} =^{def} \frac{D_{41}^2 - d_{21}^2}{D_{31}^2 - D_{21}^2} \tag{3}$$

We simulated the propagation model with $m = 4$ sources and by supposing the source messages to be narrow-band signals. Figure 3 shows the measured and the separated source signals, as provided by the JADE algorithm. The geometry of the antennas and of the sensors is depicted in Fig. 4, which also shows the estimated antennas positions. We also estimated the source carrier frequencies and amplitudes, which are, respectively: 72, 199, 206, 121 (true amplitudes 72, 198, 206, 121), 13.27, 14.84, 6.66, 17.59 (true 13.12, 14.84, 6.89, 17.68); both have been approximated with good faith.

Neural runoff models to forecast hydrogeologic risk: *Cagliari research unit*

River flow forecasts are required to provide basic information for reservoir management in a multipurpose water system optimisation framework. An accurate prediction of flow rates in tributary streams is

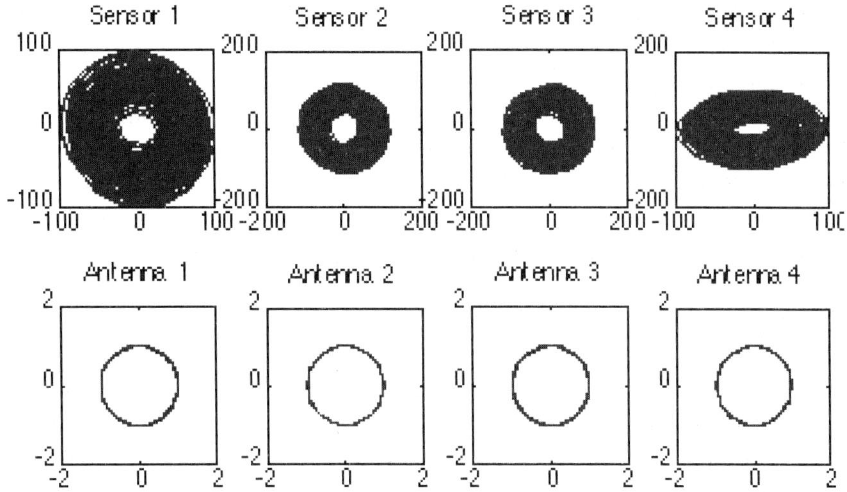

Fig. 3. Source separation (observed (top) and separated (bottom) complex signals).

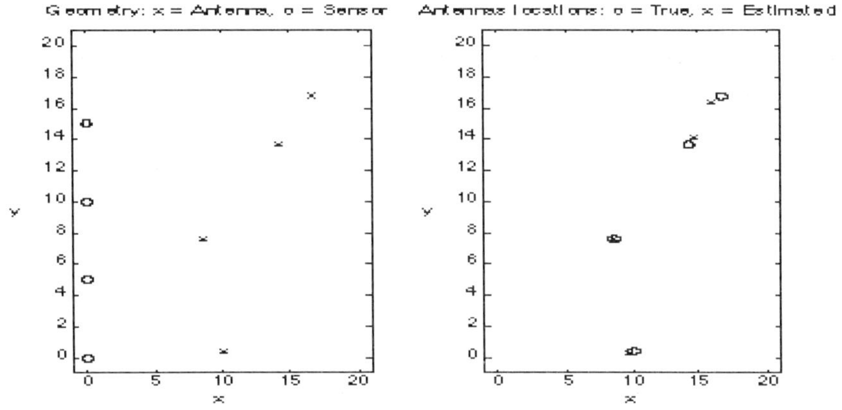

Fig. 4. Geometry of the problem (left) and estimated source positions (right).

Table 2
Prediction results (seasonal efficiency) with MLP and Locally Recurrent NN

N of prediction steps	MLP	LRNN
1	−0.2354	0.2303
2	−0.3353	0.0868
3	−0.3313	0.0622

basically required during flood events, to control flow-peak discharge, as well as to optimise management of water resources, considering extended time horizons.

Recently, the problem of forecasting the river run-off has been approached on the basis of known historical time series of the run-off flow, and by using different neural network architectures to model a run-off $\rightarrow$ run-off problem. The research group compared a MLP and Locally Recurrent Neural Network predictor on a data set measured in the "Flumendosa" (Sardegna) hydrologic basin. The data

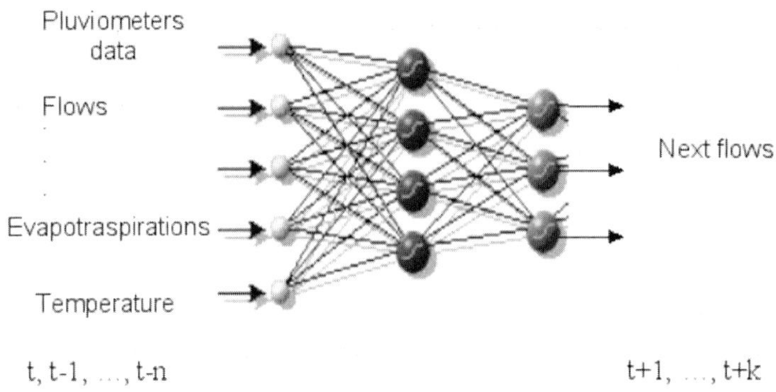

Fig. 5. structure of the MLP for the Tirso experiment.

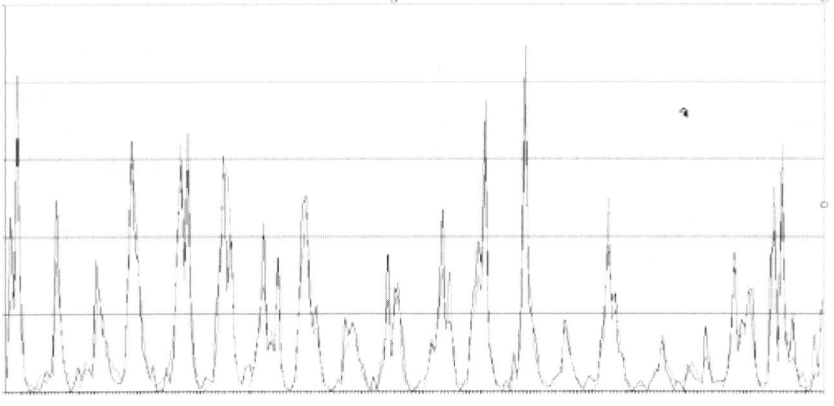

Fig. 6. MLP prediction results for the Tirso experiment.

set consists of monthly flows: the data from October 1922 to September 1977 (55 hydrological years–660 monthly values) were used for training, and the monthly flows from October 1978 to September 1997 (20 hydrological years–240 monthly values) were used for validation. Table 2 reports the prediction results obtained: the MLP results to be a poor predictor for the run-off → run-off problem, while the LRNN shows to be able to some howcpture the dynamic of the time series.

A different experiment has been performed for a rain-fall → run-off k-step forecasting problem. The data set was relative to of the Tirso river, in Sardinia (Italy) for which detailed data from several hydrometers and pluviometers along the basin were also available. A MLP with pluviometric, flow, evapo-transpiration and temperature data sequences as inputs produces an accurate flow prediction, as shown in Fig. 6.

4. Conclusions

The use of neural models to the complex, non-linear, multivariate environmental problems can represent an interesting benchmark for their possible wide application to a class of real world problems of increasing importance. The activities of the project "Neural Networks for environmental data processing" begun in Italy in the last few months. The project tries to stress the potentials of neural models in this field and to give a unitary view of the different neural modelling techniques in the perspective of their use for approach environmental problems.

The project hopefully opens an application field for neural modelling systems for which interesting international co-operation can be envisaged.

References

[1] http://www.unipg.it/burrasc/#Research

[2] M.W. Gardner et al., Neural Network Modelling And Prediction Of Hourly NOx And NO2 Concentrations In Urban Air In London,, *Atmospheric Environment* **33** 709–719.

[3] D. Bastarache, N. El-Jabi, N. Turkkan and T.A. Clair, Predicting conductivity and acidity for small streams using neural networks, *Revue canadienne de g'enie civil* **24** (1997), 1030–1039.

[4] A.M. Buch, H.S. Mazumdar and P.C. Pandey, Application of artificial neural networks in hydrological modeling, a case study of runoff simulation of a himalayan basin, *Proceedings of the International Joint Conference on Neural Networks* **1** (1993), 971–974.

[5] P. Burrascano, S. Fiori and M. Mongiardo, A Review of Artificial Neural Networks Applications in Microwave CAD, *Int. Journal of RF and Microwave CAE* **9**(3) (1999), 158–174,

[6] A. Fanni, S. Liberatore, G.M. Sechi, M. Soro and P. Zuddas, Tabu Search metaheuristic for water distribution system optimization, *Computing Tools for Modeling, Optimization and Simulation, Kluwer Academic Publishers* (2000), 279–298

[7] D.Marino and F.C. Morabito, *A Neural Network Based Urban Environment Monitoring System Exploiting Previous Knowledge*, Neural Nets, WIRN,Vietri-99, Proc. 11th Workshop, Springer Verlag, ISBN/ISSN: 1-85233-177-1, 1999, pp. 323–328.

[8] P. Campolucci, A.Uncini, F. Piazza and B.D. Rao, On-line Learning algorithms for Locally Recurrent Neural Networks, *IEEE Transactions on Neural Networks* **10** (1999).

International Journal of Applied Electromagnetics and Mechanics 13 (2001/2002) 13–18
IOS Press

Speech recognition using the extraction of particular feature by the discrete wavelet transform

Yoichi Midorikawa* and Masanori Akita
Department of Electrical and Electronic Engineering, Faculty of Engineering, Oita University, 700 Dannoharu, Oita 870-1192, Japan

Abstract. We have proposed the method for recognition of the signals polluted with noise using modification of spectral envelopes shape. However, these spectral envelopes have wrong valleys. For that reason recognition rate is not improved. There is plenty of scope for improvement. In this paper, we propose that method for extracting particular feature by the discrete wavelet transform. (Wavelet transform is widely used for the extraction of particular feature from data.) Pattern recognition rate is improved by the new method. It was shown that the wavelet analysis was one of the promising methodologies for the pattern recognition noisy signal.

1. Introduction

The pattern recognition is important for human interface with computer and machine. Speech is the most natural form of human communication. Speech recognition is used in recent electrical products (e.g. pet robots, word processor and telephone etc.). Speech recognition technology has made it possible for computers to follow human voice commands and even understand human languages. The pattern recognition (e.g. speech recognition) is one of important technology in this century. However, these pattern recognition methods have a weak point. These methods are used in the pattern recognition under the noiseless condition. Therefore, we took notice of this point. We have proposed the method for recognition of the signals polluted with color noise using modification of spectral envelopes shape. The proposed emphasis method is based on cepstral analysis method. One of the emphasis methods of spectral envelopes is performed using filtering by cepstra. We have proposed the modified rules for adding valleys are defined. These methods are applied to the enhancement system of noisy signals and the performance of the system is improved [1,2].

We have researched applications of the discrete wavelet transform [3,4]. Discrete wavelets transform is widely used for the wave and image analysis. Wavelet analysis is extremely useful tools for the noise reduction and data compression [5–8]. In this paper, we apply a wavelet transform to the pattern recognition under the noisy environment using cepstral analysis. As a result, speech recognition rate is improved by the applications of the discrete wavelet transform.

*Corresponding author: Yoichi Midorikawa, Tel./Fax: +81 97 554 7817; E-mail: ymido@cc.oita-u.ac.jp.

1383-5416/01/02/$8.00

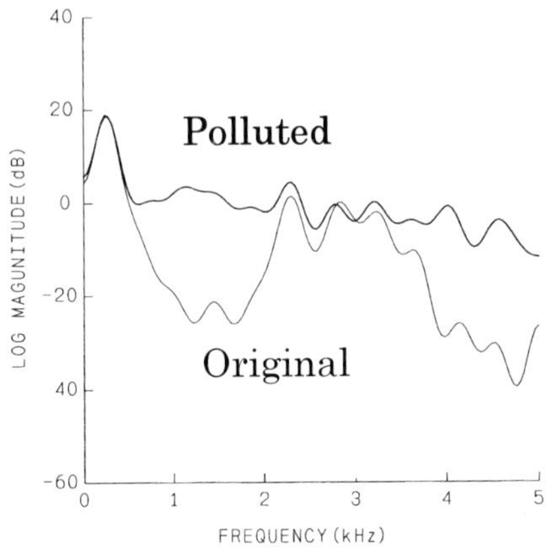

Fig. 1. Normalized spectral envelopes of clean and polluted data.

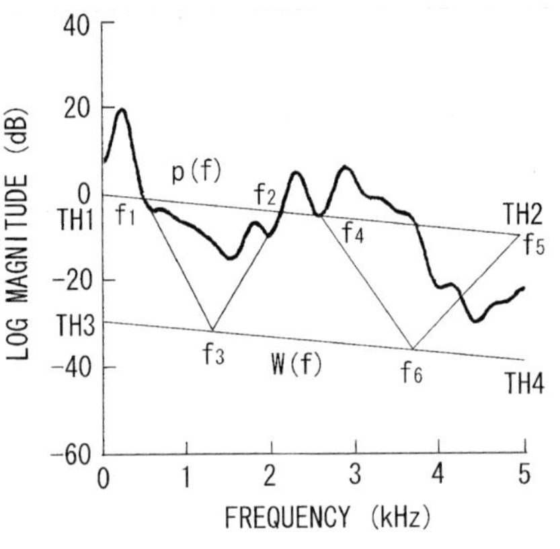

Fig. 2. Example illustrating of the method for recovering the valley of spectral envelope by rule.

2. Improvement of cepstral coefficients matrix by wavelet transformation

Figure 1 shows that normalized spectral envelopes of clean and polluted data. Previously, we have proposed the method for recognition of the signals polluted with noise using modification of spectral envelopes shape. In this case, suitable rules are defined for recovering valleys shown in Fig. 2 [2]. TH1, TH2, TH3, TH4 are thresholds.

However, this rule makes use of only frequency information. Therefore, these cepstral coefficients of neighboring frame become independent of time information. We apply two-dimensional wavelet transformation to the modification of spectral envelopes shape. Two-dimensional wavelet transformation is used for the extraction of particular feature in consideration of frequency and time domain.

We denote the cepstral coefficients of spectra modified by rules as $c_i^{(k)}$, where i is the ith order of the cepstral coefficients and k means the number of k frame.

These cepstral coefficients matrix **C** is

$$\mathbf{C} = \begin{bmatrix} c_1^{(1)} & c_2^{(1)} & \cdots & & c_n^{(1)} \\ c_1^{(2)} & c_2^{(2)} & \cdots & & c_n^{(2)} \\ & & \ddots & & \\ \vdots & \vdots & c_i^{(k)} & & \vdots \\ & & & \ddots & \\ c_1^{(m)} & c_2^{(m)} & \cdots & & c_n^{(m)} \end{bmatrix}. \tag{1}$$

The two-dimensional wavelet transformation is given by

$$\mathbf{S} = \mathbf{W} \cdot \mathbf{C} \cdot \mathbf{W}^T. \tag{2}$$

Here,

$$\mathbf{W} \cdot \mathbf{W}^T = \mathbf{I}. \tag{3}$$

where **S** is wavelet spectrum matrix, **W** is wavelet matrix [3–7], $\mathbf{W}^T$ is transpose matrix and **I** is unit matrix.

Compressed data method is given by Eq. (4). $\mathbf{S}_{i,j}$ is elements of matrix **S**. $\alpha \times \beta$ is compression rate. In this paper, these rate are $\alpha \times \beta = 0.5, 0.25$. Equation (5) is inverse wavelet transformation. $\mathbf{S}'$ is the modification of wavelet spectrum matrix by Eq. (4). $\mathbf{C}'$ is modified cepstral coefficient matrix. The pattern recognition experiments is used this modified cepstral coefficient.

$$[S'] = \begin{cases} s_{i,j} = s_{i,j} \ (i \leqslant m \times \alpha \text{ and } j \leqslant n \times \beta) \\ s_{i,j} = 0 \quad \text{(the others)} \end{cases} \tag{4}$$

$$\mathbf{C}' = \mathbf{W}^T \cdot \mathbf{S}' \cdot \mathbf{W} \tag{5}$$

3. Pattern recognition experiment

This new method is applied to the pattern recognition problem. The pattern recognition experiments by Japanese noisy digits are performed. 240 Japanese digits uttered by 8 male speakers are used. Each speaker uttered digits for three times. Two kinds of colored noise are added to the original clear speech data for making noisy data. These are expressed as Pink and Automobile. The dynamic time warping (DP) matching is used for the pattern matching [9].

The reference patterns are obtained from single speaker and the data of other speakers are recognized. In the experiments, reference patterns are obtained from clear speech data. Experimental results of the pattern recognition are shown in Table 1.

In the experimental results, recognition results of Pink and Automobile S/N10dB are improved about 10%. Recognition results of Pink and Automobile S/N0dB are the same results, in the case of wavelet base function of Haar [3–7]. However, in the case of wavelet base function of Daubechies 4–8th [3–7], recognition results of Pink and Automobile S/N0dB are improved about 5%. As a result, this new method is effective in the environment of a low noise.

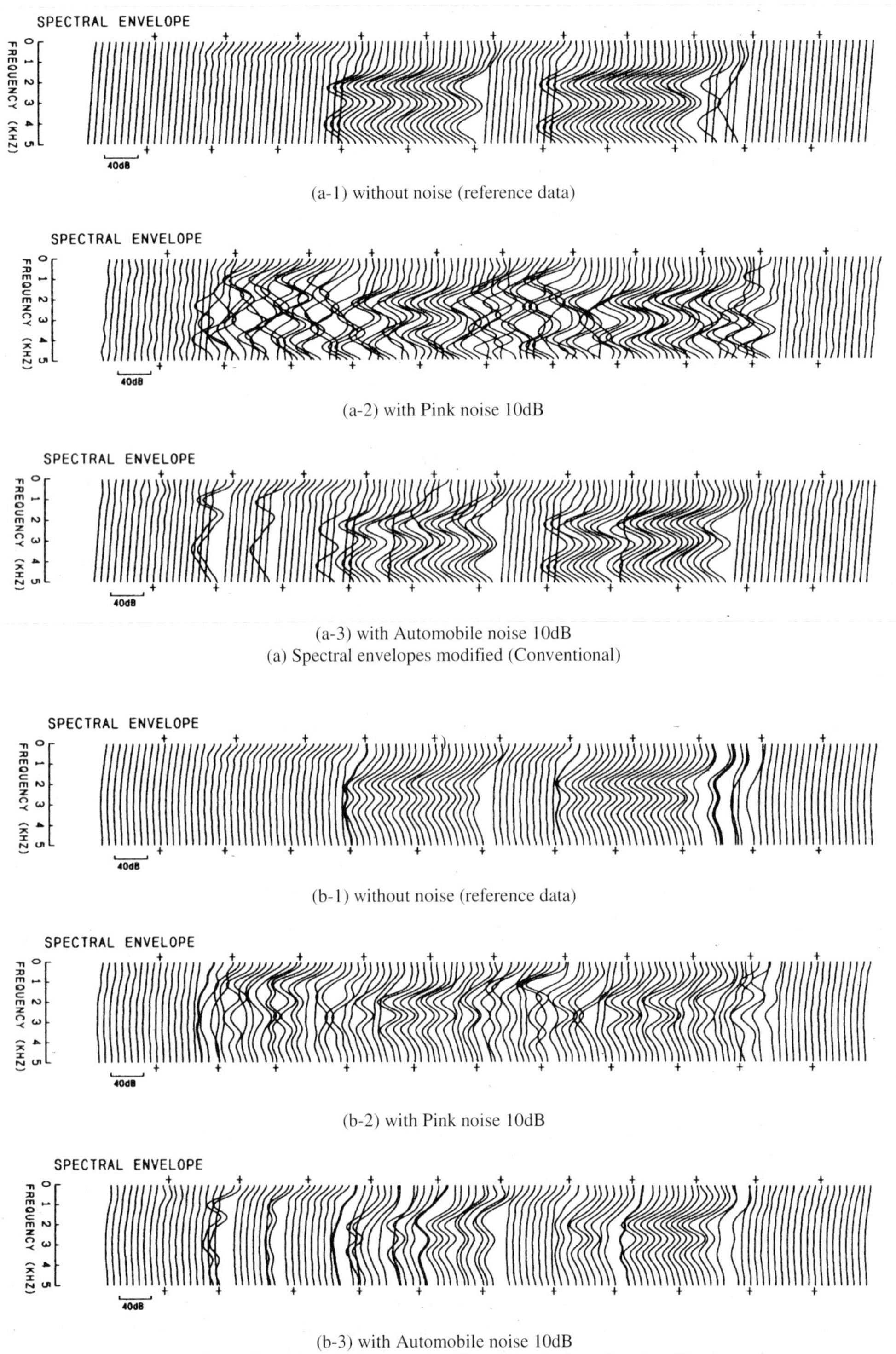

(a-1) without noise (reference data)

(a-2) with Pink noise 10dB

(a-3) with Automobile noise 10dB

(a) Spectral envelopes modified (Conventional)

(b-1) without noise (reference data)

(b-2) with Pink noise 10dB

(b-3) with Automobile noise 10dB

(b) Spectral envelopes using wavelet (New, Wavelet base function: Haar)

Fig. 3. Spectral envelopes of Japanese word /nana/(seven).

Table 1
Noisy digit recognition results. Recognition rate (%) (Threshold: TH1, TH2, TH3, TH4 = 0, −10, −30, −40 [dB])

	α	β	Wavelet base function	Pink 0dB	Pink 10dB	Auto 0dB	Auto 10dB
Conventional				19.15	54.94	16.85	49.29
New	0.5	1	Harr	20.58	56.63	16.45	51.49
	1	0.5	Harr	22.62	61.75	17.64	59.64
	0.5	0.5	Harr	23.51	63.93	16.45	61.59
	0.5	0.5	Daubechies 4	24.50	65.62	20.54	63.85
	0.5	0.5	Daubechies 6	24.01	65.42	24.37	64.07
	0.5	0.5	Daubechies 8	23.77	65.63	22.08	62.94

Figure 3 shows that the spectral envelopes of Japanese word /nana/ (seven). Figure 3(a) shows that the spectral envelopes modified by conventional method. In this case, suitable rules are defined for recovering valleys [1,2]. However, in poor environment, the incorrect valleys are sometimes added. Figures 3(a-2) and (a-3) show that the spectral envelops added wrong valleys must be removed. Figure 3(b) shows that the spectral envelopes modified by new method using wavelet transform. Figures 3(b-2) and (b-3) show that the spectral envelops is improved by the new method using wavelet transform. It is obvious that the spectral envelops added wrong valleys is removed, in the Pink noise 10 dB case shown Figs 3(a-2) and (b-2).

4. Conclusion

In this paper, we apply a wavelet transform to the speech recognition under the noisy environment using cepstral analysis. As a result, speech recognition rate is improved by the applications of the discrete wavelet transform. In the experimental results, recognition results of Pink and Automobile S/N10dB are improved about 10%. Recognition results of Pink and Automobile S/N0dB are the same results, in the case of wavelet base function of Haar. However, in the case of wavelet base function of Daubechies 4–8th, recognition results of Pink and Automobile S/N0dB are improved about 5%. As a result, this new method is effective in the environment of a low noise. Therefore, it is necessary to improve this method under the loud noisy environment. But, it was shown that the wavelet analysis was one of the promising methodologies for the pattern recognition noisy signal.

References

[1] M. Akita and S. Uchida, A Method for Improving the quality of sound waveform using analysis-synthessystem, Studies in Applied Electromagnetics in Materials 10, IOS Press, 1996, pp. 186–189.
[2] M. Akita, Enhancement of Signals Polluted with Colored Noise using Modification of Spectral Shape, Studies in Applied Electromagnetics in Materials 18, IOS Press, 2000, pp. 323–326.
[3] Y. Midorikawa, S. Hayano and Y. Saito, A Thin Film Common Mode Noise Filter and Its Evaluation by Wavelet Analysis, *IEEE Transaction on Magnetics* **32**(5) (1996), 5001–5003.
[4] Y. Midorikawa, N. Masuda, Y. Saito and H. Tohya, Discrete Wavelets Analysis of Current Distributions on the Printed Circuit Boards, *T. IEE Japan* **118-A**(7/8) (1998), (in Japanese).
[5] D.E. Newland, *Random vibrations, spectral and wavelet analysis,* (3rd ed.), Longman Scientific & Technical, 1993.
[6] S. Mallat, *A wavelet tour of signal processing,* (2nd ed.), Academic Press, 1998.
[7] H. William, *Press etc. Numerical Recipes in Fortran,* (2nd ed.), Cambridge University Press, 1992, pp. 584–599.
[8] J.K. Rogers and P.C. Cosman, Wavelet Zerotree Image Compression with Packetization, *IEEE Signal Processing Letters* **5**(5) (1998), 105–107.

[9] H. Sakoe, Two level DP-matching- a dynamic programming based pattern matching algorithm for connectedword recognition, *IEEE Trans. On Acoust., Speech & Signal Process* **ASSP-27**(6) (1979), 588–595.

International Journal of Applied Electromagnetics and Mechanics 13 (2001/2002) 19–23
IOS Press

Automatic text summarizing based on sentence extraction: A statistical approach

Bogdan Cranganu-Cretu[a,*], Zhenmao Chen[a], Tetsuya Uchimoto[b] and Kenzo Miya[a]
[a]*International Institute of Universality IIU, 1-4-6 Nezu, Bunkyo-Ku, 113-0031 Tokyo, Japan*
[b]*Nuclear Engineering Research Laboratory, Graduate School of Engineering, The University of Tokyo, Tokai-mura, Naka-gun, Ibaraki 319-1106, Japan*

Abstract. The present work describes a system for automatic summarizing of texts. Rather than focusing on abstracts ¤C a hard NLP task of not asserted effectiveness the system produces extracts through selection of most important sentences in the original text. Statistical concepts are involved in order to evaluate the degree of significance of words, groups of words and sentences. Currently both Japanese and English texts can be treated. Procedures for computing importance, information content of sentences and measures of correlation between sentences are implemented. Comments are given on the feasibility of the approach and future developments.

1. Introduction

The number of papers that are published today is ever increasing, while the Internet burst has created and continues to create huge amounts of machine-readable texts. Since it is impossible for one to keep informed and updated through extensive reading of available material, presenting the information in a condensed manner (such as summarizing) is of vital importance. This is but one of the many reasons presently sustaining the today worldwide research work on abstraction. To summarize is to reduce in complexity, and hence in length, while retaining as much as possible of the information content of the original text. Summarizing, was an early task in automation, and the first landmarks were set by Luhn [1] and Edmundson [2]. In summarization two different approaches may be used: one is to create an abstract of the text by understanding the text, the structure of the discourse and the main concepts. Thus, as a first step, a logical structure of the text is obtained. Consequently this in formation is fed to a text generator which will produce a coherent narrative abstract. The tasks above mentioned are Natural Language Processing problems, and their solution implies use of artificial intelligence techniques. Knowledge intensive methods proved successful to some extent mostly in restricted domains [3,4]. The other approach, simpler but more general, avoids the main difficulties in the NLP problem through transforming the task into summary by extraction, an approach closely related to the more general problem of information retrieval [5]. To create an extract, a system needs to identify the most important/topical/central topic(s) of the text, and return them to the reader. It is generally accepted that document extracts consisting of roughly 20% of the original text can be as informative as the full text of a document [6]. Although the summary lacks heavily in coherence (presenting what

*Corresponding author: Bogdan Cranganu-Cretu, E-mail: bogdan@jsaem.gr.jp.

Table 1
Pros and cons of extraction versus abstraction

Summaries as Extracts	Summaries as Abstracts
Goal oriented or not	Goal oriented
Slightly field dependent	Heavily field dependent
Unresolved anaphors	Narrative coherence
Use statistical techniques	Use Artificial Intelligence approach
Needs no or little background knowledge	Needs human input for background knowledge
Provides a list of main concepts in the document	Provides a true substitute for the document

are usually called unresolved anaphoras – incomplete addressed subjects due to i ncomplete selection of sentences related to the topic), the reader can form an opinion of the content of the original text [7]. The approach on text extraction – is in our opinion many folded. To name a few there are linguistic and rhetoric approach [8,9], mathematical approach, statistical approach [10,11]. If one wants to use statistics, one needs a big and homogeneously annotated corpus along with corpus analysis tools. The following work, originated from the Miya's long pursued interest [12], is a general summarizing system, based on the word frequency parameter, and relies upon computation of some statistical parameters in order to quantify the importance of the sentences. Although frequency based criterias are not new to the summarisation approach the original part of our approach lies in the combined use of information related parameters (e.g. Shanon's entropy) to provide means for sentence selection and classification.

2. The text abstraction system

In the following is presented the application that the authors have devised addressing mainly the concepts and the particular choice of parameters as well as algorithms used. The system described can consider English language text files, and relies for the morphological analysis upon freely available software on the Internet.

As pointed out from the title our system is basically a sentence extracting system. We feel that a brief comparison of pros and cons of extraction versus abstraction, as presented in Table 1, will clarify the rationale for our option. As can readily be seen, in order to obtain a system with large applicability, basically we sacrifice coherence for speed and feasibility by choosing the extraction approach.

2.1. Basic concepts in extraction

In the extraction type of approach we conducted the sentence is the basic structural unit of the text. In order to rank the sentences statistical parameters computed upon words within the text are used. On criterion of significance used is word frequency. Prior to any analysis elimination of very frequent words with no informational content is performed. Presently the approach uses just nouns, adjectives and verbs. In order to perform the needed morphological analysis our system makes use of KTEXT and PC-KIMMO text analysis software produced and distributed free on the internet by Summer Institute of Linguistics [13,14]. On the remainder of the words in the sentences parameters are computed both at word level and at sentence level.

2.2. Parameters defined in extraction process

At word level the parameter used in the analysis is the word's frequency Eq. (1):

$$w_f = \frac{n_{as}}{n_{ws}} \cdot \frac{n_{at}}{n_{wt}} \tag{1}$$

where n_{as} is the number of word's appearances in the sentence, nws the total number of words in the sentence, nat the number of word's appearances in the text and nwt the total number of words in the text. The hierarchy provided by the frequency criteria can be enhanced by means of various other criteria – commonly called heuristics. We have used Key words heuristic in which high frequency content words are positively relevant and title words heuristic in which words in title and headings throughout the document are assigned positive weights. Other criteria are: cue heuristic in which certain words (e.g. related to a certain topic that one wants to be emphasized in the summary) have positive weights whilst others have negative or zero weight;

At a sentence level we define three parameters: sentence's importance as in Eq. (2), where Ai – vector containing the number of appearances for each word in sentence I, Fi – vector containing the frequencies for each word in sentence I and n_s – number of words within sentence i:

$$im\,p_i = \mathbf{F}_i \cdot \mathbf{A}_i = \sum_{j=1}^{n_s} F_j \cdot A_{ij} \tag{2}$$

To quantify the information in a sentence, based on Shannon's principles we define Eq. (3) and (4):

$$H_i = -\sum_{i=1}^{n} P_{ij} \log_2 P_{ij} \tag{3}$$

$$P_{ij} = \frac{n_{as}}{n_{ts}} \tag{4}$$

where n_{as} – total number of appearances in sentence j of a specific word, n_{ts} – total number of words in sentence j (considering multiple appearances of a word) and n – number of distinct words in the sentence. The degree of correlation between two sentences is defined as in Eq. (5) based on the fact that similar content in word or word stems accounts for high correlation.

$$corr_{ij} = \mathbf{A}_i \cdot \mathbf{A}_j \tag{5}$$

Alternatively a location heuristic is used in some approaches. In such the sentences in first and last paragraph, the first and last sentence in a paragraph have greater importance.

2.3. Algorithms used in abstraction

Figure 1 presents schematically the type of algorithm we have used in order to select among the most important sentences in the text. Using the above-described parameters sentences are ranked according to their importance. Among the most important ones a search is performed to select, in order of their appearance, the top important sentences with the lowest degree of correlation between them, thus obtaining the main "chain" of the extract ("A","B","C" in Fig. 1). For each node in this chain a subsequent selection processed is performed in order to select the top importance sentences with the highest degree of correlation with the sentence in the node - thus forming clusters of sentences around the nodes.

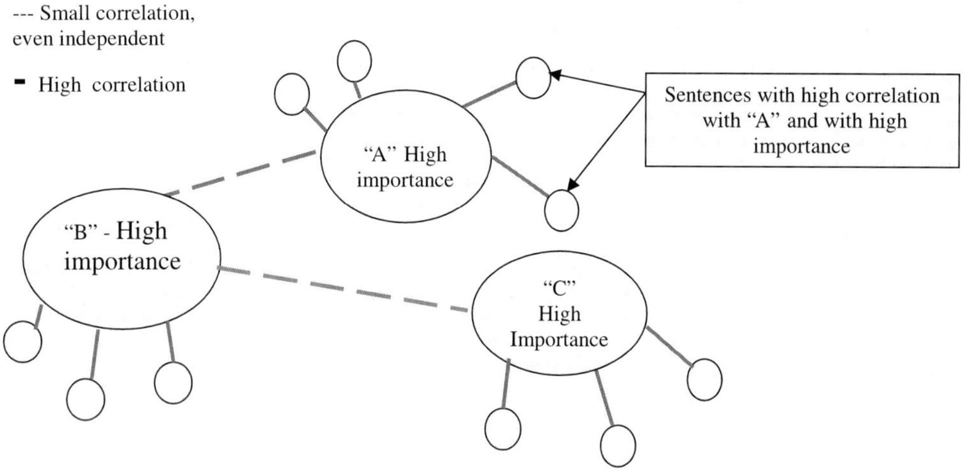

Fig. 1. Schematic description of algorithm used for text extraction.

The Titanic Disaster An Enduring Example of Money Management versus Risk Management .
What gets far less comment is that most of the problems all came from a larger systemic problem the owners and operators of steamships had for five decades taken larger and larger risks to save money risks to which they had methodically blinded themselves. The Titanic disaster suddenly ripped away the blindfolds and changed dozens of attitudes practices and standards almost literally overnight. He compares the ships of Titanic s day to the first great liner the Great Eastern built in She was designed by I K Brunel England s most celebrated engineer who got every feature he wanted. When the unsinkable Titanic was completed in she matched the Great Eastern in only one respect she too had transverse bulkheads. The Great Eastern s bulkheads were carried feet above the waterline the Titanic's only feet. However the Great Eastern s inner hull was unbroken and the engine room remained dry. The Great Eastern had defense in depth against hull breach. Various other corners ceased to be cut in safety standards as well. The lifeboat problem was exacerbated by poor procedure. Only at the last did lifeboats leave full at first many left partly empty because passengers were not queued up to them .
The first hockey players to use helmets were jeered by other athletes until it became the new custom. Sooner or later it will be pointed out that a few head injuries occur to pedestrians but I hope that it will never be made mandatory to wear a helmet to take a walk. Even today however it is still often the case that money management wins out over risk management. After helping to organize the confused lifeboat loading and personally assisting many to the last boats he remained aboard. The steward called to him to come try to swim for it but he did not answer. I hope it comforted him that this disaster was going to be so traumatic as to bring about tremendous changes despite all their costs changes that would save more lives in the long run. Were it not for the Titanic the safety standards might have continued for a long time causing a long string of smaller disasters each too small in its own right to bring about basic changes to the whole industry.

Fig. 2. Results of an essay summarization.

3. Results

The algorithm output can be tailored by selecting the number of nodes in the main chain as well as how big a cluster we want to be. This of course is in direct correlation with the original document size and optimum value tuning was up to know based on trial and error methods. In addition using a previous devised list of concepts for a certain field, thus attributing bigger importance to certain words – so not relying entirely on frequency considerations - brought to better results. Figure 2 presents results of summarization of the essay "The Titanic Disaster: An Enduring Example of Money Management vs. Risk Management" by Roy Brander [15]. The original text file considered was 14 k, and had 118 sentences. The abstract size obtained was 2.5 k and consist of 17 sentences. We compared our results with the summary obtained by a professional core, TextAnalyst 2.0, demo version. The number of identical sentences extracted from original text amounted to 40% of the summary.

4. Conclusions

We have briefly outlined the present situation in text summarization by means of extraction. Our approach is also an extraction approach type, which, although based on evaluation of words frequency goes further through the introduction of parameters to account for the degree of information present in sentences (Shannon's entropy). Based on combination of these parameters, as well as on heuristics involving words in the title we devised the present algorithms of summarization. The system is currently under development and subsequent consideration of combinations of heuristics will be implemented.

References

[1] H.P. Luhn, The Automatic Creation of Literature Abstracts, *IBM J. Res. Develop.* **2**(2) (1959), 68–73.
[2] H.P. Edmundson, New Methods in Automatic Extracting, *Journal of the Association for Computing Machinery* **16**(2) (1969), 264–285.
[3] I. Mani and M. Maybury, eds, *Advances in Automatic Text Summarization*, Cambridge MIT Press, 1999.
[4] U. Reimer and U. Hahn, Text Condensation as Knowledge Base Abstraction in IEEE Conf. On AI Applications, 1988, pp. 338–344.
[5] J. Goldstein, M. Kantrowitz, V. Mittal and J. Carbonell, Summarizing Text Documents: Sentence Selection and Evaluation Metrics, Proceedings of SIGIR'99, Berkley, USA, 1999
[6] J. Kupiec, J. Pedersen and F. Chen A trainable document summarizer, Proceedings of the 18th ACM-SIGIR Conference, 1995, pp. 68–73.
[7] C.D. Paice, Constructing literature abstracts by computer: techniques and prospects, *Information Processing and Management* **26**(1) (1990), pp. 171–186.
[8] D. Marcu , The Rhetorical Parsing of Natural Language Texts, The Proceedings of the 35th Annual Meeting of the Association for Computational Linguistics, (ACL'97/EACL'97), Madrid, Spain, July 7–10, 1997, pp. 96–103.
[9] D. Marcu, To build text summaries of high quality, nuclearity is not sufficient, *The Working Notes of the AAAI-98 Spring Symposium on Intelligent Text Summarization*, Stanford, USA, March, 1998, pp. 1–8.
[10] C.D. Paice and P.A. Jones The identification of important concepts in highly structured technical papers, *Proceedings of Sixteenth Annual International ACMM SIGIR Conference on Research and Development in Information Retrieval*, ACM Press, June, 1993, pp. 69–78.
[11] Simone Teufel and Marc Moens (1999). Argumentative classification of extracted sentences as a first step towards flexible abstracting, in: *Advances in Automatic Text Summarization*, I. Mani and M. Maybury, eds, Cambridge MITPress, 1999, pp. 155–175.
[12] K. Miya, Form and sense, (8) Analysis of creation in brain activity, *Journal of the Japan society of Applied Electromagnetics and mechanics* **8**(1) (March, 2000), 110–120.
[13] E.L. Antworth, Users guide to k-text, version 2, ftp://ftp.sil.org/software/k-text/guide.zip, 1997.
[14] E.L. Antworth, *"PC-Kimmo", a two level processor for morphological analysis*, Ocasional Publications in Academic computing, No. 16, Summer Institute of Linguistic, Dallas, 1990.
[15] http://www.cuug.ab.ca:8001~branderr/risk_essay/titanic.html

International Journal of Applied Electromagnetics and Mechanics 13 (2001/2002) 25–33
IOS Press

Application of topology optimization to electromagnetic system

Jin-Kyu Byun* and Song-Yop Hahn
School of Electrical Engineering #051, Seoul National University, Kwanak, P.O. Box 34, Seoul 151-742, Korea

Abstract. In this paper, the theories and general concept of the topology optimization in the electromagnetic system are explained, and the applications to the optimal design of electromagnetic devices and the inverse problems are shown. As a way of defining the relationship between the design variables and material properties, homogenization method or density method can be used. The mutual energy method is also explained for the design sensitivity analysis in the topology optimization.

1. Introduction

Nowadays, the topology optimization is applied to the electromagnetic systems quite often. The topology optimization was first proposed and researched actively in the field of structural engineering from the late 80's [1]. However, there have been relatively few works on the topology optimization in the electromagnetic systems until Dyck et al. adopted the universal representation of the device for the optimal design of the magnetic systems in the mid 90's [2,3]. He called his method 'Optimal Material Distribution' (OMD) method, because optimal design was attained by distributing materials in the design space according to the cost and constraints. He calculated sensitivity based upon Tellegen's theorem from circuit theory.

After the papers by Dyck, many authors proposed new ideas and methods for the topology optimization in the electromagnetic system. The authors introduced mutual energy concept and proposed new topology optimization algorithm for both electric and magnetic systems [4,5]. Wang et al. designed the C-core actuator by topology optimization method [6], and Kikuchi et al. applied homogenization method to the magnetic system and optimized the C-core in basic 3D setup [7].

In this paper, the theories and general concept of the topology optimization are explained, and the applications to the optimal design of the electromagnetic devices and the inverse problems are shown.

2. Homogenization method and density method

In the topology optimization, the design space is composed of distributed material properties, which enable the universal, gray-scale pixel-like representation of the device. This means that the topology as well as the shape of the device can be determined automatically during the optimization process. It is

*Corresponding author: Jin-Kyu Byun, Fax: +82 2 878 1452; E-mail: aiolia@dreamwiz.com.

one of the significant merits of the topology optimization, because the conventional shape optimization only changes the boundary surface between different material regions, thus limiting the design space and topology of the device.

For distributed representation of the device, the entire design region where the materials can exist is divided into small cells. These cells can coincide with the finite element method (FEM) meshes. The material properties are allowed to vary continuously between the different available solid materials.

There are several techniques for associating the design variables with the material properties in the topology optimization. The general approach is to select certain design variables in each cell that control the material properties within that particular cell with physical or mathematical relationships. The homogenization method considers microscopic perforated composite material and calculates the 'average' material property using asymptotic expansion, while density method simplifies the relationships between the design variables and the material properties as the artificial mathematical equations that can be modified at the convenience of the designer. These methods are discussed in detail in the following sections.

2.1. Homogenization method

The homogenization method was first used in the structural design by Bensoe et al. [1]. This is a mathematical method to characterize material properties of possibly complex composite material. It relies on the ability to model a micro-scale perforated or layered material for the universal, distributed representation of the device.

In the homogenization method, the material is constructed from a basic unit cell, consisting at a microscopic level of different materials. The composite medium consists of infinitely many such cells, repeated periodically through the medium. At the limit, the material density continuously varies throughout the medium. The material property of the medium can be described by effective or 'averaged' macroscopic properties which depend on the geometry of the basic cell. These 'averaged' properties can be computed using the formulas of homogenization theory.

Two kinds of the microscopic composite structures are explained in this paper. The unit cell with microscopic rectangular holes and the so-called rank-2 microscopic structure (a two-scale orthogonal laminate).

2.1.1. Microscopic rectangular holes

The contents in this section mainly follow the work by Kikuchi et al. [7]. Figure 1 shows the design domain composed of the inhomogeneous composite material with microscopic rectangular holes. The macro-scale design domain consists of a finite number of finite elements. However, each finite element can be decomposed into infinite number of unit cells, which are allowed to have rectangular perforation whose size is updated at each iteration. If the hole size of the unit cell is zero, the unit cell becomes a solid. On the other hand, if the hole size is equal to the unit cell size, that cell becomes a void. By changing the hole size, the material is varied from a solid to a hollow sector, and this allows for complete topology optimization. The final distribution defines the optimal topology in a design domain.

Hereafter, $r(x, y, z)$ will represent the macroscopic level coordinate, whereas $(r_{\text{micro}}, y_{\text{micro}}, z_{\text{micro}})$ will denote the microscopic level coordinate. The computational domain Ω is an open connected domain in R^3 with smooth boundaries. It consists of an infinite number of microstructures, each of which has a hole, as shown in Fig. 1.

The microscopic level coordinate r_{micro} is normalized as:

$$r_{\text{micro}} = \frac{r - r_0}{\alpha} \tag{1}$$

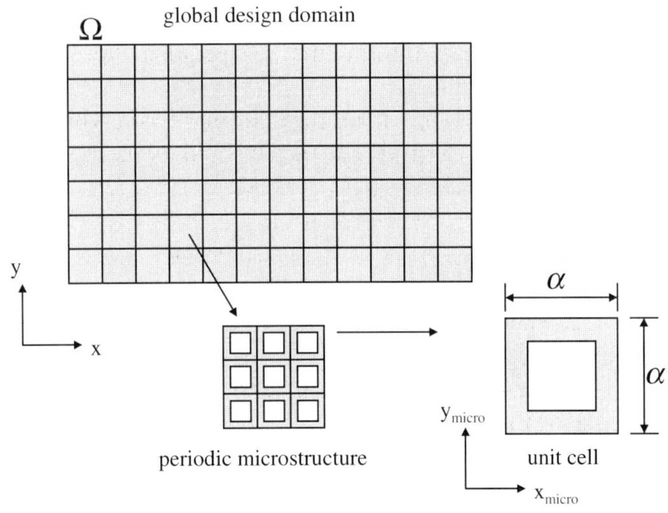

Fig. 1. Design domain and its decomposition into microstructures with rectangular holes in homogenization method.

where α is a dimension of the unit cell and r_0 denotes the coordinate shift between the global and the microscopic system origins. If the microscopic cells are of different size, then α take different values for each cell to ensure that the components of r_{micro} becomes $0 < x_{\text{micro}} < 1,\ 0 < y_{\text{micro}} < 1,\ 0 < z_{\text{micro}} < 1$.

A correspondence between the microscopic composite structure and its homogenized permeability $[\mu_H]$ can be calculated by using homogenization equation and is given by:

$$[\mu_H] = \int_{R_{\text{micro}}} [\mu] dr_{\text{micro}} + \int_{R_{\text{micro}}} [\mu] \nabla_{\text{micro}} \vec{\gamma}(r_{\text{micro}}) dr_{\text{micro}} \tag{2}$$

where $\vec{\gamma}(r_{\text{micro}})$ is the characteristic magnetic field strength in the unit cell and can be obtained from solving the following weighted residual in the microstructure:

$$\begin{aligned} &\int_{R_{\text{MICRO}}} \left[\mu_{xx} \frac{\partial \gamma_i}{\partial x_{\text{micro}}} \frac{\partial N_j}{\partial x_{\text{micro}}} + \mu_{yy} \frac{\partial \gamma_i}{\partial y_{\text{micro}}} \frac{\partial N_j}{\partial y_{\text{micro}}} + \mu_{zz} \frac{\partial \gamma_i}{\partial z_{\text{micro}}} \frac{\partial N_j}{\partial z_{\text{micro}}} \right] dr_{\text{micro}} \\ &= - \int_{R_{\text{MICRO}}} \mu_{ii} \frac{\partial N_j}{\partial r_{\text{micro}_i}} dr_{\text{micro}}, \quad (i = 1, 2, 3) \end{aligned} \tag{3}$$

where N_j represents the shape function. The homogenized value of other material properties can be computed from similar procedure.

2.1.2. Rank-2 microstructure

The cross section of the typical rank-2 microstructure is shown in Fig. 2 at different levels of magnification. The microstructure has two levels of detail (i.e. the material used to form the first level microstructure is itself a composite material defined using solid materials on an even smaller scale). The two design variables of the structure are the relative thickness of the laminations at the first level (g_1), and the relative thickness of the material between gaps at the second level (g_2). Both of these parameters are defined relative to the period of the microstructure at the corresponding level, so when $g_1 = g_2 = 1$, the composite becomes solid.

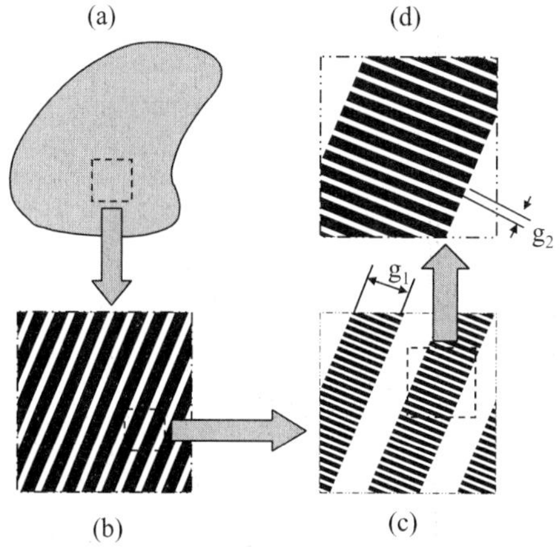

Fig. 2. Successive zooms of rank-2 microstructure.

The purpose of the two levels can be described qualitatively as follows. In the first level (Fig. 2(b)), the laminations are assumed to be aligned parallel to the magnetic field. Assuming their relative permeability is sufficiently high, these laminations will carry essentially all of the magnetic flux. Therefore, the width of the laminations will determine the magnetic flux density in the rank-1 composite making up the laminations. The second level (Fig. 2(d)) can be more accurately described as a periodically gapped core. The gaps serve to reduce the effective permeability of the composite.

2.2. Density method

The field analysis of the microstructures proposed in the previous section reveals that there is a certain relationship between the size of the microstructure and the homogenized material property. Thus there must be a correspondence between the material density and the material property, too. The class of the microstructure used in the modeling determines this relationship, and if this relationship can be approximated by an explicit mathematical equation, there is no need to solve the homogenization equation to calculate the intermediate material property.

The density method was first proposed in this context, because the microstructure is not of main concern in most engineering problems. In the density method, the material density of each finite element is the design variable, and the relationship between the material density and the material property is defined by simple mathematical equation. The density method was also called the 'engineer's approach' in [8].

To use with the finite element method, the normalized density of material is assumed to be identical over one finite element. The normalized density within each finite element controls the electromagnetic properties of the material such as permeability, permittivity, etc. For example, the relationship between the permeability and the normalized density can be written as:

$$\mu = \mu_0(1 + (\mu_r - 1)p^n) \quad (0 \leqslant p \leqslant 1,\ 2 \leqslant n \leqslant 4) \tag{4}$$

where p is the normalized density, μ_r is the relative permeability of the real material, and n is the exponent which defines the relationship between material property and normalized density. When $n > 1$, the intermediate density is penalized and as a result the normalized density is forced to either 0 or 1. The plot of μ versus p is given in Fig. 3 for various values of n. When n is larger, the intermediate

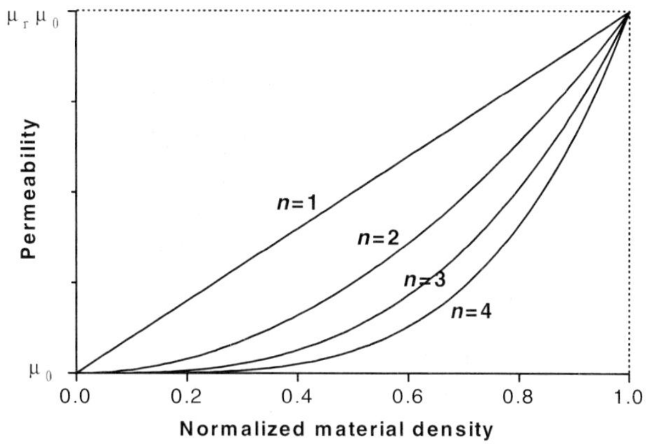

Fig. 3. Relationship between normalized density and permeability for various values of n.

density is more heavily penalized, thus eliminating 'gray' materials more efficiently. But larger value of n can also limit the design space at the same time. Usually, the value of n can be chosen between 2 and 4 considering this trade-off [9]. That is, if the intermediate density is not significant problem and the user want to access larger design space, $n = 2$ can be used. On the other hand, if the design space is sufficiently smooth and the intermediate material is major difficulty in optimization, $n = 4$ should be used. Permittivity and other material properties can be represented in a similar manner. The density method is widely used because it is easy to implement, has less design variables than homogenization method, and the intermediate density can be eliminated efficiently by adjusting the value of n.

3. Design sensitivity analysis

In the topology optimization, the number of the design variables grows very large as the number of meshes increases to get reasonable resolution. Thus, quick and efficient calculation of the sensitivity information is essential. A number of approaches can be used for the design sensitivity analysis, including: the adjoint variable method, the method based upon Telegen's theorem, and the mutual energy method. The mutual energy method is explained in this paper.

3.1. Mutual energy method

The mutual energy method is best suited for problems where the system energy or the parameters closely related to it, such as inductance or capacitance, are of main concern. It is also useful when the objective function is defined by electric or magnetic potential. No additional computation is required other than the solution of one auxiliary system in the mutual energy method. It utilizes the two mutual energy terms for the calculation of the sensitivity. For example, let's consider the arrangement of linear magnetic material, two coils, and air in region Ω as in Fig. 4. Suppose coil 1 (region Ω_1) carries a source current density $\vec{J}_1$ and coil 2 (region Ω_2) carries a source current $\vec{J}_2$ with corresponding magnetic vector potential $\vec{A}_1$ and $\vec{A}_2$.

Two mutual energy terms can be defined as the following equations.

$$L_1\left(\vec{A}_2\right) = \int_{\Omega_1} \vec{A}_2 \cdot \vec{J}_1 d\Omega, \quad L_2\left(\vec{A}_1\right) = \int_{\Omega_2} \vec{A}_1 \cdot \vec{J}_2 d\Omega \tag{5}$$

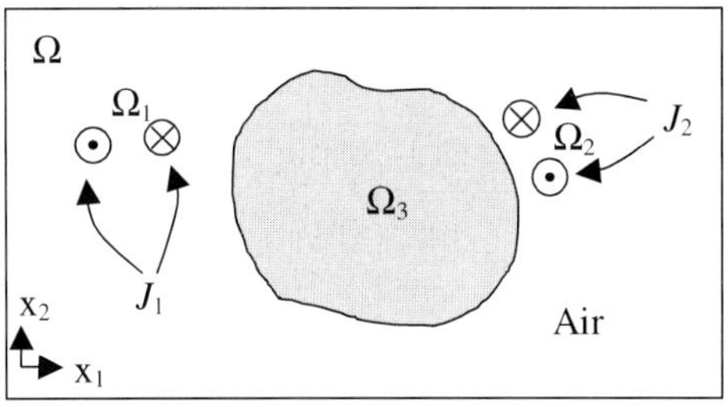

Fig. 4. Arrangement of magnetic material and two coils in region Ω.

Assuming that the current $\vec{J}_1$ is the applied source current density, and current $\vec{J}_2$ is a unit dummy current density, Eq. (5) shows the measurement of the flux linkage at Ω_2 when $\vec{J}_1$ is applied at Ω_1. By using Eq. (5), the energy functional of the system can be written in the sense of mutual energy, i.e.,

$$\chi = \frac{1}{2}a\left(\vec{A}_1, \vec{A}_2\right) - \frac{1}{2}L_1\left(\vec{A}_2\right) - \frac{1}{2}L_2\left(\vec{A}_1\right) \tag{6}$$

where $a\left(\vec{A}_1, \vec{A}_2\right) = \int_\Omega \left(\nabla \times \vec{A}_1\right) \cdot \nu \left(\nabla \times \vec{A}_2\right) d\Omega$ and ν is the reluctivity of the medium.

For the vector potential distribution $\vec{A}_1$ and $\vec{A}_2$ which satisfy Maxwell's equation, the following relations are obtained from the principle of virtual work,

$$a\left(\vec{A}_1, \vec{A}_2'\right) = L_1\left(\vec{A}_2'\right) \text{ for all } \vec{A}_2' \text{ in admissible space } U \tag{7}$$

$$a\left(\vec{A}_2, \vec{A}_1'\right) = L_2\left(\vec{A}_1'\right) \text{ for all } \vec{A}_1' \text{ in admissible space } U \tag{8}$$

so that the energy functional χ satisfies at equilibrium,

$$\chi = \frac{1}{2}a\left(\vec{A}_1, \vec{A}_2\right) - \frac{1}{2}L_1\left(\vec{A}_2\right) - \frac{1}{2}L_2\left(\vec{A}_1\right) = -\frac{1}{2}L_2\left(\vec{A}_1\right) \tag{9}$$

Now let p be the design variable, and consider χ as a function of $\vec{A}_1$, $\vec{A}_2$ and p. The first variation of χ yields [10],

$$\begin{aligned}\delta\chi\left(\vec{A}_1, \vec{A}_2, p\right) = &\frac{1}{2}a\left(\vec{A}_1, \frac{\partial \vec{A}_2}{\partial p}\delta p + \delta \vec{A}_2\right) - \frac{1}{2}L_1\left(\frac{\partial \vec{A}_2}{\partial p}\delta p + \delta \vec{A}_2\right)\\ &+\frac{1}{2}a\left(\vec{A}_2, \frac{\partial \vec{A}_1}{\partial p}\delta p + \delta \vec{A}_1\right) - \frac{1}{2}L_2\left(\frac{\partial \vec{A}_1}{\partial p}\delta p + \delta \vec{A}_1\right)\\ &+\frac{1}{2}\int_\Omega \left(\nabla \times \vec{A}_1\right) \cdot \frac{\partial \nu}{\partial p}\left(\nabla \times \vec{A}_2\right) d\Omega \delta p\end{aligned} \tag{10}$$

Since $\vec{A}_2'$ and $\vec{A}_1'$ in Eq. (8) are arbitrary variables, the first four terms in Eq. (10) can be cancelled by setting $\vec{A}_2' = \left(\partial \vec{A}_2/\partial p\right)\delta p + \delta \vec{A}_2$ and $\vec{A}_1' = \left(\partial \vec{A}_1/\partial p\right)\delta p + \delta \vec{A}_1$. Therefore, the sensitivity of χ with respect to p is given by,

$$\frac{\partial \chi}{\partial p} = \frac{1}{2}\int_\Omega \left(\nabla \times \vec{A}_1\right) \cdot \frac{\partial \nu}{\partial p}\left(\nabla \times \vec{A}_2\right) d\Omega \tag{11}$$

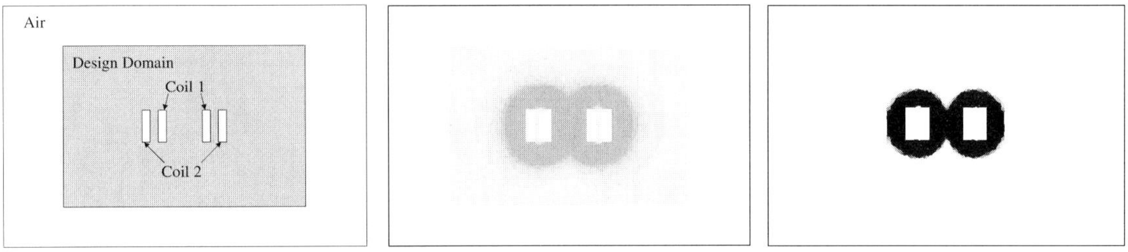

(a) 0 iterations (b) 20 iterations (c) 100 iteration

Fig. 5. Gray-scale pictures of the transformer core (shell-type).

From Eqs (9) and (11), the sensitivity of mutual energy term $L_2\left(\vec{A}_1\right)$ as to the design variable p is obtained as follows:

$$\frac{\partial L_2\left(\vec{A}_1\right)}{\partial p} = -\int_{\Omega}\left(\nabla \times \vec{A}_1\right) \cdot \frac{\partial \nu}{\partial p}\left(\nabla \times \vec{A}_2\right) d\Omega \tag{12}$$

4. Applications to the electromagnetic system

4.1. Optimal design of the electromagnetic devices

The topology optimization method can be applied to the optimal design of the electromagnetic devices such as electrical motors, electrostatic actuators, and transformers. Figure 5 shows a numerical example where the core of the shell-type transformer is optimized by the topology optimization method. The design objective is to maximize the flux linkage in the secondary coil. The total volume of the magnetic material used in the transformer core is fixed during the optimization process.

Figure 6 shows the example of SRM rotor optimization at the various iteration numbers. The design objective is to find the optimum energy profile as to the rotation of the rotor. It can be seen from the results that the topology optimization method can 'find' the optimal shape and topology of the salient-pole rotor without any prior knowledge of the basic rotor layout. However, the shape of the final device is somewhat coarse because of the limited number of the design variables in the topology optimization.

4.2. Applications to the inverse problems

The topology optimization method can also be applied to the inverse problems. As a matter of fact, the idea of representing the device by the gray-scale, pixel-like image in the topology optimization originated from the field of the inverse problems and eddy current imaging. Figure 7(a) shows the initial setup for the numerical inverse problem example where dielectric material $2(\varepsilon_r = 10)$ is included in material $1(\varepsilon_r = 1)$, and there are intermediate materials between material 1 and 2. The reconstructed normalized density using the topology optimization method is given in Fig. 7(b).

5. Conclusion

The theories and applications of topology optimization for electromagnetic system were explained in this paper. The topology optimization method has several merits compared to the conventional

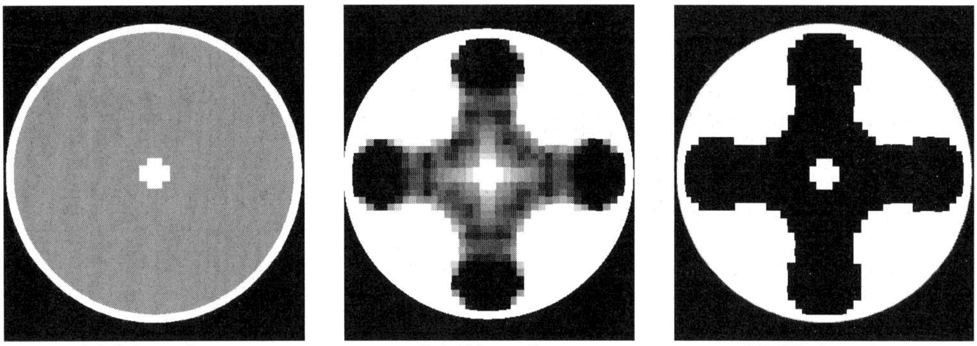

(a) 0 iterations (b) 5 iterations (c) 50 iterations

Fig. 6. The material distribution of SRM rotor.

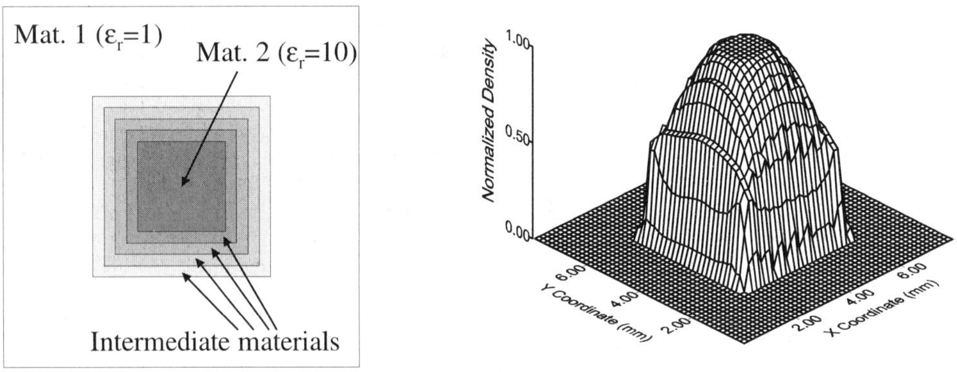

(a) Initial setup for the inverse problem (b) Reconstructed normalized density

Fig. 7. Applications to the inverse problems.

shape optimization method. First, the knowledge of the basic layout of the device is not needed before optimization, thus original topology and/or shape of the device can be designed by this method. Second, the objective function tends to be smooth in the topology optimization because of the universal, distributed representation of the device, so the use of the first-order method to search the high-dimensional design space can be justified. Third, the level of interaction and expertise needed in the design process is reduced to minimum, thus 'automated design system' can be realized.

The homogenization method or density method can be used to parametrize the design space and establish the relationship between the design variables and material properties of the electromagnetic system. The homogenization method has solid physical and mathematical background, and provides a systematic way to interpret the intermediate material density. However the number of design variables increases greatly in homogenization method and the computation time for homogenized value of material property can further delay the design process. The density method defines the relationship between normalized density and material property with simple artificial equation. The density method is easy to implement and understand, and can provide reasonable results without additional computation time for intermediate material property.

With all its merits, topology optimization method cannot be expected to replace the conventional shape optimization method all by itself. The topology optimization method performs well when the optimal topology of the device is not known or the intermediate materials exist. However, the limited resolution of the final device makes it difficult to design a detailed device. Furthermore, if the device is composed of only the solid materials, the intermediate materials generated during the optimization process need special treatment after the optimization. Thus, the main goal of the topology optimization should be to find the basic topology and layout of the device as a pre-processor module of an optimal design software suite. And then conventional shape design method can take over and determine the detailed shape and size of the device.

References

[1] M.P. Bendsoe and N. Kikuchi, Generating optimal topologies in structural design using a homogenization method, *Computer Methods in Applied Mechanics and Engineering* **71** (1988), 197–224.
[2] D.N. Dyck, *Automating the Topological Design of Magnetic Devices,* Ph. D. thesis, Dept. of Electrical Engineering, McGill University, 1995.
[3] D.N. Dyck and D.A. Lowther, Automated design of magnetic devices by optimizing material distribution, *IEEE Trans. Magn.* **32**(3) (May 1996), 1188–1193.
[4] J.K. Byun, S.Y. Hahn and I.H. Park, Topology optimization of electrical devices using mutual energy and sensitivity, *IEEE Trans. Magn.* **35**(5) (September 1999), 3718–3720.
[5] J.K. Byun, J.H. Lee, I.H. Park, H.B. Lee, K. Choi and S.Y. Hahn, Inverse problem application of topology optimization method with mutual energy concept and design sensitivity, *IEEE Trans. Magn.* **36**(4) (July 2000), 1144–1147.
[6] S. Wang, Y. Kim and K. Park, Topology optimization of electromagnetic systems, *Proceedings of the KIEE Spring Annual Conference 2000 (Korea),* April 2000, pp. 65–69.
[7] J. Yoo, N. Kikuchi and J.L. Volakis, Structural optimization in magnetic devices by the homogenization design method, *IEEE Trans. Magn.* **36**(3) (May 2000), 574–580.
[8] H.P. Mlejnek and R. Schirrmacher, An engineer's approach to optimal material distribution and shape finding, *Computer Methods in Applied Mechanics and Engineering* **106**(1–2) (1993), 1–26.
[9] R.J. Yang, Multidiscipline topology optimization, *Computer & Structures* **63**(6) (1997), 1205–1212.
[10] S. Nishiwaki, M.I. Frecker, S. Min and N. Kikuchi, Topology optimization of compliant mechanisms using the homogenization method, *Int. J. Numer. Meth. Engng.* **42** (1998), 535–559.

International Journal of Applied Electromagnetics and Mechanics 13 (2001/2002) 35–43
IOS Press

Computer simulation of magnetic levitation of thin steel plate

Kazuo Marunaga[a], Yasuo Oshinoya[b] and Kazuhisa Ishibashi[c,*]
[a]*Graduate Student, Course of Mechanical Engineering, Graduate School of Engineering, Tokai University, Japan*
[b]*Department of Mechanical Engineering, Future Science and Technology Joint Research Center, Tokai University, Japan*
[c]*Department of Mechanical Engineering, Faculty of Engineering, Tokai University, 2-28-4 Tomigaya, Shibuya-ku, Tokyo, 151-0063 Japan*

Abstract. This paper describes how to simulate magnetic levitation of a thin steel plate by using electromagnets. Since the levitation system is essentially unstable, the steel plate is levitated by controlling the attractive force produced by the electromagnets with feedback signals: the displacement and velocity of the steel plate and the current of the electromagnet. The motion equation of the steel plate is solved by the finite difference method. The attractive magnetic force by electromagnets is obtained by using the finite element method.

1. Introduction

A thin steel plate is used for various industrial products. So as to maintain quality of the surface of the steel plate in manufacturing process, we have proposed a method of magnetic levitation of the steel plate – with free edges. Since the attractive magnetic levitation system is essentially unstable, the feedback control is indispensable and fundamental investigation of the control system has been carried out experimentally [1].

In these days, computer simulation has been widely used in solving engineering problems. Computer simulation is useful to examine the levitation system because we can modify easily the system in order to obtain better system. We have reported how to simulate time response of the steel plate with free edge when impulsive disturbance is given to the steel plate under steady state [2]. In this paper, we study how to simulate magnetic levitation of a steel plate with free edge taking gravity into consideration.

2. The motion equation of the steel plate

The magnetic levitation system is shown in Fig. 1. The size of a steel plate is as follows: length $l_x = 800$ mm, width $l_y = 600$ mm, thickness $h = 0.3$ mm. Five electromagnets are used for levitating

*Corresponding author: Kazuhisa Ishibashi, Tel.: +81 3 3467 2211 ext. 381; E-mail: isibasi@keyaki.cc.u-tokai.ac.jp.

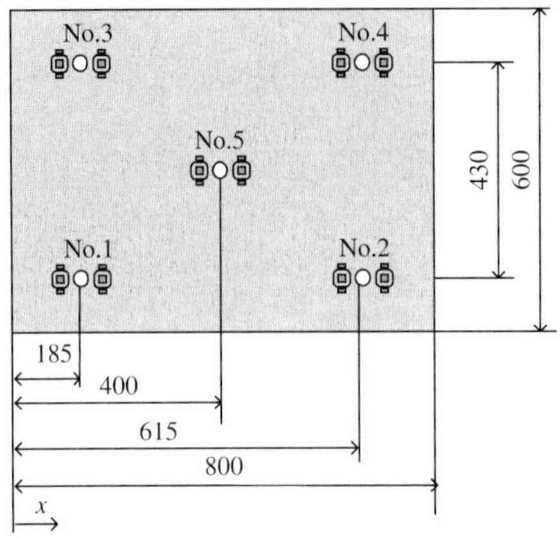

Fig. 1. Magnetic levitation control system.

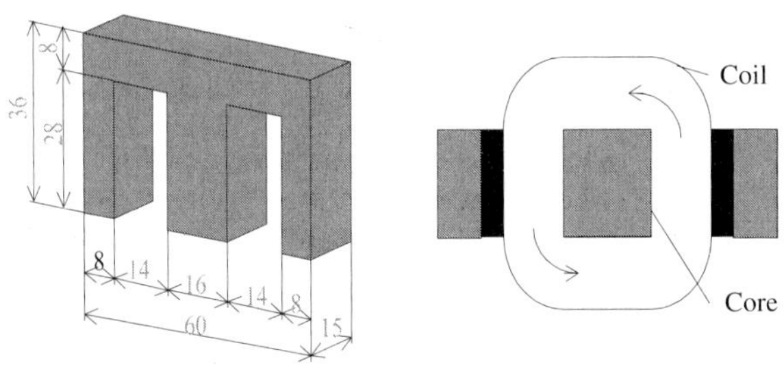

Fig. 2. Dimension of electromagnet.

the steel plate. The configuration of the electromagnet is shown in Fig. 2. The number of coil turns is 1005 and the resistance and inductance of the coil are 9.51 Ω and 0.277 H, respectively.

The attractive magnetic force is obtained with the help of the finite element method using "EM solution" [3]. When the current of the electromagnet is 0.54 A, that is for supporting the steel plate by the distance 5 mm apart from the electromagnet, the attractive magnetic force is given as shown in Fig. 3. From this figure, it is found that the attractive magnetic force is given to a small area of the steel plate. So, in this simulation, it is supposed that the attractive magnetic force is given to a point on the steel plate.

The relationship between the attractive magnetic force and electric current of electromagnet is shown in Fig. 4 when the distance between the electromagnet and the steel plate is 5 mm. The relationship between the attractive magnetic force and the distance between the electromagnet and the steel plate is shown in Fig. 5 when the electric current of the electromagnet is 0.54 A.

The steel plate vibrates transversely in the lengthwise direction when the length is much larger than the width. We simulate the magnetic levitation system when the steel plate vibrates in the lengthwise direction. The attractive magnetic force given to the steel plate is given by integrating with respect to the widthwise direction. A two-dimensional analysis model for the magnetic levitation control system is shown in Fig. 6. The magnetic actuator consists of two electromagnets and a sensor. The sensor is placed between the two electromagnets.

The motion equation of the steel plate is given as follows [4].

$$\rho h \ell_y \frac{\partial^2}{\partial t^2} + C_3 D \frac{\partial}{\partial t} \frac{\partial^4}{\partial x^4} w + ED \frac{\partial^4}{\partial x^4} w = \rho h \ell_y g + F_m$$

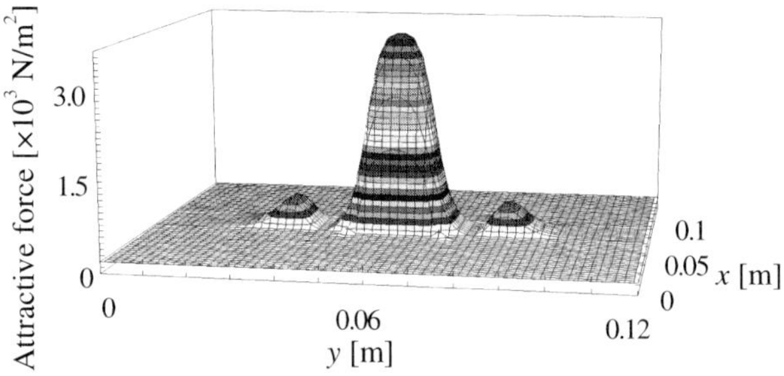

Fig. 3. Distribution of attractive force.

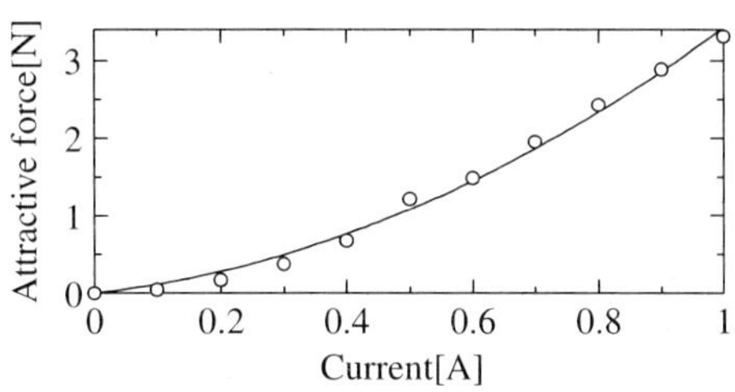

Fig. 4. Relationship between current and attractive force.

$$D = 3l_y h/12 \tag{1}$$

w: Displacement of steel plate [m]
F_m: Levitation force given to steel plate per unit length [N/m]
ρ: Density of the steel [kg/m^3]
C_s: Internal damping coefficient [N·s/m]
E: Young's modulus [N/m^2]
D: Geometrical moment of inertia [m^4]
g: gravitational acceleration [m/s^2]
t: Time [s]
x: Position of the steel plate along the length [m]

where $\rho = 7.27 \times 10^3$ kg/m^3, $E = 1.97 \times 10^{11}$ N/m^2, $g = 9.8$ m/s^2 and $C_s = 3.75 \times 10^8$ N·s/m^2.

The boundary conditions are given as

$$\frac{\partial^2 w}{\partial x^2} = 0, \quad \frac{\partial^3 w}{\partial x^3} = 0 \tag{2}$$

The motion equation of the steel plate is solved with the help of the finite difference method applying Crank-Nicolson Method [5]. Then the motion equation of the steel plate is written as follows.

$$\frac{v_{n+1}}{\Delta t} + \frac{C_s D}{2\rho h l_y}\frac{\partial^4}{\partial x^4}v_{n+1} = \frac{v_n}{\Delta t} - \frac{C_s D}{2\rho h l_y}\frac{\partial^4}{\partial x^4}v_n - \frac{ED}{\rho h l_y}\frac{\partial^4}{\partial x^4}w_n + g + \frac{F_m}{\rho h l_y} \tag{3}$$

$$w_{n+1} = w_n + v_{n+1}\Delta t \tag{4}$$

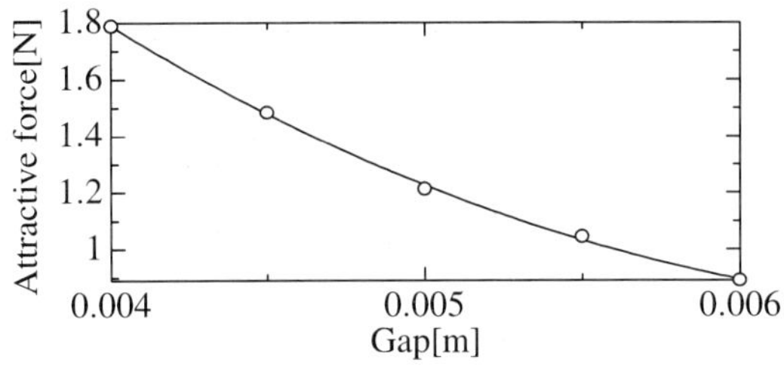

Fig. 5. Relationship between gap and attractive force.

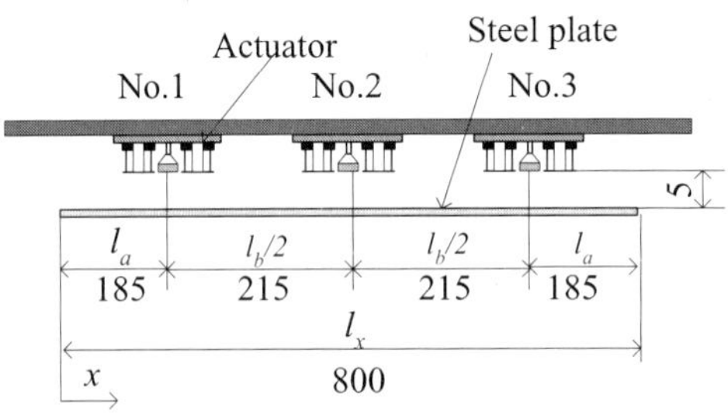

Fig. 6. Two-dimensional analysis model of the magnetic levitation control system.

where v is velocity of the steel plate [m/s], Δt is the time interval [s] and n is number of the time intervals. From Eq. (3), when v and w at time corresponding to n are known, we can get v at time corresponding to $n+1$. Then by substituting v to Eq. (4), we can get w at time corresponding to $n+1$. We can get vibration of steel plate by those repetitions.

The finite difference approximation of the terms in Eqs (1) and (2) becomes as follows.

$$\frac{\partial^2 w}{\partial x^2} = \frac{w(i+1,n) - 2w(i,n) + w(i-1,n)}{\Delta x^2}$$

$$\frac{\partial^3 w}{\partial x^3} = \frac{1}{2}\left\{\frac{w(i+2,n) - 3w(i+1,n) + 3w(i,n) - w(i-1,n)}{\Delta x^3} + \frac{w(i+1,n) - 3w(i,n) + 3w(i-1,n) - w(i-2,n)}{\Delta x^3}\right\}$$

$$\frac{\partial^4 w}{\partial x^4} = \frac{w(i+2,n) - 4w(i+1,n) + 6w(i,n) - 4w(i-1,n) + w(i-2,n)}{\Delta x^4}$$

where Δx is space interval and i is number of the space intervals. The finite difference approximation corresponding to v is given similarly.

In finite difference approximation of Eq. (1), fictitious points over the edge are required. The fictitious points are determined by utilizing the boundary condition [6].

In this analysis, the steel plate is divided into 80 sections, time interval is 0.00001 [s].

3. Vibration of steel plate in impulsive disturbance

We simulate vibration of the steel plate when impulsive disturbance is given at the center of the steel plate. Since the levitation system is essentially unstable, the steel plate is levitated by controlling the

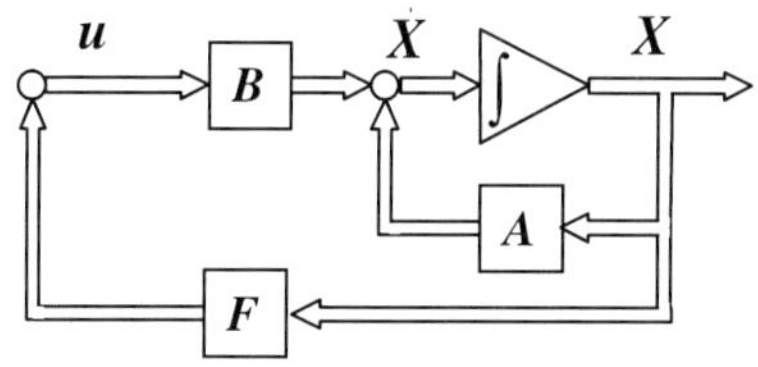

Fig. 7. Block diagram of Levitation system.

attractive force produced by the electromagnets.

In the system shown in Fig. 7, state variables are defined as general coordinate, W_1, W_2, W_3, General velocity W_1, W_2, W_3, and the current of the electromagnet i_1, i_2, i_3. The control inputs are e_1, e_2, e_3.

Then, the state equation is given as

$$\begin{aligned} \dot{\boldsymbol{X}} &= \boldsymbol{A}\boldsymbol{X} + \boldsymbol{B}\boldsymbol{u} \\ X &= [W_1\ W_2\ W_3\ \dot{W}_1\ \dot{W}_2\ \dot{W}_3\ i_1\ i_2\ i_3]^T \\ \boldsymbol{u} &= [e_1\ e_2\ e_3]^T \end{aligned} \tag{5}$$

The relationship between i and e is given as

$$e_i = R_i i_i + L_i \frac{d}{dt} i_i \quad (i = 1, 2, 3) \tag{6}$$

where R is resistance and L is inductance of the electromagnet.

The feedback gain F is defined as

$$F = \begin{pmatrix} k_{11}\ k_{12}\ k_{13}\ k_{14}\ k_{15}\ k_{16}\ k_{17}\ k_{18}\ k_{19} \\ k_{21}\ k_{22}\ k_{23}\ k_{24}\ k_{25}\ k_{26}\ k_{27}\ k_{28}\ k_{29} \\ k_{31}\ k_{32}\ k_{33}\ k_{34}\ k_{35}\ k_{36}\ k_{37}\ k_{38}\ k_{39} \end{pmatrix}.$$

3.1. *Magnetic levitation control by local control*

In the feedback control system shown in Fig. 8, the general coordinate W is defined as the distance between the surface of the actuator and the steel plate w. Feedback signals to the electromagnets come from the sensor placed between two electromagnets.

We simulate the magnetic levitation. The magnetic levitation system contains two set of the feedback control system. We call the magnetic levitation control 'local control'.

The feedback gains of the magnetic levitation system are set as follows.

$$k_w = 170000, \quad k_v = 4500, \quad k_i = 450$$

The actuators are placed at the position where the attractive force acts at each node position of the first elasticity vibration mode of the steel plate (Case A), 2 cm inside the node position (Case B) and 2 cm outside the node position (Case C).

The results of simulation are shown in Fig. 9. Time history of the vibration of the steel plate at the position of the actuator and that at the center of the steel plate are shown. In addition, the shape of the steel plate in steady state is contained. From these results, we notice that the position of the electromagnets affects the characteristic of the magnetic levitation control.

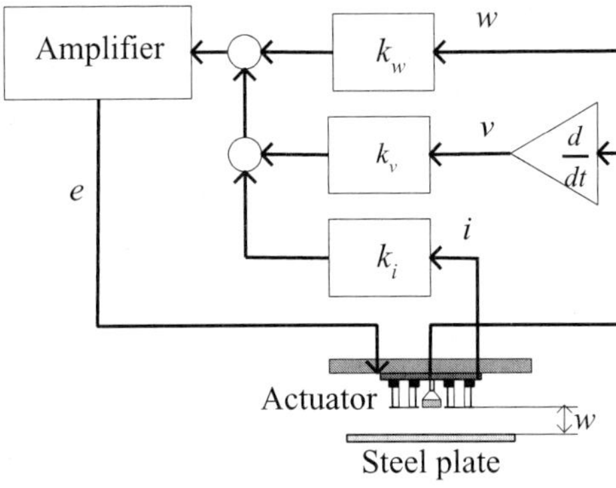

Fig. 8. Feedback control system.

In order to investigate the adequacy of simulation, we compare the shape of the steel plate in steady state, that is the state when the vibration disappears, with that bent by gravity.

The shape of the steel plate bent by gravity is given as follows.

$$0 \leqslant x \leqslant l_a$$

$$x_1 = x$$

$$w = -\frac{\rho g}{ED}\left\{\frac{1}{24}x_1^4 + \left(-\frac{1}{6}l_a^3 + \frac{1}{12}l_b^3 + \frac{1}{4}l_a^3 l_b\right)x_1 + \left(-\frac{1}{24}l_a^4 + \frac{1}{12}l_a l_b^3 + \frac{1}{4}l_a^3 l_b\right)\right\}.$$

$$l_a \leqslant x \leqslant l_a + l_b$$

$$x_2 = x - l_a$$

$$w = -\frac{\rho g}{ED}\left\{\frac{1}{24}x_2^4 - \frac{1}{12}l_b x_2^3 + \frac{1}{4}l_a^2 x_2^2 + \left(\frac{1}{12}l_b^3 + \frac{1}{4}l_a^2 l_b\right)x_2\right\}$$

$$l_b \leqslant x \leqslant l_x$$

$$x_3 = x - l_a - l_b$$

$$w = -\frac{\rho g}{ED}\left(\frac{1}{24}x_3^4 - \frac{1}{6}l_b x_3^3 + \frac{1}{4}l_a^2 x_3^2 + \frac{3}{4}l_a^2 l_b x_3\right)$$

The values obtained by these equations are almost the same as those shown in Fig. 9(a). It is confirmed that simulation by the Crank-Nicolson method is adequate.

In order to suppress the first elasticity vibration mode, three set of feedback control system shown in Fig. 8 are used. The actuators are placed as shown in Fig. 6. The feedback gains of the added feedback control system are the same as those in two set of the feedback control system.

The results of simulation are shown in Fig. 10. Time history of the vibration of the steel plate at the position of the side actuator and that at the position of the center actuator are shown. In addition, the shape of the steel plate in steady state is contained.

From these results, it is noticed that the magnetic levitation system using three set of the feedback control system can depress the elasticity vibration effectively.

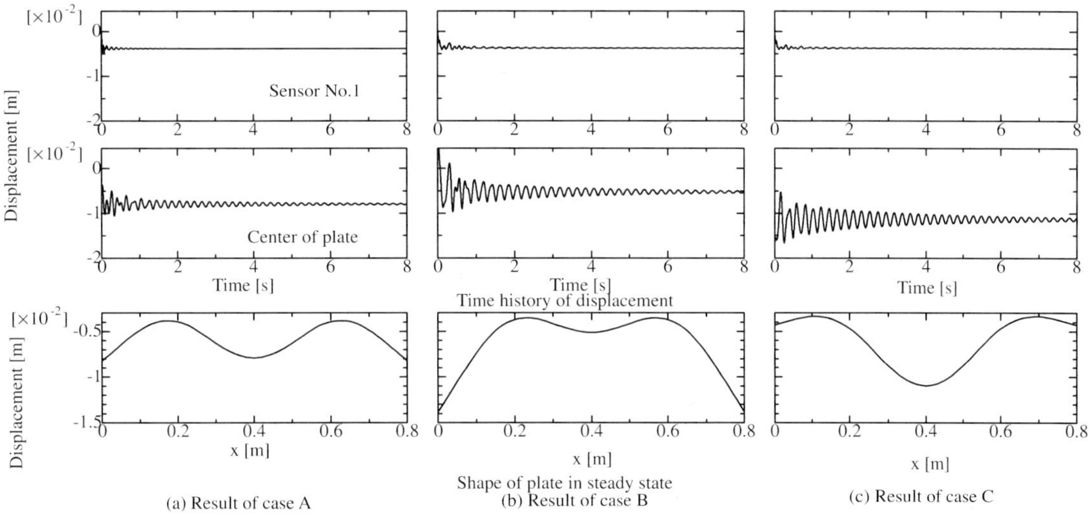

Fig. 9. Result of simulation.

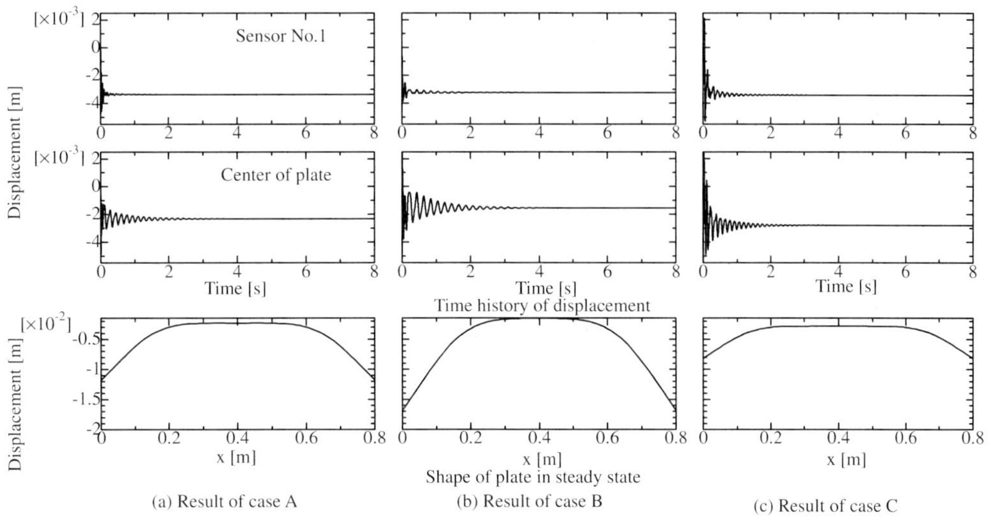

Fig. 10. Result of simulation.

3.2. Magnetic levitation control by optimal control

We compare the control performance of local control with optimal control [7] in order to examine the effect of the control system with 3 actuators. The actuators are placed as shown in Fig. 6.

Displacement of the steel plate from stable position $\Delta w(x,t)$ is supposed to be represented by using the mode shape $\phi(x)$ and magnitude $W(t)$ as

$$\Delta w(x,t) = \sum_{i=1}^{3} \phi_i(x) W_i(t). \tag{7}$$

Each mode shape is given as

$$\phi_1(x) = 1$$

$$\phi_2(x) = \frac{\sqrt{3}}{l_x}(2x - l_x)$$

$$\phi_3(x) = \cos\frac{\lambda_x}{l_x}x + \cosh\frac{\lambda_x}{l_x} + \frac{\sin\lambda_x + \sinh\lambda_x}{\cos\lambda_x - \cosh\lambda_x}\left(\sin\frac{\lambda_x}{l_x}x + \sinh\frac{\lambda_x}{l_x}xx\right)$$

where λ_x is the solution of the equation of

$$\cosh\lambda_x \cos\lambda_x = 1. \tag{8}$$

Substituting Eq. (7) into Eq. (1), multiplying both sides of Eq. (1) by $\phi_j(x)$ $(j = 1, 2, 3)$ and integrating Eq. (1) from $x = 0$ to $x = l_x$, we get the state equations for each mode.

The feedback gain $\boldsymbol{F}$ is obtained with the help of the optimal control theory [7].

The performance index is given as

$$J = \lim_{tf\to\infty}\int_0^{tf}(\boldsymbol{X}^T\boldsymbol{Q}\boldsymbol{X} + \boldsymbol{u}^T\boldsymbol{R}\boldsymbol{u})dt \tag{9}$$

where $\boldsymbol{X}$ is state variable, $\boldsymbol{u}$ is the control input and $\boldsymbol{Q}$ and $\boldsymbol{R}$ are the weighting matrices given as

$$\boldsymbol{Q} = \begin{bmatrix} q_1 & 0 & \ldots & 0 \\ 0 & q_2 & \ddots & \vdots \\ \vdots & \ddots & \ddots & 0 \\ 0 & \ldots & 0 & q_9 \end{bmatrix} \quad \text{and} \quad \boldsymbol{R} = \begin{bmatrix} r_1 & 0 & 0 \\ 0 & r_2 & 0 \\ 0 & 0 & r_3 \end{bmatrix}.$$

The control input $\boldsymbol{u}$ which makes the performance index minimum is given as

$$\boldsymbol{u} = -\boldsymbol{R}^{-1}\boldsymbol{B}^T\boldsymbol{P}\boldsymbol{X} = -\boldsymbol{F}\boldsymbol{X} \tag{10}$$

where $\boldsymbol{F}$ is feedback gain, $\boldsymbol{P}$ is the unique positive-definite solution of following Riccati's equation.

$$\boldsymbol{A}^T\boldsymbol{P} + \boldsymbol{P}\boldsymbol{A} - \boldsymbol{P}\boldsymbol{B}\boldsymbol{R}^{-1}\boldsymbol{B}^T\boldsymbol{P} + \boldsymbol{Q} = 0 \tag{11}$$

In simulation, supposing $q_n = 1$ and $r_n = 0.00001$ with $n = 1, 2, 3$, the feed back gain is obtained as follows.

The feedback gain for optimal control is $k_{11} = k_{33} = 176000$, $k_{12} = k_{32} = -63200$, $k_{13} = k_{31} = 60900$, $k_{14} = k_{36} = 3320$, $k_{15} = k_{35} = -1510$, $k_{16} = k_{34} = 958$, $k_{17} = k_{39} = 259$, $k_{18} = k_{38} = 0.655$, $k_{19} = k_{37} = 1.48$, $k_{21} = k_{23} = -81800$, $k_{22} = 174000$, $k_{24} = k_{26} = -2000$, $k_{25} = 4180$, $k_{27} = K_{29} = 0.495$, $k_{28} = 450$.

The feedback gain for local control is $k_{11} = k_{33} = 18000$, $k_{22} = 170000$, $k_{14} = k_{36} = 3300$, $k_{25} = 4200$, $k_{17} = k_{39} = 260$, $k_{28} = 450$ and others $= 0$.

The results of the simulation are shown in Fig. 11. In these figures, (a) is the results by the optimal control and (b) is those by the local control. Time history of the vibration at the position under the side sensor (sensor No. 1 and 3 shown in Fig. 6) and at the position under the center sensor (sensor No. 2 shown in Fig. 6) are shown. In addition, the shape of the steel plate in steady state is contained.

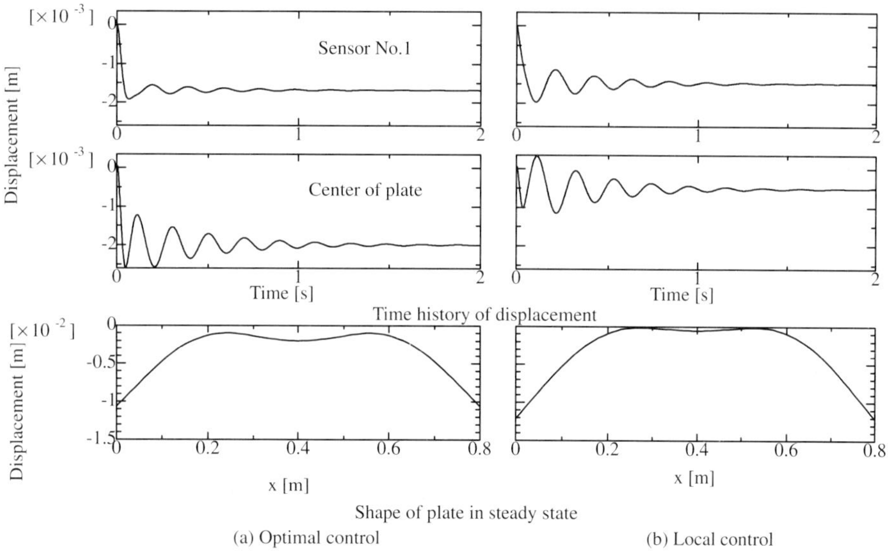

Fig. 11. Result of simulation.

From these results, it is noticed that the control quality by the optimal control is better than that by the local control in vibration and flatness of the steel plate. However, when we employ the local control, we can adjust the control characteristic easily because the actuator is controlled independently by the feedback signals from the sensor.

4. Conclusion

We have studied computer simulation of the attractive magnetic levitation system of the thin steel plate. The attractive force by the electromagnets are controlled by feedback signals the displacement and velocity of the steel plate and the current of the electromagnets. We simulated the displacement of the steel plate when a disturbance is given to the levitation system. It is found that the vibration of the steel plate is affected considerably by gravity in levitating the steel plate. It is necessary to simulate the magnetic levitation by taking influence of gravity into consideration in order to study how to construct the magnetic levitation system.

References

[1] Y. Oshinoya, Study on electromagnetic levitation conveyance of a steel plate with free edges by using electromagnets and permanent magnets, *Journal of Japan Society Applied Electromagnetics and Mechanics (in Japanese)* **6**(4) (October 1998), 364–371.
[2] K. Ishibashi and Y. Oshinoya, Simulation of Magnetic Levitation of a Steel Plate, NON-LINEAR ELECTROMAGNETIC SYSTEMS (ISEM'99) in: *Studies in Applied Electromagnetics and Mechanics,* (Vol. 18), P.D. Barba and A. Savini, eds, IOS Press, 2000, pp. 489–492.
[3] A. Kameari, Local force calculation in 3D FEM with edge elements, *International Journal of Applied Electromagnetics in Materials* **3** (1993), 231–240.
[4] R.W. Clough and J. Penzien, *DYNAMICS OF STRUCTURES,* McGraw-Hill, 1975, pp. 301–302.
[5] C.F. Gerald and PO. Wheatly, *APPLIED NUMERICAL ANALYSIS,* Addison-Wesley Publishing, 1989, pp. 555–558.
[6] S. Timoshenko and S. Woinowsky-krieger, *Theory of Plates and Shells,* McGraw-Hill, 1959, pp. 351–362.
[7] K. Ogata, *MODERN CONTROL ENGINEERING,* Prentice-Hall, 1997, pp. 915–059.

International Journal of Applied Electromagnetics and Mechanics 13 (2001/2002) 45–51
IOS Press

Characteristics of the forced vibration with magnetic damping

Yoshikazu Tanaka* and Tomoyoshi Horie
Department of Mechanical System Engineering, Kyushu Institute of Technology 680-4 Kawazu, Iizuka, Fukuoka 820-8502, Japan

Abstract. It is required to clarify the characteristics of forced vibration with magnetic damping for an earthquake-resistant design of conductive thin shell structures located in an electromagnetic field. In this paper, the characteristics of forced vibration are examined using finite element tightly coupled analysis and single-degree-of-freedom coupled model. The results show that the change of resonance frequency is different from that of general viscous damping problems for some combinations of material properties and magnetic fields. It is pronounced when the eddy current decay rate is close to the natural period of the structure.

1. Introduction

A mechanical vibration of thin shell structures located in a strong magnetic field such as fusion reactor components and magnetically levitated vehicles may be suppressed by magnetic damping effect caused by electromagnetic and structural coupled phenomenon. Therefore, it is important to evaluate the electromagnetic and structural coupled effect or magnetic damping effect. Especially, it is required to clarify the characteristics of forced vibration with magnetic damping from the view point of an earthquake-resistant design.

Over the past few years, several attempts have been made on the magnetically damped vibration. Experiments and benchmark analyses on bending and torsion of cantilevered plates in magnetic fields have been made [1–3]. The critical magnetic viscous damping ratio has been proposed by Takagi, which can be used for the approximate solution method for electromagnetic mechanical coupling analysis [4]. The authors have proposed the coupling intensity parameter as the evaluation parameter for the magnetic damping characteristics and its validity has been verified [5]. But the characteristics of forced vibration with magnetic damping have not been examined yet.

In general damping or viscous damping problem, the characteristics of vibration are based on the second order differential equation. The characteristics of the magnetically damped vibration, however, are based on the third order differential equation and the absolute value of the eigenvalue is not constant. Thus the resonance characteristics of magnetic damping should be different from those of general damping. In this paper, the unusual change of the resonance frequency which is important for an earthquake-resistant design is discussed, and the resonance characteristics for the case when the eddy current decay rate is close to the natural period of structure are also discussed.

*Corresponding author: Yoshikazu Tanaka, Department of Mechanical System Engineering, Kyushu Institute of Technology, 680-4 Kawazu, Iizuka, Fukuoka 820-8502, Japan; Tel.: +81 948 29 7757; Fax: +81 948 29 7751; E-mail: yoshi@solid.mse.kyutech.ac.jp.

2. Evaluation method of the forced vibration with magnetic damping

2.1. The equation of motion with magnetic damping

The equation of motion for the coupled problem is expressed using displacement u and the normal component T of the current vector potential as

$$m\ddot{u} + C_s T + ku = F^{ex}\,, \tag{1}$$

where F^{ex} and $C_s T$ mean the external force and the coupling term by the electromagnetic force respectively. The equation of the eddy current becomes

$$U\dot{T} + C_e\dot{u} + RT = \dot{B}^{ex}\,, \tag{2}$$

where $U, R, \dot{B}^{ex}$, and $C_e\dot{u}$ mean the inductance, resistance, change of external magnetic field, and coupling term by the electromotive force respectively. Elimination of T from Eqs (1) and (2) yields the third order differential equation [6]

$$mU\dddot{u} + mR\ddot{u} + (kU - C_eC_s)\dot{u} + kRu = U\dot{F}^{ex} + RF^{ex} - C_s\dot{B}^{ex}\,. \tag{3}$$

The roots of the characteristic equation of Eq. (3) are three real numbers for discriminant $D \geqslant 0$, and a real number γ_e and a pair of complex numbers $-\gamma_s \pm \omega_s i$ for $D < 0$. The discussion of the resonance characteristics below are made for $D < 0$. Then, the coupled eigenvalues are expressed as [5]

$$-\gamma_e = -\frac{R}{3U} + \left\{b - \sqrt{b^2 + a^3}\right\}^{\frac{1}{3}} + \left\{b + \sqrt{b^2 + a^3}\right\}^{\frac{1}{3}}\,, \tag{4}$$

$$\begin{aligned}-\gamma_s \pm \omega_s i = &-\frac{R}{3U} - \frac{\left\{b - \sqrt{b^2 + a^3}\right\}^{\frac{1}{3}} + \left\{b + \sqrt{b^2 + a^3}\right\}^{\frac{1}{3}}}{2} \\ &\pm \frac{-\left\{b - \sqrt{b^2 + a^3}\right\}^{\frac{1}{3}} + \left\{b + \sqrt{b^2 + a^3}\right\}^{\frac{1}{3}}}{2}\sqrt{3}\, i\,,\end{aligned} \tag{5}$$

where

$$a = -\frac{1}{9}\frac{R^2}{U^2} + \frac{1}{3}\frac{k}{m} - \frac{1}{3}\frac{C_eC_s}{mU}\,, \qquad b = -\frac{1}{27}\frac{R^3}{U^3} - \frac{1}{3}\frac{k}{m}\frac{R}{U} - \frac{1}{6}\frac{R}{U}\frac{C_eC_s}{mU}\,.$$

2.2. The resonance curve under harmonic force with magnetic damping

If the external force is given by $F^{ex} = F_0 \sin(\omega_f t)$ and the external magnetic field $\dot{B}^{ex}$ is neglected, the resonance curve obtained from Eq. (3) is expressed by

$$\frac{A}{u_0} = \sqrt{\frac{\gamma_0^2 + \omega_f^2}{(\gamma_e^2 + \omega_f^2)\{(\gamma_s^2 + \omega_s^2 + \omega_f^2)^2 - 4\omega_f^2\omega_s^2\}}\frac{k}{m}}\,, \tag{6}$$

where u_0, γ_0 and ω_0 mean static deflection, uncoupled eigenvalue of eddy current and natural angular frequency respectively. Although it is possible to obtain the resonance frequency from Eq. (6), it is a

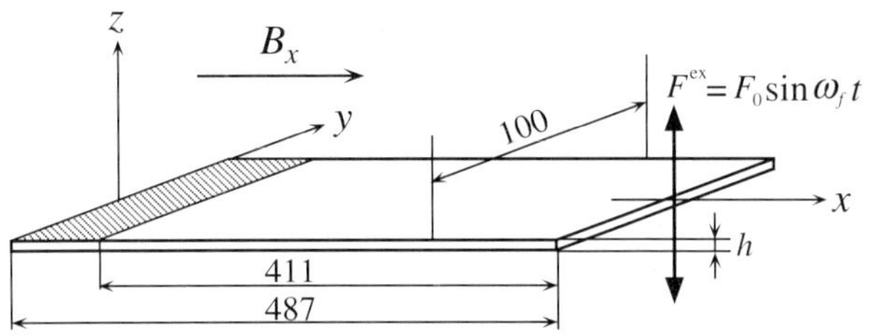

Fig. 1. A bending plate with a harmonic force in a steady magnetic field B_x.

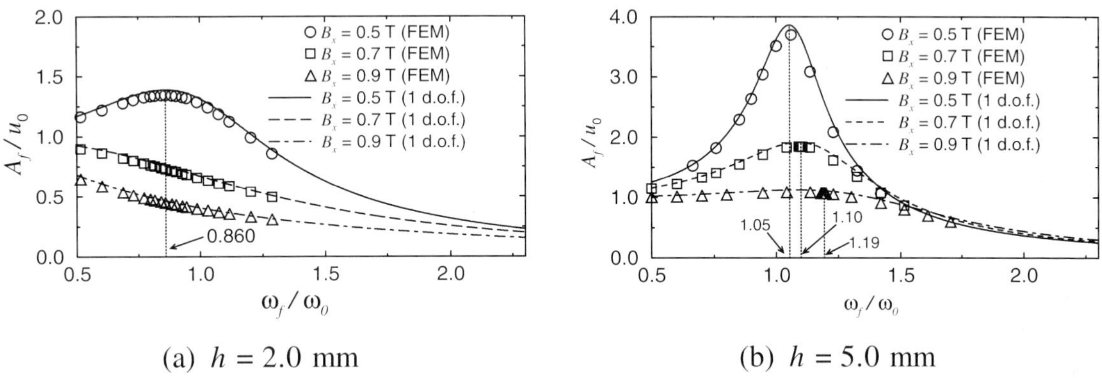

Fig. 2. Resonance curve under the harmonic force.

little complicated to discuss the change of the resonance frequency analytically because of the cubic roots in Eqs (4) and (5). Then, $\gamma_e \approx \gamma_0$ is assumed in Eq. (6) because of the small change of γ_e with the steady magnetic field. Upon differentiating Eq. (6) by ω_f, the resonance frequency becomes

$$\omega_f = \sqrt{\omega_s^2 - \gamma_s^2}\,. \tag{7}$$

Since γ_e, γ_s, ω_s are the function of the steady magnetic field B, Young's modulus E, mass density ρ, electric conductivity κ and thickness h, the resonance curve Eq. (6) and the resonance frequency Eq. (7) are also the function of these parameters [6].

3. Results of the forced vibration with magnetic damping

3.1. Analysis model

The resonance characteristics of a cantilevered copper plate in a magnetic field shown in Fig. 1 are examined by changing the steady magnetic field B_x and thickness h. Though the geometry of the plate is the same as TEAM problem 12 [2], a harmonic force is applied at the free end of the plate.

3.2. Results with different thickness

Figure 2 shows the amplitude A at the free end of the plate with either 2.0 or 5.0 mm thickness as a function of the frequency ω_f of the harmonic force. The lines are the results of 1 d.o.f. model Eq. (6)

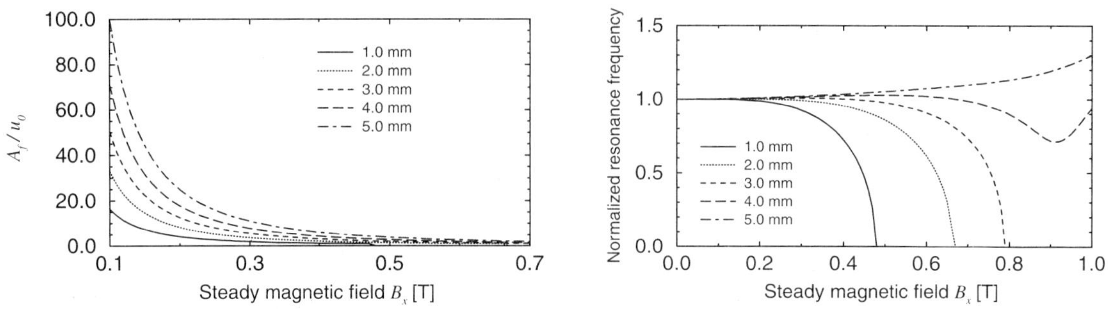

(a) Normalized amplitude (b) Normalized resonance frequency

Fig. 3. Change of resonance frequency and amplitude under the harmonic force.

and the symbols are those of finite element tightly coupled analyses. The latter results are obtained by the coupled finite element analysis code that uses the isoparametric 8-node shell element and which is developed by the authors [7]. The characteristics for 2.0 mm thickness are similar to those of the general damping. On the other hand, the characteristics for 5.0 mm thickness are different from those of the general damping; the resonance frequency increases with larger damping and the resonance curves are crossing each other. Since the results of 1 d.o.f. model Eq. (6) agree well with those of the finite element tightly coupled analysis, 1 d.o.f. model Eq. (6) is verified. Then, the resonance characteristics are examined using Eq. (6). Since the results of Eq. (7) also agree with the resonance frequencies of Eq. (6) the validity of the approximation of Eq. (7) is also verified.

The relation between the normalized amplitude at the resonance frequency and the steady magnetic field obtained from Eq. (6) is shown in Fig. 3(a). The amplitude becomes smaller with the stronger steady magnetic field because of the magnetic damping effect. The relation between the normalized resonance frequency and the steady magnetic field obtained from Eq. (7) is shown in Fig. 3(b). The resonance frequency of the thick plate is increased, which is in contrast with the general viscous damping. This may be the feature of the resonance characteristics based on the third order differential equation. Since the ratio of the eddy current decay rate divided by the natural period of structure R_T is 0.076 for 5.0 mm thickness and 0.012 for 2.0 mm thickness, the feature is pronounced when R_T is larger. Further consideration will be made in Section 3.3.

The coupled eigenvalues γ_s and ω_s are shown in Fig. 4 as a function of the steady magnetic field. Since $|\gamma_s|$ increases with B_x for 2.0 mm thickness, the resonance frequency $\sqrt{\omega_s^2 - \gamma_s^2}$ is decreased as shown in Fig. 3(b). When B_x becomes larger than 0.66 T, no resonance frequency is obtained because $|\gamma_s| > \omega_s$. In the case of 4.0 mm thickness, ω_s increases with B_x, and $|\gamma_s|$ does not exceed ω_s. $|\gamma_s|$ is approaching ω_s when $B_x < 0.9$ T, but it is separating away from ω_s when $B_x > 0.9$ T. Therefore, the resonance frequency $\sqrt{\omega_s^2 - \gamma_s^2}$ is decreasing when $B_x < 0.9$ T, but it is increasing when $B_x > 0.9$ T as shown in Fig. 3(b).

3.3. Results when the eddy current decay rate is close to the natural period of structure

The influence of ratio of the eddy current decay rate and the natural period of structure on the characteristics of magnetically damped vibration is discussed.

Figure 5(a) shows the resonance curve of 2.0 mm thickness, $E = 1.10\times10^{13}$ Pa, $\rho = 8.91\times10^3$ kg/m^3, $\kappa = 5.81 \times 10^7$ S/m, R_T of which is 0.121. Figure 5(b) also shows the resonance curves for $E = 1.10\times10^{14}$ Pa, R_T of which is 0.510. Compared to Fig. 2(b), in which $E = 1.10\times10^{11}$ Pa ($R_T = 0.076$),

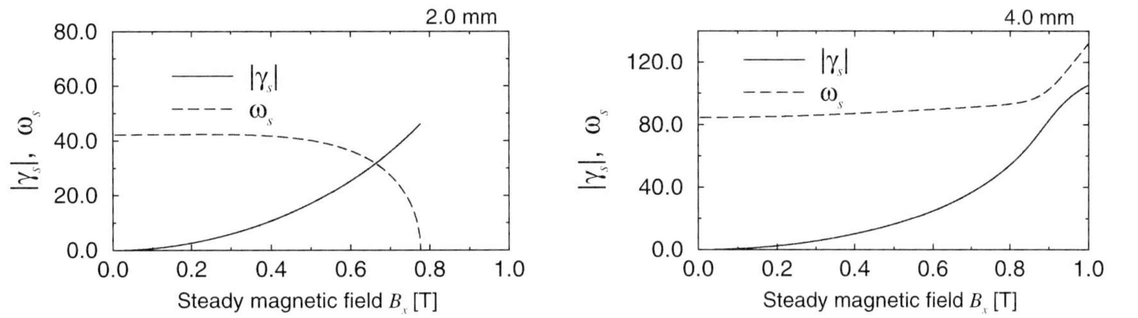

(a) $h = 2.0$ mm (b) $h = 4.0$ mm

Fig. 4. Change of coupled eigenvalues with magnetic field for different plate thickness.

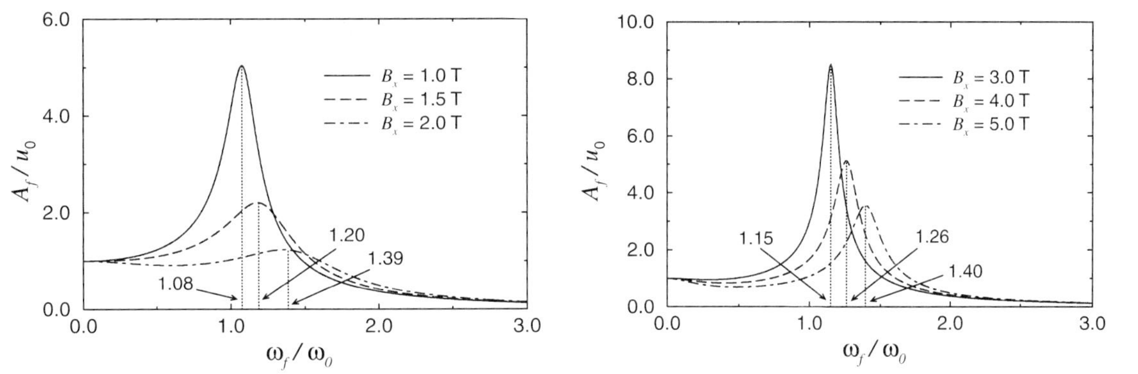

(a) $E = 1.10\ \ 10^{13}$ Pa (b) $E = 1.10\ \ 10^{14}$ Pa

Fig. 5. Resonance curves of the forced vibration with different Young's modulus.

the resonance frequency becomes much larger when the eddy current decay rate is close to the natural period of structure.

Equation (7) can be rewritten using Eqs (4) and (5) and R_T as

$$\sqrt{\omega_s^2 - \gamma_s^2} = \sqrt{\frac{k}{m}\left(1 - \frac{1}{8\pi^2 R_T^2}\right) - \frac{C_e C_s}{mU} + \frac{1}{2}\gamma_e^2}\,. \tag{8}$$

Then, the resonance frequency with the magnetic damping is affected by $-\dfrac{C_e C_s}{mU}$ and γ_e. Figure 6 shows the relation between R_T and each term of Eq. (8) with increasing κ for $B_x = 0.2$ T and $h = 2.0$ mm. The 1st term is close to $\frac{k}{m}$, the 2nd term is constant and the 3rd term is decreasing when R_T becomes large. As the result, the resonance frequency, which is summation of each term, becomes large and is close to $\sqrt{\frac{k}{m} - \frac{C_e C_s}{mU}}$ for large R_T.

Since γ_e becomes small with large R_T as shown in Fig. 6 the decay of eddy current is slow. To discuss the effect of the change of γ_e to the magnetically damped vibration, the coupled finite element analyses are performed for small and large R_T. Results of deflection w and electromagnetic force F_z in the direction of z at the free end for $R_T = 0.012$, $B_x = 0.5$ T and $\omega_f = 43.5$ rad/s(near the resonance

International Journal of Applied Electromagnetics and Mechanics 13 (2001/2002) 53–58
IOS Press

Elastic vibration of steel plate with magnetic force for restraining horizontal position in magnetic levitation

Atsushi Igusa[a], Yasuo Oshinoya[b] and Kazuhisa Ishibashi[c,*]
[a]*Graduate Student, Course of Mechanical Engineering, Graduate School of Engineering, Tokai University, Japan*
[b]*Department of Mechanical Engineering, Future Science and Technology Joint Research Center, Tokai University, Japan*
[c]*Department of Mechanical Engineering, Faculty of Engineering, Tokai University, 2-28-4 Tomigaya, Shibuya-ku, Tokyo, 151-0063 Japan*

Abstract. It has been confirmed that a thin steel plate can be levitated by controlling the current ofelectromagnets. However, the plate gets out of position in the case of high-speed conveyance of the steel plate if there is no force for restraining the plate from slipping horizontally. Then, electromagnets for controlling the horizontal position are installed. In this paper, we study how the elastic vibration of the steel plate is affected by the magnetic field produced by the electromagnets. The magnetic force pulls back the steel plate to its center. With the magnetic field produced by the electromagnets, the frequency of the vibration increases and the vibration is attenuated more rapidly. In addition, the tensile force makes the steel plate flat.

1. Introduction

The authors have proposed a magnetic levitation control system for a thin steel plate and confirmed experimentally the realization of the levitation system [1]. However, even if the levitation control is applied for the steel plate, the steel plate gets out of position in high speed conveyance because there is no horizontal restraining force. Then, electromagnetic actuators are installed in order to control the horizontal position of the steel plate and the effectiveness is examined experimentally [2].

In order to construct better levitation system, we have studied how to simulate the magnetic levitation because we can modify easily the system in computer simulation. We have reported a simulation method for the magnetic levitation [3]. In this paper, we study how the elastic vibration of the steel plate is affected by the magnetic field produced by the actuators controlling the horizontal position.

*Corresponding author: Kazuhisa Ishibashi, Tel.: +81 3 3467 2211 ext. 381; E-mail: isibasi@keyaki.cc.u-tokai.ac.jp.

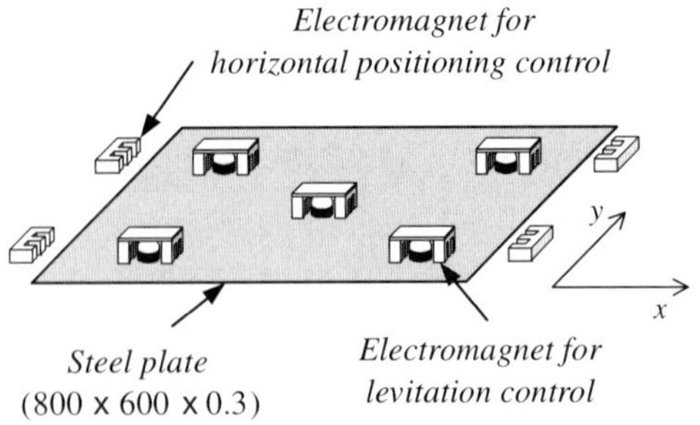

Fig. 1. Conveyance system of thin steel plate.

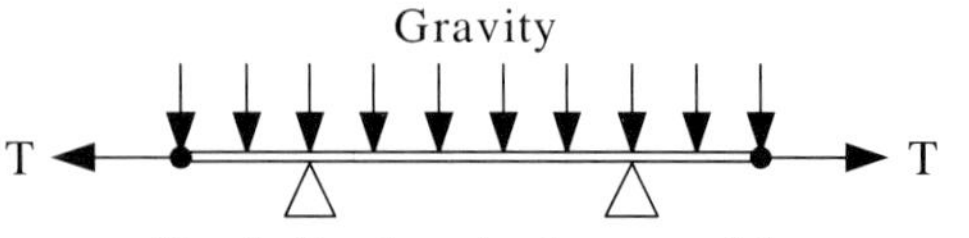

Fig. 2. Overhanging beam model.

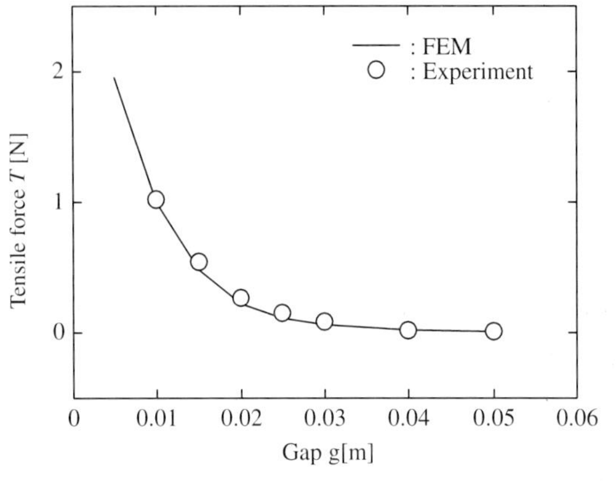

Fig. 3. Tensile force given to the edge of the plate.

2. Motion equation of thin steel plate

The conveyance system of thin steel plate is shown in Fig. 1. The size of a steel plate is as follows: length $l_x = 800$ mm, width $l_y = 600$ mm, thickness $h = 0.3$ mm. Five electromagnets are used for levitating the steel plate. At the same time, the horizontal position of the steel plate is controlled by four electromagnets for horizontal positioning control when the steel plate is conveyed to x direction. We investigate how the characteristics of elastic vibration of the steel plate are affected by the electromagnets for horizontal positioning control.

The steel plate vibrates transversely in the lengthwise direction when the length is much larger than the width. We obtain the characteristics of the steel plate when the steel plate vibrates in the lengthwise direction. In this case, the motion equation of the steel plate is given as follows.

$$\rho h l_y \frac{\partial^2}{\partial t^2} w + \frac{h^3 l_y}{12} C_s \frac{\partial}{\partial t} \frac{\partial^4}{\partial x^4} w + \frac{h^3 l_y}{12} E \frac{\partial^4}{\partial x^4} w - T \frac{\partial^2}{\partial x^2} w = \rho h l_y g + f_m \tag{1}$$

x: Position of the steel plate along the length [m]
w: Displacement of steel plate [m]
f_m: Levitation force given to steel plate per unit length [N/m]
ρ: Density of the steel [kg/m^3]
C_s: Internal damping coefficient [N·s/m]
E: Young's modulus [N/m^2]

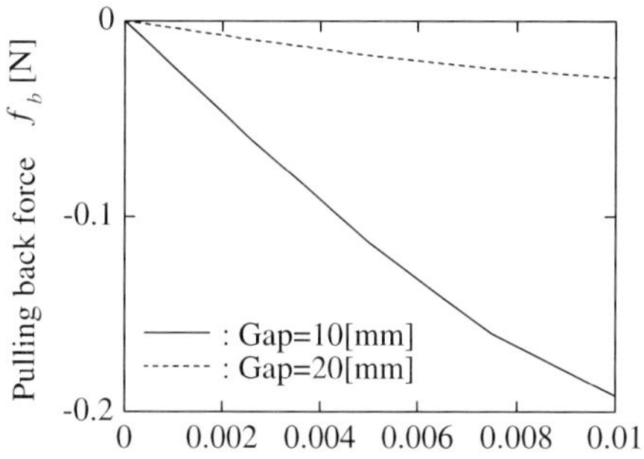

Fig. 4. Magnetic force pulling back to center.

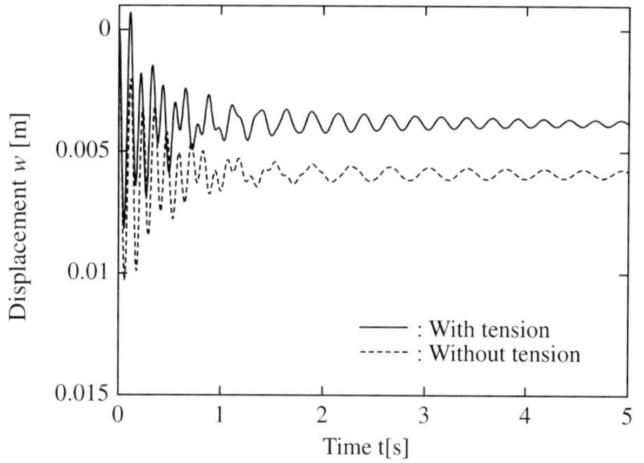

Fig. 5. Time history of the vibration at the edge of the steel plate.

T: Tensile force [N]
g: gravitational acceleration [m/s^2]
t: Time [s].

where $\rho = 7.27 \times 10^3$ kg/m^3 and $g = 9.8$ m/s^2. Young's modulus E and internal damping coefficient C_s is obtained experimentally by data of free vibration as 2.30×10^{11} N/m^2 and 2.55×10^8 Ns/m^2, respectively.

The magnetic force f_o given to the steel plate per unit area is obtained with the help of the finite element method and the magnetic force f_m is obtained by integrating f_o with respect to y.

3. Vibration of steel plate by impulsive disturbance

In order to investigate how the elastic vibration of the steel plate is affected by the magnetic field produced by the electromagnets controlling the horizontal position, we solve the vibration of the steel plate when the impulsive disturbance is given at the center of the steel plate. In this investigation, we introduce a modified conveyance system.

Firstly, we solve the vibration of the overhanging beam model shown in Fig. 2. Each supporting point is positioned at 0.18 m from each edge of the steel plate. The boundary condition at the free edge is given as

$$\partial^2 w/\partial x^2 = 0, \quad (h^3 \ell_y/12)E(\partial^3 w/\partial x^3) = T(\partial w/\partial x) + kw$$

with a spring constant k [N/m]. The boundary condition at the supporting point is given as $w = 0$.

The tensile force T is given by electromagnets. If the steel plate is moved up or downward, the electromagnet pulls back the steel plate to its center just as spring action. The tensile magnetic force T is obtained as shown in Fig. 3. The abscissa is the gap g between the edge of the steel plate and the surface of the electromagnet. The magnetic force f_b pulling back to its center is shown in Fig. 4. The abscissa is the movement distance s from the center of the magnet. The spring constant k is obtained from the slope given in Fig. 4.

The motion Eq. (1) of the steel plate is solved with the help of the finite difference method applying Crank-Nicolson Method [4]. In finite difference approximation of Eq. (1), fictitious points over the

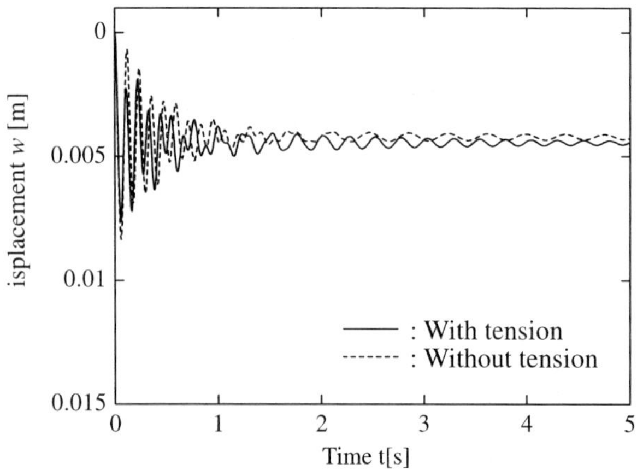

Fig. 6. Time history of the vibration at the center of the steel plate.

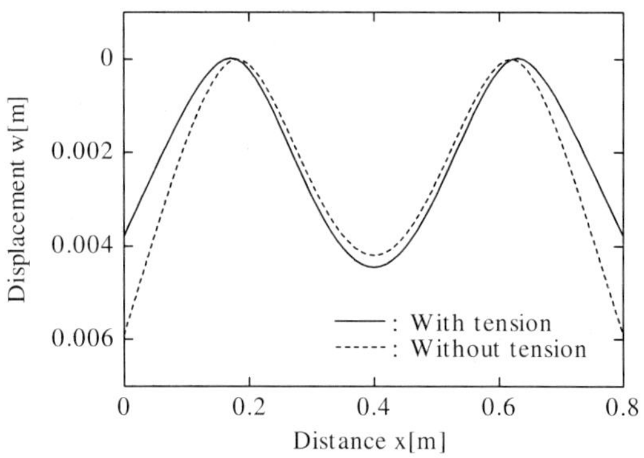

Fig. 7. Shape of the steel plate in the case of overhanging beam model.

edge are required. The fictitious points are determined by utilizing the boundary condition [5]. In this numerical analysis, the steel plate is divided into 80 sections and the time interval is 0.0001 [s].

The gap between the edge of the steel plate and the electromagnet is supposed to be 10 mm. Figure 5 shows the time history of the vibration at the edge of the steel plate. Figure 6 shows the time history of the vibration at the center of the steel plate. Figure 7 shows the shape of the steel plate in the steady state. The ordinate is the displacement of the steel plate from the position at $t = 0$.

Next, we solve the vibration of the magnetic levitation beam model shown in Fig. 8. Two electromagnets are placed at the same position, where the attractive force acts, as the supporting point in the case of the overhanging beam model.

We simulate motion of the steel plate when impulsive disturbance is given at the center of the steel plate. Since the levitation system is essentially unstable, the steel plate is levitated by controlling the attractive force produced by the electromagnet. The feedback control system is shown in Fig. 9. Feedback signals to the electromagnet come from the sensor.

In the control system shown in Fig. 9, the state variables are the gap w between the electromagnet and the steel plate, the velocity of the steel plate v given by differentiating the gap w with respect to time t and the current of electromagnets i. Then, the state equation is given as

$$\dot{\boldsymbol{X}} = \boldsymbol{A}\boldsymbol{X} + \boldsymbol{B}\boldsymbol{u} \tag{2}$$

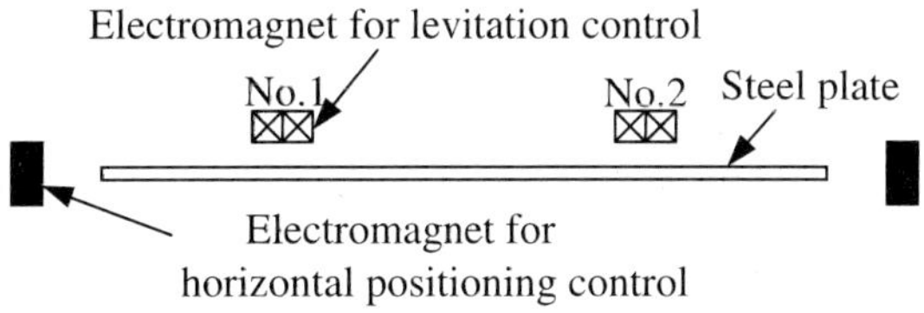

Fig. 8. Magnetic levitaion beam model.

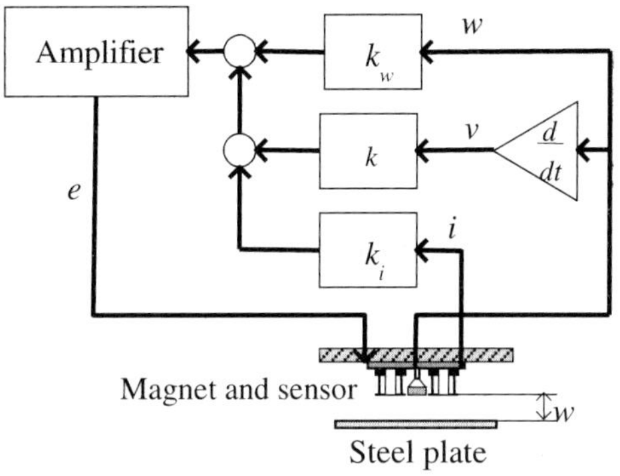

Fig. 9. Feedback control system for levitating steel plate.

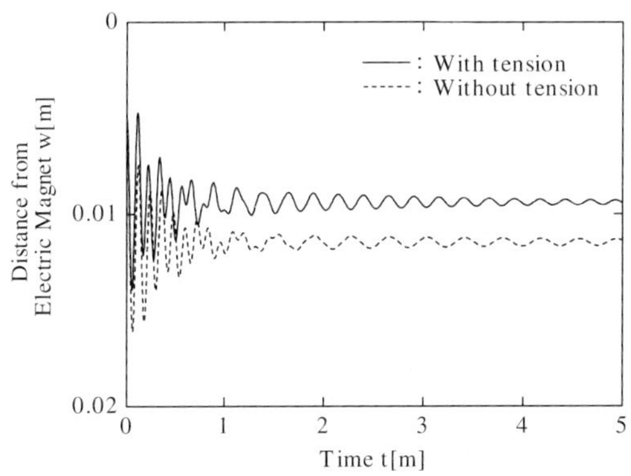

Fig. 10. Time history of the vibration at the edge of the steel plate.

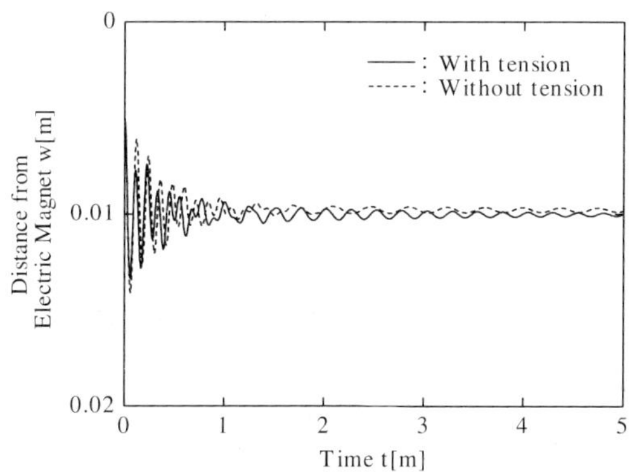

Fig. 11. Time history of the vibration at the center of the steel plate.

where $\boldsymbol{X} = [w\ v\ i]^T$, $\boldsymbol{u} = [e]^T$.

The voltage for supplying the electromagnet current is given as

$$e = Ri + L\frac{d}{dt}i \tag{3}$$

where the resistance R and the inductance L of the electromagnet are 19 Ω and 0.54 H, respectively.

The attractive magnetic force is supposed to be

$$f_m = f_a + \frac{\partial f}{\partial w}\Delta w + \frac{\partial f}{\partial i}\Delta i \tag{4}$$

where $\partial f/\partial i = 5.3$ N/A, $\partial f/\partial x = 670$ N/m and $f_a = 0.19$ N.

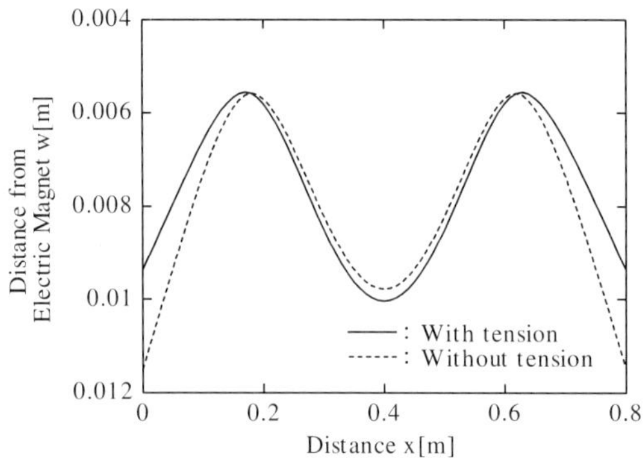

Fig. 12. Shape of the steel plate in the case of magnetic levitation beam model.

The feedback gains of the magnetic levitation system are set as follows.

$$k_w = 170000, \ k_v = 4500, \ k_i = 450$$

The motion equation of the steel plate is solved with the help of the finite difference method applying Crank-Nicolson Method. The gap between the edge of the steel plate and the electromagnet is supposed to be 10 mm as in the case of the overhanging beam model. In this numerical analysis, the steel plate is divided into 80 sections and the time interval is 0.0001 [s]. The numerical analysis is carried out in the same way as [3].

The results of the time history of the vibration are shown in Figs 10 and 11. The shape of the steel plate is shown in Fig. 12 when the vibration disappears.

By comparing these results with those in the case of the overhanging beam model, it is found out that the results of the simulation is almost the same. The ordinate of the former is the displacement of the steel plate from the position at t =0 and that of the latter is the distance between the steel plate and the electromagnet. It is confirmed that the simulation of the magnetic levitation is adequate.

4. Conclusion

We have examined how the elastic vibration of the steel plate is affected by the magnetic field produced by the electromagnet for controlling the horizontal position. The electromagnet pulls back the steel plate to its center and gives the tensile force to the steel plate. By the tensile force given to the steel plate by the electromagnet, the frequency of the vibration of the steel plate increases and the vibration is attenuated more rapidly. In addition, the tensile force makes the steel plate flat.

References

[1] Y. Oshinoya, Study on Electromagnetic Levitation Conveyance of a Steel Plate with Free Edges by Using Electromagnets and Permanent Magnets, *Journal of the Japan Society Applied Electromagnetics and Mechanical Engineers (in Japanese)* **6**(4) (1998), 364–371.

[2] Y. Oshinoya, Basic Research on Contactless Position Control of a Steel Plate in the Horizontal Direction, *Journal of the Japan Society Applied Electromagnetics and Mechanical Engineers (in Japanese)* **6**(4) (1998), 358–363.

[3] M. Higashi, A. Igusa, Y. Oshinoya and K. Ishibashi, Computer Simulation of Magnetic Levitation of Steel Plate, *Proceedings of School of Engineering Tokai University (in Japanese)* **40**(1) (2000), 119–124.

[4] C.F. Gerald and P.O. Wheatly, *APPLIED NUMERICAL ANALYSIS,* Addison-Wesley Publishing, 1989, pp. 555–558.

[5] S. Timoshenko and S. Woinowsky-krieger, *Theory of Plates and Shells,* McGraw-Hill, 1959, pp. 351–362.

International Journal of Applied Electromagnetics and Mechanics 13 (2001/2002) 59–64
IOS Press

Control- performance improvement for magnetic fluid type semi-active damper

Toshiko Nakagawa[a], Koji Itoda[a,*], Tadashi Furukawa[a] and Shigeru Futami[b]
[a]*Tokyo Denki University, Tokyo, Japan*
[b]*Yasukawa Electric Corporation, Japan*

Abstract. In the previous paper 1, a novel type of semi-active damper that is composed of a cylinder injected with Magnetic Fluid (M.F.) and a magnetic flux controller was proposed for improving the ride comfort of trains. In this paper, at first, a modified model considering nonlinear characteristic of M.F. such as a Non-Newtonian-Fluid is proposed. Next, in order to enhance the control effects, two kinds of improvements are made in the semi-active damper. One is to change the voltage control input to an electromagnet into the current control input. The other is to expand the control range for the M.F. damper by means of feeding a bias current to the electromagnet. Finally, the effects of the improved semi-active damper are shown by various digital simulations and vibration experiments.

1. Introduction

In railways engineering, to lighten a carriage is a method for improving energy consumption. However, the method induces the vibration of a carriage and, as the result, the ride comfort becomes worse. To set up active dampers to the carriage is an effective method for controlling vibrations. However, such kind of active dampers sometimes fall into disorder by active force of pneumatic or hydraulic actuators once their controller fails.

This study introduces a novel semi-active damper which is composed of a cylinder injected with Magnetic Fluid (M.F.) and a nonlinearly controlled electromagnet. Generally, M.F. has the very interesting property that the viscosity can be easily changed within the positive values by an outside magnetic field. Therefore, a carriage with this kind of semi-active damper will move stably at any time.

2. Basic model of the system

Figure 1 shows the device which is composed of two masses, two types of springs, an electromagnet with a nonlinear controller and a cylinder filled with M.F. The viscosity of the fluid is controlled by the electro magnet. Supposing that only the whole viscosity (C; $C_0 + U$) of M.F. is changed by the control input U, the device shown in Fig. 1 is modeled as Fig. 2. In addition, based on some experimental

*Corresponding author: Koji Itoda, Tel.: +81 3 5280 3181 (ext.2137); Fax: +81 3 5280 3573; E-mail: 00gme06@ed.cck.dendai.ac.jp.

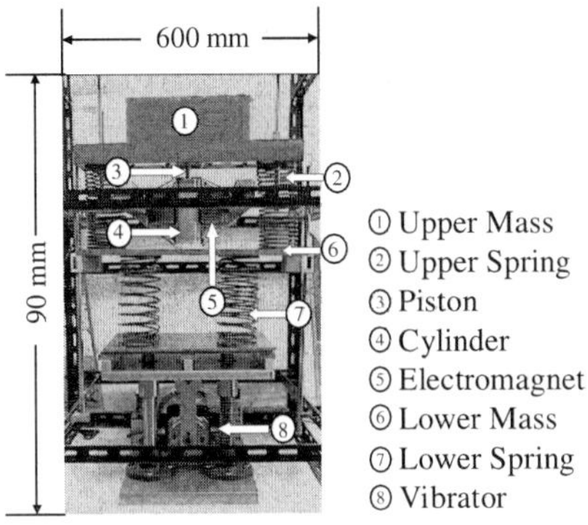

Fig. 1. Experimental device.

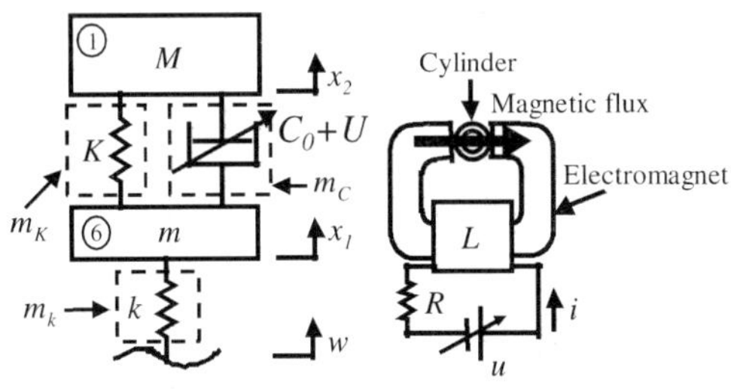

Fig. 2. Device model.

results [1] between the magnet current i and the control input U, the nonlinear dynamic equations for the device are formulated by,

$$U = bi \quad (b > 0) \tag{1}$$

$$m\ddot{x}_1 = -k(x_1 - w) + K(x_2 - x_1) + C_0(\dot{x}_2 - \dot{x}_1) + U(\dot{x}_2 - \dot{x}_1) \tag{2}$$

$$M\ddot{x}_2 = -K(x_2 - x_1) - C_0(\dot{x}_2 - \dot{x}_1) - U(\dot{x}_2 - \dot{x}_1) \tag{3}$$

3. Modification of the basic model

Although the masses of the damper and the springs are not considered at the basic system (Fig. 2), it actually exists. In addition, M.F. has a characteristic as Non-Newtonian-Fluid. These nonlinearities are not neglected. Iflarge control effects can be expected, a compensator should be designed considering them. Therefore, in this chapter, the values of system constants are modified and new terms are added to the Eqs (1)–(3) in order to model the real characteristics approach as well as possible. Now, a frequency response at the magnet current $i = 0$ [A] is shown in Fig. 3(a). Based on the result, the system constants are determined as follows

$$K = 674.1 \text{ [N/m]}, \quad k = 1286.9 \text{ [N/m]}, \quad C_0 = 42.8 \text{ [Ns/m]}$$

These are different from the original values. These differences come from the fact that the proper system contains the masses of the damper and springs. Consequently, in the modified model, these are adopted as the system constants at $i = 0$ [A].

Next, the frequency responses at $i = 1.5$ [A], $i = 3.0$ [A] are showed in Figs 3(b) and (c). From these figures it is presumed that the damping coefficient (C) increases depending on the magnet current and the Eqs (1)–(3) do not represent the real system, because the M.F. has nonlinear characteristics as Non-Newtonian-Fluid. Therefore the corrective term Δf should be added to the Eqs (1)–(3). Finally, the modified equations are determined as follows [2].

$$m\ddot{x}_1 = -k(x_1 - \omega) + K(x_2 - x_1) + C_0(\dot{x}_2 - \dot{x}_1) + \Delta f \tag{4}$$

$$M\ddot{x}_2 = -K(x_2 - x_1) - C_0(\dot{x}_2 - \dot{x}_1) - \Delta f \tag{5}$$

$$\Delta f = \{(\dot{x}_2 - \dot{x}_1) - h(x_2 - x_1)\}U \tag{6}$$

$$U = b'i \tag{7}$$

where, $b' = 3.27$ [Ns/Am], $h = 31.6$ [1/s].

These are also different from the original values ($b = 6.91$, $h = 0$). In Figs 3(a), (b) and (c), the solid lines show the characters obtained by adapting both the modified Eqs (4)–(7) and the corrected values. The dot lines shows the characters obtained by Eqs (1)–(3). These absolutely correspond to the experimental data.

4. Improvement of control system

The effect of the proposed control method [1] has been already confirmed by various digital simulations and vibration experiments. In this paper, the control method is improved to obtain better effects with respect to 2 points as follows.

4.1. Improvement of a power supply device

Previously, the input to the electromagnet was a type of voltage. This time the power supply device is changed from a voltage source to a current source, because the control input U could be handled directly with current as shown in Fig. 2.

4.2. Spread of control range by bias current

The control input U is permitted to be positive or negative in the usual control system. But when the steady current $i_0 = 0$ [A], it can not be negative actually, because the viscosity should be always positive. This time, the steady value of the damping constant is shifted from $C_0(i = 0)$ to $C_0'(i = 1.5)$ with the bias current $0 = 1.5$ [A]. As the result, it is possible that the input varies from a positive value to a negative one. Therefore, the control input works effectively.

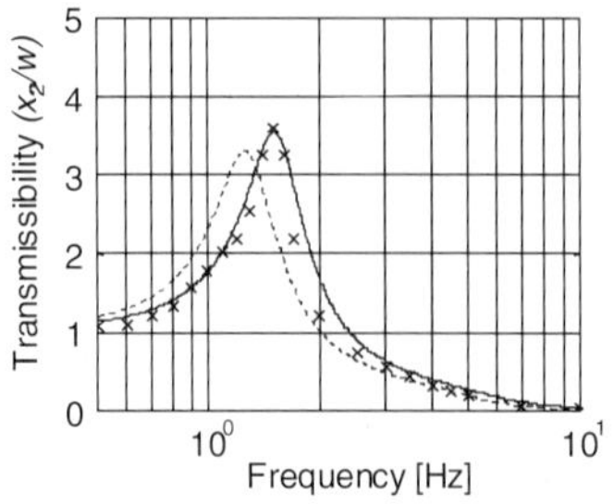

(a) i =0A

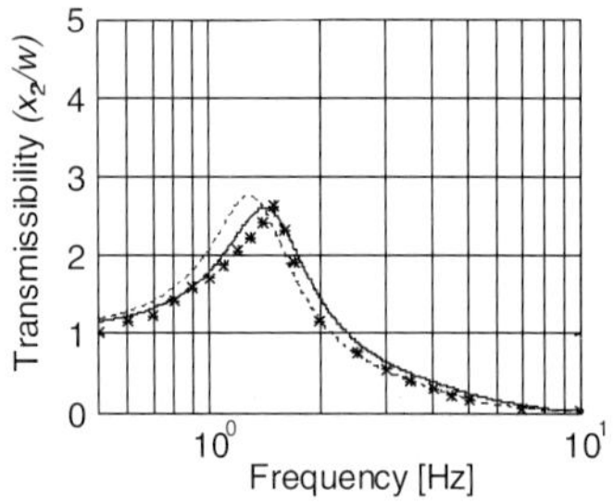

(b) i =1.5A

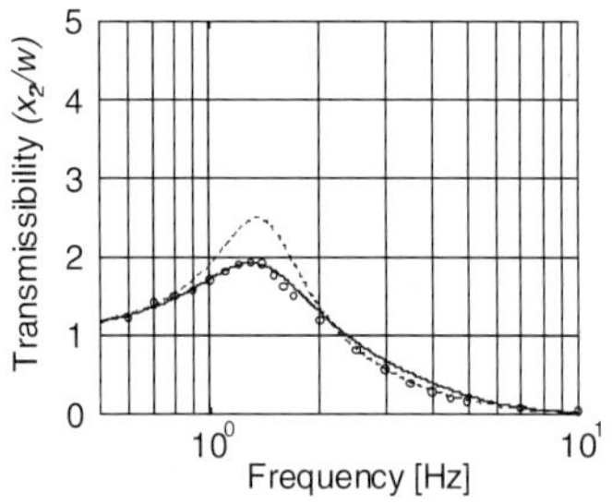

(c) i =3A

Fig. 3. Frequency response to i.

5. Compensator based on the modified model

The compensator is renewed for the modified model (the Eqs (4)–(7)). The method of nonlinear control by Hamilton Jacobi inequalities is applied in this study.

The step is as follows. The suitable frequency weighting function $W(s)$ for shaping the state variable x_2 is introduced to this system. The bias current i_0 is chosen as the new operating point. The control input U is determined following the procedure in [1]. As the result, U is given by

$$
\begin{aligned}
U = -\{(\dot{x}_2 - \dot{x}_1) - h(x_2 - x_1)\} \times \{(1.10 \times 10^{-3})\dot{x}_1 - (1.12 \times 10^{3})\dot{x}_2 \\
+(1.07 \times 10^{4})x_1 + (1.59 \times 10^{3})x_2 + (3.89 \times 10^{5})x_{w1} - (7.46 \times 10^{4})x_{w2}\}
\end{aligned} \tag{8}
$$

where, variable x_{w1} and x_{w2} are new state variables of the augmented system by $W(s)$.

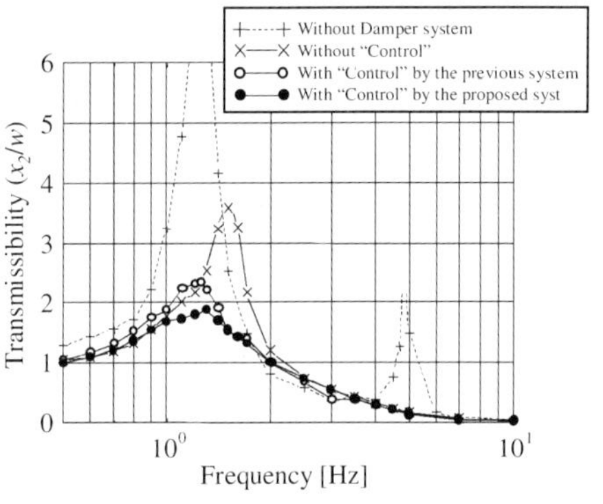

Fig. 4. Frequency response.

6. Experimental results

In this chapter, vibration experiments under a disturbance of a simple harmonic motion or a disturbance composed of harmonic motions of different frequencies were conducted to conform the damping effects of the compensator in Chapter 5.

6.1. Disturbance of a simple harmonic motion

Figure 4 shows the results of the vibration experiments under a disturbance of a simple harmonic motion (amplitude $A = 1$ cm, frequency range 0.5 to 10.0 Hz).

A broken line by + is the frequency responses without the damper device. A line by x shows the characteristics without the proposed control. A line by ∘ is the result with the unmodified controller in the previous system [1]. A line by • is the result with "Control" by the proposed system in this paper.

In the case without the damper device, the fiercely vibrations of the 1st and the 2nd order of vibration modes ($f_1 = 1.25$, $f_2 = 4.7$ Hz) reveal. In the case without any control, the vibration in the vicinity of 4.7 Hz is restricted enough, but the resonance peak in the vicinity of 1.25 Hz is still high. In case with the control by the previous unmodified system, the resonance peak is restricted very much. In case with the controller by the proposed system in this paper, the proposed system provides the sufficient effects.

6.2. Disturbance composed of two motions of different frequencies

Under the assumption that a disturbance has a complex waveform, the vibration experiments are conducted. The disturbance is; the amplitude $A = 0.5$ [cm], the frequencies f_1, f_2 are combinations of 0.5, 0.8, 1.5, 2.0, 3.0, and 5.0 Hz.

Figures 5(a) and (b) show the frequency responses under the condition without the proposed control and the condition with the proposed control. These figures reveal that the control can damp the vibration enough.

7. Conclusion

The followings were certificated in this paper.

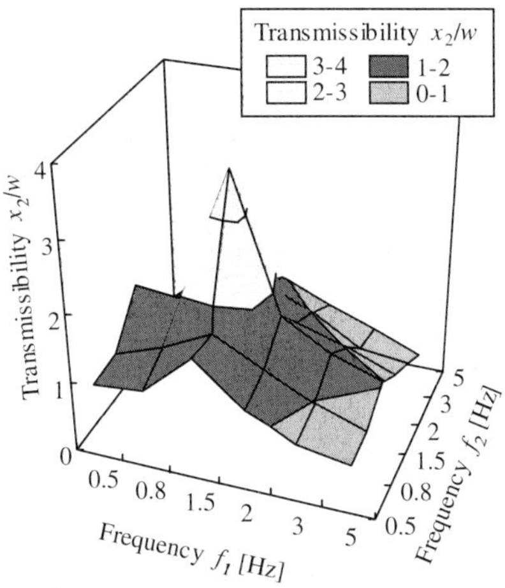

(a) Without "Control"

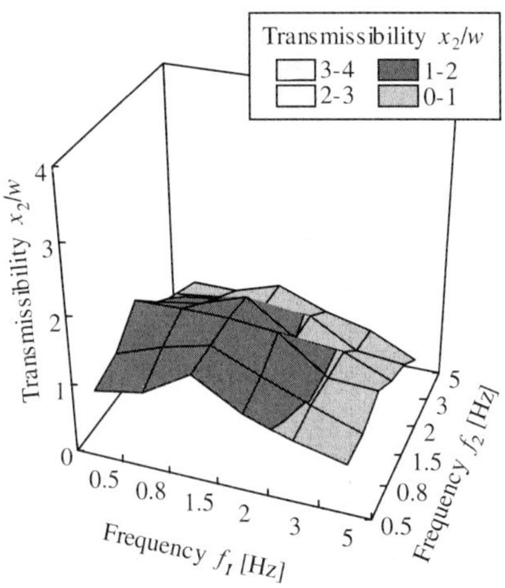

(b) With "Control"

Fig. 5.

(1) The corrective term was added to the basic model to represent real behaviors of the M.R. damper system more accurately and the corresponding compensator was designed newly.
(2) The power supply device was changed from a voltage type to a current one and the range of the control input is expanded by adapting a bias current.
(3) The efficacy of the proposed control method was confirmed by many vibration experiments.

References

[1] T. Nakagawa and A. Yamada, Design for a Novel M.R.F. Semiactive Damper and Certification of the Nonlinear Controller's Effects, *IEEE Transaction on Magnetics* **35**(5) (1999), 3604–3606.

[2] T. Nakagawa, T. Furakawa and A. Yamada, Study on Performance Improvement of a Magnetic Fluid Type Semi-Active Damper, *IEEJ IAS* **121-D**(5) (May 2001).

[3] A.J. von der Schaft, L_2 gain analysis of nonlinear system and state feedback H^∞ control, *IEEE Trans. On AC-37-6,* 1992, pp. 770–784

International Journal of Applied Electromagnetics and Mechanics 13 (2001/2002) 65–70
IOS Press

Numerical study on sticking of paper in electrostatic transfer process for electrophotography

Nobuyuki Nakayama* and Yu Yimei
Research and Development Center, Document Products Company, Fuji Xerox Co., Ltd., 430 Sakai Nakai-machi Ashigarakami-gun Kanagawa, 259-0157, Japan

Abstract. A numerical study was carried out on the electrostatic sticking of a charged paper to photoreceptor in electrostatic transfer process for electrophotography. Numerical procedures to predict sticking of the paper were presented which consists of (1) determination of initial angle of the paper, (2) estimation of electric field intensity, (3) estimation of charge on paper created by Paschen's air breakdown, (4) estimation of electrostatic force acting on the paper and (5) calculation of deformation of paper. Deformation of the charged paper in electric field was numerically simulated according to the procedures and the dependencies of paper sticking on photoreceptor size and initial angle of paper were quantitatively clarified. It is clarified that the smaller and softer photoreceptor improves sticking of paper.

1. Introduction

In electrostatic transfer process for electrophotography as shown in Fig. 1, it is desired to transfer most toner particles on photoreceptor to proper positions on charged paper without defects and to transport paper precisely [1]. In the process, transfer roller is pressed to photoreceptor and the paper pass out between photoreceptor and transfer roller. The paper is charged by Paschen's air breakdown in post-transfer area and it sometimes stick to photoreceptor by the electrostatic force. In the system using large photoreceptor, charged paper is easy to stick to photoreceptor. The limit diameter of photoreceptor to avoid sticking is experimentally estimated at 110 mm in typical transfer process. It is necessary to clarify the relation between the paper sticking behavior and transfer conditions.

In the present study, the electrostatic force applied to a charged paper in post-transfer area was estimated by the calculation of electric field intensity and charge density on paper. Then the deformation of paper was calculated and the effects of transfer conditions on the sticking were investigated and discussed.

2. Numerical method

The paper sticking behavior can be predicted by the deformed profile of charged paper in the transfer field. The model of a paper in post-transfer area is described in Fig. 2, where transfer roller with radius,

*Corresponding author: Nobuyuki Nakayama, Tel.: +81 465 80 2157; Fax: +81 465 81 8983; E-mail: Nobuyuki.Nakayama@fujixerox.co.jp.

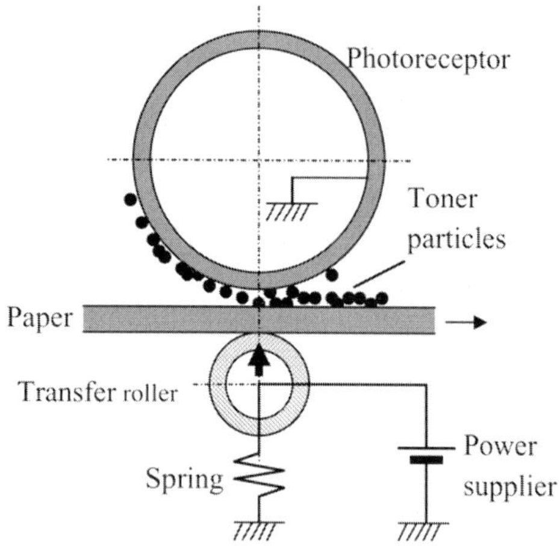

Fig. 1. Electrostatic transfer process for electrophotography.

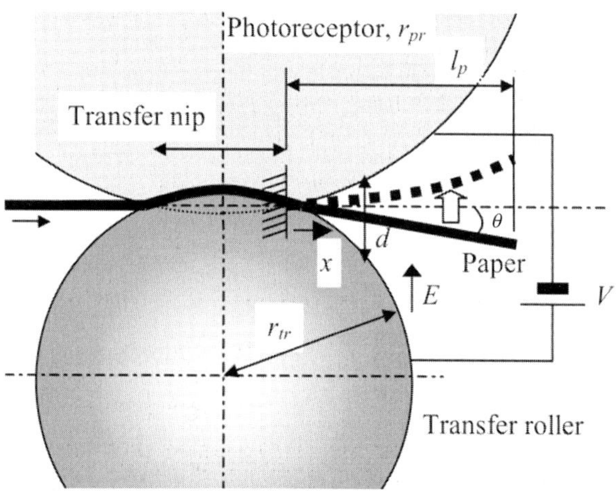

Fig. 2. Model of the paper in post-transfer area.

r_{tr}, is harder than the photoreceptor with radius, r_{pr}. In the numerical study, leading part of the paper in post-transfer area is focused and is modeled as a plate fixed at the trailing edge. The deformation of paper is calculated using finite element method [2] according to following steps: (1) Determine initial angle of paper, θ, (2) estimate electric field intensity in post-transfer area, E, (3) estimate charge density on paper, σ, (4) calculate electrostatic force and (5) calculate deformation of paper.

Initial angle of paper is determined by the mechanical deformation of photoreceptor and transfer roller, which were calculated by finite element method. The example of calculated profile of paper is shown in Fig. 3. In the case of Fig. 3(a), rigid transfer roller is harder than the photoreceptor and the angle of paper is -6.23 degree. In the case of Fig. 3(b), softer transfer roller was introduced and the absolute angle of paper became smaller. The paper angle depends on the hardness of photoreceptor and transfer roller.

Electric field can be estimated by solving following Gauss's theorem and charge conservation rule.

$$\nabla \times \boldsymbol{D} = \rho \tag{1}$$

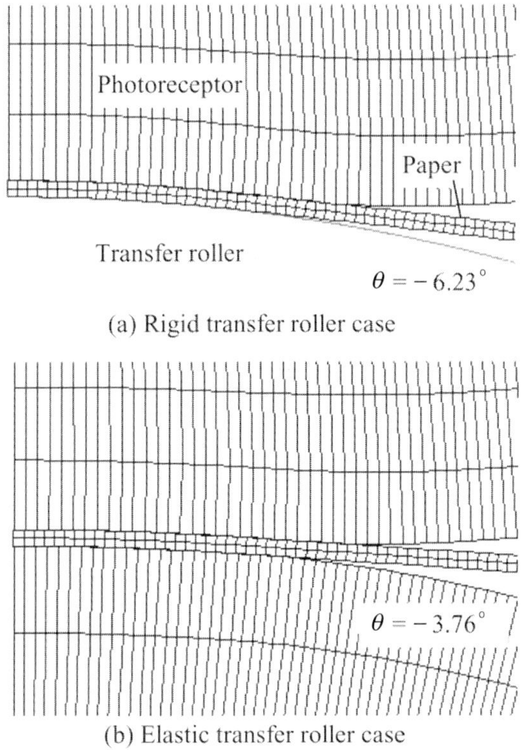

Fig. 3. Deformed profile of paper in transfer nip.

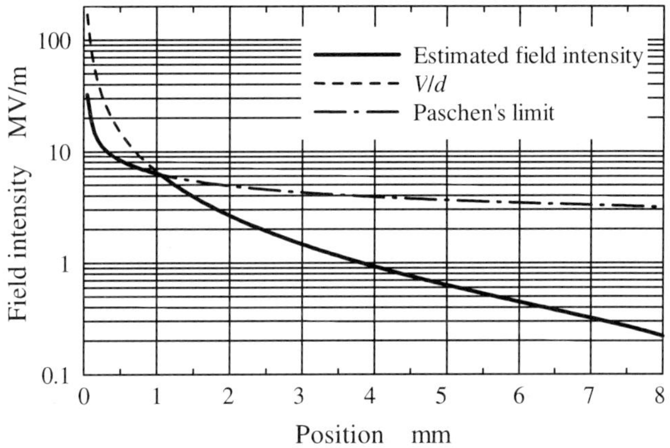

Fig. 4. Electric field intensity.

$$\nabla \times \boldsymbol{J} = -\frac{\partial \rho}{\partial t} \tag{2}$$

Where, $\boldsymbol{D}$ is electric flux density, $\boldsymbol{J}$ is current density and ρ is charge density. In the present study, layer-structured one-dimensional model of transfer area is introduced [3]. In this model, photoreceptor, air gap, paper and transfer roller are considered as piled up layers. In each time step in the field calculation, voltage difference in each layer is determined by solving Eq. (1). Then the current density

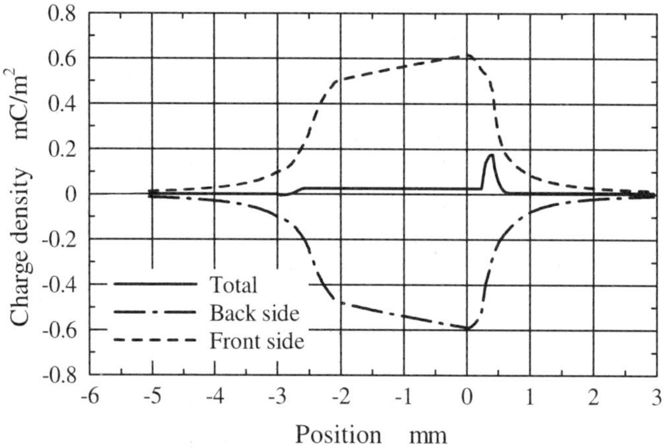

Fig. 5. Charge density on the paper.

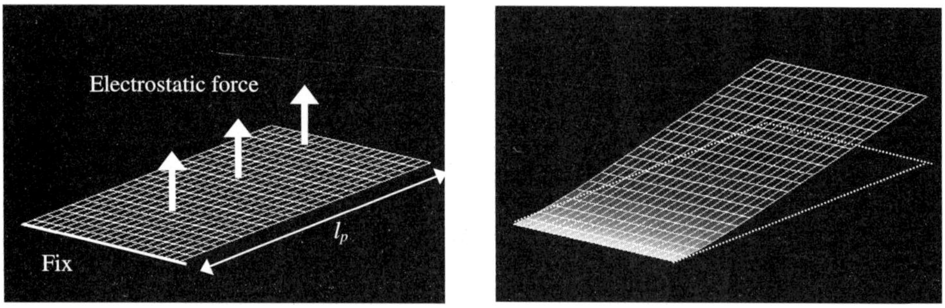

(a) Model of paper

(b) Deformed profile of paper

Fig. 6. Model and calculated deformation of paper.

and increment of charge is estimated from Eq. (2). By starting the calculation at the certain point in pre-transfer area and by solving Eqs (1) and (2) repeatedly with changing air gap thickness, voltage differences and charges on boundaries of layers are determined.

In the post-transfer area, the calculated electric field intensity is approximately equal to the value V/d, where V is applied voltage difference between photoreceptor and transfer roller, and d is thickness of the air gap. However, maximum field intensity is limited by Paschen's air breakdown limit. The example of electric field intensity as a function of position, x, is plotted in Fig. 4. In the field calculation, the charge density on paper created by air breakdown is also estimated. That is, if the field intensity exceeds Paschen's limit then the proper amount of charge is generated on the surface of each boundary of air gap layer. The estimated charge densities are plotted in Fig. 5. Maximum total charge density on paper, which flowed in from the transfer roller and photoreceptor in transfer nip, is estimated at 0.2 mC/m^2 in this case. The charge is reduced less than 1 μC/m^2 in the area of $x > 1$ mm by the Paschen's air breakdown.

Finally, electrostatic force acting on the paper can be calculated by a product of electric field intensity, E and charge on paper, σ. Then the deformation of paper is calculated using finite element method. The model and calculated deformation of paper are described in Fig. 6. From the calculated deflection of paper as is plotted in Fig. 7, gap between paper and photoreceptor is evaluated. If the gap is negative, it is supposed that the paper sticks to photoreceptor.

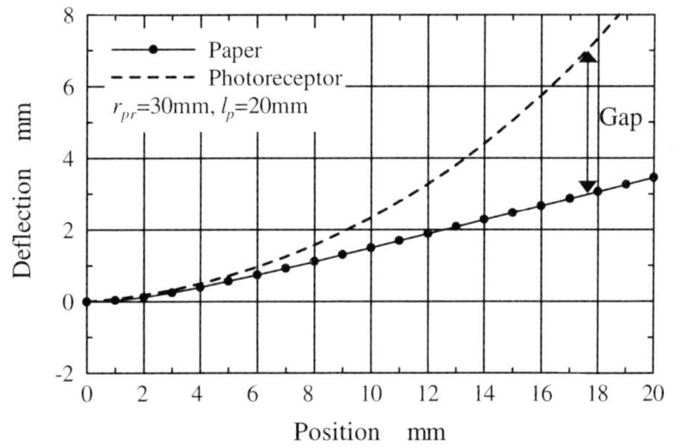

Fig. 7. Deflection of paper.

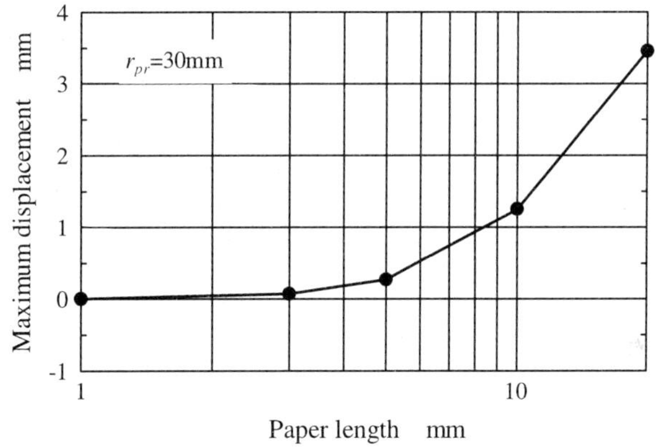

Fig. 8. Maximum displacement of paper.

3. Result

Calculated maximum displacements at the leading edge of papers are plotted in Fig. 8. Maximum displacement depends on stiffness of paper, which is a function of paper length. If the paper length is small, stiffness of the paper is large enough to prevent deformation. On the other hand, longer paper is easy to deform because of its small stiffness.

Calculated gap as a function of paper length, lp, for various photoreceptor radius are shown in Fig. 9. The gap depends on paper length as is predicted from the results shown in Fig. 8. The gap also depends on photoreceptor radius because the field intensity is lower and the initial gap is larger for the smaller photoreceptor radius.

Comparing Figs 9 (a) with (b), it is recognized that the initial angle of paper has significant effect on the gap. In case of $r_{pr} = 60$ mm in Fig. 9(a), paper of 10 mm length is supposed to stick to photoreceptor. Sticking of the paper is improved by increase in radius of photoreceptor. Photoreceptor with 30 mm radius is required to prevent sticking. In the case of Fig. 9(b) where initially inclined paper is assumed, sticking of paper is prevented even for 60 mm radius photoreceptor. To obtain the initial paper inclination, photoreceptor must be softer than the transfer roller.

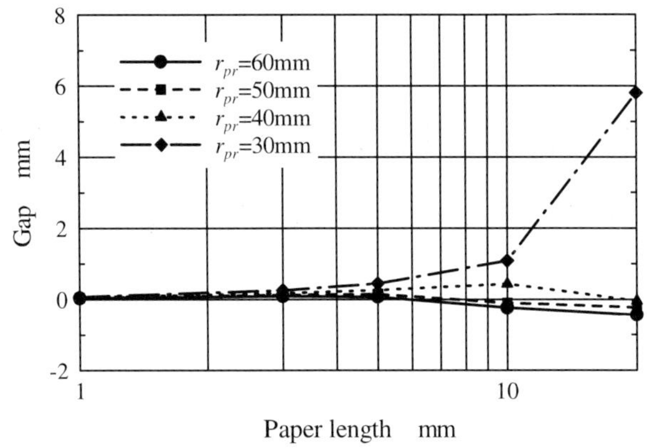

(a) θ=0°

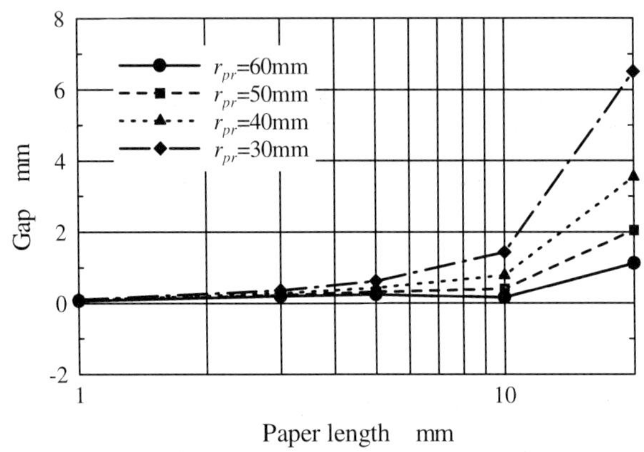

(b) $\theta = -2°$

Fig. 9. Gap between photoreceptor and deformed paper.

4. Conclusion

Numerical study was carried out on the electrostatic sticking of charged paper to photoreceptor or transfer roller in electrostatic transfer process for electrophotography. Electrostatic force applied to a charged paper in post-transfer area was estimated by the calculation of electric field intensity and charge density on paper using one-dimensional model of transfer area. Then the deformation of paper was calculated and the effects of radius of photoreceptor and initial posture of paper on the sticking were clarified. It is good for prevent sticking to engage smaller and softer photoreceptor.

References

[1] E.M. Williams, *The physics and technology of xerographic processes,* Krieger Publishing Co., 1993.

[2] N. Nakayama, Y. Toi and K. Yuge, Deformation analysis of charged dielectric thin film under electric field, *Seisan-Kenkyu* **42**(3) (1990), 209–212.

[3] T. Ito and H. Kawamoto, Simulation of biased transfer roller process in electrophotography, *Trans. Jpn. Soc. Mech. Eng. C* **65**(637) (1999), 3559–3566.

International Journal of Applied Electromagnetics and Mechanics 13 (2001/2002) 71–78
IOS Press

Effect of magnetic field on a tuned liquid damper using a magnetic fluid

Yasuhiro Ohira, Hiroshi Houda and Tatsuo Sawada*
Department of Mechanical Engineering, Keio University, 3-14-1 Hiyoshi, Kohoku-ku, Yokohama 223-8522, Japan

Abstract. The frequency response of a tuned liquid damper (TLD) which uses a magnetic fluid (called a "tuned magnetic fluid damper" (TMFD)) is investigated. A single-degree-of-freedom tuned mass damper (TMD) analogy is used to define the effective mass, stiffness and damping and these parameters are then estimated from experimental results. A numerical simulation is used to examine the TMFD and appropriate magnetic field conditions for effective damping are predicted. In particular, the performance of the TMFD was improved by switching the applied magnetic field at a point near the resonant frequency of the undamped structure.

1. Introduction

A tuned liquid damper (TLD) is a passive mechanical damper that uses a sloshing liquid to dampen oscillations. Many researchers in the civil engineering field have examined the performance of TLDs. Sun et al. [1] measured liquid motion in TLDs with different tank shapes and different liquid viscosities and described the complex non-linear characteristics using a single-degree-of-freedom tuned mass damper (TMD) analogy. Modi and Munshi [2] performed a parametric study focused on enhancing the energy dissipation efficiency of a rectangular tuned liquid damper through introduction of a two-dimensional obstacle. They observed a significant increase, up to 60%, in energy dissipation when using the obstacle. Yamamoto and Kawahara [3] performed a numerical study using a modified ALE method to clarify the effectiveness of structural control by TLDs. Abé et al. [4] proposed a new active TLD which uses a magnetic fluid to enhance the performance of the TLD and studied its feasibility experimentally. A magnetic fluid is a stable medium which is essentially a fluid but also has magnetic properties comparable to those of a solid magnet. It can be confined, positioned, shaped and controlled by the application of magnetic fields.

In the present paper we describe further experiments for this tuned magnetic fluid damper (TMFD), expanding on the study by Abé et al. We also develop an analytical model for the TMFD using an analogy with a tuned mass damper (TMD) in which the total mass of the magnetic fluid in the container is separated into an effective mass and a non-effective mass. The TMFD parameters and the magnetic fluid hydrodynamic force are calculated from experimental results and then a numerical simulation is carried out in order to test this model of and examine the effectiveness of the TMFD.

*Correspondence to: Tatsuo Sawada, Department of Mechanical Engineering, Keio University, 3-14-1 Hiyoshi, Kohoku-ku, Yokohama 223-8522, Japan. Tel.: +81 45 566 1504; Fax: +81 45 566 1495; E-mail: sawada@mech.keio.ac.jp.

1383-5416/01/02/$8.00

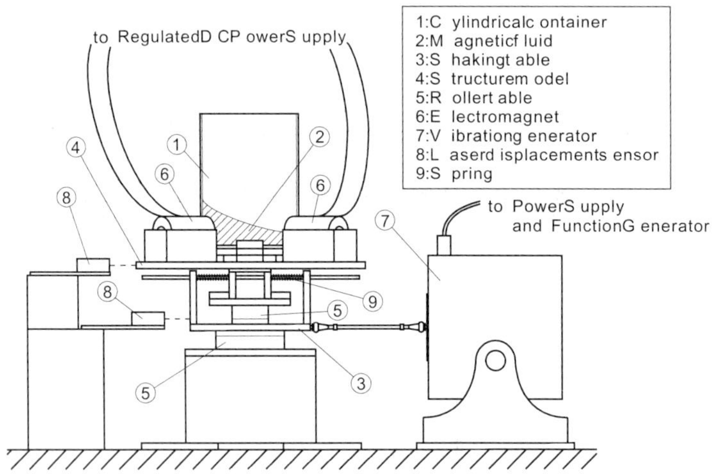

Fig. 1. Experimental apparatus.

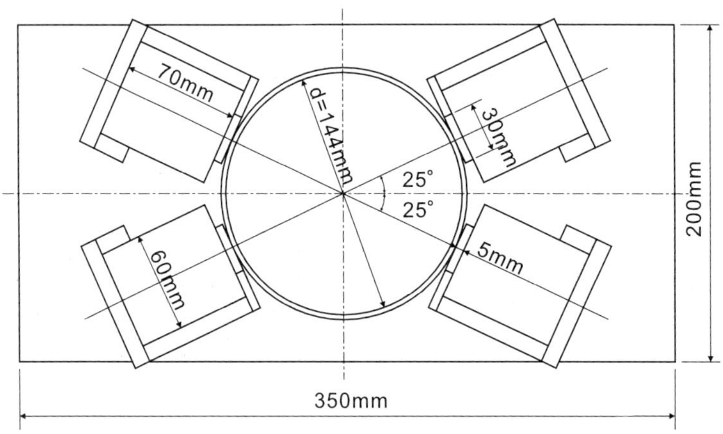

Fig. 2. Arrangement of electromagnets.

2. TMFD excitation experiment

2.1. Experimental apparatus

Figure 1 shows a schematic diagram of the experimental apparatus. The TMFD is a cylindrical container, inner diameter 144 mm, containing a water-based magnetic fluid (W-40), depth 31 mm, density and kinematic viscosity at 23°C, 1382 kg/m^3 and 10.6×10^{-6} m^2/s, respectively. The cylindrical container is provided with four electromagnets positioned on two sides as shown in Fig. 2. The TMFD is placed on a roller table (the cylinder and roller table are referred to as the "structure model") that is further supported on a vibrating table. The roller table is held in place by two springs having a stiffness of 1329 N/m. The vibrating displacement of the structure model and the vibrating table are measured by two laser displacement transducers, respectively. The sampling frequency for the measurements was fixed at 50 Hz in all experiments.

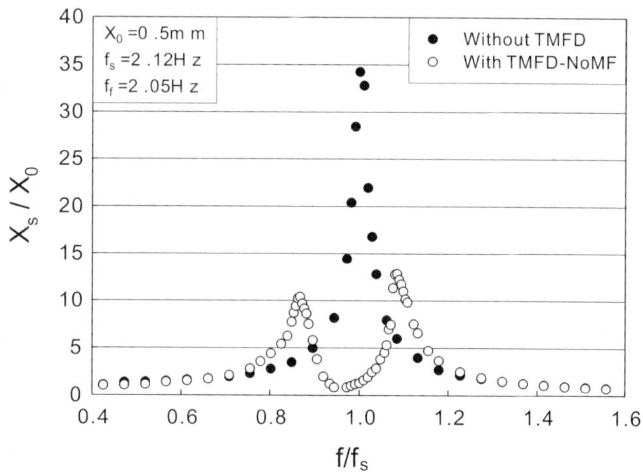

Fig. 3. Frequency responses of the structure model.

2.2. Frequency response

Figure 3 shows frequency responses for the structure model without the magnetic fluid (Without TMFD) and for the structure model with the magnetic fluid but without a magnetic field (With TMFD-NoMF). In Fig. 3, f_s is the first resonant frequency of the structure model without TMFD and f_f is that of the liquid sloshing for TMFD-NoMF. The displacement of the structure model is X_s, the displacement of the vibrating table is X_0, and f is the excitation frequency of the vibrating table. When the structure model is intensively vibrated in a region around the resonant frequency f_s, the displacement of the structure is clearly suppressed by using the magnetic fluid (TMFD-NoMF case). The suppression is particularly strong near the resonant frequency f_s. Notably, there are two almost symmetrical peaks that occur at opposite sides of f_s. These two peaks are very small in comparison with the peak for the result without TMFD. Thus, there is a large damping effect achieved only by installing the magnetic fluid (TMFD-NoMF case) in the structure model even if there is no magnetic field applied.

Frequency responses with and without magnetic field are compared in Fig. 4. The intensity of the applied magnetic field is varied in proportion to the velocity of the structure $\dot{x}_s$ with a proportional constant defined as K_B. In particular, when $\dot{x}_s = 0.1$ m/s, the value of the magnetic field intensity is 47.7 mT, with the result that the rate of change of the magnetic field intensity K_B is (47.7 mT)/(0.1 m/s) = 0.477 T/(m/s). ϕ indicates the phase delay between the application of the magnetic field and the vibration of the structure model. For example, the intensity of the magnetic field of the right electromagnet pair increases with $\dot{x}_s$ when $\phi = 3\pi/2$ whereas the intensity of the magnetic field of the left electromagnet pair increases with $\dot{x}_s$ when $\phi = \pi/2$. In short, $\phi = 3\pi/2$ and $\phi = \pi/2$ indicate that the phase of vibration of the velocity and magnetic field is the same or opposite, respectively.

When a magnetic field is applied out of phase with the velocity of the structure model ($\phi = \pi/2$), a reduction of the left peak is noticeable while the right peak actually increases. We believe that these results are connected with an increase in the apparent viscosity caused by the magnetic field and the existence of the magnetic body force. These results indicate that an active TMFD may be achieved by using magnetic fields appropriately.

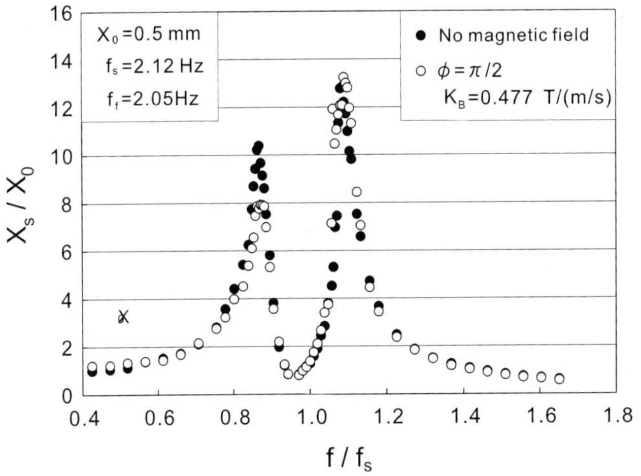

Fig. 4. Frequency responses of the structure model with TMFD.

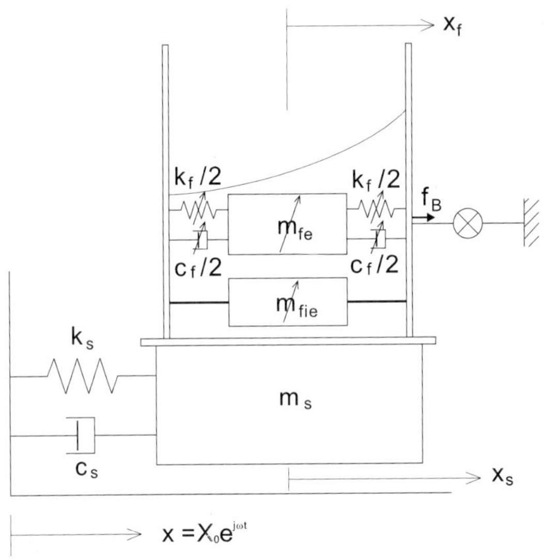

Fig. 5. Analytical model of TMFD.

3. Modeling TMFD using a TMD analogy

3.1. Definition of effective mass, stiffness and damping

Figure 5 shows an analytical model of the TMFD using a TMD analogy. When a magnetic field is applied, motion of the magnetic fluid is restrained. Thus, conceptually, we can separate the internal fluid mass into a non-effective mass m_{fie} and an effective mass m_{fe} [5]. The non-effective mass is considered to be rigidly attached to the moving structure model. Therefore, the additional mass of the vibrating system becomes m_{fie}, and the total mass of the structure model becomes $m_s + m_{fie}$, where m_s is the net mass of the structure model including the cylindrical container, electromagnets, roller table and other components. This model is illustrated in Fig. 5. Here, k_s and c_s are stiffness and damping in relation to the structure model, respectively, and f_B is a hydrodynamic magnetic fluid force caused by an applied magnetic field. The TMFD parameters are defined as follows:

$$\mu_f = \frac{m_{fe}}{m_f}, \quad k_f = m_{fe}{\omega_{fe}}^2, \quad \zeta_f = \frac{c_f}{2m_{fe}\omega_f} \tag{1}$$

where μ_f, k_f and ζ_f are effective mass ratio, effective stiffness and effective damping ratio, respectively. In these equations, m_f is a total fluid mass, that is, $m_f = m_{fie} + m_{fe}$, $\omega_f (= 2\pi f_f)$ and ω_{fe} are the first resonant angular frequency of the sloshing with the total fluid mass m_f and with the effective fluid mass m_{fe}, respectively, and c_f is the effective damping of m_{fe}. These TMFD parameters are varied according to displacement of the structure x_s. Considering the present model as a single-degree-of-freedom vibration structure with an auxiliary system and the steady-state displacement of the base structure as $x(t) = X_0 \exp(j\omega t)$, where $\omega = 2\pi f$, the equations of motion for the TMFD are

$$\begin{aligned}(m_s + m_{fie})\,\ddot{x}_s = &-k_s\left(x_s - X_0 e^{j\omega t}\right) - c_s\left(\dot{x}_s - jX_0\omega e^{j\omega t}\right) + k_f\left(x_f - x_s\right)\\ &+c_f\left(\dot{x}_f - \dot{x}_s\right) + f_B\end{aligned} \tag{2}$$

$$m_{fe}\ddot{x}_f = -k_f\left(x_f - x_s\right) - c_f\left(\dot{x}_f - \dot{x}_s\right) \tag{3}$$

3.2. Sloshing hydrodynamic force

The sloshing hydrodynamic force F_f is represented in a single-degree-of-freedom vibration system as:

$$F_f = m_{fe}\ddot{x}_f \tag{4}$$

Then, letting X_f and X_s be the amplitude of x_f and x_s, respectively, F_f is expressed in the following dimensionless form:

$$\left|\frac{F_f}{\rho\pi R^2 h X_s \omega^2}\right| = \left|\frac{m_{fe}}{\rho\pi R^2 h}\cdot\frac{-X_f\omega^2}{X_s\omega^2}\right| = \mu_f\frac{X_f}{X_s} \tag{5}$$

where ρ, R and h are fluid density, radius of the container and fluid depth, respectively. When $x_s = X_s e^{j(\omega t+\varphi_s)}$, $x_f = X_f e^{j(\omega t+\varphi_f)}$, and φ_s and φ_f are the phase delay from the vibration of the vibrating table, the linearized equation of motion for Eq. (3) becomes

$$m_{fe}\ddot{x}_f + c_f\dot{x}_f + k_f x_f = \left(k_f + jc_f\omega\right) X_s e^{j(\omega t+\varphi_s)} \tag{6}$$

and Eq. (5) can be solved. Consequently, the following relation is obtained:

$$\left(\frac{X_f}{X_s}\right)^2 = \frac{1 + 4{\zeta_f}^2{\lambda_f}^2}{\left(1 - {\lambda_f}^2\right)^2 + 4{\zeta_f}^2{\lambda_f}^2} \tag{7}$$

where $\lambda_f = \omega/\omega_f$. Using Eqs (5) and (7), the sloshing hydrodynamic force is given by

$$\left|\frac{F_f}{\rho\pi R^2 h X_s \omega^2}\right| = \mu_f\sqrt{\frac{1 + 4{\zeta_f}^2{\lambda_f}^2}{\left(1 - {\lambda_f}^2\right)^2 + 4{\zeta_f}^2{\lambda_f}^2}} \tag{8}$$

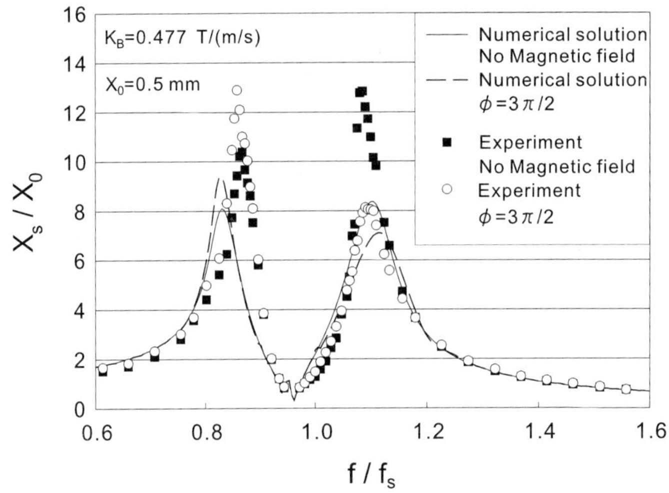

Fig. 6. Comparison of numerical and experimental frequency responses of the structure model.

3.3. Calculation of TMFD parameters

In order to solve Eqs (2) and (3) it is still necessary to obtain μ_f, k_f and ζ_f. We calculated these values from experimental results under no magnetic field in accordance with Eq. (7). The sloshing hydrodynamic force was measured by a load cell attached to the cylindrical container. Various excitation experiments were carried out, changing amplitude X_0 and frequency f of the vibrating table. The TMFD parameters were then calculated by fitting the experimental results to Eq. (8). We used a method for least-squares estimation for non-linear parameters [6] in the calculation and obtained the following empirical expressions for TMFD parameters as a function of X_s:

$$\mu_f = 2.05 {X_s}^{0.16} \tag{9}$$

$$\zeta_f = 0.0154 \log X_s + 0.134 \tag{10}$$

$$k_f = 185 {X_s}^{0.131} \tag{11}$$

4. Numerical simulation and results

We then proceeded to carry out a numerical simulation in order to solve Eqs (2) and (3) using the 4-th order Runge-Kutta method. In this simulation, μ_f, k_f and ζ_f are expressed by Eqs (9), (10) and (11), respectively. However the hydrodynamic magnetic fluid force f_B in Eq. (2) is still undetermined. Various experiments were carried out to measure f_B under various applied magnetic fields. In these experiments, the frequency response tends to resemble that of a one-degree-of-freedom vibration system and we thus obtained empirical expressions for f_B on the basis of a solution of a one-degree-of-freedom vibration system by trial and error. These expressions are classified into several formulas according to the intensity and frequency of the magnetic field. The detailed classification procedure and expressions of f_B are provided in [7].

Figure 6 shows some typical results of the numerical simulations compared with the experimental results. The numerical solutions are in fairly close agreement with the experimental results. As we

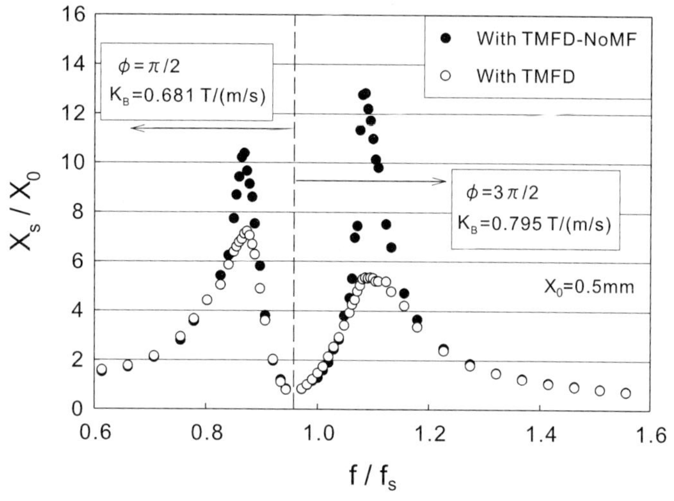

Fig. 7. Experimental results when switching the magnetic field condition at a point near the resonant frequency.

discussed in Section 2.2 with regard to the $\phi = \pi/2$ magnetic field condition, two peaks occur in the frequency response, similar peaks occur for the $\phi = 3\pi/2$ condition as shown in Fig. 6. When the magnetic field is applied in the $\phi = 3\pi/2$ condition, the right peak becomes smaller in both experimental and numerical results, however the left peak actually increases. Since there is a limit in the number of experiments that can be performed quickly, we used the numerical simulation to further analyze these peaks for various different conditions. Consequently, it became clear that reduction of the peak in either the high (right) or low (left) frequency region is possible by applying an appropriate magnetic field.

Based on these numerical simulations, in order to attempt to reduce both peaks, we conducted a further experiment, in which we switched the magnetic field condition at a point near the resonant frequency. Figure 7 shows the resulting frequency response. Using the experimental parameters predicted by the numerical simulation, we see that both peaks actually decrease.

We suggested that the differences between the numerical results and the experimental results may be due to incorrect estimation of the TMFD parameters and hydrodynamic magnetic force because of a phenomenon called "swirling". When a cylindrical container of liquid is subject to a harmonic lateral vibration, the free surface may rotate harmonically or non-harmonically around the central axis of the container in a specific forcing frequency range. The direction of rotation is not fixed and may change irregularly. In the present experiment, swirling occurred at some excitation frequencies.

5. Concluding remarks

The characteristics of the TMFD are examined experimentally and numerically. Various TMFD parameters are determined based on a single-degree-of-freedom TMD analogy and these parameters are estimated based on several trial experiments for measuring hydrodynamic sloshing forces. Numerical simulations using the 4-th order Runge-Kutta method are then carried out to support the experimental results. The numerical results are in fairly good agreement with the experimental results. In the frequency response curves using a TMFD, there are two dominant peaks on opposite sides of the resonant frequency. The peaks can be reduced by application of a magnetic field, however, both peaks could not be reduced with a single set of magnetic field conditions. Based on the numerical simulations, we conducted a further experiment in which we switched the magnetic field condition at a point near the resonant frequency and obtained reduction of both peaks.

The effectiveness of the TMFD is confirmed, however, there are several remaining problems. In particular, swirling can occur in a cylindrical container and produce a circumferential sloshing force and has a very complicated influence on the TMFD or TLD. Further, we used a simple magnetic field that varied sinusoidally and only altered its intensity and phase delay from the vibration of the structure model. A more suitable magnetic field condition may improve the TMFD performance.

Acknowledgements

This work was partially supported by Keio Gijuku Fukuzawa Memorial Fund for the Advancement of Education and Research. We also thank Mr. K. Sadoyama, a student of Keio University, for his help in the experiments.

References

[1] L.M. Sun et al., *Earthquake Eng. Struct. Dyn.* **24** (1995), 967–976.
[2] V.J. Modi and S.R. Munshi, *J. Fluids Struct.* **12** (1998), 1055–1071.
[3] K. Yamamoto and M. Kawahara, *Compt. Struct.* **71** (1999), 435–446.
[4] M. Abé, Y. Fujino and S. Kimura, *Proc. SPIE* **3329** (1998), 620–623.
[5] G.W. Housner, *Bull. Seismol. Soc. Amer.* **47** (1957), 15–35.
[6] D.W. Marquardt, *J. Soc. Indust. Appl. Math.* **11** (1963), 431–441.
[7] Y. Ohira, Application of a magnetic fluid to the tuned liquid damper, Master Thesis, Keio University, 2001, (in Japanese).

International Journal of Applied Electromagnetics and Mechanics 13 (2001/2002) 79–83
IOS Press

A study on the vibration analysis of a maglev vehicle – A theoretical investigation of the effect of magnetic damping on a vibration control system

Ken Watanabe[a,*], Yasuhiro Ohta[b], Masao Nagai[b] and Takayoshi Kamada[b]
[a]*Railway Technical Research Institute, 2-8-38 Hikari-cho, Kokubunji-shi, 185-8540 Tokyo, Japan*
[b]*Tokyo University of Agriculture and Technology, 2-24-16 Naka-cho, Koganei-shi, 184-8588 Tokyo, Japan*

Abstract. The superconducting Maglev system is conceptualized as the next generation high-speed public transportation system. For practical use, it is important to improve the ride comfort particularly in high-speed running. The Maglev vehicle is composed of lightweight car bodies and bogies which are mounted with SCMs (superconducting magnets), on-board refrigerating system, etc. The vibration isolation performance is lower than in conventional railways, since the mass ratio of sprung mass to un-sprung mass in the Maglev vehicle is lower than in the typical railway car. In this magnetically levitated system, the passive electromagnetic damping between the SCMs and ground coils is very small. Therefore, it is important to add active electromagnetic damping to the primary suspension between the SCMs and ground coils, and to adjust the secondary suspension between the car body and bogie.

This paper examines lateral and rolling vibration control of the Maglev vehicle using LQ (linear quadratic) control theory. Moreover, the estimated electromagnetic damping, which interacts between the SCMs and the guideway, is also considered in the model to improve the ride comfort.

1. Introduction

The superconducting Maglev vehicle is composed of lightweight car bodies and bogies mounted with SCMs (superconducting magnets) and on-board refrigerating systems. The vibration isolation performance is lower than in conventional railways, since the mass ratio of sprung mass to un-sprung mass in the Maglev vehicle is higher than in the typical railway car. In this system, the electromagnetic damping between the SCMs and ground coils is very small.

This paper focuses on the reduction of the lateral and rolling vibrations of the Maglev vehicle using a vibration control system designed based on LQ (linear quadratic) control theory. Moreover, the case considering the effect of the electromagnetic damping is investigated.

*Corresponding author: Ken Watanabe, Tel.: +81 42 573 7299; Fax: +81 42 573 7300; E-mail: ken_w@rtri.or.jp.

1383-5416/01/02/$8.00

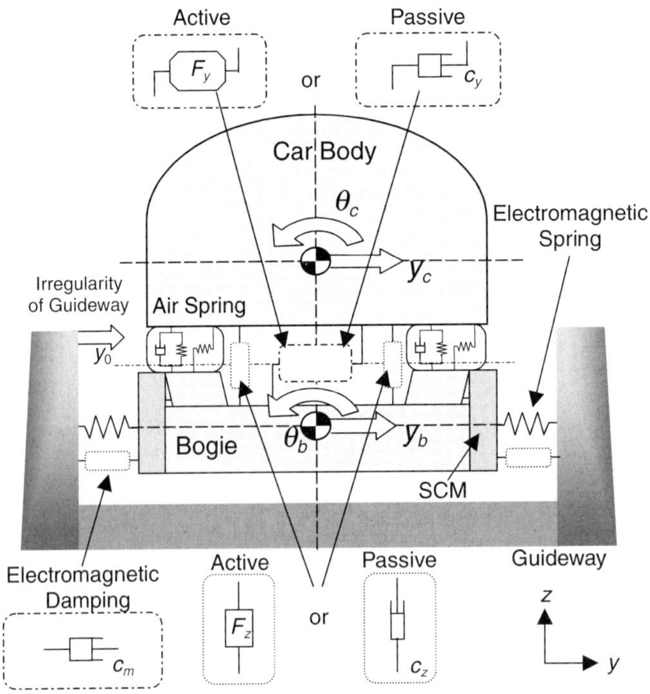

Fig. 1. Simulation model.

2. Simulation model

Figure 1 shows the 4-degree of freedom simulation model, which consists of the car body and the bogie. The car body and bogie are connected in secondary suspension, which consists of air springs and vertical and lateral dampers. Vehicle vibrations are excited by the irregularities of the guideway through the primary suspension of the electromagnetic springs. The primary suspension is generated by the interaction between the SCMs and levitation coils of the guideway.

In this paper, we examine the following two methods to improve the ride comfort.

(1) LQ control theory applied to the secondary suspension.
(2) LQ control theory applied to the secondary suspension, with electromagnetic damping added to the primary suspension.

We adopt lateral actuators instead of lateral dampers and regard the output force of the actuators as control input. We also adopt vertical actuators instead of vertical dampers and regard the output forces of the actuators as control input. We assume these actuators to be ideal ones that have no time lag in response.

3. Control system design

The objective of this control system is to reduce the lateral acceleration and rolling angle of the car body and relative displacement between the bogie and guideway. In addition, the lateral relative displacement of the minimum gap between the bogie and guideway must be within 40 mm [1].

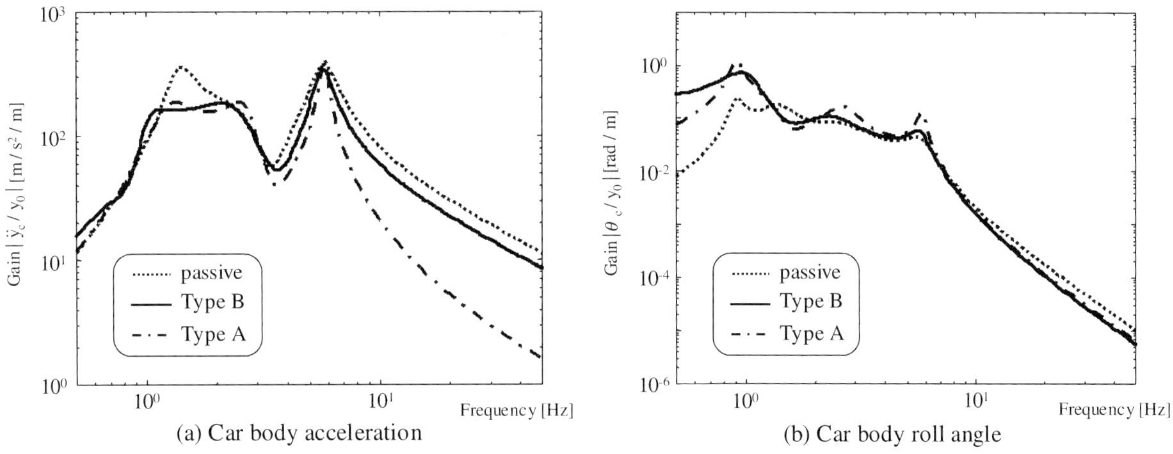

Fig. 2. Comparison of vibration characteristics with different patterns of weighting factors.

The state variables of the system are as follows:

$$x = [\dot{y}_c \, y_c \, \dot{y}_b \, y_b \, \dot{\theta}_c \, \theta_c \, \dot{\theta}_b \, \theta_b]^T \tag{1}$$

The state equation of the system is as follows:

$$\dot{x} = Ax + Bu + Wy_0 \tag{2}$$

For the system given in Eq. (2), according to LQ control theory, the cost function is expressed as:

$$J = \int_0^\infty \left[\left(\frac{\ddot{y}_c}{q_1}\right)^2 + \left(\frac{y_b - y_0}{q_2}\right)^2 + \left(\frac{\theta_c}{q_3}\right)^2 + \left(\frac{\theta_b}{q_4}\right)^2 + \left(\frac{u}{r}\right)^2 \right] dt \tag{3}$$

The objective is to minimize the above cost function J, where, u is control input that is determined by state feedback as:

$$u = -Kx \tag{4}$$

4. Simulation results

4.1. Without electromagnetic damping

(a) Lateral actuators

First, we control the lateral acceleration of the car body. For the cost function given in Eq. (3), we set the weighting factors of the rolling angles q_3 and q_4 to infinity and design the controller. We call this case "Type A". Figure 2 shows a comparison between Type A and the passive system without control. From this figure, we confirm that the gain of the lateral acceleration of the car body near 1.3 Hz is lower in Type A than in the passive system. However, the gain of the rolling angular acceleration of the car body increases near 1 Hz, since we do not evaluate the rolling angles in the controller.

Second, we consider the effects of the rolling angles on the lateral acceleration. We call this case "Type B". The simulation results of Type B are shown in Figs 2 and 3. We confirm that the gain of the rolling angles is improved, except for frequencies lower than 1.2 Hz.

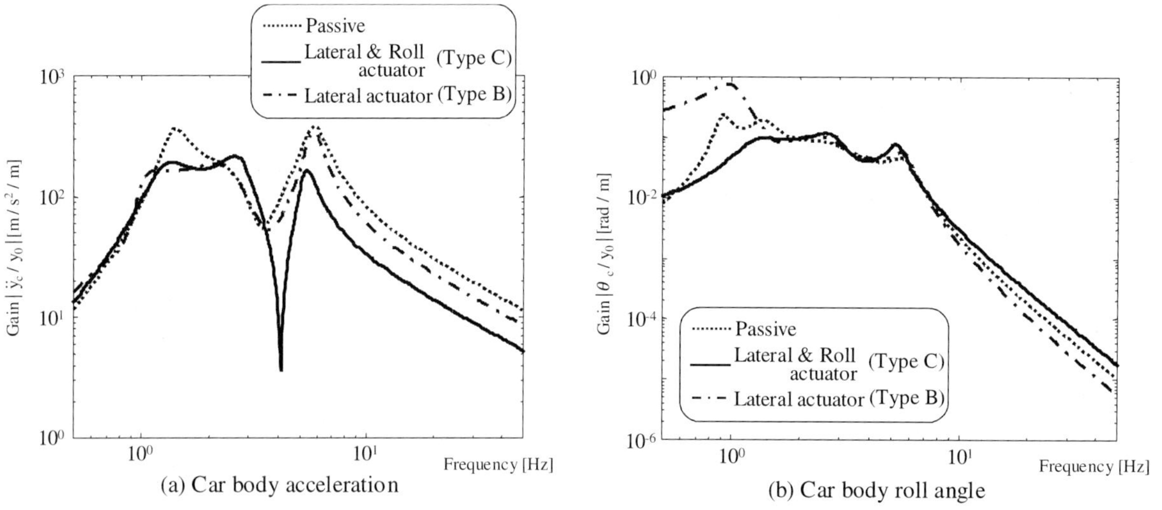

Fig. 3. Comparison of vibration characteristics with different patterns of actuators.

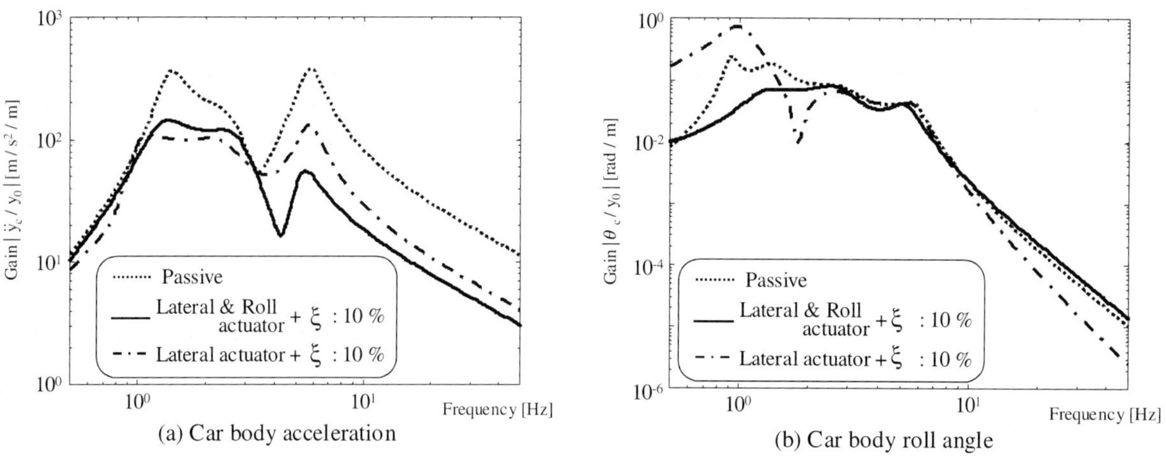

Fig. 4. Comparison of vibration characteristics considering electromagnetic damping.

(b) Lateral and rolling actuators

As we mentioned above, the improvement of the gain of the rolling angles is insufficient with the lateral actuators alone. Therefore, we add rolling actuators to control the rolling angle of the car body. We call this case "Type C". The cost function is expressed as:

$$J = \int_0^\infty \left[\left(\frac{\ddot{y}_c}{q_1}\right)^2 + \left(\frac{y_b - y_0}{q_2}\right)^2 + \left(\frac{\theta_c}{q_3}\right)^2 + \left(\frac{\theta_b}{q_4}\right)^2 + \left(\frac{u_{la}}{r_l}\right)^2 + \left(\frac{u_{ve}}{r_v}\right)^2 \right] dt \tag{5}$$

where, u_{la} and u_{ve} are control inputs of the lateral and rolling actuators, respectively.

Figure 3 compares the vibration characteristics of different patterns of actuators. From this figure, it is clarified that the peak of the lateral acceleration of the car body in the passive system near 1.3 Hz is reduced in Type C. It is also important to note that the gain of the rolling angular acceleration of the car body for frequencies less than 1.2 Hz is reduced in Type C.

4.2. With electromagnetic damping

In the superconducting Maglev system, there is small electromagnetic damping between the SCMs and guideway. There are several methods of generating electromagnetic damping in the primary suspension. One of them is the electromagnetic damping control system with the application of the distributed-type linear generator [2]. This system generates lateral force and rolling moment by changing the current phase of the on-board power collecting coils. There is a merit in which this system needs no exclusive damping coils.

In this study, the lateral force and rolling moment are assumed as parameters of each damping coefficient. We add 10% damping coefficient to the primary suspension, and control the lateral and rolling actuators of the secondary suspension. Figure 4 shows a comparison of the vibration characteristics considering electromagnetic damping. From this figure, it is effective for the improvement of the ride comfort to combine the electromagnetic damping and the actuators. Especially, it is more effective for the car body acceleration in the area of 6 Hz, which is near the natural frequency of lateral electromagnetic spring.

5. Conclusions

It is shown that the lateral acceleration and rolling angle of the car body on the superconducting Maglev vehicle can be effectively controlled using LQ theory applied to the secondary suspension. It is also found that adding electromagnetic damping to the primary suspension further enhances the control system performance.

References

[1] M. Azakami, The Development of Maglev Bogie System on the First Train Set for Yamanashi Test Line, *RTRI Report* **10**(1) (1996), (Japanese).

[2] T. Murai, H. Hasegawa, T. Yamamoto and S. Fujiwara, Active Magnetic Damper Using Linear Generator, *T. IEE Japan* **119-D**(11) (1999), (Japanese).

IOS Press

Numerical simulation of biased roller transfer by one-dimensional quasi-steady electric field analysis

Tomoyuki Ito[a] and Hiroyuki Kawamoto[b]
[a]*Fuji Xerox Co., Ltd., 430 Sakai, Nakai-machi, Ashigarakami-gun, Kanagawa, Japan*
Tel.: +81 465 80 2161; Fax: +81 465 81 8983; E-mail: Tomoyuki.Ito@fujixerox.co.jp
[b]*Department of Mechanical Engineering, Waseda University, 3-4-1, Okubo, Shinjuku, Tokyo, Japan*
Tel.: +81 3 5286 3914; E-mail: kawa@mn.waseda.ac.jp

Abstract. A stable and accurate numerical simulation technique is proposed to investigate performance of the transfer system using a biased transfer roller (BTR) in electrophotography. A one-dimensional layer-structured model was used to calculate distribution of electric field and current under air discharge. Concentration of generated ozone was also calculated. As results of calculations for a typical black and white laser printer, it was shown that (1) a transfer system using a semi-conductive BTR under constant current operation was robust against variation in resistance of media, and (2) concentration of ozone in the BTR system was two orders lower than that in the system using a conventional corona device.

1. Introduction

In the electrophotographic system used in copiers and laser printers, electrostatic images generated by charging and exposing the surface of the photoreceptor is developed by charged toner particles. Generated toner images are transferred to paper and fused by pressure and heat [1]. In the biased transfer roller (BTR) system that is one of the transfer systems, as shown in Fig. 1(a), a voltage is applied through a resistive elastomer layer of the roller inducing discharge to form a transfer field, the electric field to apply electrostatic force to toner particles. Compared to the conventional systems using the corona discharge, the BTR system is able to realize higher image quality and lower ozone generation. However, in the BTR system, parameters must be appropriately tuned for stable operation under variation of environmental and operational parameters.

For prediction of the transfer field, various theoretical studies have been carried out most of which were excessively simplified for accurate predictions. In 1993 a transfer field simulation based on a quasi-steady analysis was reported [4], but the method was limited to a belt-type photoreceptor and its discharge analysis was unstable. Even though multi-dimensional finite difference approaches are possible, the analysis tends to be excessively time consuming for parameter tuning for development of products.

Although the BTR system is known as "ozone-less", generation of ozone cannot be eliminated and quantitative evaluation of its concentration is still important. Regarding ozone formation in roller discharge system, Kawamoto carried out a theoretical and analytical study of a charger roller [3].

1383-5416/01/02/$8.00

5. Conclusion

In order to clarify the mechanism and predict performance of the BTR transfer system in the electrophotographic copier and printer, a generalized numerical simulation technique that is stable and accurate was proposed. Assuming that the electric flux line is perpendicular to the surface of each layer, a one-dimensional layer-structured model was constructed to conduct a quasi-steady analysis. At each time step, effect of conductive current is calculated by Runge-Kuitta method and the set of equations is solved considering air discharge subsequently. Based on the discharge current, concentration of generated ozone is calculated also.

Using the technique, calculations using parameters of a typical black and white laser printer was carried out. By comparing the result with that of previously proposed method and experiment, stability and accuracy of the calculation was shown. Calculated ozone concentration was two orders lower than that in the conventional corona device. It was also indicated that a transfer system using a semi-conductive BTR under constant current operation was robust against variation in resistivity of media. Both the results agree with empirically known facts. This technique can be utilized in development of electrophotographic products, for quick and low cost parameter tuning and trouble shooting.

References

[1] E.M. Williams, *The physics and technology of xerographic processes,* Krieger Publishing Co., 1993.

[2] G.V. Nadis, Modeling of plasma chemical processes in stable corona discharge at thin wires, *J. Phys. D: Appl. Phys.* **25** (1992), 477.

[3] H. Kawamoto, Modeling of ozone formation by contact charger roller, *J. Imaging Sci. Tech.* **39** (1995), 267.

[4] M.C. Zaretsky, Performance of electrically biased transfer roller in Kodak ColorEdgeTM CD copier, *J. Imaging Sci. Tech.* **37**(2) (1993), 187.

[5] R.M. Schafert, *Electrophotography,* (2nd ed.), Focal Press, London, 1975, pp. 514.

International Journal of Applied Electromagnetics and Mechanics 13 (2001/2002) 93–97
IOS Press

Response of a prestressed orthotropic thick plate subjected to an impulsive load

Naotake Nozawa
Department of Mechanical Engineering, Hachinohe National College of Technology, 16-1 Uwanotai, Tamonoki, Hachinohe City, Aomori 039-1192, Japan
Tel.: +81 178 27 7266; Fax: +81 178 27 7275; E-mail: nozawa-m@hachinohe-ct.ac.jp

Abstract. This paper concerns the dynamic response of an infinitely long thick plate. The plate is simply supported at the edges and subjected to initial in-plane stresses parallel and perpendicular to the edges. The plate is an orthotropic plate and rests on a viscoelastic foundation, where an impulsive line load excites it. Dynamic coefficients have been calculated and displayed graphically for several values of the initial stress parameters.

1. Introduction

Given the number of airline crashes in recent years, aircraft safety has become an issue of global relevance. For space development, spacecraft safety must also take into account the fact that parts will be strike each other while being assembled in space and that the structure will be struck by particles of cosmic dust and small meteors. Debris from man-made satellite larger than 10 cm in diameter numbers more than eight thousand objects in the earth's orbit, and the possibility of pieces of debris striking a structure must be considered. However, restrictions for the mass of launch vehicle are a problem one should not ignore, as spacecraft is generally designed to the limit. In order to increase the reliability of the limits in design, one must determine how an external dynamic force moves parts. In relation to this problem, Schonberg and Tullos [1] are experimenting with the resistance to penetration by orbital debris impacts using a dual-wall structure. For the movement of parts of spacecraft, Saito et al. [2], Padovan [3] and Chonan [4] are studying the response of plate bands with elastically restrained edges to traveling line loads. However, to the author's knowledge, there were only paper of Chonan et al. [5] and Nozawa et al. [6] on the impulse response of plates under such circumstances.

It is in this context, then, that this paper offers theoretical analysis of the transient response of a plate regarded as part of a spacecraft. The plate is simply supported at the edges and subjected to initial in-plane stresses that are parallel and perpendicular to the edges. The plate is an orthotropic plate and resting on a viscoelastic foundation. Governing equations are solved using methods for Laplace transforms with respect to time and methods for Fourier transforms with respect to axial space variables. Dynamic coefficients (= maximum dynamic displacement / static displacement) have been calculated and are graphically displayed for several values of the initial stress parameters.

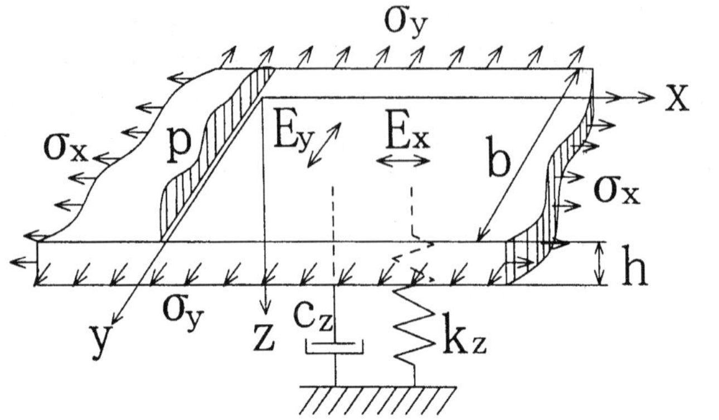

Fig. 1. Geometry of the plate.

2. Formuration of problem

The plate is an orthotropic and an elastic plate. Figure 1 shows an infinitely long thick plate with resting on an elastic foundation. The x-direction is the neutral plane in the direction of the length of the plane, the y-direction is the width of plate, and the z-direction is the line perpendicular to the plate. The plate has in-plane prestresses σ_x and σ_y in the x- and y-directions. The plate is an orthotropic and an elastic plate. And it is assumed that the plate is simply supported along the edges $y = 0$, b, and subjected to an impulsive line load $p(x, y, t)$ applied to the plate at $x = 0$. The displacement of the plate in the z-direction is denoted by $w(x, y, t)$, and the rotation angles in the xz and yz-planes are denoted by $\Psi_x(x, y, t)$ and $\Psi_y(x, y, t)$; the angle of rotation is assumed to be clockwise. In this case, with the application of Hamilton's principle, the equations for the plate are given as:

$$\left[(\kappa Gh + \sigma_x h)\frac{\partial^2}{\partial x^2} + (\kappa Gh + \sigma_y h)\frac{\partial^2}{\partial y^2} - \rho h\frac{\partial^2}{\partial t^2} - c\frac{\partial}{\partial t} - k_z\right] w$$
$$+ \left[-\kappa Gh\frac{\partial}{\partial x}\right] \Psi_x + \left[-\kappa Gh\frac{\partial}{\partial y}\right] \Psi_y + p(x, y, t) = 0, \tag{1}$$

$$\left[\kappa Gh\frac{\partial}{\partial x}\right] w + \left[(E_x + \sigma_x)\frac{h^3}{12}\frac{\partial^2}{\partial x^2} + (G + \sigma_y)\frac{h^3}{12}\frac{\partial^2}{\partial y^2} - \kappa Gh - \frac{\rho h^3}{12}\frac{\partial^2}{\partial t^2} - \right.$$
$$\left.\left(c\frac{\partial}{\partial t} + k_x\right)(\frac{h}{2})^2\right] \Psi_x + \left[(E_v + G_{xy})\frac{h^3}{12}\frac{\partial^2}{\partial x \partial y}\right] \Psi_y = 0, \tag{2}$$

$$\left[\kappa Gh\frac{\partial}{\partial y}\right] w + \left[(E_v + G)\frac{h^3}{12}\frac{\partial^2}{\partial x \partial y}\right] \Psi_x$$
$$+ \left[(G + \sigma_x)\frac{h^3}{12}\frac{\partial^2}{\partial x^2} + (E_y + \sigma_y)\frac{h^3}{12}\frac{\partial^2}{\partial y^2} - \kappa Gh - \frac{\rho h^3}{12}\frac{\partial^2}{\partial t^2} - \left(c\frac{\partial}{\partial t} + k_y\right)\left(\frac{h}{2}\right)^2\right] \Psi_y = 0. \tag{3}$$

Here, E_x, E_y, and E_v are Young's modulus of the x- and y- directions and Poisson's effects, G is the shear modulus, κ is the shear coefficient of plane, b is the plate width, h is the thickness of the plate, ρ the mass density, and t is time. The normal stiffness of the foundation under the unit area of plate is denoted by k_z and the corresponding horizontal stiffness in the x- and y- directions by k_x and k_y, respectively; c is the coefficients of external damping. $p(x, y, t) = [\sum p\sin(n\pi y/b)]\delta(x)f(t)$, where $\delta(x)$ is Dirac delta function and $f(t)$ is step function of time. Applying the methods for Laplace transforms with respect to t and Fourier transforms with respect to x allows solution of the resulting equations. And using the inverse Fourier transform and the inverse Laplace transform

$$w^* = \frac{1}{2\pi i}\int_{\varsigma - i\infty}^{\varsigma + i\infty} \Gamma_1(x^*, s^*)\Lambda(s^*) \exp{(s^* t^*)} ds^* \tag{4}$$

Where

$$w^* = \frac{E_x}{p\sin(\pi y/b)} w$$

Here, w^* is the non-dimensional displacement.

The resulted Eq. (4) is calculated the using the numerical inversion method for the Laplace transform presented by Weeks [7].

3. Numerical results

As a numerical example, a thick strip plate subjected to a step line load is presented in detail. In this case, the time dependent part of the loading is the Heaviside unit step function.

Physical parameters used are as follows: $\kappa = \pi^2/12$, $\alpha = h/b$, $\varepsilon_y = \sigma_y/E_y$, $E_v/E_x = 0.3$, $G = 0.35 \times E_x$, $\zeta_z = \{k_z h/E_x\}$, $c = 0$.

Figure 2 shows the dynamic coefficients of the displacement of the plate as a function of the in-plane prestress for the width (y- direction). The dynamic coefficients are the ratio of the maximum dynamic displacement w^*_{max} to w^*_{st} is labeled K_{w} such that:

$$K_W = w^*_{\text{max}}/w^*_{\text{st}}.$$

The dynamic coefficient K_w displays no variation to the in-plane prestress in the x-direction, and when the x- and y- directions vary at the same time, the result is considered to be only the y-direction prestress.

Figure 2 shows the dynamic coefficient of displacement (K_W) where $\zeta_z = 0$, and the thickness-to-width ratio α is 0.04, 0.06, 0.08, 0.09, 0.10, 0.12. Given these results, the dynamic coefficient is small with a smaller thickness-to-width ratio and the increase is uniform; as α increases, it increases slightly and then decreases again, so the maximum value is roughly where α equals 0.09. As shown by the increase in the ratio E_y/E_x (modulus ratio) for the Young's modulus in the direction of the width to the Young's modulus in the x-direction, the maximum value for the dynamic coefficient appears in the compression stress.

Figure 3 show the dynamic coefficient of the displacement as a function of increasing stiffness with $\alpha = 0.06$. Comparison to values in Fig. 2(b) indicates that the dynamic coefficient increases slightly with an increase in stiffness. And, the maximum value of the dynamic coefficient appears in the compression stress with the modulus ratio (E_y/E_x). Figure 4 illustrate the results when $\alpha = 0.08$. Compared to Fig. 2(c), the dynamic coefficient increases with an increase in stiffness. In addition, the maximum value for the dynamic coefficient appears in the compression stress with an increase in the modulus ratio (E_y/E_x).

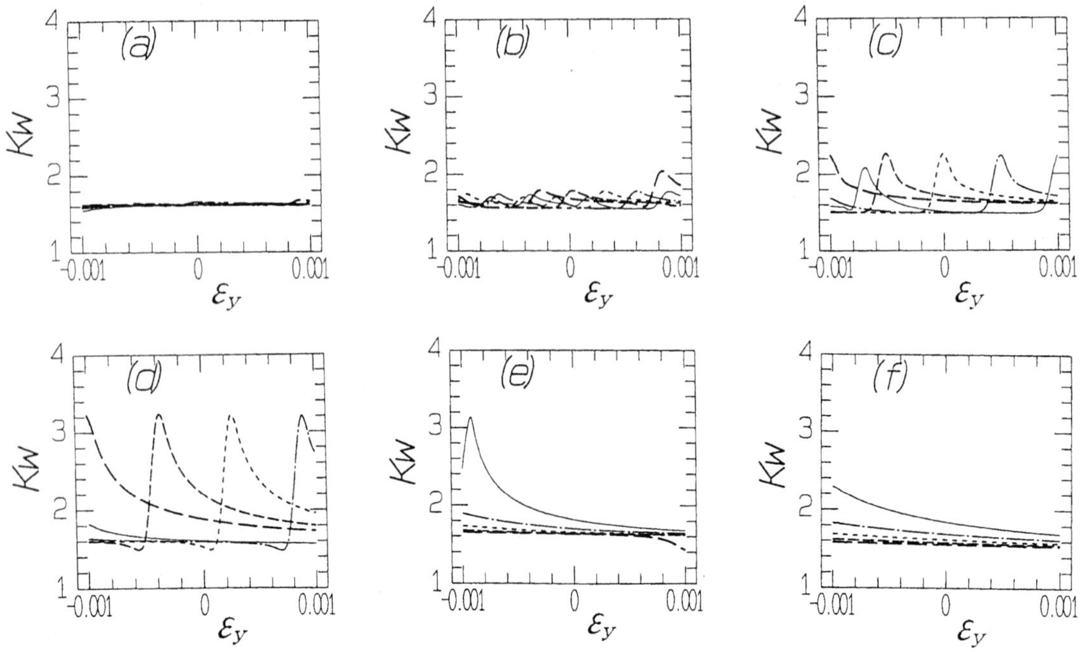

Fig. 2. Variations in the dynamic coefficient of displacement K_W with increasing initial stress. (a) $\alpha = 0.04$; (b) $\alpha = 0.06$; (c) $\alpha = 0.08$; (d) $\alpha = 0.09$; (e) $\alpha = 0.10$; (f) $\alpha = 0.12$; $\zeta_z = 0$. ——, $E_y/E_x = 0.8$; - - - - - $E_y/E_x = 0.9$; - - - - - - $E_y/E_x = 1.0$; – – –, $E_y/E_x = 1.1$; —— $E_y/E_x = 1.2$.

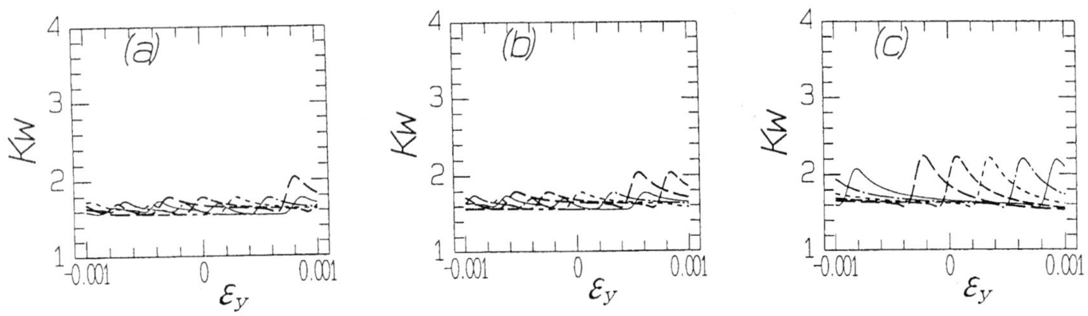

Fig. 3. Variations in the dynamic coefficient of displacement k_W with increasing initial stress. $\alpha = 0.06$. (a) $\zeta_z = 0$; (b) $\zeta_z = 0.000001$; (c) $\zeta_z = 0.00001$. ——, $E_y/E_x = 0.8$; - - - - - $E_y/E_x = 0.9$; - - - - - - $E_y/E_x = 1.0$; – – –, $E_y/E_x = 1.1$; —— $E_y/E_x = 1.2$.

4. Conclusions

A theory of transient vibration has been developed for the response of an infinitely long plate with thick bands and in-plane stress when it is subjected to an impulsive step line load. The results obtained can be summarized as follows:

(1) The dynamic coefficient of the displacement is a maximum value when α is roughly equal to 0.09.
(2) As the ratio E_y/E_x for the Young's modulus in the direction width to the Young's modulus parallel to edges increases, the maximum value for the dynamic coefficient appears in the compression

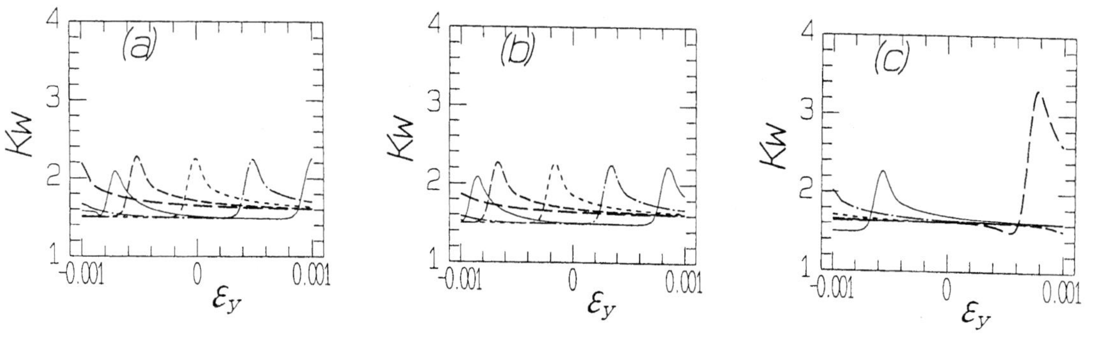

Fig. 4. Variations in the dynamic coefficient of displacement K_W with increasing initial stress. $\alpha = 0.08$. (a) $\zeta_z = 0$; (b) $\zeta_z = 0.000001$; (c) $\zeta_z = 0.00001$. ——, $E_y/E_x = 0.8$; - - - - - $E_y/E_x = 0.9$; - - - - - - $E_y/E_x = 1.0$; – – –, $E_y/E_x = 1.1$; — — $E_y/E_x = 1.2$.

stress when α are 0.08∼0.10.

(3) The dynamic coefficient of the displacement increase according to the plate stiffness.

Acknowledgments

The author would like to cordially thank Professor Dr. S. Chonan of Tohoku University for assistance with the initiation of this work and further discussion as it progressed.

References

[1] W.P. Schonberg and R.J. Tullos, *AIAA Journal* **29**(12) (1991), 2207.
[2] H. Saito, S. Chonan and O. Kawanobe, *J. Sound Vib.* **71**(2) (1980), 191.
[3] J. Padovan, *Int. J. Eng. Sci.* **19**(9) (1981), 1203.
[4] S. Chonan, *J. Sound Vib.* **93**(3) (1984), 427.
[5] S. Chonan and N. Nozawa, *Tran. of the JSME C* **52**(476) (1986), 1203.
[6] N. Nozawa and S. Chonan, *Tran. of the JSME C* **53**(492) (1987), 1663.
[7] W.T. Weeks, *J. Association Comput. Mach* **13**(3) (1966), 419.

International Journal of Applied Electromagnetics and Mechanics 13 (2001/2002) 99–106
IOS Press

Design of switchable transformer control system employing the micro-controller 80C552

H.A. Dharmawan* and S.A.M. Ali
School of Electrical and Information Engineering, University of South Australia, Adelaide, South Australia 5095

Abstract. Research on a switchable transformer has proven that series and parallel configurations used in the primary and secondary windings can decrease the amounts of total losses which are made up of core losses and winding losses. Both configurations are integrated through some Solid State Relays (SSRs). When utilising this transformer however, some complications arise. The main concern is timings of switching from one configuration to another, and timings of triggers which should be applied to the SSRs. This paper proposes the use of the micro-controller 80C552 as the main control system to overcome the problems. The timings are controlled through the combination use of timer T0, T2, capture and compare interrupts.

1. Introduction

The applications of micro-controllers are becoming more and more popular, as it can solve complicated tasks using their smart digital processing. If compared with analog control circuits, micro-controller based control systems provide easy designs and higher reliability. Integrated components inside micro-controller chips enable users to design a control system efficiently and reduce complications due to wiring components.

The micro-controller proposed in this research is 80C552 which is a single chip 8-bit micro-controller from Philips Semiconductors. This type of micro-controller provides some additions to the popular 80C51 architecture. One of the very useful additional hardware parts is timer T2 and its related registers which provide functions such as capture logic and compare logic. In the design of our control system, many timing sequences rely on this timer, though timer T0 is also used for measuring time intervals.

The application of the micro-controller in this research, with some extra circuits, has also given a chance to extend the system to be able to monitor and protect the switchable transformer. The control system works with the configurations specified by users. These configurations can be modified and stored to an external memory. In addition, an LCD and a 3x4 keypad are added to make it easier for users to communicate with the micro-controller.

*Corresponding author: H.A. Dharmawan, Tel.: +61 8 83023346; Fax: +61 8 83023384; E-mail: dhaha001@students.unisa.edu.au.

1383-5416/01/02/$8.00

Table 1
Capture interrupts and their trigger sources

Interrupts	Trigger sources (phase 1)
CT0	Negative zero crossing of primary voltage
CT1	Positive zero crossing of primary voltage
CT2	Negative zero crossing of secondary current
CT3	Positive zero crossing of secondary current

Table 2
Information stored in the compare registers

Registers	Information stored in the registers
CM0	Timing to trigger phase 1
CM1	Timing to trigger phase 3
CM2	Timing to trigger phase 2

2. The use of SSRs

In our design, we use eighteen surface mounting SSRs from Crydom to connect tappings of windings into series or parallel configuration. The SSRs are optically isolated and feature an internal snubber circuit for high dv/dt applications and inductive loads. These encapsulated devices contain inverse-parallel SCR output devices suitable for AC switching.

Timings of trigger sequences to turn on the SSRs hold a crucial role in the design of the micro-controlled switchable transformer. Right timings are required to achieve lowest harmonic generations as an effect of switching actions. Attempts have been made to turn on these devices at zero crossing of current.

3. Timing control using capture and compare interrupts

The use of capture and compare interrupts is a good strategy to achieve precise timings of triggers. Compared with using normal procedures inside a main program, service procedures of these interrupts have the following advantages,

- Triggers to the SSRs can be generated without disturbances from any commands or normal procedures in the main program, unless interrupted by interrupts having higher priority.
- The controller would handle the service procedures swiftly as a corresponding external trigger comes.

There are four capture interrupts available inside 80C552 i.e. CT0, CT1, CT2 and CT3. These capture interrupts in the design are activated by the signals as shown in Table 1. The other interruptions used are compare interrupt CM0, CM1 and CM2. The combination of timer T2 and compare interrupts gives a chance to utilise these three interrupts to send triggers to the SSRs. In relation to these interrupts, there are three accompanied register CM0, CM1 and CM2 which are connected to timer T2 and hold values as indicated in Table 2. Once a match between values of timer T2 and these registers occurs, the controller will activate the corresponding compare interrupt.

Subroutines of interruption CT0 and CT1 have some commands to enable compare interrupt CM0, CM1, CM2 and to run timer T2. Besides, interruption CT0 and CT1 also have a job to run timer T0 which will measure phase differences between voltage and current signals in phase 1. When a zero crossing of current condition occurs, interruption CT2 (for falling edge) or CT3 (for rising edge) is generated, and

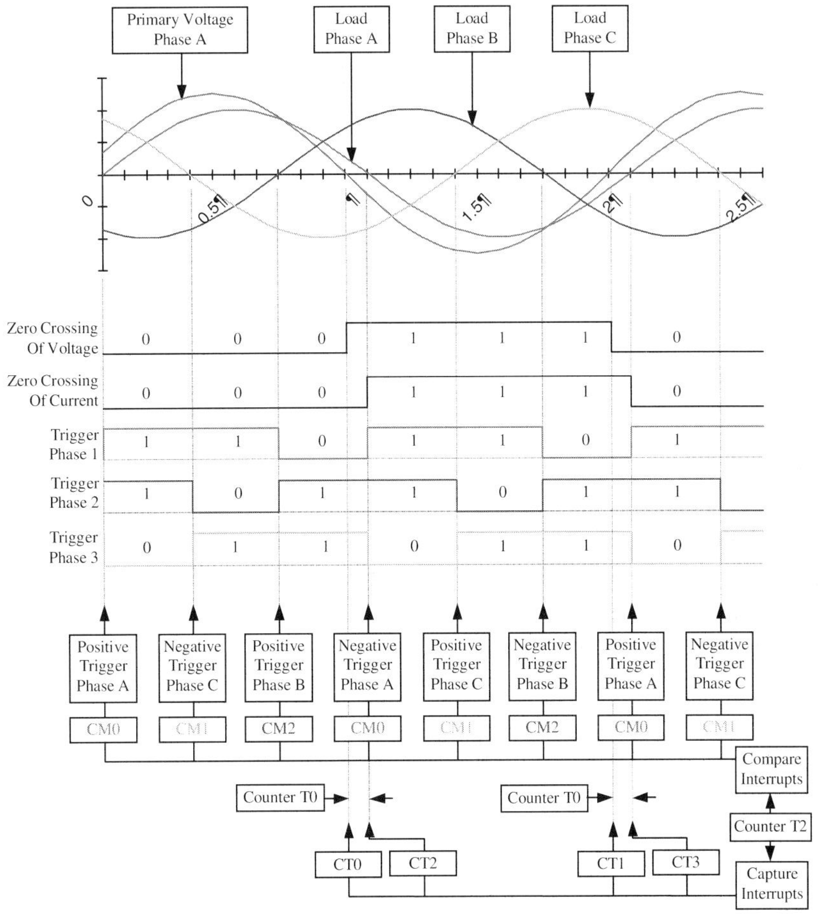

Fig. 1. Trigger sequences generated from the interruptions.

the subroutine inside these interruptions would stop T0 and save its value. An example of sequences when triggers are sent to the SSRs is depicted in Fig. 1. In this three-phase system, it is assumed that the three-phase current is symmetrical.

4. Capture and compare interrupt handlers

Logical commands inside service procedures of capture interrupt CT0, CT1, CT2 and CT3 are as follows

Interrupt CT0

```
reset CT0 interrupt flag
reset timer T0
reset timer T2
assign all compare registers(for rising edge)
reset all compare interrupt flags
enable all compare interrupts
run timer T2
run timer T0
```

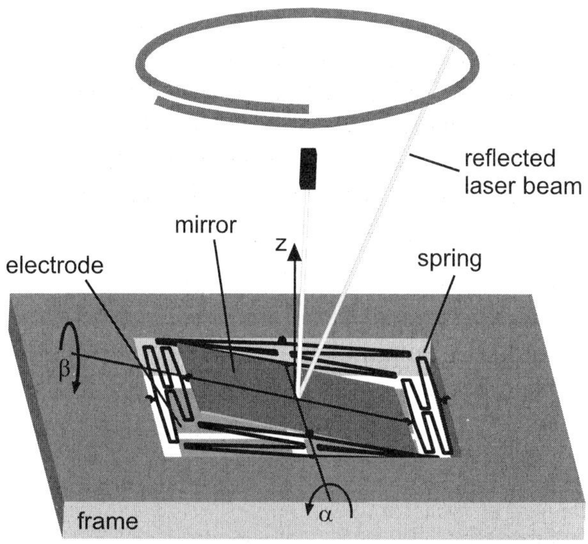

Fig. 1. Micromirror with laser beam.

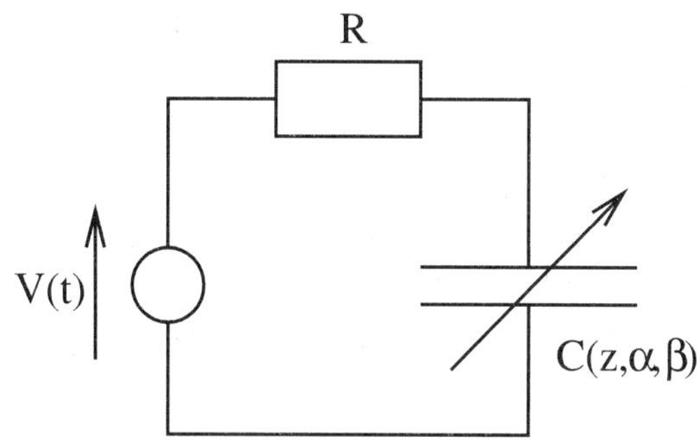

Fig. 2. Electrical network of one electrode.

3. Inverse dynamics of the mechanical subsystem

The given motion, the mirror shall move along in $_0E^3$, can be defined in the configuration space $_0R^n$ (n – dof. of the mechanical subsystem) as nominal trajectory (denoted by $q_0^\kappa(t)$) using inverse kinematics. Let r be the number of the generalized coordinates to be controlled (q^{κ_1}), then a submanifold $_0R^{(n-r)}$ is defined by

$$f^{\kappa_1}(q,t) := q^{\kappa_1} - q_0^{\kappa_1}(t) = 0, \quad |\{\kappa_1\}| = r. \tag{1}$$

If no electrically produced forces act on the mechanical subsystem, the well-known Lagrange's motion equations can be devided into two parts due to Eq. (1) ($g_{\kappa\lambda}$ - metric, $\Gamma_{\kappa\lambda\varrho}$ - Christoffel symbols of the first kind, Q_κ^{mech} - generalized mechanical forces)

$$g_{\kappa_1\lambda_1}\ddot{q}^{\lambda_1} + g_{\kappa_1\lambda_2}\ddot{q}^{\lambda_2} + \Gamma_{\kappa_1\lambda_\varrho}\dot{q}^\lambda\dot{q}^\varrho = Q_{\kappa_1}^{mech} + R_{\kappa_1}, \tag{2a}$$

$$g_{\kappa_2\lambda_1}\ddot{q}^{\lambda_1} + g_{\kappa_2\lambda_2}\ddot{q}^{\lambda_2} + \Gamma_{\kappa_2\lambda_\varrho}\dot{q}^\lambda\dot{q}^\varrho = Q_{\kappa_2}^{mech}, \tag{2b}$$

where R_{κ_1} are the reaction forces needed to hold the nominal trajectory. Using Eq. (2a,2b) with Eq. (1), R_{κ_1} may be calculated. If the $\mathfrak{o}R^n$-machine of Eq. (2a, 2b) would be kinematically controlled by Eq. (1) then R_{κ_1} would be the correct corresponding reaction forces. However, using these forces to control the mechanical subsystem dynamically, the solution will gradually leave the nominal trajectory because of numerical errors or unknown disturbances. Hence, an augmented PD-controller (see [3,4]) consisting of a nonlinear feedforward (based on the reaction forces) and a linear feedback (characterized by symmetric, positive definite gain matrices $K_{\kappa_1\lambda_1}, C_{\kappa_1\lambda_1}$) according to

$$\begin{aligned} R_{\kappa_1} := {} & g_{\kappa_1\lambda_1}\ddot{q}_0^{\lambda_1}(t) + \Gamma_{\kappa_1\lambda\varrho_1}\dot{q}^{\lambda}\dot{q}_0^{\varrho_1} + \Gamma_{\kappa_1\lambda\varrho_2}\dot{q}^{\lambda}\dot{q}^{\varrho_2} - Q_{\kappa_1} + \\ & + g_{\kappa_1\lambda_2}g^{\lambda_2\kappa_2}[Q_{\kappa_2} - g_{\kappa_2\lambda_1}\ddot{q}_0^{\lambda_1}(t) - \Gamma_{\kappa_2\lambda_\varrho}\dot{q}^{\lambda}\dot{q}^{\varrho}] - K_{\kappa_1\lambda_1}\dot{\varepsilon}^{\lambda_1} - C_{\kappa_1\lambda_1}\varepsilon^{\lambda_1} \end{aligned} \tag{3}$$

with $\varepsilon^{\lambda_1} := q^{\lambda_1} - q_0^{\lambda_1}(t)$ is introduced. This control law is asymptotically stable, i.e. the deviation ε of the motion from its nominal trajectory decreases asymptotically.

4. Electrically produced control forces

If the electromechanical interaction is only caused by capacitors, the forces produced electrically in an EMS read generally

$$Q_{\kappa}^{el} = -\frac{1}{2}\partial_{\kappa}c_{\mu v}(q^{\lambda})q^{\mu}q^{v}, \tag{4}$$

where $\partial_\kappa = \partial/\partial q^\kappa$ and $c_{\mu\nu}$ is derived from the capacities C^{ij} using the fundamental loop matrix $A^i{}_\mu$ of the representing graph of the electrical network ($c_{\mu\nu} = A^i{}_\mu A^j{}_\nu C_{ij}$). The control forces to drive the mirror along the nominal trajectory shall exclusively be produced electrically ($R_{\kappa_1} := Q_{\kappa_1}^{el}$). Using a convenient principal-axes transformation $y^\mu = T^\mu{}_\nu(q^\lambda)q^\nu$, T - orthogonal, and $(y^\mu)^2 = x^\mu$, the control forces read with Eq. (4)

$$R_{\kappa_1} = b_{\kappa_1\mu}(q^{\lambda})x^{\mu}, \ x \geqslant 0. \tag{5}$$

This is a linear equation system (usually underdetermined because of $r < m$, m - dof. of the electrical subsystem) with additional constraints. It can be solved e.g. by non-linear optimization using the electrical energy as cost function.

If $c_{\mu\nu}$ has diagonal form then an orthogonal transformation is not needed. Because of the topology of the electrical network of the micromirror, the electrically produced control forces read

$$\begin{aligned} Q_{\kappa_1}^{el} = -\frac{1}{2} & \left[\partial_{\kappa_1}(1/C_I)(q^I)^2 + \partial_{\kappa_1}(1/C_{II})(q^{II})^2 + \right. \\ & \left. + \partial_{\kappa_1}(1/C_{III})(q^{III})^2 + \partial_{\kappa_1}(1/C_{IV})(q^{IV})^2\right], \end{aligned} \tag{6}$$

where e.g. q^I denotes the charge at the first quadrant's capacitor and C_I its capacity. With $x^\mu := (q^\mu)^2$, $\mu = I \dots IV$, Eq. (6) is an underdetermined linear system

$$R_{\kappa_1} := Q_{\kappa_1}^{el} = -\frac{1}{2}\partial_{\kappa_1}c_{\mu\mu}x^{\mu}, \ x^{\mu} \geqslant 0, \tag{7}$$

which has to be solved for every integration step τ_i under the given constraints. Because the electrical energy as appropriate cost function is also linear in x^μ ($W_{el} = \frac{1}{2}c_{\mu\nu}q^\mu q^\nu = \frac{1}{2}c_{\mu\mu}x^\mu$), Eq. (7) can be

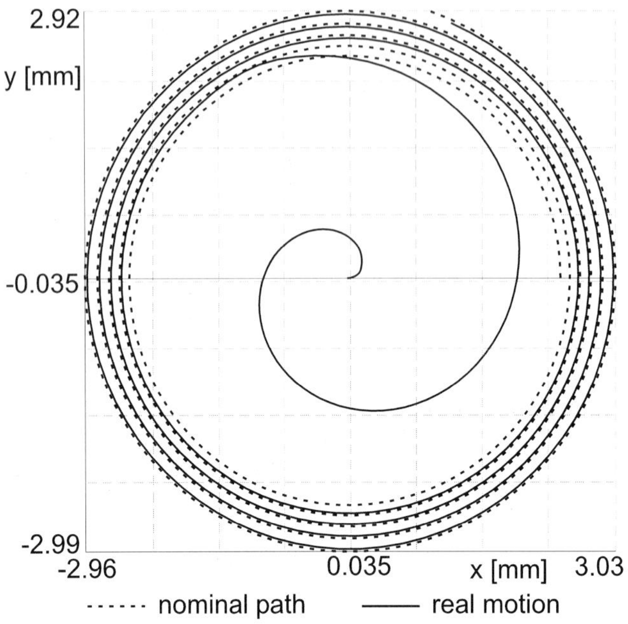

Fig. 3. Laser spot motion.

considered as equality and inequality constraints of a linear programming problem. Solving Eq. (7) for x^{μ}_{opt}, the nominal values of the corresponding generalized electrical coordinates (mesh charges) can simply be computed for every integration step τ_i. Finally, an interpolation yields a time history $q^{\mu}_0(t)$ of the mesh charges.

5. Inverse dynamics of the electrical subsystem

The time history $q^{\mu}_0(t)$ forms a nominal trajectory in the configuration space ${}_{\mathfrak{o}}R^m$ of the electrical subsystem. Because of the components used in the electrical network, the motion equations of the electrical subsystem read

$$r_{\mu v}\dot{q}^{v} + c_{\mu v}q^{v} + v_{\mu 0} = 0, \tag{8}$$

where $r_{\mu\nu}$ contains the matrix of resistances, $c_{\mu\nu}$ is derived from the capacities, and $v_{\mu 0}$ comprehends the voltage sources. The electrical subsystem has a clearly faster dynamical behaviour than the mechanical subsystem. Hence, the electrical subsystem can be considered as quasi static (term with $\dot{q}^{\nu}$ in Eq. (8) is neglected) and the nominal values of the mesh voltages $v_{\mu 0}$ can be calculated from the nominal mesh charges $q^{\nu}_0(t)$:

$$v_{\mu 0}(t) = -c_{\mu v}(q^{\lambda})q^{v}_0(t). \tag{9}$$

Injecting finally the resulting time history in the voltage source components of the electrical network, the necessary charges can be produced at the capacitors. These charges are suitable to originate mechanical forces in such a way that the real motion of the mirror approximates asymptotically to its nominal motion.

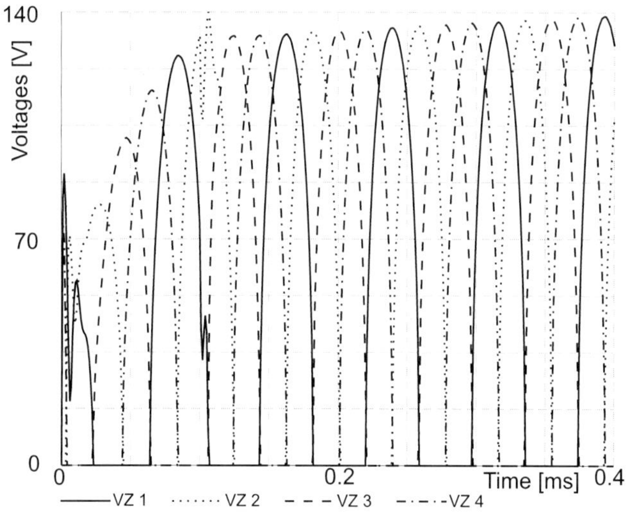

Fig. 4. Computed voltages.

6. Results

In a test setup called "Panorama projection", the mirror shall carry out a swaying motion with increasing amplitude in such a way, that the laser spot describes a spiral in an xy-parallel plane (see dotted line in Fig. 3. Near the resonance frequency, this could be done using sinusoidal voltages. In other cases, special voltages have to be used. The solid line in Fig. 3 shows the laser spot motion if the mirror starts in xy-parallel position. The time histories of the voltages needed to cause this motion were computed using the three steps of inverse dynamics shown in this paper. They are plotted in Fig. 4. After about 0.1 ms the starting process is finished and the nominal trajectory is reached (the voltage amplitudes increase slowly). Hence, the asymptotic stability of the controller was able to be verified.

7. Conclusion

Based on a unified mathematical approach for modelling electromechanical systems, a method to control dynamically the motion of an electrically driven micromirror was shown. The use of inverse dynamics, carried out in three steps for this task, was described. In the first step, an augmented PD-control law consisting of a nonlinear feedforward and a linear feedback was used. Finally, simulation results verifying the global asymptotic stability of the controller were given.

Acknowledgements

All research work have been done within the Collaborative Research Centre Sonderforschungsbereich 379 at the Chemnitz University of Technology supported by the Deutsche Forschungsgemeinschaft. The authors would like to acknowledge the cooperation with the other research groups of the SFB 379.

References

[1] O. Enge, P. Maißer and G. Kielau, Modelling and Simulation of Electromechanical Systems in the Case of Micromirrors, in: *International Symposium on Non-Linear Electromagnetic Systems (ISEM'99), Pavia, Italy, May, 10–12, 1999, proc.*, P. Di Barba, A. Savini, eds, IOS Press, Amsterdam, 2000, pp. 563–566.

[2] P. Maißer, O. Enge, H. Freudenberg and G. Kielau, Electromechanical Interactions in Multibody Systems Containing Electromechanical Drives, *Journal Multibody System Dynamics* **1**(3) (1997), 281–302.

[3] P. Maißer and U. Jungnickel, Stability of Controlled Motion in Diving Simulations, in: *XVI International Symposium on Biomechanics in Sports (ISBS'98), Konstanz, Germany, July, 21–25, 1998, proc.*, H.J. Riehle, M.M. Vieten, eds, Universitätsverlag Konstanz, 1998, 367–370.

[4] R.M. Murray, Z. Li and S.S. Sastry, *A Mathematical Introduction to Robotic Manipulation*, CRC Press, Boca Raton, 1994.

International Journal of Applied Electromagnetics and Mechanics 13 (2001/2002) 113–120
IOS Press

Optimal piezo-electro-mechanical coupling to control plate vibrations

Silvio Alessandroni[a], Francesco dell'Isola[b,*] and Fabrizio Frezza[c]
[a]*Dott. in Meccanica Teorica ed Applicata, Universitá di Roma "La Sapienza", 00184 Roma, Italia*
E-mail: salessan@vt.edu
[a]*Department of Engineering Science and Mechanics, Virginia Polytechnic Institute and State University, Blacksburg, VA 24061-0219, USA*
[b]*Dip. Ing. Strutturale e Geotecnica, Universitá di Roma "La Sapienza", 00184 Roma, Italy*
E-mail: francesco.dellisola@uniromal.it
[c]*Dip Ing. Elettronica, Universitá di Roma "La Sapienza", 00184 Roma, Italy*
E-mail: frezza@die.ing.uniromal.it

Abstract. A new way of coupling electrical and mechanical waves, using piezoelectric effect, is presented here. Since the energy exchange between two systems supporting wave propagation is maximum when their evolution is governed by similar equations, hence, an optimal electro-mechanical coupling is obtained by designing an electric network which is "analog" to the mechanical structure to be controlled. In this paper, we exploit this idea to enhance the coupling, between a Kirchhoff-Love plate and one possible synthesis of its circuital analog, as obtained by means of a set of piezoelectric actuators uniformly distributed upon the plate. It is shown how this approach allows for an optimal energy exchange between the mechanic and the electric forms independent of the modal evolution of the structure. Moreover, we show how an efficient electric dissipation of the mechanical energy can be obtained adding dissipative elements in the electric network.

1. Introduction

It is known that a deformation field of a piezoelectric material generates an electric field in it and vice versa (see [1]). However, it is not possible to obtain an efficient energy exchange between the electrical and mechanical waves propagating in a macroscopic structure of piezoelectric material, because the propagating speed of an electric field, in such a material (i.e., light speed) is much larger than the propagating speed of any deformation field (i.e., sound speed) [2]. This means that the electric field, generated by the deformation in a piezoelectric (PZT) material, is not supported and efficiently transmitted by the material itself, so that the energy is not efficiently transformed from one form to the other (see [3]). Since light speed can be considered infinite when compared to sound speed, we can accept the lumped hypotheses for the used electrical components and we synthesize, in Section 1, an electric circuit, the equations of which are analogous to the finite-difference equations for the deflection field of a plate. These circuits are called 'circuital analogs' of the considered mechanical systems [4]. They were used as analogical computers before the incoming of the digital ones for the design of mechanical structures

*Corresponding author: Francesco dell'Isola, E-mail: francesco.dellisola@uniroma1.it.

1383-5416/01/02/$8.00

(see [5]). In Section 2 the interconnection between the plate and the circuital analog, using piezoelectric actuators, is considered and the model of the coupled electromechanical system is developed by means of a suitable homogenization procedure. We explicitly remark that no mathematical convergence result is obtained or used there. The identification procedure, which we developed, is purely heuristic and may deserve a more careful mathematical analysis similar to that developed in [6,7]. In Section 3, the analysis of the coupled system will show that the maximum energy transfer from the mechanical to the electrical form is independent of the modal evolution of the plate. Finally, considering the presence of dissipative elements in the analog, an efficient mechanical damping is obtained.

2. Electric analog synthesis

Let us consider the Kirchhoff-Love plate dimensionless equation:

$$\nabla^2\nabla^2 u = \alpha\ddot{u} = 0, \quad \alpha = \frac{3\rho l_0^4}{h^2 E t_0^2} \quad \forall u \in D \tag{1}$$

where u and $\nabla^2\nabla^2$ are the dimensionless deflection field and the double-Laplacian operator defined on the bidimentional domain D, the superscript dot represents the dimensionless time derivative, l_o and t_o are the characteristic length and time, ρ is the plate mass density per unit volume, h is the thickness of the plate, and E is the Young modulus of the material constituting the plate. Moreover, two different kinds of boundary conditions (BC) are chosen [2]:

- Completely-clumped BC:

$$\begin{cases} u|_{\partial D} = 0 \\ \frac{\partial u}{\partial n}|_{\partial D} = 0 \end{cases} \tag{2}$$

  where n is the normal to ∂D in the tangent plane to D
- Simply-supported BC:

$$\begin{cases} u|_{\partial D} = 0 \\ \frac{\partial^2 u}{\partial n^2}|_{\partial D} = 0 \end{cases} \tag{3}$$

The electric analog is found sampling the domain D by a uniform-step grid and associating the electric tension of each node of the circuital analog to the deflection-velocity field in the nodes of the grid. Indeed let us introduce the Laplace transform of the field u. Therefore Eq. (1) becomes:

$$\nabla^2\nabla^2 u + \alpha s^2 u = 0 \tag{4}$$

where s is the Laplace variable. Then we replace the spatial differential operator $\nabla^2\nabla^2$ with its centred finite difference approximation. This means that we need to synthesize a lumped circuit in which the nodes correspond to the nodes of the grid and the Kirchhoff laws in terms of their voltage drop referred to ground become

$$\begin{aligned}(\frac{16}{s\varepsilon^4} + \alpha s)V[i,j] &+ \frac{1}{s\varepsilon^4}(V[i+2,j] - 6V[i+1,j] - 6V[i-1,j] + V[i-2,j]) \\ &+ \frac{1}{s\varepsilon^4}(V[i,j+2] - 6V[i,j-1] + V[i,j-2]) \\ &+ \frac{1}{s\varepsilon^4}(V[i+1,j+1] - 6V[i-1,j+1] - 6V[i+1,j-1] + V[i-1,j-1]) = 0\end{aligned} \tag{5}$$

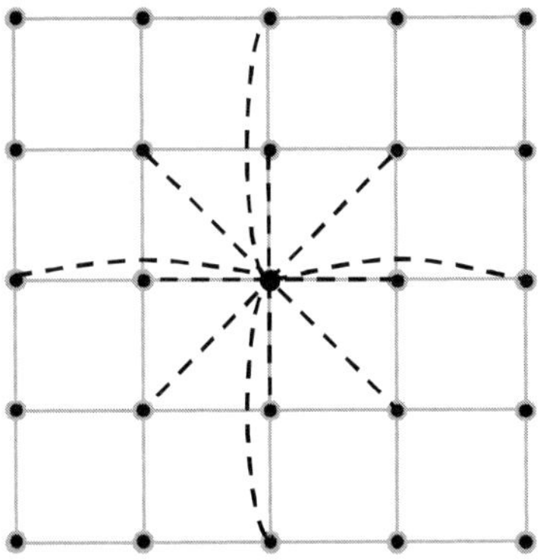

Fig. 1. Involved nodes for the equation at ij node (dashed lines).

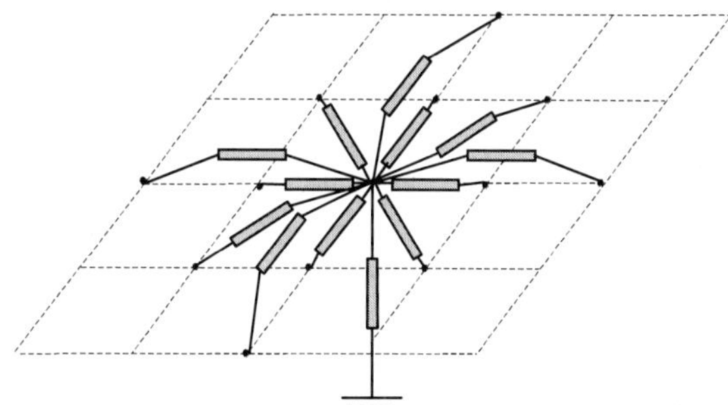

Fig. 2. Circuit-elements topology.

where $V[i,j]$ is the dimensionless L-transform of the voltage drop v at node ij, α has to be identified in terms of electric immittances and ε is the dimensionless sampling step used.

The evolution Eq. (5) present a coupling among the ij node and its adjacent ones according to the Fig. 1.

This implies the existence of a non-zero admittance edge in the electric analog between the ij node and all those to which it is connected. Thus, we can consider the topology in Fig. 2 for the circuital element of the analog limiting ourselves to the use of two-terminal networks (TTN) depicted as block boxes

The balance equations for the current I at node ij of the circuit in Fig. (2) are the following:

$$\sum_{m=-2}^{2}(I[i+mj]+I[i,j+m])+\sum_{n=-1}^{1}\sum_{m=-2}^{2}I[i+kj+k]+I_G[i,j]=0 \tag{6}$$

where $I_G[i,j]$ is the current flowing from the node ij to the ground. The analog is synthesized assuming the connections between the electrical nodes constituted only by TTNs. Therefore, using the general expressions for the constitutive relation of a linear TTN, for each edge we obtain

$$I[i+p,j+q]=Y[i,j;i+p,j+q](V[i+p,j+q]-V[i,j])\ p,q=-2,\cdots 2 \tag{7}$$

where $Y[i,j;i+q,j+q]$ are the edge electric admittances. The values for the admittances are finally obtained equating the coefficients of Eq. (5) to those of Eq. (6) once transformed by using Eq. (7). In this way, the element of a possible electric analog of Kirchhoff-Love plate can be synthesized as shown by Fig. 3, where L and C are given by

$$L=\alpha\varepsilon^4R_0t_0,\quad C=\frac{t_0}{R_0}, \tag{8}$$

and R_0 is a characteristic dimensionless resistance. The presence of negative inductances (see Fig. 3) is due to the assumption that the electric edges are constituted only by two-terminal networks and implies the presence of active elements able in their synthesis. Indeed, it is easy to realize negative admittances using the Antoniou circuit [8], the scheme of which is presented in Fig. 4.

In order to analyze the coupled system introduced in the next section, it is useful to consider the homogenized form of the circuital equations dependent on the time integral of the electric potential denoted by Ψ:

$$\nabla^2\nabla^2\Psi+\alpha\ddot{\Psi},\quad \alpha=\frac{LC}{t_o^2\varepsilon^4}\ \forall\Psi\in D. \tag{9}$$

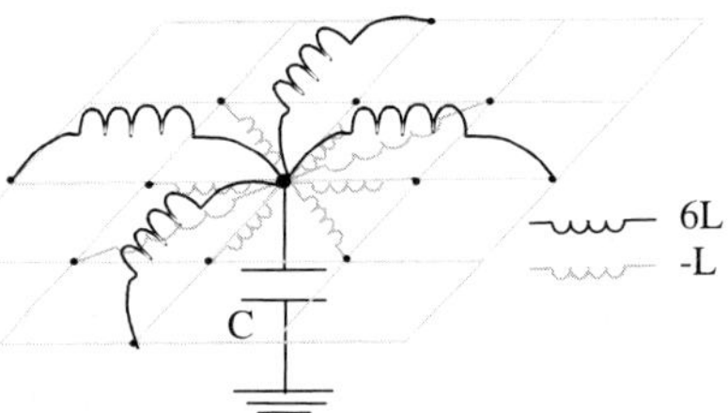

Fig. 3. Analog circuital element.

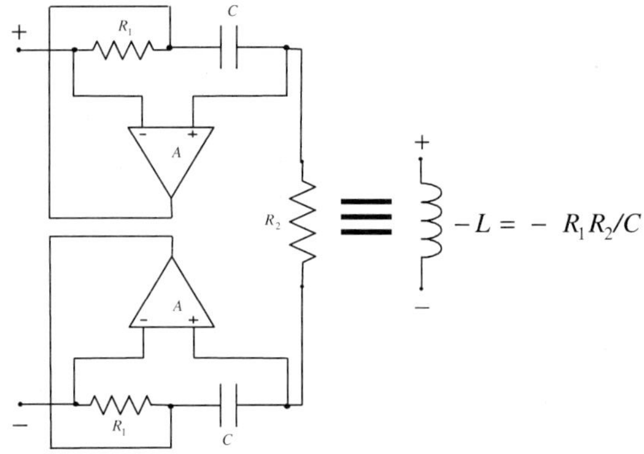

Fig. 4. Negative inductance.

The previos equation allows us to state that the circuit defined by Fig. (3) is electric analog of Kirchhoff-Love plate if

$$LC = \frac{3\rho}{h^2 E} l_0^4 \varepsilon^4. \tag{10}$$

Let us explicitly remark that, once mechanical properties of the plate and the grid step are given, the condition Eq. (10) determines the value of the product LC.

3. Piezoelectro-mechanical plate

The electro-mechanical coupling between the plate and its circuital analog is realized using square piezoelectric actuators of negligible thickness which are centered at the sampling nodes of the plate (see Eq. (2)) and interconnect the nodes of the electric analog to the ground. Indeed, as electric components, they can be regarded as TTN of mainly capacitive admittance.[1]

Since the plate is described using a homogenized model while the circuit is a lumped device, it is convenient to describe the behavior of PZT actuators, which interconnect the two system, by the following constitutive relations, in which mechanical distributed variables, assumed constant on the domain of the actuator, and lumped electric quantities are used:

$$\begin{bmatrix} M_{a_{xx}} \\ M_{a_{yy}} \\ Q \end{bmatrix} = \begin{bmatrix} g_{mm} & g_{12} & -\frac{g_{em}}{b} \\ g_{21} & g_{mm} & -\frac{g_{em}}{b} \\ bg_{em} & bg_{em} & g_{ee} \end{bmatrix} \begin{bmatrix} u_{,xx} \\ u_{,yy} \\ v \end{bmatrix} \tag{11}$$

Where the bending-moment tensor components $M_{a_{xx}}$, $M_{a_{yy}}$, and the electric charge Q stored into the actuator are related the plate curvatures $u_{,xx}$, $u_{,yy}$ and the voltage across the actuator terminals by the characteristic actuator coefficients g. The constant b represents the edge length of the actuator.

In order to characterize the actuator from an electric point of view, let us consider the time derivative of the third relation of Eq. (11)

$$i = g_{ee}\dot{v} + sbg_{em}\nabla^2\dot{u} \tag{12}$$

[1] We neglected the resistences introduced by the electrodes since they contribute only as a minor effect in the behaviour of the piezoelectric patch as a sensor/actuator (see [8]).

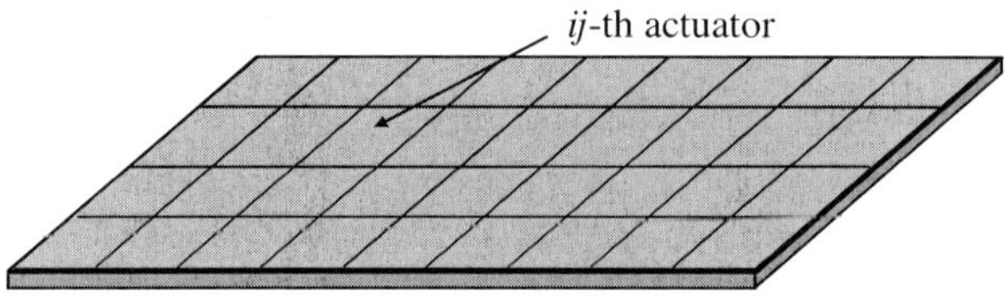

Fig. 5. Plate with actuators.

where i represents the current flowing in the actuator. Equation (12) shows how the actuator behaves like a capacitance in parallel connection with a current generator, the impressed value of which is related to the curvature of the actuator. Therefore, the use of the actuators as the capacitances of the electric analog is fully justified. The additional impressed currents modify Eq. (9) adding an electro-mechanical coupling term. In conclusion, the electric-coupled equation for the piezoelectro-mechanical (PEM) plate is

$$\nabla^2\nabla^2\Psi + \alpha\ddot{\Psi} + \beta_e\nabla^2\dot{u} = 0, \ \ \beta_e = \frac{Lbg_{em}}{\varepsilon^4}\frac{l_0^7}{v_0 t_0^2} \tag{13}$$

where v_0 is a characteristic voltage drop.

From a mechanical point of view, it is easy to consider the presence of the actuator layer as an additional term to the bending moment tensor of the plate, which is given by the first two relations of Eq. (11). Neglecting the contribution due to the purely-mechanical bending-stiffness of the actuator, the plate Eq. (1) is modified by an additional coupling term and gives the mechanical coupled equation of the PEM plate

$$\nabla^2\nabla^2 u + \alpha\ddot{u} - \beta_m\nabla^2\dot{\Psi} = 0, \ \ \beta_m = \frac{3g_{em}}{2bh^3E}v_0. \tag{14}$$

In the coupled system, the value of the circuital capacitance equals the actuator purely electric coefficient g_{ee}; therefore, the analogy condition given in Eq. (10) becomes the following condition for the circuit inductance L:

$$L = \frac{\varepsilon^4}{l_0^4}\frac{3\rho}{h^2 g_{ee}E} \tag{15}$$

Moreover the equality between the coupling coefficients β in Eq. (13) and Eq. (4) can be obtained setting the characteristic voltage v_0 as follows

$$V_0 = L_0\frac{b}{t_0}\sqrt{\frac{2h\rho}{g_{ee}}}. \tag{16}$$

Therefore, the PEM plate equations are:

$$\begin{cases} \nabla^2\nabla^2\Psi + \alpha\ddot{\Psi} + \beta\nabla^2\dot{u} = 0 \\ \nabla^2\nabla^2 u + \alpha\ddot{u} - \beta\nabla^2\dot{\Psi} = 0, \end{cases} \quad \alpha = \frac{3\rho l_0^4}{h^2 E t_0^2}, \ \ \beta = \frac{3 l_0 g_{em}}{bh^3 E} \tag{17}$$

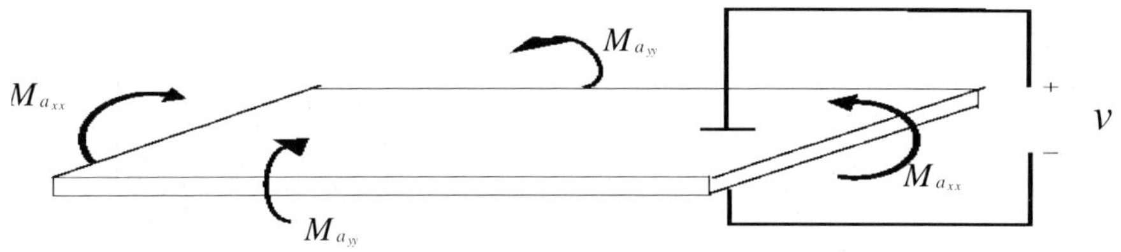

Fig. 6. Bending actuator.

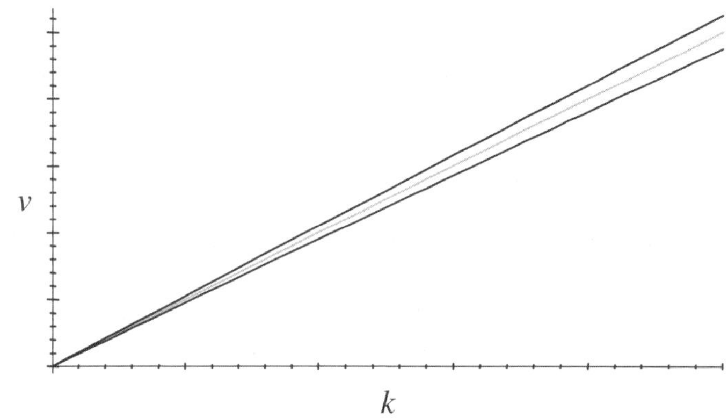

Fig. 7. Dispersion relations.

4. Vibration in PEM plate

In order to observe the coupling properties in the piezo-electro-mechanical plate, let us consider a monochromatic electro-mechanical wave propagating in it

$$u = A_e^{j(k\cdot r-\omega t)}, \quad \Psi = B_e^{j(k\cdot r-\omega t} \tag{18}$$

where k and r are the wave vector and the position vector, respectively, and ω is the angular frequency. Replacing Eq. (18) into the Eq. (17), the dispersion relations for the PEM plate are obtained:

$$v_{p_1} = \frac{\omega_1}{k} = \frac{\beta}{2\alpha}\left(\sqrt{1+\frac{4\alpha}{\beta^2}}+1\right), \quad v_{p_2} = \frac{\omega_2}{2\alpha}\left(\sqrt{1+\frac{4\alpha}{\beta^2}}-1\right) \tag{19}$$

In Fig. 7, the phase speeds v_p are plotted for the uncoupled case (gray line, $\beta = 0$) and for the coupled one (black lines), in which it is apparent how the coupling generate a split of the dispersive relation around the uncoupled case.

Using the dispersion relations, it can be easily derived also the relation between the electric and mechanical wave-amplitudes A and B which proves that the coupling is independent of the wave angular frequency and the wavelength:

$$A = iB. \tag{20}$$

Therefore, the electric and the mechanical components show a phase difference of $\pi/2$ and are always equal in modulus. This means that, for monochromatic waves in PEM plates (as expected because of

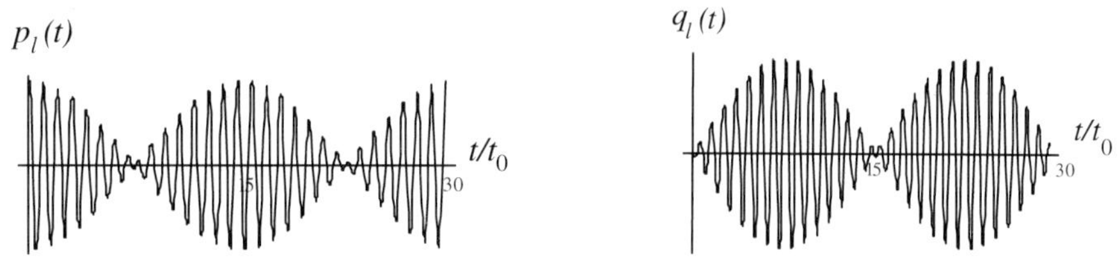

Fig. 8. Mechanical and electrical time-evolution for the generic e-m mode.

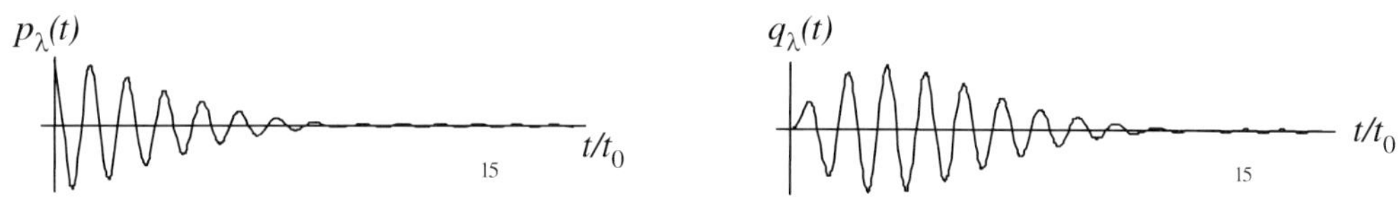

Fig. 9. Mechanical and electrical time-evolution for the generic e-m damped mode.

the results found in [9–11]) all the energy flows from one form to the other, thus realizing an optimal coupling.

Finally, a modal analysis is performed representing the solution u, Ψ on the basis m_λ constituted by the eigenvectors of the problem Eq. (1) with simply supported boundary conditions:

$$u = \sum_\lambda p_\lambda(t) m_\lambda, \quad \Psi = \sum_\lambda q_\lambda(t) m_\lambda \tag{21}$$

where $p_\lambda(t)$ and $q_\lambda(t)$ are the mechanical and electric Fourier coefficients, respectively. The time evolution for any mode, is presented in Fig. 8 considering a purely mechanic deformation of the PEM plate at time $t = 0$, and the characteristic frequency equal to the natural frequency of the uncoupled-system mode:

Figure 8 proves again how the energy goes back and forth from the mechanic to the electrical forms. If dissipative elements are introduced in the circuit-analog, then the mechanical energy converted into the electric form by the actuators can be damped. Hence, a non-vanishing purely resistive admittance in parallel connection with the actuator is added and the PEM plate relation Eq. (17) become

$$\begin{cases} \nabla^2\nabla^2\Psi + \alpha\ddot{\Psi}\beta\nabla^2\dot{u} = 0 \\ \nabla^2\nabla^2 u + \alpha\ddot{u} - \beta\nabla^2 P\dot{s}i + \gamma\dot{\Psi}, \end{cases} \quad \alpha = \frac{3\rho l_0^4}{h^2 E t_0^2}, \quad \beta = \frac{3 l_0 g_{em}}{b h^3 E}, \quad \gamma = \frac{L}{t_0 R \varepsilon^4} \tag{22}$$

where R is the resistance.

Performing a modal analysis, as done for the non-dissipative case, an optimal resistance value can be found which is able to optimize the damping of the plate as shown in Fig. 9.

The simulations have been performed considering an aluminum plate of 1 m of edge, and 1 mm of thickness with a grid of 10×10 commercial actuators, and an initial mechanical deformation of one per cent of the plate edge length. Let us remark that values for the circuital inductances of few Henry and voltages on the actuator less then 50 volts have been found, exhibiting the feasibility of the PEM plate.

5. Conclusions

This paper has demonstrated the possibility of obtaining an efficient vibration damping for a plate using a grid of piezoelectric actuators connected to the circuital analog of the plate. The approach used permits to obtain an optimum energy exchange and a vibration damping independent of the vibration mode of the plate. Moreover, the values found for the components used in the simulations show the technical feasibility of the PEM plate.

Acknowledgements

F.d. I. and S.A. wish to thank the Department of Engineering Science and Mechanics of the Virgina Polytechnic Institute and State University for the warm hospitality and research grants provided in the last two years.

References

[1] A.C. Eringen and G.A. Maugin, *Electrodynamics of Continua*, I-II, New York Springer, 1990.
[2] R. Courant and D. Hilbert, *Methods of mathematical physics Vol.II Partial differential equations* Interscience Publishers, New York London, 1962.
[3] F. Bardati, G. Barzilai and G. Gerosa, Elastic wave Excitation in PZT slabs, *IEEE Transaction on sonics and ultrasonics* **SU-15** (4) (October, 1968), 193–202.
[4] C.T. Molloy, Four Pole Parameters in Vibration Analysis in Mechanical impedance Methods for Mechanical Vibrations in: *Colloquium on Mechanical Impedance Methods for Mechanical Vibrations,* presented at the ASME Annual Meeting, New York, N.Y, sponsored by Shock And Vibrating Committee Applied Mechanics Division, The American Society of Mechanical Engineers, New York, edited by R. Plunkett, December 2, 1958.
[5] G. Kron, Equivalent circuits of the elastic field, *J. App, Mech.* **11** (1944), 149–161.
[6] K.H. Hoffmann and N.D. Botkin, Homogenization of Von Karman plates excited by piezoelctric patches, *ZAMM Z. Angew. Math. Mech.* **89**(9) (2000), 579–590.
[7] N.D. Botkin, Homogenization of an equation describing linear thin plates by piezopatches, *Commun. Appl. Anal.* **3**(2) (1999), 271–281.
[8] K.W. Wang, Structural vibration suppression via parametric control actions piezoelectric materials with real-time semi-active networks, in: *Wave Motion, Intelligent Structures and Nonlinear Mechanics*, (Vol 1), A. Guran and D.J. Inman, eds, Series on Stability, Vibration and Control of Structures, World Scientific Singapore, 1995, pp. 112–134.
[9] F. dll'Isola and S. Vidoli, Bending-Waves Damping in Truss Beams by Electrical Transmission Line with PZT Actuators, *Archive of Applied Mechanics* **68** (1998), 626–636.
[10] S. Vidoli, F. dell'Isola, Modal coupling in one-dimensional electro-mechanical structured continua, *Acta Mechanica* **141** (2000), 1–2.
[11] S. Vidoli, F. dell'Isola, Vibration control in plates by uniformly distributed actuators interconnected via electric networks, *European Journal of Mechanics/A*, to appear.

International Journal of Applied Electromagnetics and Mechanics 13 (2001/2002) 121–128
IOS Press

Horizontal noncontact positioning control for a magnetic levitation plate

Takanori Sekihara[a], Yasuo Oshinoya[b,*] and Kazuhisa Ishibashi[c]
[a]*Graduate school of Tokai University, Japan*
[b]*Department of Mechanical Engineering, Faculty of Engineering, Future Science and Technology Joint Research Center, Tokai University, 1117 Kitakaname, Hiratsuka-shi, Kanagawa, Japan*
[c]*Department of Mechanical Engineering, Faculty of Engineering, Tokai University, Japan*

Abstract. When sheet steel travels at high speed in a processing facility, rollers are usually used as supporters. However, when rollers are used, deterioration caused by cracking is possible. The authors have proposed a magnetic levitation control system for sheet steel and confirmed the feasibility by an experiment on digital control. However, there is no horizontal restraining force in transport in the case of using only levitation control. Therefore, we proposed to install electromagnets in order to control the horizontal position. For the basic examinations, the sheet steel plate (length 80 0mm, width 600 mm, thickness 0.3 mm) was reinforced by light carbon pipes in this study. To verify the usefulness of the proposed noncontact positioning control system of sheet steel, a digital control experiment was performed on transport. As a result, good positioning control performance for the levitated steel plate was obtained.

1. Introduction

In recent years, innovative magnetic levitation techniques have been developed and applied to various contactless modes of transport. These techniques also make a large contribution toward promoting innovation in technology. For example, applications to contactless suspension of a traveling thin steel plate, which is widely used in the manufacture of various industrial products, especially steel works, are possible. When steel plate travels at high speed, rollers are usually used as supporters. However, when rollers are used, deterioration caused by cracking and exfoliation in the process of coating or gilding is possible. Electromagnetic levitation techniques can solve many problems in these conveyance modes of steel plates [1–3]. The authors dealt with theoretical features of magnetic levitation control for steel plates and carried out an experiment on digital control using nontraveling plates [4].

In this study, the authors proposed a horizontal noncontact positioning control system for a magnetically levitated steel plate [5]. A zinc-coated steel plate is levitated and positioned in a contactless manner by attractive forces of electromagnets which are controlled by feedback signals from gap sensors. For the basic examinations, sheet steel was reinforced by light carbon pipes in this study. The electromagnetic attractive control forces of the noncontact positioning control actuators are applied to the two forced edges of the levitated steel plate from the horizontal direction. By assuming that the interaction of

*Corresponding author: Yasuo Oshinoya, Tel.: +81 463 58 1211 ext. 4308; Fax: +81 463 59 8293; E-mail: ossy@keyaki.cc.u-tokai.ac.jp.

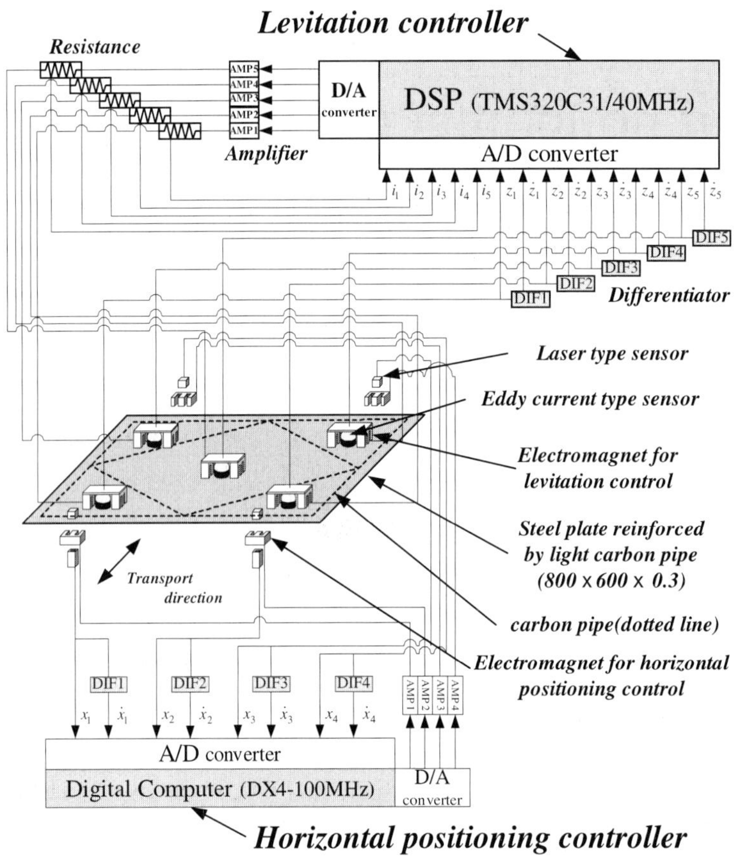

Fig. 1. Electromagnetic levitation control system with horizontal positioning controller.

levitation control with horizontal positioning control could be disregarded, each control system was independently designed as a one-degree-of-freedom system by applying the optimal control theory. To verify the usefulness of the proposed system, a digital control experiment was performed. Levitated steel plates subjected to horizontal random excitation and during transport were considered. The disturbance was found to be sufficiently suppressed and the offset displacement during transport to be reduced as a result of using the proposed horizontal positioning controller.

2. Control system

A schematic illustration of the electromagnetic levitation control system with a horizontal positioning controller is shown in Fig. 1. A zinc-coated steel plate (length 800 mm, width 600 mm, thickness 0.3 mm, material SS40 steel) is levitated and positioned in a contactless manner by attractive forces of electromagnets which are controlled by feedback signals from gap sensors. The dotted lines in the steel plate shown in Fig. 1 are pipes made of light carbon fiber. The total weight of the carbon pipes is about 8% of the steel plate mass = 1.2 kg.

The vertical displacement of the plate, its velocity and the current through the magnet are acquired using five eddy-current-type displacement sensors, five differentiators and five resistors connected in

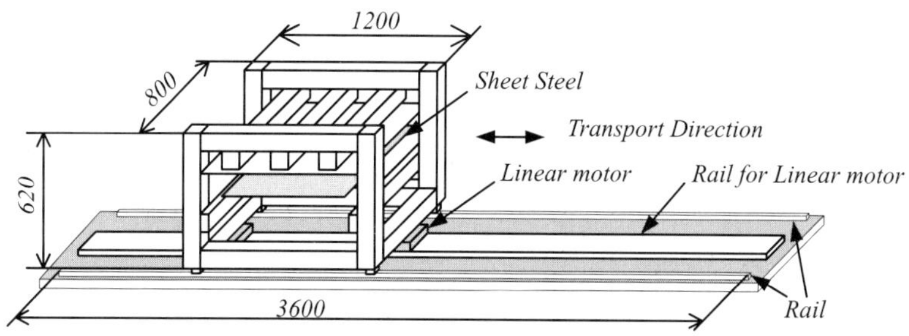

Fig. 2. Transport system of the plate.

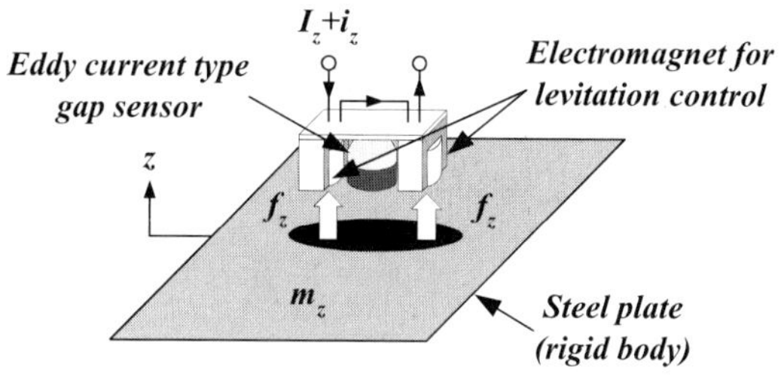

Fig. 3. Theoretical model of levitation control of the steel plate.

series to the magnetic circuit, respectively. The total number of acquired values is fifteen. These values are converted from analog to digital in the sampling interval of 0.2 ms to be processed in a digital signal processor. The control forces (voltage) are calculated in the DSP and are fed to the five electromagnets through the D/A converter and five power amplifiers. The horizontal displacement of the plate and its velocity are acquired using four laser-type displacement sensors and four differentiators, respectively. The control forces are calculated on a computer using these eight values (sampling interval is 0.2 ms). These sensors and electromagnets are installed in the frame shown in Fig. 2. The steel plate is carried about 1.9 m with this frame, using the linear motor.

3. Modeling and fundamental equations

3.1. Electromagnetic levitation control system

In this study, the displacement, the velocity and the coil current detected at one electromagnet position were used for the control of the electromagnet. Consequently, the electromagnetic levitation system was modeled as a one-degree-of-freedom system, as shown in Fig. 3.

In an equilibrium levitation state, magnetic forces are determined so as to balance with gravity. The equation of small vertical motion around the equilibrium state of the steel plate subjected to magnetic forces is expressed as

$$m_z \ddot{z} = 2 f_z, \tag{1}$$

where $m_z = m/5$ [kg], z: vertical displacement [m], $f_z(t)$: dynamic magnetic force [N].

The inductance of the electromagnetic coil is expressed as a sum of the component inversely proportional to the gap between the steel plate and magnet and the component of leakage inductance. If deviation

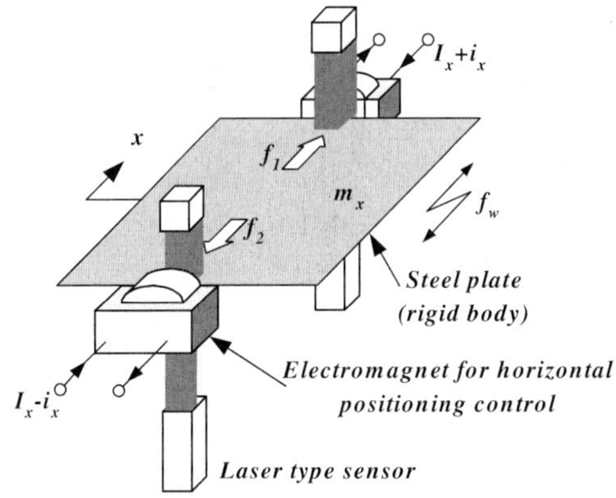

Fig. 4. Theoretical model of horizontal positioning control of the steel plate.

around the static equilibrium state is very small, the characteristic equations of the electromagnet are linearized as

$$f_z = \frac{2F_z}{Z_0}z + \frac{2F_z}{I_z}i_z \tag{2}$$

$$\frac{d}{dt}i_z = -\frac{L_{eff}}{L_z} \cdot \frac{I_z}{Z_0^2}\dot{z} - \frac{R_z}{2L_z}i_z + \frac{1}{2L_z}v_z \tag{3}$$

$$L_z = \frac{L_{eff}}{Z_0} + L_{lea}, \tag{4}$$

where F_z: magnetic force in the equilibrium state [N], Z_0: gap between steel plate and electromagnet in the equilibrium state [m], I_z: current in the equilibrium state [A], i_z: dynamic current of the coupled magnets [A], L_z: inductance of the magnet coil in the equilibrium state [H], R_z: resistance of the coupled magnet coils [Ω], vz: dynamic voltage of the coupled magnets [V], L_{lea}: leakage inductance of the magnet coil [H].

3.2. Horizontal positioning control system

Horizontal positioning control was carried out using the electromagnets installed at two edges of the steel plate. The horizontal displacement was measured by the banded laser beam sensor, keeping 5 mm of clearance between the edge of the steel plate and each electromagnet surface.

For the basic examination, the horizontal motion of the steel plate was modeled to have one-degree-of-freedom in the transport direction, as shown in Fig. 4. Therefore, the same attractive forces were generated from two electromagnets placed at one side of the steel plate. The equation of small horizontal motion around the equilibrium state of the steel plate subjected to the same static magnetic forces from the electromagnets at two edges is expressed as

$$m_z\ddot{x} = f_1 - f_2 + f_w = f_x + f_w \tag{5}$$

$$f_x = \frac{4F_x}{X_0}x - \frac{4F_x}{I_x}i_x, \tag{6}$$

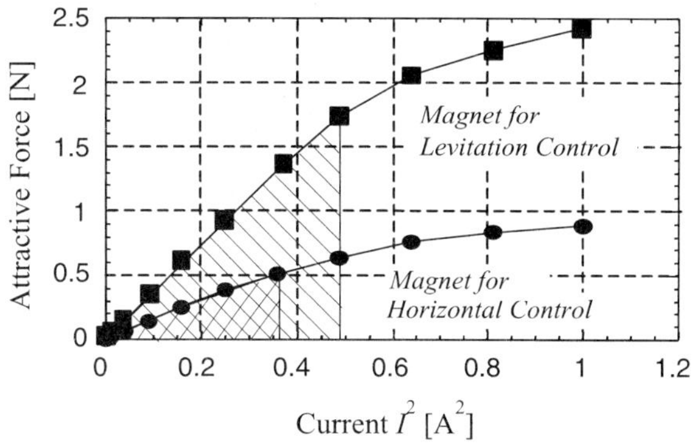

Fig. 5. Characteristics of electromagnetic attractive forces for levitation and horizontal positioning control.

where $m_x = m/2$ [kg], x: horizontal displacement [m], $f_x(t)$: dynamic magnetic force [N], $f_w(t)$: disturbance [N], F_x: magnetic force in the equilibrium state [N], X_0: gap between steel plate edge and electromagnet in the equilibrium state [m], I_x: current of the magnet in the equilibrium state [A], i_x: dynamic current of the magnet [A].

4. Design of controller

In this study, by assuming that the interaction of the levitation system with the horizontal positioning system could be disregarded, each control system was independently designed. Using the state vector, the Eqs (1)–(4) and (5), (6) are written as the following state equations:

$$\dot{\boldsymbol{z}} = \boldsymbol{A}_z \boldsymbol{z} + \boldsymbol{B}_z v_z$$
$$z = [z\ \dot{z}\ i_z]^T, \quad \boldsymbol{A}_z = \begin{bmatrix} 0 & 1 & 0 \\ \frac{4F_z}{m_z Z_0} & 0 & \frac{4F_z}{m_z I_z} \\ 0 & -\frac{L_{eff}}{L_z}\cdot\frac{I_z}{Z_0^2} & -\frac{R_z}{2L_z} \end{bmatrix}, \quad \boldsymbol{B}_z = \begin{bmatrix} 0\ 0\ \frac{1}{2L_z} \end{bmatrix}^T \tag{7}$$

$$\dot{\boldsymbol{x}} = \boldsymbol{A}_x \boldsymbol{x} + \boldsymbol{B}_x v_x + \boldsymbol{D}_x f_w$$
$$\boldsymbol{x} = [x\ \dot{x}]^T, \quad \boldsymbol{A}_x = \begin{bmatrix} 0 & 1 \\ \frac{4F_x}{m_x X_0} & 0 \end{bmatrix}, \quad \boldsymbol{B}_x = \begin{bmatrix} 0\ -\frac{4F_x}{m_x I_x R_x} \end{bmatrix}^T, \quad \boldsymbol{D}_x = \begin{bmatrix} 0\ \frac{1}{m_x} \end{bmatrix}^T. \tag{8}$$

To design their controllers based on optimal control theory, the following criterion functions are chosen.

$$J_z = \lim_{t_f \to \infty} \int_0^{t_f} (\boldsymbol{z}^T \boldsymbol{Q}_z \boldsymbol{z} + r_z v_z^2) dt \tag{9}$$

$$J_x = \lim_{t_f \to \infty} \int_0^{t_f} (\boldsymbol{x}^T \boldsymbol{Q}_x \boldsymbol{x} + r_x v_x^2) dt \tag{10}$$

$\boldsymbol{Q}_z = \begin{bmatrix} q_z & 0 & 0 \\ 0 & q_{zs} & 0 \\ 0 & 0 & q_{zi} \end{bmatrix}$, $\boldsymbol{Q}_x = \begin{bmatrix} q_x & 0 \\ 0 & q_{xs} \end{bmatrix}$, q and r are weighting coefficients.

Table 1
Weighting coefficient and feedback gain for levitation control

Weighting coefficient			Feedback gain		
q_z	q_{zs}	r_z	f_{bz}	f_{bzs}	f_{bzi}
1	1	1	– 11239	– 187	– 43

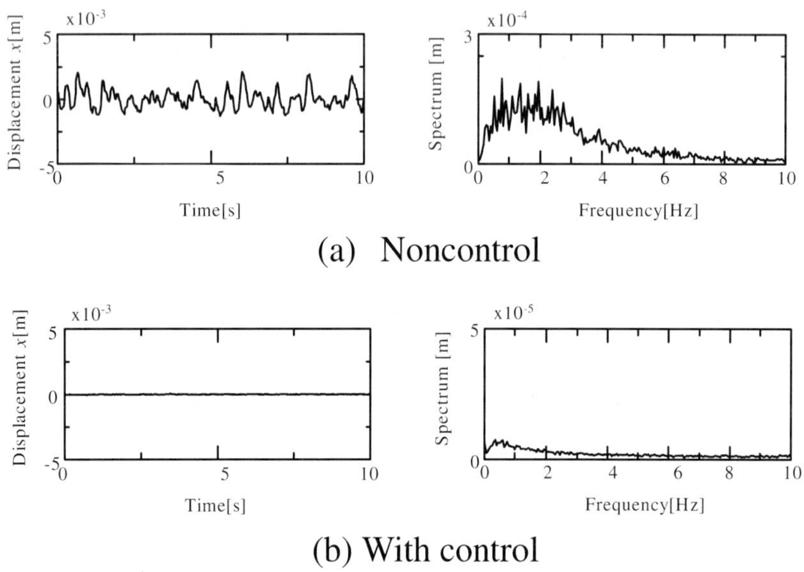

Fig. 6. Time response and spectrum of horizontal displacement of the steel plate subjected to random excitation.

According to the optimal control theory, the optimal state feedback gain is determined by solving the Riccati equation. The optimal voltages of electromagnets are obtained as

$$v_z = [-f_{bz} \quad -f_{bzs} \quad -f_{bzi}]\boldsymbol{z} \tag{11}$$

$$v_x = [-f_{bx} \quad -f_{bxs}]\boldsymbol{x}. \tag{12}$$

5. Control experiments

5.1. Specifications of the experimental apparatus

Levitation system: $m = 1.2$ kg, $Z_0 = 5$ mm, $I_z = 0.45$ A, $F_z = 1.18$ N, $R_z = 20.6\ \Omega$, $L_z = 2.21 \times 10^{-1}$ H, $L_{eff} = 1.16 \times 10^{-4}$ Hm. Horizontal positioning system: $X_0 = 5$ mm, $I_x = 0.50$ A, $F_x = 0.40$ N, $R_x = 9.50\ \Omega$.

The characteristics of electromagnetic attractive forces used for levitation and horizontal positioning control are shown in Fig. 5.

5.2. Consideration for random excitation in nontransport

To verify the usefulness of the proposed system, a digital control experiment for a nontransported steel plate was performed. The steel plate levitated using feedback gain values shown in Table 1 was subjected

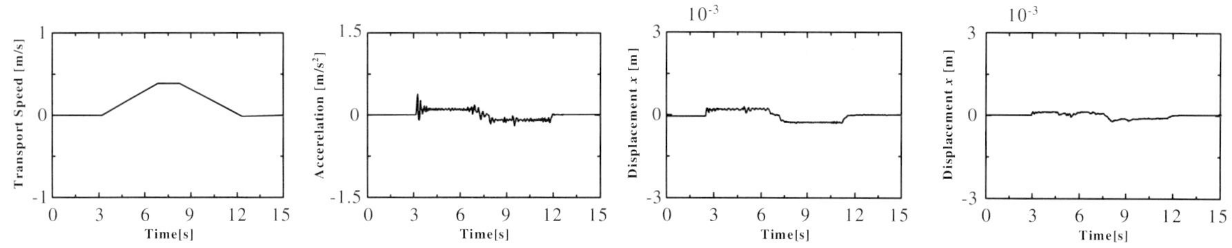

command value observed value in the case of using gain No.1 in the case of using gain No.2

(a) Transport acceleration and deceleration = ±0.1m/s^2

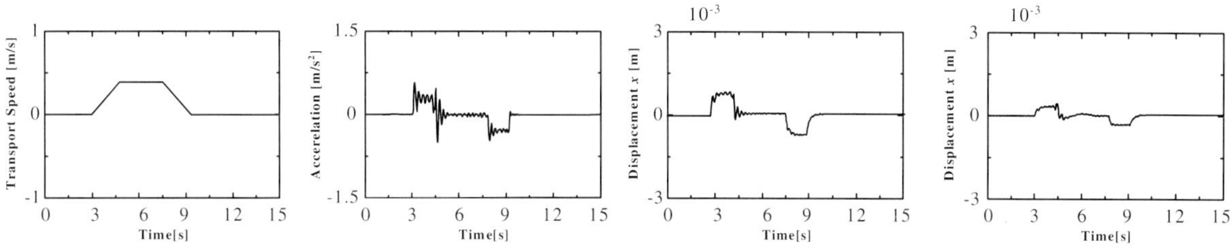

command value observed value in the case of using gain No.1 in the case of using gain No.2

(b) Transport acceleration and deceleration = ±0.3m/s^2

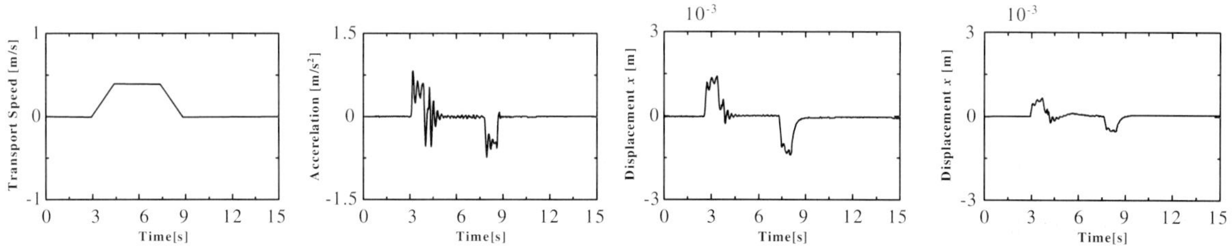

command value observed value in the case of using gain No.1 in the case of using gain No.2

(c) Transport acceleration and deceleration = ±0.5m/s^2

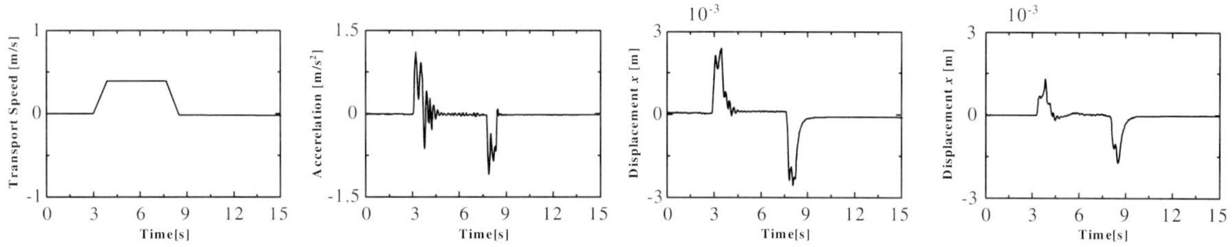

command value observed value in the case of using gain No.1 in the case of using gain No.2

(d) Transport acceleration and deceleration = ±0.7m/s^2

Fig. 7. The command values of transport speed, its observed acceleration and horizontal displacements of the plate.

to horizontal random excitation by a shaker. The disturbance is transferred to the plate through a soft coil spring connected by a universal joint. The experimental results are shown in Fig. 6. In Fig. 6(a), the horizontal displacement and its spectrum in the case of no horizontal positioning control are shown. Figure 6(b) shows control results. The validity of modeling was confirmed, because the horizontal disturbance was sufficiently suppressed upon using horizontal positioning control in the steady state.

Table 2
Weighting coefficient and feedback gain for horizontal positioning control

	Weighting coefficient			Feedback gain	
	q_x	q_{xs}	r_x	f_{bx}	f_{bxs}
No. 1	1	100	1.0×10^{-3}	− 1857	− 322
No. 2	100	10	1.0×10^{-5}	− 3831	− 887

5.3. *Consideration during transport*

The experimental results in the case of the transport mode are shown in Fig. 7. The command values of the transport speed of the linear motor, its observed accelerations and the horizontal displacements of the steel plate when using two different feedback gains, No. 1 and 2 in Table 2, are shown. Each figure indicates the results for the maximum transport speed of the steel plate of 0.4 m/s, with the transport acceleration and deceleration of (a) $\pm$ 0.1 m/s^2, (b) $\pm$ 0.3 m/s^2, (c) $\pm$ 0.5 m/s^2, and (d) $\pm$ 0.7 m/s^2 (command all). Zero horizontal displacement x indicates the equilibrium position. When the transport acceleration and deceleration increase as from (a) to (d), the offset horizontal displacement increases in the case of using feedback gain No. 1. However, there is a marked difference in the control effect when using feedback gain No. 2. It is possible to reduce the horizontal displacement by adjusting the feedback gains. Consequently, it was shown that the present noncontact horizontal positioning control system for the levitated steel plate was very effective.

6. Conclusion

This study was aimed at developing a control system that would enable noncontact horizontal positioning of a levitated steel plate. The electromagnetic forces were controlled via signals from laser sensors installed at the control positions. Digital control experiments with disturbance and during transport were carried out. Horizontal positioning control of a levitated steel plate was realized as a result of using the proposed control system.

References

[1] Y. Oshinoya and T. Shimogo, Electro-Magnetic Levitation Control of a Traveling Elastic Plate, *Proc. of Int. Conf. on Advanced Mechatronics,* Tokyo, May 1989, pp. 845–850.
[2] Y. Kurita, A. Sakurai, Y. Hamazaki, H. Ueda, K. Kato and H. Kondo, Suppression of Vibration on a Thin Steel Plate Using a Controlled Electromagnets, *Proc. of the 1991 Asia-Pacific Vibration Conference,* (Vol. 1), Melbourne, November 1991, pp. 63–68.
[3] K. Matsuda, M. Yoshihashi, Y. Okada and A.C.C. Tan, Self-Sending Active Suppression of Vibration of Flexible Steel Sheet, *Trans. ASME Journal of Vibration and Acoustics* **118** (July 1996), 469–473.
[4] Y. Oshinoya, S. Kobayashi and K. Tanno, Optimal Electromagnetic Levitation Control of a Thin Rectangular Steel Plate with Two Opposite Edges Reinforced by a Beam, *Trans. of Japan Society of Mechanical Engineers* (*in Japanese*) **62**(600-C) (August 1996), 127–133.
[5] Y. Oshinoya, Study on Electromagnetic Levitation Conveyance of a Steel Plate with Free Edges by Using Electromagnets and Permanent Magnets, *Journal of The Japan Society Applied Electromagnetics and Mechanics* (*in Japanese*) **6**(4) (October 1998), 364–371.

International Journal of Applied Electromagnetics and Mechanics 13 (2001/2002) 129–136
IOS Press

Magnetic levitation control with real-time vibration analysis using finite element method

Masanori Sase and Susumu Torii*
Musashi Institute of Technology, 1-28-1 Tamazutsumi, Setagaya, Tokyo 158-8557, Japan

Abstract. The authors propose the magnetic levitation control of thin steel plates using the real-time finite element analysis of the vibration. This paper describes the concept of the proposed levitation system, the finite element simulation of presuming the displacement, and the applicability from the viewpoint of calculation time.

1. Introduction

There is a requirement to improve the quality of the thin steel plates, in the steel making process. The thin steel plates used for the body of automobile requires high quality paint for the improvement of a fine sight. It is required that steel plates prevent occurring of the wounds because of flatting the surface. Current conveyance system uses the rollers with contact in the manufactural process of the product. This system, however, has the problem of causing the wounds in surface of the plates. Contact-less transportation schemes are thus required in the manufacturing processes, and the magnetic levitation technology is one of the most expected ones.

The thin steel plates vibrate more easily than the thick ones because of their high flexibility. The analysis of the vibration modes [1] and the application of Hrobust control theory [2] are the conventional countermeasure of the problem. High order controller is successfully used in the both system, however it is not always available in the real application. One gap sensor for each electromagnet is required without exception and causes the high cost of the total system.

This paper proposes the magnetic levitation control of the thin steel plates using the real-time finite element analysis of the vibration. There are interactions of supporting force between the plural electromagnets and the consideration of dynamic vibration is therefore necessary. This system aims at the stable levitation by presuming the vibration of the plate and observing the force of the electromagnet. This system can omit the gap sensor. This paper shows the concept of the proposed levitation system, the technique of analyzing vibration using finite element method, and simulation of presuming the displacement.

*Corresponding author: Susumu Torii, E-mail: torii@ee.musashi-tech.ac.jp; Fax: +81 3 5707 2212.

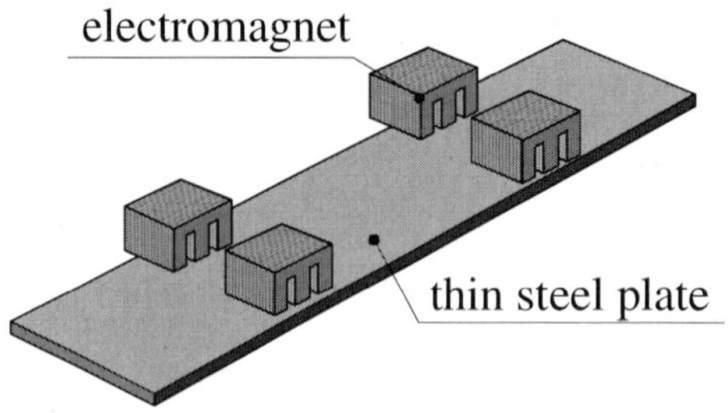

Fig. 1. The concept of the thin steel plates transported system with magnetic levitation technology.

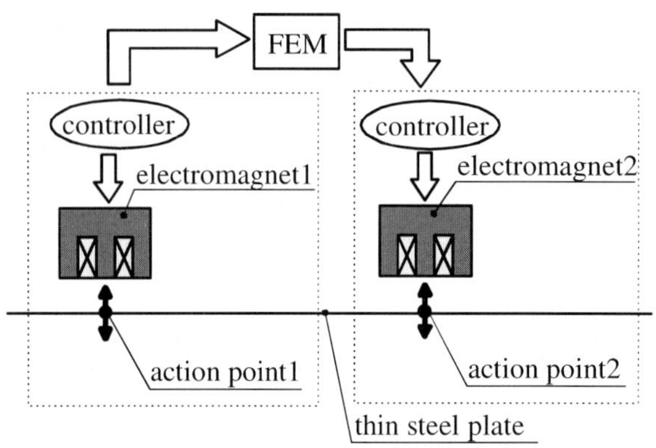

Fig. 2. The concept of magnetic levitation control using real-time finite element vibration analysis.

2. The concept of the proposed system

In case of the steel plate transporting system with magnetic levitation technology, the object of the transportation is very thin. It is difficult to support the weight of the thin steel plate with one electromagnet in magnetic levitation system because of the effect of magnetic saturation. Thus, plural electromagnets support the thin steel plate in the magnetic levitation system. The Fig. 1 shows the concept of the thin steel plates transported system with magnetic levitation technology. When the magnetic levitation control system support the thin steel plate by the electromagnets, vibration on the plate occurs on action point because of the magnetic levitation technology is unstable as the levitate direction. When the thin steel plate is levitated by the plural electromagnets, there are interactions of supporting forces as the action points. It is necessary to consider the dynamic vibration.

The proposed system analyses the effect on the other point by vibration occurred at the action point on real time using finite element method. Figure 2 shows the concept of magnetic levitation control using real-time finite element vibration analysis. The electromagnet2 controls in consideration for the effect of vibration occurred in the point that the electromagnet1 acts in Fig. 2. This system has the flexible ability to deal with the various shapes of steel sheets because of the flexible nature of small domain analysis like finite element method.

At the time of the levitation by the plural electromagnets, it is necessary to combine the gap sensor and the electromagnet. The proposal system, however, analyze the effect of the other oint by vibration which occurs at the action point on real time. The proposal system presumes the displacement of the other point. Therefore, this system can omit the gap sensor on other point. Figure 3 shows the composition of the system with reduced number of gap sensors. Figure 4 shows the block diagram of the system

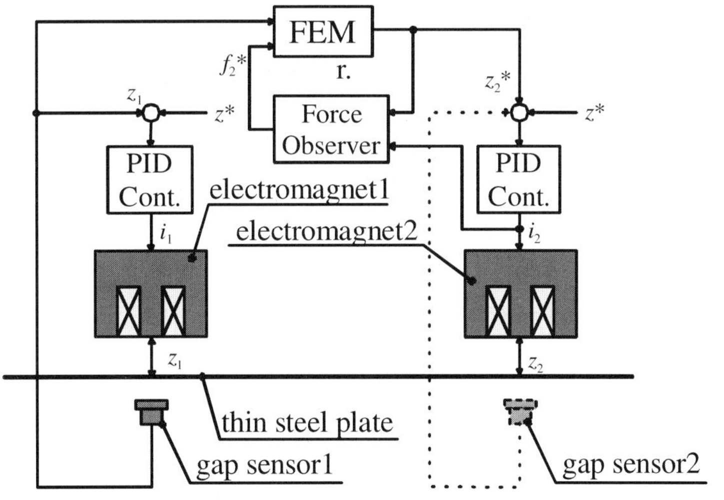

Fig. 3. The reduction of the number of gap sensors is realized by using finite element analysis and force observer.

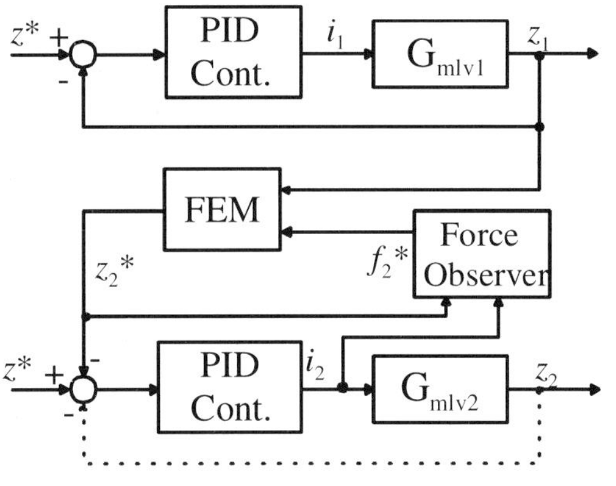

Fig. 4. The block diagram of the number of gap sensors is realized using finite element analysis and force observer.

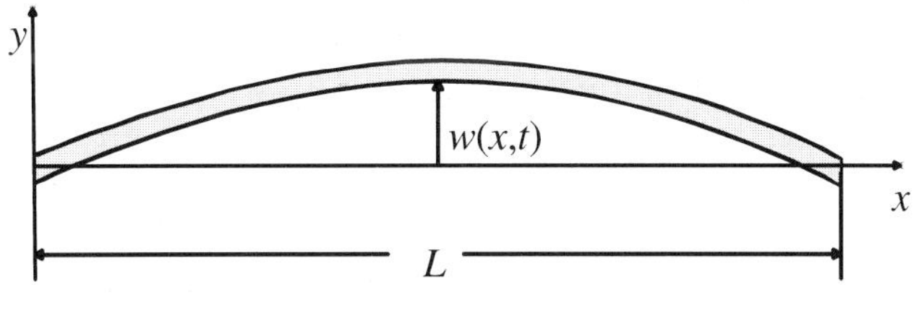

Fig. 5. Coordinate of the girder.

with reduced number of gap sensors. This system omits the portion of dotted line in Figs 3 and 4. The gap sensor1 considers only effect of electromagnet1. Therefore, this system considers the effect of the electromagnet2 using force observer.

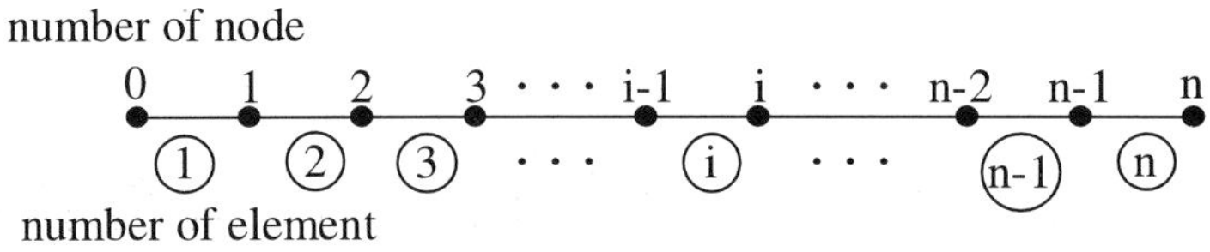

Fig. 6. Model of the girder disperses for n parts.

3. Vibration analysis using finite element method

The thin steel plate is regarded as the girder because of the vibration of thin steel plate is considered to be in only one direction. Figure 5 shows the coordinates of the girder. L is length of girder in Fig. 5. The Eq. (1) shows the motional equation of the girder.

$$m\frac{\partial^2 z}{\partial t^2} + CI\frac{\partial^5 z}{\partial t \partial x^4} + EI\frac{\partial^4 z}{\partial x^4} = f \tag{1}$$

z: displacement [m], m: mass per unit length [kg/m], C: damping coefficient $[\mathrm{Ns/m^2}]$, E: Young's modulus $[\mathrm{N/m^2}]$, I: polar moment of inertia of area $[\mathrm{m^4}]$ f: external force per unit length [N/m].

The vibration analysis of much flexibility calculates the matrix of dynamic motional equation successively. The Eq. (1) is dispersed in the direction of x creates matrix of motional equation. Figure 6 shows the model of the girder disperses for n parts. The Eq. (2) shows the matrix of motion equation.

$$[M]\{a(t)\} + [C]\{V(t)\} + [K]\{z(t)\} = \{F\} \tag{2}$$

$[M]$: mass matrix, $[C]$: damping matrix, $[K]$: stiffness matrix, $\{F\}$: vector of external force, $\{z(t)\}$, $\{v(t)\}$, $\{a(t)\}$: vector of displacement, velocity and acceleration.

The target girder has an isotropy, same quality, and the uniform section. The model of Fig. 6 is partition element at equal intervals. The i-th element matrix creates the matrix of whole of the girder. The Eqs (3), (4) and (5) show mass, damping, and stiffness matrix of whole of the girder on one end fixed. The l is length of element in Eqs (4) and (5).

$$m\begin{bmatrix} 1 & 0 & \cdots & 0 \\ 0 & \ddots & \ddots & \vdots \\ \vdots & \ddots & \ddots & 0 \\ 0 & \cdots & 0 & 1 \end{bmatrix} \tag{3}$$

$$\frac{CI}{l^4}\begin{bmatrix} 6 & -4 & 1 & 0 & \cdots & \cdots & \cdots & 0 \\ -4 & 6 & -4 & 1 & 0 & & & \vdots \\ 1 & -4 & 6 & -4 & 1 & 0 & & \vdots \\ 0 & \ddots & \ddots & \ddots & \ddots & \ddots & \ddots & \vdots \\ \vdots & \ddots & \ddots & \ddots & \ddots & \ddots & \ddots & 0 \\ \vdots & & 0 & 1 & -4 & 6 & -4 & 1 \\ \vdots & & & 0 & 1 & -4 & 5 & -2 \\ 0 & \cdots & \cdots & \cdots & 0 & 1 & -2 & 1 \end{bmatrix} \tag{4}$$

$$\frac{EI}{l^4}\begin{bmatrix} 6 & -4 & 1 & 0 & \cdots & \cdots & \cdots & 0 \\ -4 & 6 & -4 & 1 & 0 & & & \vdots \\ 1 & -4 & 6 & -4 & 1 & 0 & & \vdots \\ 0 & \ddots & \ddots & \ddots & \ddots & \ddots & \ddots & \vdots \\ \vdots & \ddots & \ddots & \ddots & \ddots & \ddots & \ddots & 0 \\ \vdots & & 0 & 1 & -4 & 6 & -4 & 1 \\ \vdots & & & 0 & 1 & -4 & 5 & -2 \\ 0 & \cdots & \cdots & \cdots & 0 & 1 & -2 & 1 \end{bmatrix} \tag{5}$$

The Eq. (2) is differential equation. This differential equation is reduced to the function of displacement using explicit time-domain solution. The speed v and the acceleration a are expressed using the displacement $z(t)$, the displacement before the Δt second $z(t-\Delta t)$ and the speed before the Δt second $v(t-\Delta t)$. The Eqs (6) and (7) show the expression. The Δt is operation cycle in the Eqs (6) and (7). The Eq. (8) shows the resultant equation of motion using only the variables before one time step.

$$v(t) = \frac{z(t) - z(t-\Delta t)}{\Delta t} \tag{6}$$

$$a(t) = \frac{z(t) - z(t-\Delta t)}{\Delta T^2} - \frac{v(t-\Delta t)}{\Delta t} \tag{7}$$

$$\begin{aligned} &[\overline{K}]\{w(t)\} = \{\overline{F}\} \\ &[\overline{K}] = \left[\frac{[M]}{\Delta t^2} + \frac{[C]}{\Delta t} + [K]\right] \\ &\{\overline{F}\} = \{F\} + [M]\left(\frac{\{z(t-\Delta t)\}}{\Delta t^2} + \frac{\{v(t-\Delta t)\}}{\Delta t}\right) + [C]\frac{\{z(t-\Delta t)\}}{\Delta t} \end{aligned} \tag{8}$$

Solving the Eq. (8) gives the presumed vibration of the girder. The inverse matrix $[\overline{K}]^{-1}$ is practicable to create previously because of $[\overline{K}]$ is the composition by known quantity. Figure 7 shows the flowchart of calculation. Figure 7 is flowchart of calculation in the system of Fig. 3. This calculation presumes displacement z_2 by inputting the displacement z_1 and the power of the electromagnet2 creating $\{\overline{F}\}$ after creating $[\overline{K}]^{-1}$. The calculation in the operation cycle calculates from creating $\{\overline{F}\}$ to presuming displacement in Fig. 7.

4. The simulation of presuming displacement

The calculation time is important in order to perform analyzing vibration on real time. Plenty of time is expected to be needed for the calculation using the finite element method, and the amount strictly depends on the number of divided elements. The calculation time for presuming the displacement is examined by performing the vibration analysis of the girder in the several cases of different number of division.

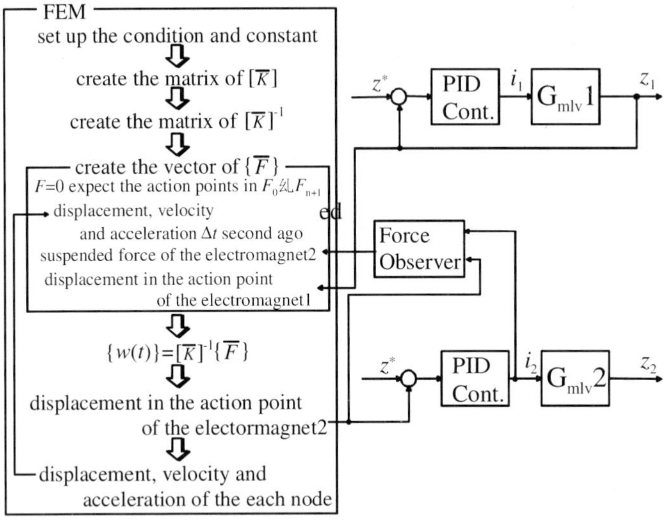

Fig. 7. The flowchart of calculation using magnetic levitation control using finite element vibration analysis.

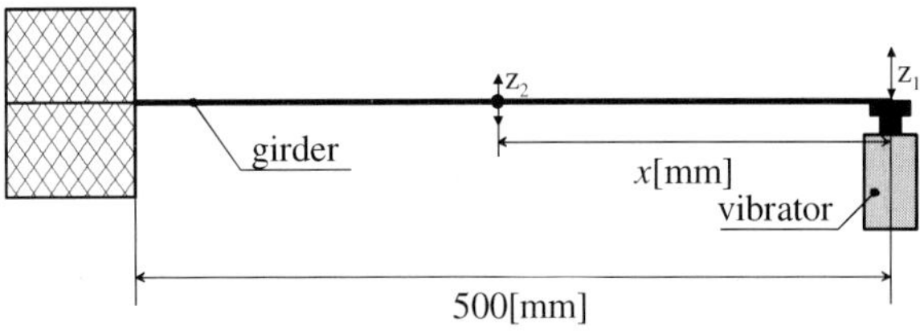

Fig. 8. The model of analyzing vibration with finite element method.

Figure 8 shows the model of analyzing vibration. One end of the girder is fixed, and forced vibration is applied to a point near the other end. The thin steel plate is 0.8 millimeters in thickness, 500 millimeters in length, and 100 millimeters in width. Here, analyzing vibration presumes z_2 of the distance from attached vibrator to x millimeters point by the inputted displacement z_1 of attached vibrator point generated from vibrator. The z_1 is the sinusoidal wave of one-millimeter amplitude, and the frequency is changed from 10 to 50 hertz in 10-hertz step. Figure 9 shows as an example the result of the vibration analysis in $x = 250$ millimeters, frequency 30 hertz, and number of element is 200. The displacement z_2 vibrates center on $z = -1.5$ millimeters from -1 millimeters because of its static flexure. In the beginning of time response, the amplitude of the waveform is large and occurring undulation because of the effect of gravity.

As mentioned previously, the number of elements and time of calculation is important in order to perform analyzing vibration on real time. Mean square errors are calculated to examine the change of presuming displacement by number of elements. The change of the time of calculation by finite elements was measured using various CPUs. Figure 10 shows the mean square error by change of the number of elements. Figure 11 shows the time of calculation by change of the number on elements. The presuming displacement scarcely changes in Fig. 10 even if the number of elements increased over 120. The operation cycle is 0.2 milliseconds in this simulation. The calculation is finished as 0.2 milliseconds, which is the typical minimum control cycle, only when the CPU is Athlon 1.4 GHz and the number of elements is 100 in Fig. 11. However, as for the operation cycle, more than 0.2 milliseconds is possible

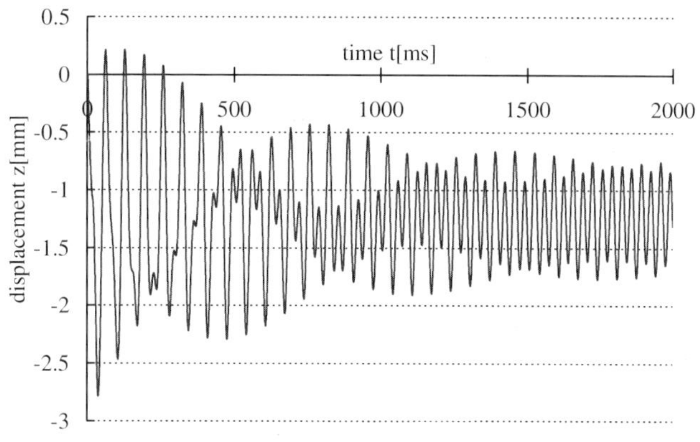

Fig. 9. The result of analyzing vibration ($x = 250$ [mm], frequency 30 [Hz], the number of element is 200).

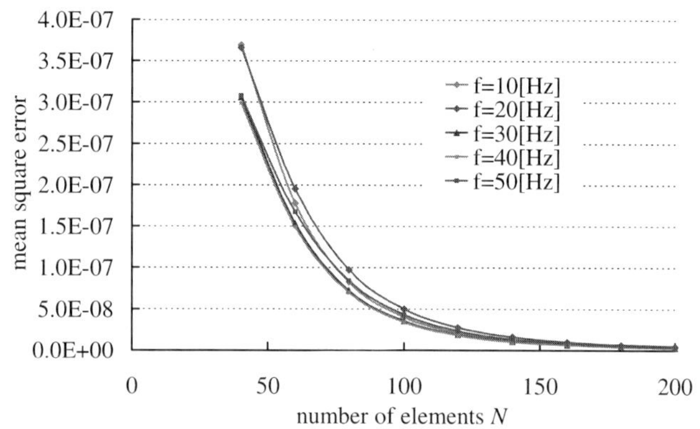

Fig. 10. The mean square error in various frequencies to the number of divided elements.

for levitation. The analyzing vibration on real time is possible by the speed of calculation on present CPU.

The operation cycle is the sampling time for calculating the speed and acceleration in the vibration analysis. The delaying the operation cycle affects the result of analysis. The change of the presuming the displacement by the sampling time was measured. Figure 12 shows the presuming the displacement by change of the sampling time. This result is maximum of the error to subtract from the time response at the $\Delta t = 0.2$ milliseconds on $t = 1800$ to 2000 milliseconds in various sampling time. The frequency of 20 hertz and 50 hertz has the large change in presuming the displacement. The resonance frequency of the one end fixed and the other end simply supported girder is $f_1 = 16.34$ hertz, $f_2 = 52.94$ hertz, $f_3 = 110.45$ hertz. The change of presuming the displacement by change of sampling time is large in the vibration in the neighborhood of the resonance frequency. Therefore, the vibration in the neighborhood needs not to be generated in the real-time vibration analysis system.

5. Conclusion

The required improvement of the quality of the thin steel plate in process is achieved by using the magnetic levitation technology. The proposed real-time finite element analyses implemented in the

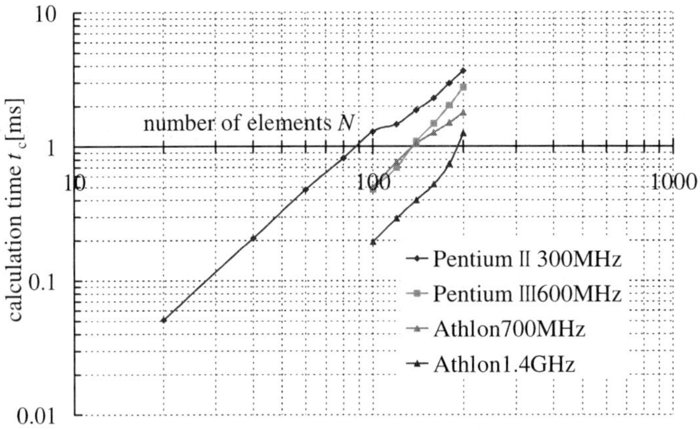

Fig. 11. The calculation time depends on the types of CPU and their clock cycles.

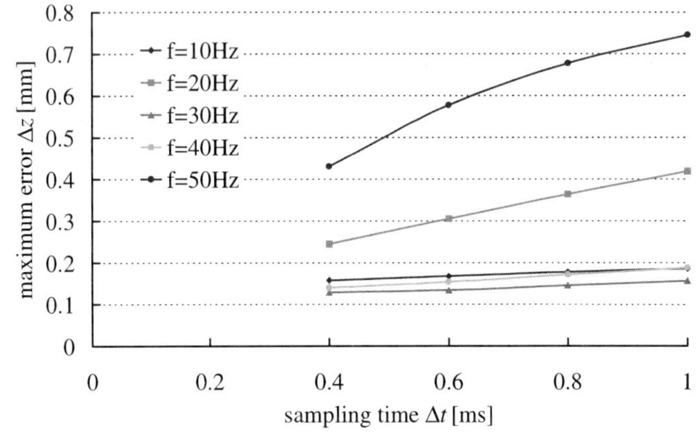

Fig. 12. The maximum error in various frequencies to the sampling time.

controller of the electromagnets are the universal solutions to solve the problem in the EMS system for elastic body. The interactions between the magnets as well as the modal vibration of the plates are taken into account in the analysis, and the reduction of the numbers of gap sensors is achieved using the force observer together.

The feasibility of the implementation was confirmed by this study, because the finite element calculation is proved to be available within the control period if the CPU of enough speed is used. Future works should be aimed first upon the experimental confirmation of the accuracy of the estimation. After establishing the presumption, the existing control system of thin steel plate EMS will be replaced to the proposed system with reduced number of gap sensors, which will realize the sufficiently lower cost.

References

[1] Y. Oshinoya and T. Shimogo, *Electro-Magnetic Levitation Control of an Elastic Plate*, Proceedings of the 11th International Conference on Magnetically Levitated Systems and Linear Drives, (Maglev'89), 1989, pp. 435–440.

[2] Y. Muraguchi, M. Watada, S. Torii, D. Ebihara, K. Nakaya and T. Nakagawa, *Design of an $H\infty$ robust control system for a magnetically levitated thin iron plate*. Advanced Computational and Design Techniques in Applied Electromagnetic Systems, Elsevier Science b.v, 1995, pp. 447–450.

International Journal of Applied Electromagnetics and Mechanics 13 (2001/2002) 137–144
IOS Press

Tilting control of railway vehicle using electric actuators

Tilting control method using vehicle body information in real time

Hidehisa Yoshida* and Masao Nagai
Tokyo University of Agriculture and Technology, Koganei-shi, 184-8588 Tokyo, Japan

Abstract. When a railway vehicle passes through a curve, reducing the lateral acceleration affecting its passengers is important in order to improve the ride comfort. Tilting control by feedback, which can maintain the vehicle inclination close to a desired value, is examined in this paper; reduction of car body vibration is also anticipated with this control. Specifically, the curvature is obtained by collecting data in real time from the car body traveling through a curve, and this information is utilized in the proposed tilting control method. The effects of the control scheme are verified by computer simulation using a dynamic model and experiment with model equipment.

1. Introduction

A need for car body inclination while traveling on curved sections of the railway has already been recognized in Japan because of its many mountainous areas. When a railway vehicle travels through a curve at excess speed, over centrifugal acceleration that cannot be fully reduced by cant acts on the car body, with the passengers affected by lateral acceleration as the result. This raises a need for a method to decrease the effect of the over centrifugal acceleration on the train by controlling the tilt of the car body.

At present, there mainly exist two methods for determination of the required tilting control amount. The first method, which is employed in Japan, uses a vehicle programmed with curve data of the travel course and operates tilting control according to position signals received from the ground. The second method, used mainly in Europe, does not depend on curve information. It instead calculates the inclination quantity from the over centrifugal acceleration computed by measuring the lateral acceleration of the non-tilting truck part of the train. In both systems, the basis for obtaining the target inclination is feedforward control.

This paper proposes a method that determines the necessary tilting quantity by feedback control. It will allow improved control of the car body inclination to the target value and additionally reduce the car body vibration. Specifically, the curvature of the track will be calculated from real time conditions of the traveling car body, and tilting control will be executed based on it. The effect of the control will be verified by both computer simulation and experiment using a physical model.

*Corresponding author: Hidehisa Yoshida, Fax: +81 42 385 7204; E-mail: hyoshida@cc.tuat.ac.jp.

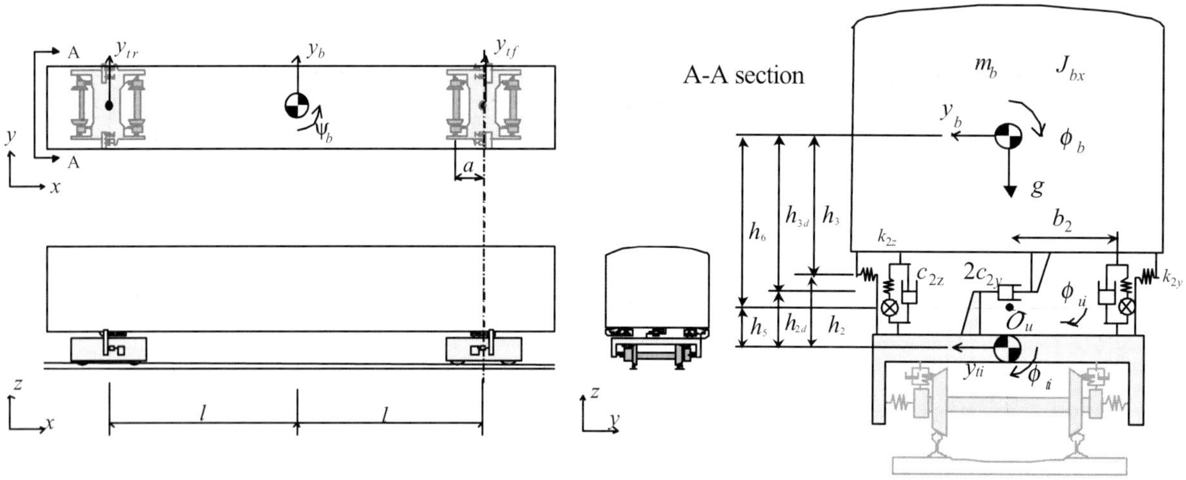

Fig. 1. Vehicle model (lateral, roll, yaw).

2. Vehicle model and control system design

2.1. Vehicle model

Figure 1 shows the railway vehicle model used in this report. The model is a single-car-body model assumed to be uniform, rigid, and symmetric with respect to the yz-plane. For simplicity, it is assumed that the lateral disturbance directly reaches the car truck, and that the gravitational force and centrifugal acceleration directly affects the car body. By balance forces, equations of motion in the lateral, roll, and yaw directions are obtained:

$$m_b\ddot{y}_b + 2k_{2y}\{y_b + h_3\phi_b - y_t\} + 2c_{2y}\{\dot{y}_b + h_{3d}\dot{\phi}_b - \dot{y}_t\} - m_b\{\alpha_u - g\phi_b\} = 0 \quad (1)$$

$$\begin{aligned} &J_{bx}\ddot{\phi}_b + 2h_3k_{2y}\{y_b + h_3\phi_b - y_t\} + 2h_{3d}c_{2y}\{\dot{y}_b + h_{3d}\dot{\phi}_b - \dot{y}_t\} \\ &+2b_2^2k_{2z}\{\phi_b - (\phi_u + \phi_t)\} + 2b_2^2c_{2z}\{\dot{\phi}_b - \dot{\phi}_t\} - m_bh_3(\alpha_u - g\phi_b) = 0 \end{aligned} \quad (2)$$

$$\begin{aligned} &J_{bz}\ddot{\psi}_b + 2lk_{2y}(l\psi_b + h_2\phi_t - y_{tf}) + 2lk_{2y}(l\psi_b + h_2\phi_t - y_{tr}) + 2lc_{2y}(l\dot{\psi}_b + h_{2d}\dot{\phi}_t - \dot{y}_{tf}) + \\ &2lc_{2y}(l\dot{\psi}_b + h_{2d}\dot{\phi}_t - \dot{y}_{tr}) + 2b_2k_{2z}(\psi_b - \psi_{tf}) + 2b_2k_{2z}(\psi_b - \psi_{tr}) + 2b_3c_{2z}(\dot{\psi}_b - \dot{\psi}_{tf}) + \\ &2b_3c_{2z}(\dot{\psi}_b - \dot{\psi}_{tr}) = 0 \end{aligned} \quad (3)$$

Here, α_u is the centrifugal acceleration affecting the car body, and ϕ_u the tilt angle which will be the control input of the tilting control. For further simplification, the air spring that is the secondary suspension system is dynamically modeled as a parallel spring-damper model with a rolling natural frequency of 1 Hz.

2.2. Lateral acceleration and tilting control target

The centrifugal acceleration $\alpha(t)$ and gravitational acceleration g influence a railway vehicle traveling on a curved section of the railway.

$$\alpha(t) = \rho(t)V^2 \quad (4)$$

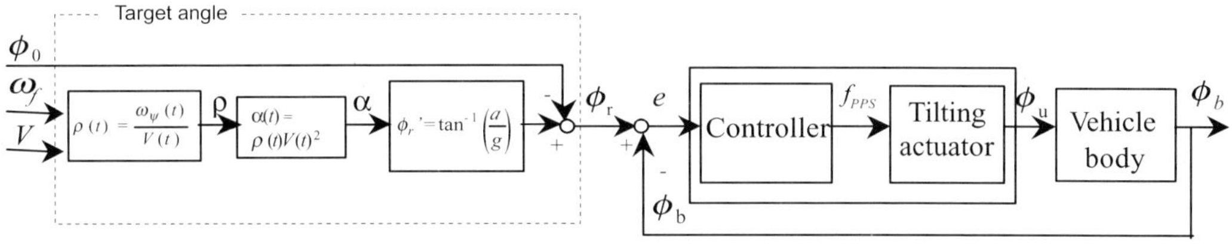

Fig. 2. Block diagram of proposed control system.

The orbital plane of the curve is designed with cant ϕ_0 so that the passengers would not perceive this centrifugal acceleration. Each curve additionally has cruising speed (Valence speed) set. When the vehicle exceeds this speed, over centrifugal acceleration $\alpha_u(t)$ acts on it as a disturbance.

$$\alpha_u(t) = \rho(t)V^2 \cos\phi_0 - g\sin\phi_0 \tag{5}$$

The value above expresses steady-state acceleration caused by the centrifugal force. In addition to the steady-state acceleration originating from the over centrifugal acceleration, the passengers in the rolling car body also would be affected by a vibrational acceleration component. Therefore, the lateral acceleration $\alpha_d(t)$ perceived by the passengers can be described as the following equation.

$$\alpha_d(t) = \{\alpha_u(t) + \ddot{y}_b\}\cos\phi_b - g\sin\phi_b \tag{6}$$

The often-recommended peak value of the lateral acceleration $\alpha_d(t)$ felt by the passengers is $0.8\ \mathrm{m/s^2}$ for frequencies under 1 Hz. This value is called a riding comfort coefficient and used as a riding comfort standard after structural and practical constraints are taken into account. To make the effect of tilting control clear, the goal of tilting control in this study is set as reducing the lateral acceleration $\alpha_d(t)$, caused by the over centrifugal acceleration $\alpha_u(t)$, to zero.

2.3. Selecting the tilting method

Tilting control is composed of an upper control layer which estimates the target tilting quantity and a low order control layer computing the required output stroke of the actuators for obtaining the desired tilting amount.

In this study, the target roll angle for eliminating $\phi_r(t)$ is determined in real time by estimating the track curvature from the car body conditions. A tilting method with a servo system that follows the target roll angle is now proposed. The block diagram of this tilting control system is shown in Fig. 2.

2.4. Determining the target roll angle

First, the yaw angle velocity of the car body is either detected using a sensor or calculated from other information. When a vehicle makes a turn, the relationship between its running speed $V(t)$ in the tangential direction and the curvature $\rho_\psi(t)$ of the track can be expressed as:

$$\rho(t) = \omega_\psi(t)V \tag{7}$$

In order to cancel the lateral acceleration α_d affecting the riders, the target car body roll angle ϕ_r can be obtained in real time in the following manner.

$$\phi_r(t) = \tan^{-1}(\alpha(t)/g) - \phi_0 \left(= \sin^{-1}(\alpha(t)/\sqrt{\alpha(t)^2+g^2}) - \phi_0\right) \tag{8}$$

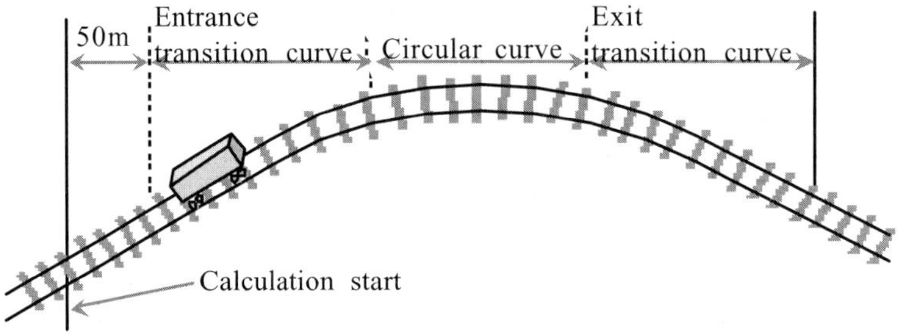

Fig. 3. Test curve condition.

2.5. Obtaining the desired roll angle

It is possible to control tilting by using the desired tilt angle as feedforward control input. This study, however, deals with feedback tilting control by a servo system for minimizing the deviation from the target roll angle ϕ_r to compensate for possible error introduced by disturbances. The error between the target roll angle ϕ_r and actual car body roll angle ϕ_b can be written as the following equation.

$$e = \phi_r - \phi_b \tag{9}$$

The output stroke of the actuators for tilting the car body is termed as tilt angle ϕ_u. From Eqs (1)–(3), the state equations become:

$$\begin{aligned} &\frac{d}{dt}x = Ax + Bu + Wx_d \\ &x = [y_b\ \phi_b\ \psi_b\ \dot{y}_b\ \dot{\phi}_b\ \dot{\psi}_b]^T, \quad u = [\phi_u], \\ &x_d = [\rho\ y_{tf}\ y_{tr}\ \phi_{tf}\ \phi\ \psi_{tf}\ \psi_{tr}\ \dot{y}_{tf}\ \dot{y}_{tr}\ \dot{\phi}_{tf}\ \dot{\phi}_{tr}\ \dot{\psi}_{tf}\ \dot{\psi}_{rt}]^T \end{aligned} \tag{10}$$

The state equations are then expanded by adding the integrated value of the roll angle error from Eq. (9) as a state quantity.

$$x_{e1} = \int e_1 dt \left(= \int (\phi_r - \phi_b) dt \right) \tag{11}$$

$$\begin{aligned} &\frac{d}{dt}\begin{bmatrix} x \\ x_{e1} \end{bmatrix} = \begin{bmatrix} A & 0 \\ C & 0 \end{bmatrix}\begin{bmatrix} x \\ x_{e1} \end{bmatrix} + \begin{bmatrix} B \\ D \end{bmatrix} u + \begin{bmatrix} W \\ W_{e1} \end{bmatrix} x_d \\ &\frac{d}{dt}x_n = A_n x_n + B_n u + W_n x_d \end{aligned} \tag{12}$$

By supposing the initial value of the disturbance to be zero, the inclination quantity can be determined by optimum control. The cost function can be described as the following equation; the tilting angle ϕ_u is computed by multiplying the integrated value of the roll angle error by a weight factor.

$$J = \int_0^\infty (q x_{e1}^2 + u^2) dt \tag{13}$$

$$u = -K x_n \tag{14}$$

Table 1
Test curve condition

Curve Number		1	–
Maximum curvature	ρ_{max}	1/400	1/m
Cant in steady state cornering	C	105	mm
Length of entrance transition curve		80	m
Length of circular curve		140	m
Length of exit transition curve		80	m
Track gauge	G	1067	mm
Curve direction		right	–

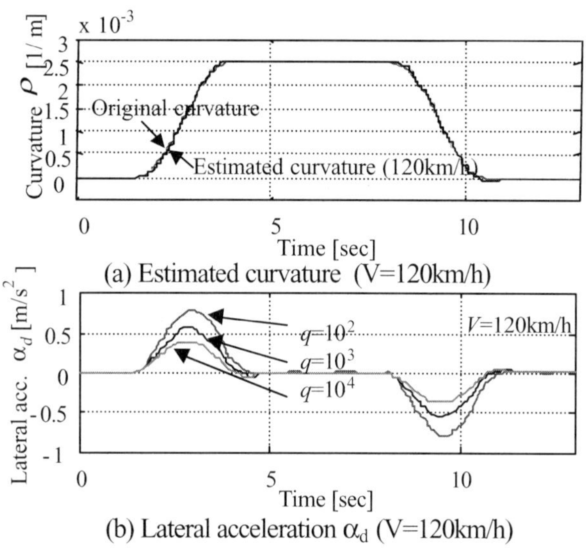

Fig. 4. Tilting control performance of simulation.

3. Examining the tilting control performance

3.1. Simulation conditions

The effects of tilting control by the proposed control method are now examined. The railway vehicle parameters for the calculation uses those of conventional express railway vehicle. The simulation recreates the vehicle traveling through the curve shown in Fig. 3 and Table 1 at the speed of 120 km/h. Over centrifugal acceleration of $1.813\ \mathrm{m/s^2}$ acts on the vehicle under this condition, exceeding twice the value of the riding comfort coefficient.

3.2. Effects of the tilting control

Figure 4 displays the resulting time responses under the simulated conditions. The estimated curvature and actual curvature are shown in Fig. 4(a), whereas the resulting lateral acceleration α_d is graphed in Fig. 4(b). The tilting control using the estimated curvature held the lateral acceleration to about 30% of the maximum acceleration caused by the disturbance, and the acceleration also has converged while on the circular curve section, proving the effect of the control process.

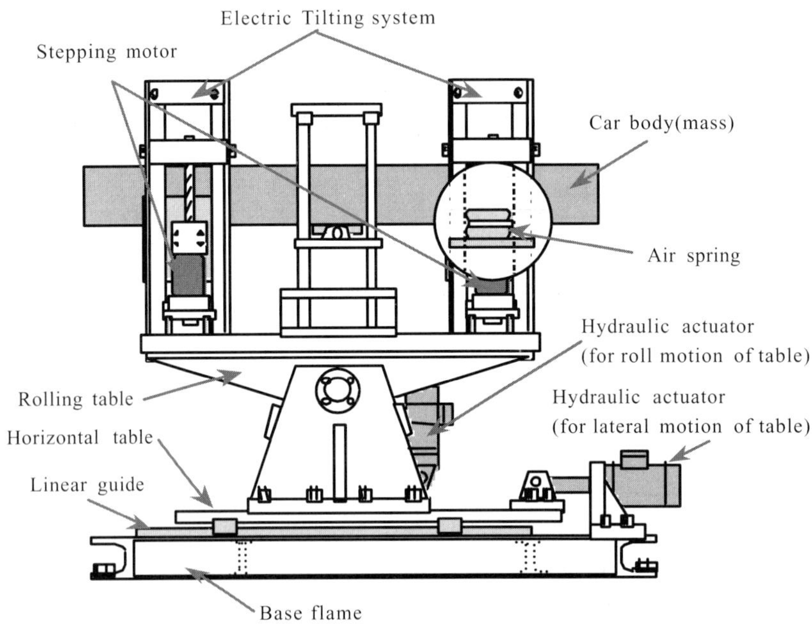

Fig. 5. Experimental set-up of tilting system.

4. Verification by experiment with a model

4.1. Test equipment model

A model experiment equipment setup, which simulates a train traveling through a curve, is shown in Fig. 5. This device can examine both the lateral and roll motions. Assuming that desired car body roll angle ϕ_r can be estimated, tilting control performance is now examined.

In this equipment, the tilting angle is produced by actuators on the right and left operating in opposite phases with stepping motors as their driving force. When the car body section of the equipment becomes level with the ground, that indicates the over centrifugal acceleration produced by the excess speed is canceled.

The output rotation angle of the stepping motors is proportional to the number of input pulses. The stepping motors have a maximum output torque that can be produced for the input pulse speed. In this paper, therefore, the actuator is modeled with the pulse speed as the input and the tilting angle the output.

$$\phi_u = a \int_0^T f_{PPS} dt \tag{15}$$

4.2. Feedback tilting control using the car body roll angle

Figure 6 shows a block diagram for feedback tilting control employed in the experiment. A gyro type sensor was used to detect the absolute car body roll angle. Since the goal of the tilting control is to make the lateral acceleration α_d zero, the target roll angle ϕ_r for this equipment is zero also. The controller here utilizes a PID control method, and its output is defined as the number of pulses $A_{fb}(t)$ per sampling time dt for reducing the deviation e of Eq. (9). The proportional gain is k_P, integral gain

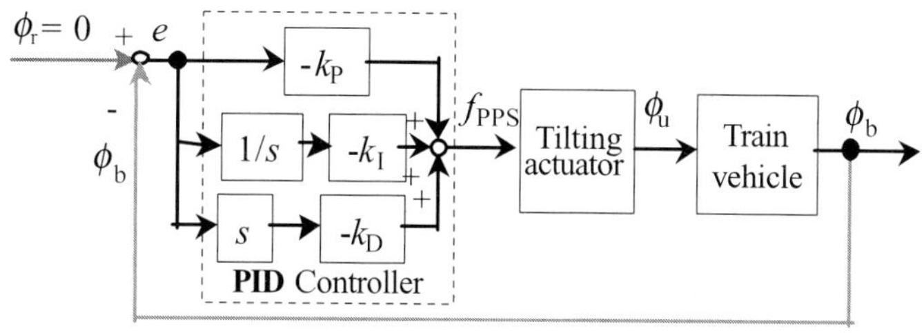

Fig. 6. PID control for experiment.

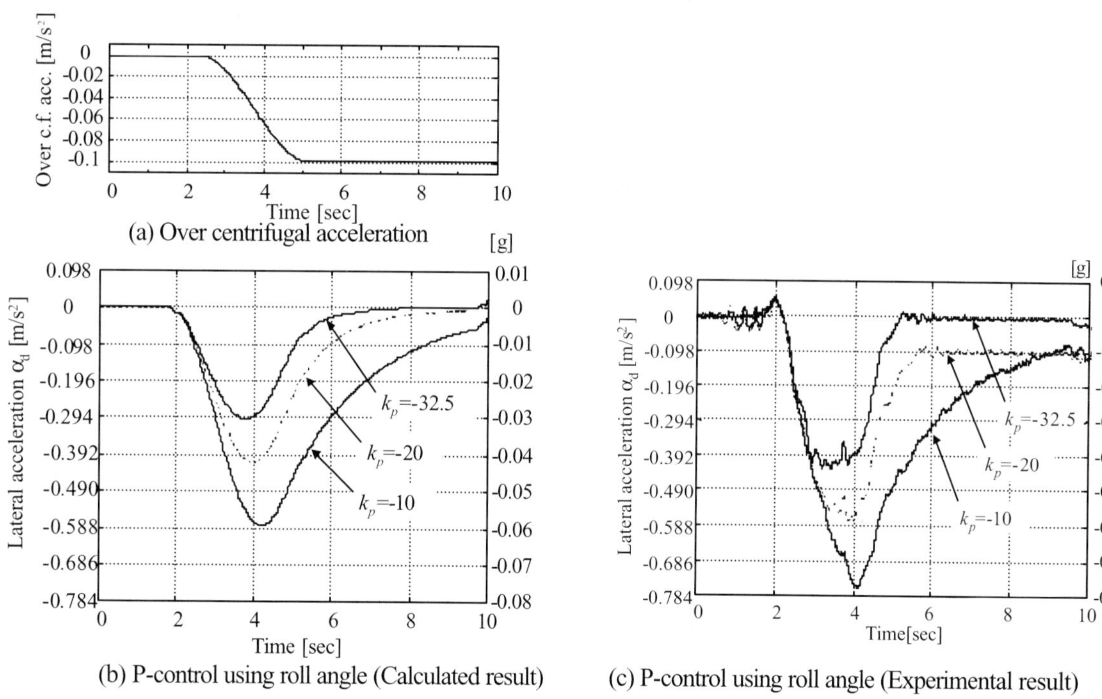

Fig. 7. Roll angle feedback tilting control (P-control).

k_I, and derivative gain k_D for this PID controller with respect to the deviation e. Pulse speed $f_{PPS}(t)$ which is an inclination command for the actuator can be calculated from Eq. (17). Eqs (15)–(17) can be rearranged to give Eq. (18).

$$A_{fb}(t) = -k_p e_{e1} - k_p \int e_{e1} dt - k_p \frac{d}{dt}\{e_{e1}\} \tag{16}$$

$$f_{PPS}(t) = \frac{d}{dt}\{A_{fb}(t)\} \tag{17}$$

$$\phi_u = -k_p e_{e1} - k_p \int e_{e1} dt - k_p \frac{d}{dt}\{e_{e1}\} \tag{18}$$

4.3. Experimental results

The results of tilting control with the vehicle running at 100 km/h on a curve of Fig. 3 and Table 1 are now described. As an example, Fig. 7 describes the results of proportional control of the deviation of the car body roll angle during 10 seconds that takes to pass the straight path, entrance transition curve, and then circular curve. A response of the over centrifugal acceleration α_u acting on the car body is shown in (a). The lateral acceleration α_d calculated by the computer simulation is displayed in (b). α_d from the experiment using the model is graphed in (c). These results confirm the effect of the tilting control that reduces lateral acceleration to zero during a curve against the disturbance of the over centrifugal acceleration. As shown in the figure, the effect increases with the K_P gain. Also, maximum values of the lateral acceleration during the transition curve sections were reduced by approximately 50% without the motors stepping out. Thus, the experimental results show the effect of the control scheme, confirming the merits of this method.

5. Conclusions

A procedure for car body tilting control was proposed on the basis of roll angle feedback whose process included collecting data from a yaw rate sensor on the car body, deriving the curvature of the track, and computing a target car body roll angle. A control scheme was first constructed, and computer simulation with a dynamic model and experiment using electric actuators were performed. Finally, the following conclusions were obtained.

1. Tilting control was shown to be possible by feedback of the deviation of the car body roll angle from the target value for required performance, which is determined in real time by using the estimated curvature.
2. Tilting control performance of the above control method was verified and its effectiveness proven by experiment employing electric actuators.

References

[1] H. Yoshida and M. Nagai, Tilting Control of Railway Vehicle using Electric Actuators 1st report, in Japanese, *JSME* **66–645** (C) (2000), 1629–1635.

[2] Roger Goodall, Active railway suspensions: Implementation status and technological trends, *Vehicle System Dynamics* **28** (1997), 87–117.

[3] The Japan Soc. of Mechanical Engineers ed., The dynamics of the railway vehicle in Japanese, The Co.,Ltd. electric railcar workshop, 1994, pp. 15–32.

[4] The Japan Soc. of Mechanical Engineers ed., Dynamics and control of the railway vehicle system, The Co.,Ltd. YOUK-ENDO, 1999, pp. 141–155.

International Journal of Applied Electromagnetics and Mechanics 13 (2001/2002) 145–149
IOS Press

Tracking control of a two-stage servo head arm for a hard disk drive

Masanori Murai*, Junji Tani and Jinhao Qiu
Institute of Fluid Science, Tohoku University, 2-1-1 Katahira, Aoba-Ku, 980-8577 Sendai, Japan

Abstract. This paper describes a new type of two stage servo head arm for magnetic disk drives. Four PZT actuators were bonded to the stainless steel head arm. The PZT actuators were used for accurate follow-up of the track and also vibration control of the head arm. The PID control method was used in the design of controllers for vibration control and tracking follow-up control of the arm. The characteristics of a prototype of the servo system were measured and its effectiveness was verified by experiment.

1. Introduction

Recently, the areal recording density of hard disk drives (HDD) is increasing at a rate of about 100% per year. It is necessary to increase the track density as well as the line density for higher areal density. Therefore, a great number of researchers have proposed several types of two stage tracking servo systems for high density magnetic disk drive systems [1–7]. These actuators are categorized into three groups.Those in the first group are often called as the first generation. They are assembled on head mounting blocks and piezoelectrically or electromagnetically driven to swing the head suspension arms. These actuators can be fabricated with conventional machining technology. However, actuators in this group are influenced by resonance of head arms and have narrow control frequency bandwidth. The second group is often called as the second generation. The microactuators are made using micromachine technology and inserted between the heads arm and the head sliders [4,5]. These microactuators are not influenced by the resonance of the head arms and good control performance were obtained [5]. However, since these actuators are almost of the same size as the sliders and the suspension payload, the space between the disks in a drive are increased. The third group is often called as the third generation. The microactuators are embedded on the head sliders and the magnetic head elements are electrostatically moved by the microactuators [6,7]. These microactuators are not influenced by the resonance of the head arms and much smaller than the second generation actuators. However, they have small displacement and weak force.

This paper describes a new type of two stage servo system for magnetic disk drives. The head arm of the two stage servo system has the second stage actuators for the tracking control. Four PZT actuators were bonded to the suspension part of the metallic head arm. The PZT actuators were used for accurate follow-up of the track and for control of some vibration modes of the head arm. The PID control theory was used in the design of the two dimensional controller for the simultaneous vibration control and

*Corresponding author: Masanori Murai, Tel.: +81 22 217 5251; Fax: +81 22 217 5251; E-mail: murai@ifs.tohoku.ac.jp.

1383-5416/01/02/$8.00

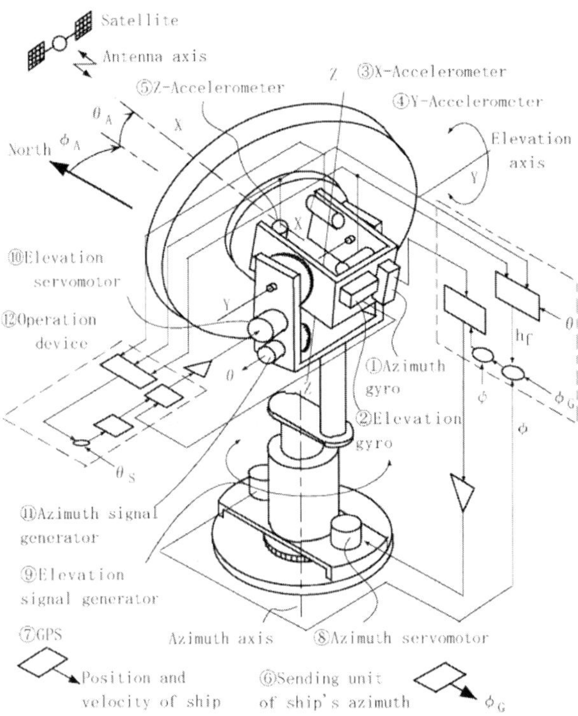

Fig. 1. Apparatus of antenna system.

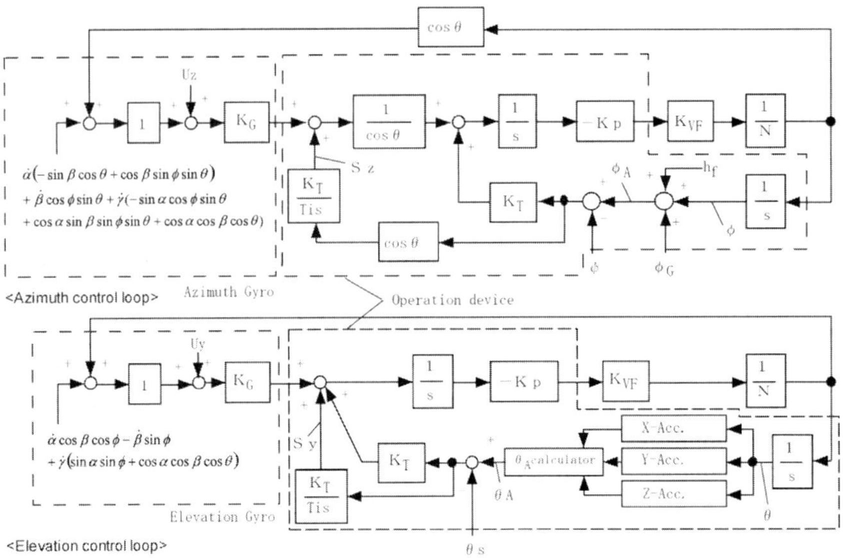

Fig. 2. Control block diagram.

and elevation gyro ② around the Elevation and Azimuth axes. This means that angular velocities of the antenna about the y and z axes relative to an inertial space are constantly zeroed even when the ship rolls, pitches and yaws (α, β, γ). This control results in the antenna axis being continually directed at

Table 1
Pointing errors in scorsby table test

ϕ (deg)	θs (deg)	P.E (mean)	P.E (maximum)
0	10	1.02	1.87
45	10	0.85	1.67
90	10	0.60	1.22
0	45	0.60	1.34
45	45	0.61	1.32
90	45	0.95	4.31

the satellite. The servo is a time-lag of first order system with 8 Hz frequency.

With this system however, azimuth angle ϕ_A and inclination angle θ_A must be controlled in order to move the antenna axis in the intended direction and to compensate for antenna axis error due to gyro drift (Uz, Uy). Antenna azimuth angle ϕ_A is calculated by using the ship azimuth angle ϕ_G, azimuth signal generator output ϕ and the compensation value of inclination h_f and is expressed as follows.

$$\phi_A = h_f + \phi_G + \phi \tag{1}$$

where

$$\tan h_f = g_y / g_z \sin \theta \tag{2}$$

g_y: the output of Y-accelerometer ④
g_z: the output of Z-accelerometer ⑤
θ: the output of elevation signal generator ⑪ (output is zero when antenna axis is parallel to deck)
ϕ: the output of azimuth signal generator ⑨ (output is zero when antenna axis is in the bow direction)

Antenna inclination angle θ_A between the horizontal plane and satellite is also obtained by accelerometer outputs ③ ④ ⑤ and is expressed as follows.

$$\tan \theta_A = g_x / (g_y \ \sin \varepsilon + g_z \cos \varepsilon) \tag{3}$$

where

$$\tan \varepsilon = g_y / g_z \tag{4}$$

g_x : the output of X-accelerometer ③

Errors between these angles and the set angles such as inclination angle θ_S and the set azimuth angle ψ to the servo system are added in pointing the antenna towards the satellite, with anglest θ_S and ψ to the satellite obtained by GPS ⑦. With this control, gyro drifts are compensated for by outputs S_Z and S_y of the integration device. These calculations are performed by micro-computer ⑫.

3. Experimental results

Pointing error of the antenna axis to the satellite was evaluated by using a 3-axis Scorsby Table with pitch, roll, yaw and antenna control signals input into a personal computer.

Figure 3 shows the results of the ship under motion. The azimuth angle ψ to the satellite is 90 degrees and inclination angle θs to the satellite is 45 degrees. Rolling angle of the ship is 25 degrees; pitching angle, 15 degrees; and yawing angle, 8 degrees. The measuring time is ten minutes. The results show a pointing error of less than 5 degrees and that the motion of both azimuth and inclination angles is stable. Table 1 depicts the results for the other directions. It can be seen that the antenna pointing error is less than the control accuracy objective described in this paper (i.e., pointing error of less than 10 degrees during 90 percent of usage time).

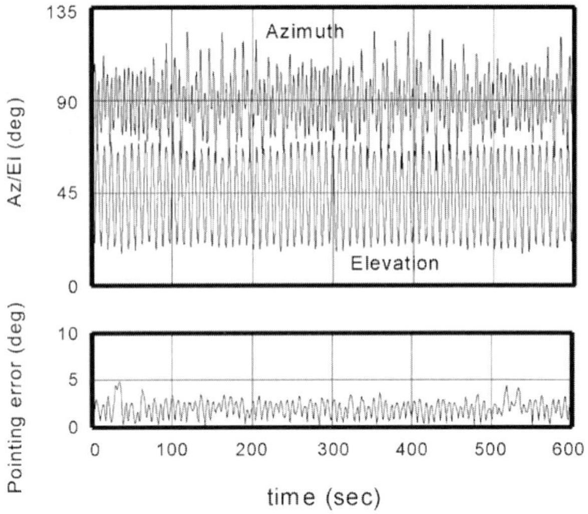

Fig. 3. Pointing errors, $\Psi 90^\circ$ $\theta s 45^\circ$.

4. Conclusion

A compact antenna direction control device is presented. The device employs a strap-down type 2-axis Az/El system with gyros and acceleration sensors but no level sensor. A control method is presented in which two servomotors control and maintain gyro outputs at zero. The advantage of this device is that the system is compact and simple. In addition, it requires no complicated calculations. This work is applicable to receiving apparatus such as CS and BS as well as for INMARSAT communication.

Acknowledgements

The author would like to express his sincere thanks to Professor K. Nagaya of GUNMA Univ. for his continuous encouragement and to Mr. Y. Kamiya, TOKIMEC for his useful discussion.

References

[1] H. Chiba, A. Yasuda, H. Ando, K. Yamashita and M. Komamiya, *A satellitebroadcasting reception system for vessels using a two-axis controlledantenna*, IEICE SANE88-24.

[2] K. Sanada, H. Ando and Y. Kanai, A satellite broadcasting reception system for ship using a two-axis antenna rotating mechanism, *JIN* **77** (May, 1987), 91–101.

[3] O. Suzuki, A. Yasuda and H. Yamashita, Onboard tracking mechanism by low speed Az-El mount rotator, *IEICE* **J80-B-II**(8) (August, 1997), 714–721.

[4] T. Shiokawa, Y. Karasawa and H. Yuki, Reduction method of pointing error of EL/AZ mount for maritime satellite communications *IEICE* **J69-B**(8) (August, 1986), 833–841.

[5] T. Shiokawa, Y. Karasawa and H. Yuki, A compact shipborne antenna system for future maritime satellite communications, *IEEE Tokyo Section DenshiTokyo* **26** (1987), 25–31.

International Journal of Applied Electromagnetics and Mechanics 13 (2001/2002) 155–162
IOS Press

Repulsive magnetic bearing stabilized by the motion control of magnets with a piezoelectric actuator

Takeshi Mizuno[a] and Mitsunori Aizawa[b,*]
[a]*Department of Mechanical Engineering, Saitama University*
[b]*Advantest Corporation*

Abstract. A magnetic bearing system using forces of repulsion between permanent magnets was developed. In the developed magnetic bearing, the radial motions of the rotor are passively supported by repulsive forces. The motion in the axial direction, which is inherently unstable, is stabilized by using the motion control of the permanent magnets for passive radial suspension with a piezoelectric actuator. The developed magnetic bearing suspended a rotor without any mechanical contact. The characteristics of the suspension system were studied both theoretically and experimentally.

1. Introduction

One of the principal methods of achieving contact-free levitation is to use forces of repulsion between permanent magnets [1]. This levitation has several advantages:

- No energy is required to generate levitation force.
- It is stable in the direction of repulsive force (normal direction).

Its disadvantages are

- It is unstable in the lateral direction.
- It must be combined with either some mechanical guidance or active magnetic suspension for completely contactless levitation.
- Damping in the levitation direction is small.

In consequence of these problems, this type of levitation system has not been used so widely as the suspension using controlled DC electromagnets. However, the possibility of new levitation systems using permanent magnets is worth being investigated because the recent improvement of magnetic material makes the flux density of permanent magnets so high.

We proposed to introduce the motion control of permanent magnets into this type of levitation system as a mean of compensating the above-mentioned disadvantages [2–5]. A Levitation system with performance of positioning and vibration control in the normal direction was realized by inserting an

*Corresponding author: Department of Mechanical Engineering, Saitama University Shimo-Okubo 255, Saitama 338-8570, Japan. Tel.: Int+48 858 3455; Fax: Int+48 856 2577; E-mail: mizar@mech.saitama-u.ac.jp.

1383-5416/01/02/$8.00

actuator between a levitation magnet and its base [2]. Stabilization in the lateral direction was achieved by moving a magnet for support like an inverted pendulum [3]. The latter principle was applied to a repulsive magnetic bearing system [4,5]. In the developed system, the radial motions of a rotor were passively supported by repulsive forces between ring-shape permanent magnets; the axial motion was actively controlled by the motion control of the permanent magnets with a pair of voice coil motors.

This paper investigates the applicability of piezoelectric actuators to the repulsive magnetic bearing system instead of voice coil motors. Piezoelectric actuators have several advantages:

- high-speed response,
- small heat dissipation,
- no necessity of winding coil.

These characteristics are distinguishable from those of electromagnets that are ordinarily used in conventional active magnetic bearings. The third characteristic is attractive in the fields where simplicity of structure is of primary importance.

In spite of these advantages, there are few works in which piezoelectric actuators are applied to magnetic suspension. Kanemitsu et al. [6] use them for the lateral vibration control of a flexible rotor. Oka et al. [7] develop a hanging-type magnetic levitation system using a piezoelectric actuator. In both the cases, they are operated to adjust the attractive force of magnet.

In this research a repulsive-type magnetic bearing system is developed which is characterized by using a piezoelectric actuator for the motion control of levitation magnets. The characteristics of the bearing are studied both theoretically and experimentally.

2. Basic model

2.1. Basic configuration

The levitation system using forces of repulsion between permanent magnets is stable in the levitation direction but unstable in the lateral directions. The lateral motions of the levitated object can be controlled by moving a permanent magnet of support in the lateral directions like an inverted pendulum [3]. Figure 1 shows a basic configuration of magnetic bearings using this principle of levitation. The four-degree-of-freedom motions in the radial directions of the rotor are passively supported by repulsive forces between permanent magnets of ring shape; the single-degree-of-freedom motion in the axial direction, which corresponds to the lateral direction, is actively controlled by moving the ring-shape permanent magnets for radial support. For this motion control, a piezoelectric actuator is used in this research.

2.2. Modeling

Figure 2 shows a physical model of the system illustrated by Fig. 1. It is assumed for simplicity that the rotor moves only in the axial direction. The gravitational force acting on the rotor and the lateral forces between the permanent magnets are balanced in the equilibrium states. The effects of the lateral forces are modeled as a spring with a negative stiffness. For small deviations from the equilibrium, the equation of motion becomes

$$m_a \ddot{x}_a(t) = k_l(x_a(t) - x_b(t)), \tag{1}$$

where

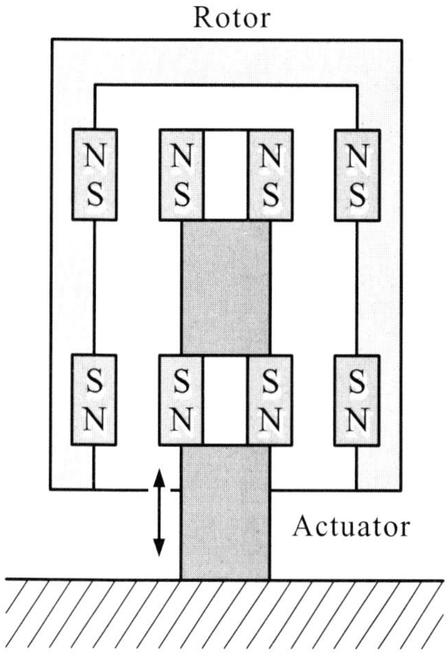

Fig. 1. Axially active and radially passive magnetic bearing.

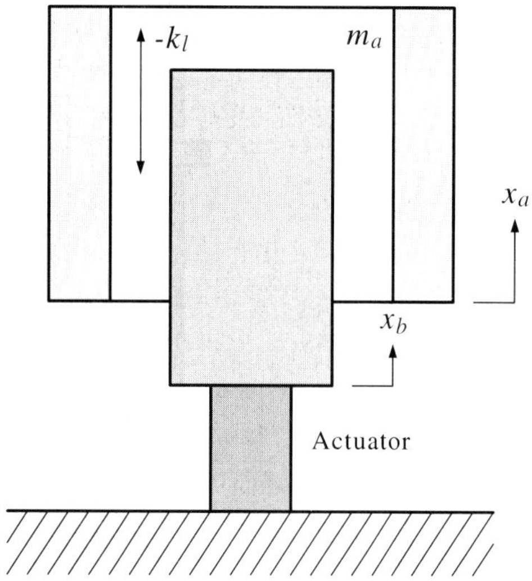

Fig. 2. Basic model.

m_a: mass of the levitated object,
k_l: stiffness of the vertual spring in the lateral direction,
x_a: displacement of the rotor,
x_b: displacement of the magnets driven by an actuator.
It is assumed that the actuator is controlled to follow a signal inputted to the driver circuit so that

$$x_b(t) = k_e u(t), \tag{2}$$

where
u: input voltage to the driver circuit,
k_e: gain of the driver circuit.
Substituting Eq. (2) into Eq. (1) gives

$$m_a \ddot{x}_a(t) = k_l x_a(t) - k_l k_e u(t). \tag{3}$$

Equation (3) indicates that the system is unstable without control $(u(t) = 0)$.

3. Control system design

3.1. PD control

The PD control is a fundamental control scheme for stabilization. The control input is represented by

$$u(t) = K_P x_a(t) + K_D \dot{x}_a(t) + v(t), \tag{4}$$

where
K_P: gain of displacement feedback,

As shown in Fig. 8, the same property as observed in the experimental result can be indicated by the theoretical calculation. However, the rotor vibration observed in Fig. 4 can't be confirmed from the theoretical analysis conducted in this section, since the system has the stable eigenvalues.

For the reason, the further study must be conducted concerning the effect of the small looseness between the axial bearing and the rotor shaft, since it is known that there exists a few cases that the self-excited rotor vibration can be induced [7]. Moreover, the transient current is induced in the induction coil because of the impact between the flange and the fixed stopper labeled with Open and Close, which is assumed another reason for the vibration. These are the future work.

On the other hand, two vibration suppression methods are given from the view of the practical application, one of which is to increase the friction force between the rotor and the axis, and the other is to increase the capacitance of C_1 in the circuit.

5. Concluding remarks

In this paper, the unexpected vibration of the exposure control system in security camera is investigated. It is found that the unexpected vibration of the exposure control system can not be observed for every sample, but suddenly occurs at a few samples. Therefore, it has been very difficult to observe the vibration itself.

First, some experiments are conducted to examine the dynamic properties of the exposure control system at the normal condition. Moreover, the vibration of the exposure control system observed in the abnormal condition is investigated. As a result, the vibration of the exposure control system is excited with the 20 times of the natural frequency of the rotor-spring system. Second, theoretical analysis at the normal condition is made to give useful information on the excitation mechanism of the vibration. Then, it is confirmed that the motion of the system at normal condition is in good agreement with that of the experimental results. Further discussion must be conducted concerning the reasons for the vibration.

References

[1] T. Ozawa, Actuator for Camera, Japanese Patent, 01-267622, 1989.

[2] T. Emura, Modal Analysis of Lens Units for Video Camcorders, *Konica Technical Report* **10** (1997), 83.

[3] H. Takano et al., Mechanism for Supporting Rotor of Direct Drive Motor Built in a Lens Barrel for Zoomlens, *Trans. IEE Japan* **116-D** (1996), 785.

[4] H. Takano et al., Performance Analysis for Direct Drive Motor Built in a Lens Barrel for Zoomlens, *Trans. IEE Japan* **115-D** (1995), 478.

[5] L. Cao et al., Mid-Frequency Oscillation and High Frequency Stability in Stepping Motors, *Trans. IEE Japan* **117-D** (1997), 1154.

[6] S. Hu et al., Reverse Rotation in PM Motors with Stepping Drive, *Trans. IEE Japan* **114-D** (1994), 1162.

[7] H. Kawamoto, Rotor Dynamics of a Laser Scanner Motor Supported by Air Bearings, *J. Imaging. Sci. Tech.* **41-6** (1997), 565.

International Journal of Applied Electromagnetics and Mechanics 13 (2001/2002) 171–180
IOS Press

Maximum power control for a photovoltaic power generator not requiring detailed system information

Kenji Takahara*, Youichi Yamanouchi and Hideki Kawaguchi
Department of Electrical and Electronic Engineering, Faculty of Engineering, Muroran Institute of Technology, 27-1 Mizumoto-cho, Muroran, Hokkaido, Japan, 050-8585

Abstract. The purpose of the present study is to construct a system to maintain the output power of the photovoltaic power generation system (PV system) at its maximum power levels. The switch duty ratio of the DC-DC converter of the PV system and its output power are selected as the input and output values, respectively. The characteristics of the PV system, which are described by the characteristic curves, are changed according to changes in external environmental conditions, such as atmosphere, loads and so on. In this study, the characteristic curves are considered to be approximated by quadratic functions and their gradients by linear functions. The change in the switch duty ratio is determined by the fuzzy controller, based on the characteristic curves and their gradients estimated by an adaptive algorithm.

1. Introduction

The effective introduction of natural energy, photovoltaic power generation and so on, are desirable for the stable supply of energy and for the protection of the earth's environments. The power generation efficiency of a PV system is low and its output power changes greatly in relation to environmental conditions, such as radiation, temperature and so on. Therefore, it is necessary to use an appropriate control method that makes the PV system track its maximum power automatically, in order to obtain energy from the system effectively. The hill-climbing method [1] is generally used as a MPPT controller for a PV system. It is a simple method that searches the optimal operating point by barely changing an input value with a certain step-size around the operating point, until the PV output is maximized, without any description about the PV system. But it may not be able to respond to the rapid changes in environmental conditions if the change of the input value is not chosen properly. Shortening the arrival time of an operating point to maximum power point was attempted by approximating the optimal operating points, which differed according to solar radiation, with a linear function [2]. However, the changes in external environmental conditions sometimes reduced the efficacy of the approximation. The use of fuzzy logic was proposed to determine the change in input values [3]. In order to make appropriate fuzzy membership functions, it is therefore necessary to appreciate detailed information concerning the PV system.

*Corresponding author: Kenji Takahara, Tel.: +81 143 46 5508; Fax: +81 143 46 5501; E-mail: takahara@elec.muroran-it.ac.jp.

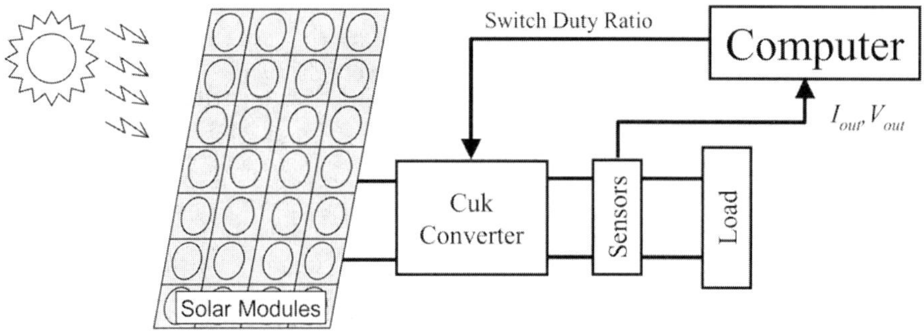

Fig. 1. Outline of the independent PV system.

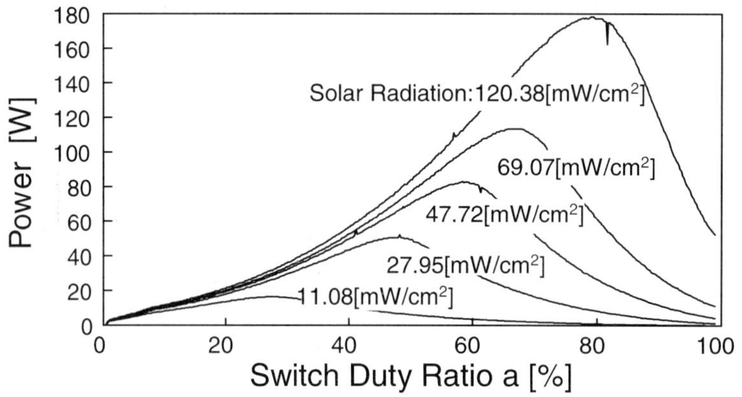

Fig. 2. Characteristics of output power.

A control method that directly yielded the value of optimal operating points was proposed, by approximating the output characteristics of the PV system to 4 th order polynomials, the parameters of which were estimated by an adaptive algorithm [4]. Its control system worked well, even without any information of the PV system, because its changing characteristics were seized as the innovated parameters of mathematical models by adaptive law, monitoring the input and output values. However, their approximation might be insufficient, because the output characteristic curves physically disaccord with the polynomial expressions. An adaptive hill-climbing method [5] was proposed, that determined the change in width of the input value in proportion to the estimated gradient values of the output characteristic curves by an adaptive algorithm. This method rapidly tracked the maximum power point of the PV system, coping with the effect of changing external environmental conditions, because it does not receive the influence of noise much, in order to obtain the gradients of the characteristic curves. However, there were cases when operating points barely fluttered around the maximum power points, because spontaneous overshoots sometimes occurred in the peaks of the characteristic curves.

In this study, the differential coefficients of quadratic functions, which approximate the characteristic curves around its peak, will also be used with estimated gradient values as linear functions for MPPT control in order to decrease the overshoots and fluttering. Furthermore, fuzzy logic will be introduced to determine the input value, based on two gradient values of the characteristic curves, to cope with the effect of changing external environmental conditions.

Table 1
Rule table of the fuzzy controller

<table>
<tr><th>S2 \ S1</th><th>NB</th><th>NM</th><th>NS</th><th>ZO</th><th>PS</th><th>PM</th><th>PB</th></tr>
<tr><td>NB</td><td rowspan="7">NB</td><td rowspan="7">NM</td><td colspan="3" rowspan="2">ZO</td><td rowspan="7">PM</td><td rowspan="7">PB</td></tr>
<tr><td>NM</td></tr>
<tr><td>NS</td><td>NS</td><td>NS</td><td>NS</td></tr>
<tr><td>ZO</td><td>ZO</td><td>ZO</td><td>ZO</td></tr>
<tr><td>PS</td><td>PS</td><td>PS</td><td>PS</td></tr>
<tr><td>PM</td><td colspan="3" rowspan="2">ZO</td></tr>
<tr><td>PB</td></tr>
</table>

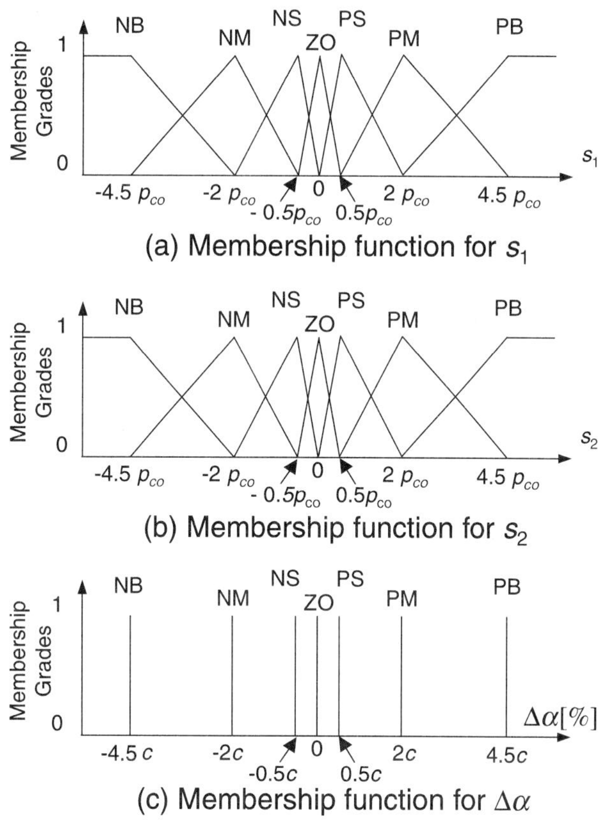

Fig. 3. Membership functions.

2. Materials and method

2.1. Independent PV system

Figure 1 shows an outline of the independent PV system. This system is composed of PV modules, a DC-DC converter, a computer and an ohmic load. Five modules of crystal silicon solar batteries (LA441K63 made by KYOCERA CORP. Co. Inc.) are used, which are connected in parallel. These

Case1		All Covered
Case2		Left three Modules Covered
Case3		Lower Left Covered
Case4		Lower Half Covered

Fig. 4. Experimental conditions.

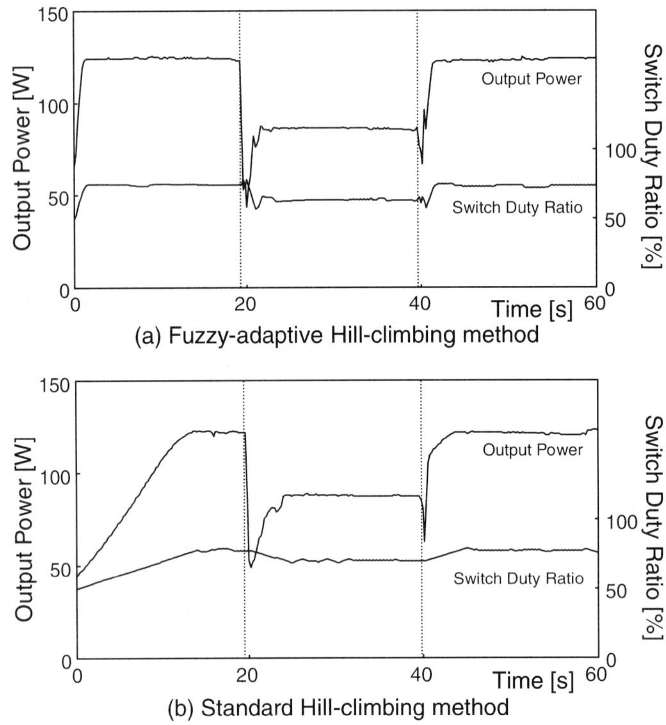

Fig. 5. Responses under conditions of sudden radiation changes when covering the whole solar modules (Average solar radiation: $81.0\ [\mathrm{mW/cm^2}]$, average temperature: $6.2\ [^\circ\mathrm{C}]$).

modules are established towards due south outside and at an angle of elevation of 43 degrees to equal the latitude (Lat. 43°N) of Muroran. The DC-DC converter is a cuk converter. The switch duty ratio of the DC-DC converter can be set by the computer.

2.2. Characteristic curve of PV system and control concept

Figure 2 shows the characteristics of output power in the PV system, which was measured under various radiation levels, when the switch duty ratios were changed continuously from 0 to 100%. The shapes of the output characteristic curves are convex like mountains and the peak of these curves are maximum power points of the PV system. This figure shows that the values of the maximum power points of the PV system and the optimal switch duty ratio, which produce maximum power, are different

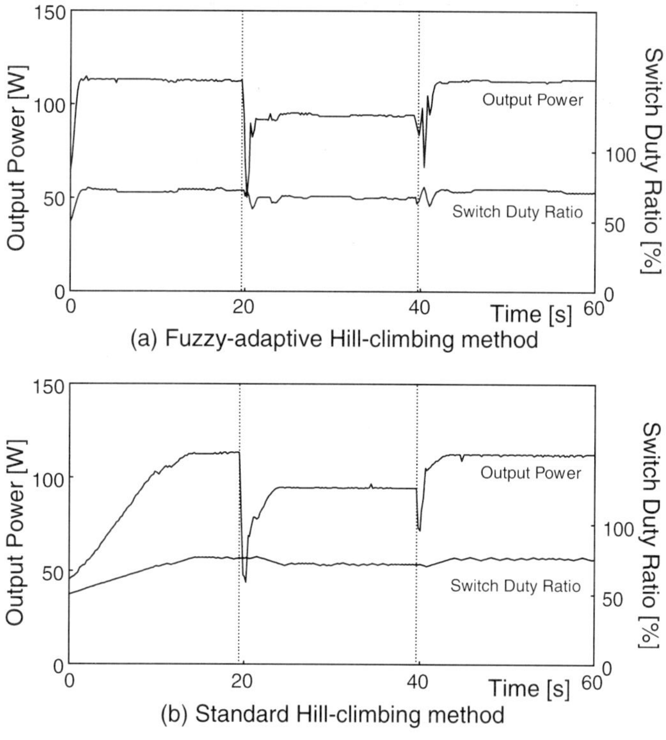

Fig. 6. Responses under conditions of sudden radiation changes when covering the left three panels of the solar modules (Average solar radiation: 70.6 $[\mathrm{mW/cm^2}]$, average temperature: 7.1 $[^\circ\mathrm{C}]$).

if radiation levels are different. Furthermore, the characteristics of the PV system are changed by changes in external conditions such as cell temperature, load and so on. Therefore, it is not easy to determine the optimal operating points of such a PV system with many variable characteristics.

The following illustrates the MPPT control method that determines the change in width of input value in accordance with the shapes of the characteristic curves using adaptive algorithms and fuzzy logic.

3. Fuzzy-adaptive hill-climbing method

The gradient values of the characteristic curves of the DC-DC converter output are zero at the maximum power points as shown in Fig. 2. Their gradient values are positive if the switch duty ratio is smaller than the optimal switch duty ratio. On the other hand, their gradient values are negative if the switch duty ratio is larger than the optimal switch duty ratio. And the farther the operating points are from the optimal operating points, the larger the absolute values of the gradient. Therefore, the approximate values of the optimal operating points are estimated based on the gradient values of the characteristic curves around the operating points. Firstly, the gradient values of the characteristic curves are estimated as follows.

At first, the switch duty ratio $\alpha(k)$ and the output power of the cuk converter $P(k)$ are selected as input and output values, respectively. The following linear and quadratic functions are used as descriptions of the characteristic curve.

$$P(k) = a_0 + a_1\alpha(k) \tag{1}$$

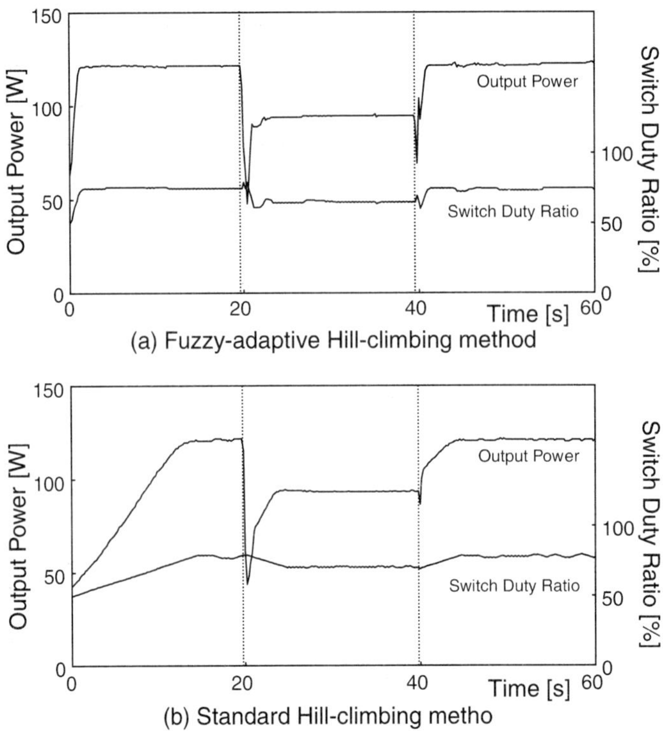

Fig. 7. Responses under conditions of sudden radiation changes when covering the lower left of the solar modules (Average solar radiation: $81.0\ [\mathrm{mW/cm^2}]$, average temperature: $5.2\ [^\circ\mathrm{C}]$).

$$P(k) = b_0 + b_1\alpha(k) + b_2\alpha^2(k) \tag{2}$$

The mathematical models, which correspond to the above models, are assumed in order to identify the characteristics of the PV system. The parameters of those identified models are estimated as $\hat{a}_0(k), \hat{a}_1(k), \hat{b}_0(k), \hat{b}_1(k)$ and $\hat{b}_2(k)$, by a constant-trace adaptive algorithm [6]. Two gradient values around the operating points are obtained as the follows based on the estimated parameters.

$$s_1(k) = \hat{a}_1(k) \tag{3}$$

$$s_2(k) = \hat{b}_1(k) + 2\hat{b}_2(k)\alpha(k) \tag{4}$$

The value change of the switch duty ratio of the DC-DC converter are determined by the following fuzzy logic with those gradient values.

The fuzzy rules based on the outlines of the characteristic curves are shown in Table 1. The variables $s_1(k)$, $s_2(k)$ and the change in the switch duty ratio $\Delta\alpha(k)$ are fuzzified by the membership functions as in Fig. 3. Here, c is a constant value, and $p_{co}(k)$ is calculated by Eq. (5) using P_{ref}, which is a rated power.

$$P_{c0}(k) = P(k)/P_{ref} \tag{5}$$

The crisp value of the change in the switch duty ratio is obtained by defuzzifying the reasoned result using a product-sum-gravity method [7]. The switch duty ratio is determined by Eq. (6).

$$\alpha(k) = \alpha(k-1) + \Delta\alpha(k) \tag{6}$$

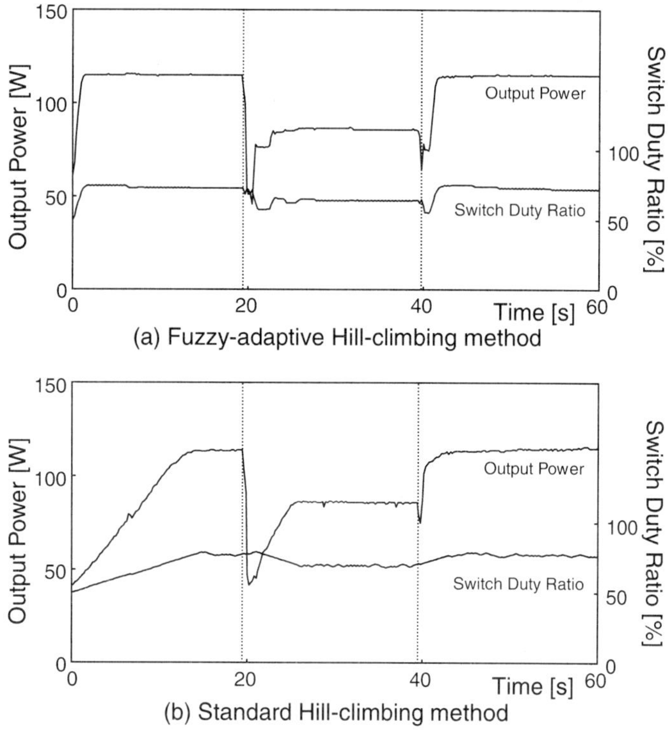

Fig. 8. Responses under conditions of sudden radiation changes when covering the lower half of the solar modules (Average solar radiation: 76.0 [$\mathrm{mW/cm^2}$], average temperature: 6.8 [$^\circ$C]).

In order to confirm the efficacy of the proposed control system, it was applied to some experiments with various conditions in the follows.

4. Experiments and results

The chosen sampling interval was 100 ms in the experiments. The initial value of the switch duty ratio of the DC-DC converter was set 50%. And the initial value of the gain matrix for the adaptive algorithm was chosen as $\Gamma(0) = 100000\mathbf{I}$. The value of P_{ref} in the Eq. (5) was chosen as 140 W and c as 1.0. All the experiments were carried out at times of fine weather. Then, the characteristics of the PV system were measured just before and after each experiment, and were used for the evaluation of the control results.

4.1. Response under conditions of sudden radiation changes

First, radiation changes are considered as one of environmental changes. Hence, the performance of the proposed system was tested during rapid changes in radiation.

Making a shadow on the whole or a part of the solar modules using gauze changed solar radiation rapidly as shown in Fig. 4. Covering the solar modules with gauze was assumed to mimic obstacles such as cloud interrupt a part of the solar radiation.

The control results under the conditions of sudden radiation changes by the shadows, which were produced with the methods of cases 1 to 4, are shown in Figs 5 to 8, respectively. The results of the

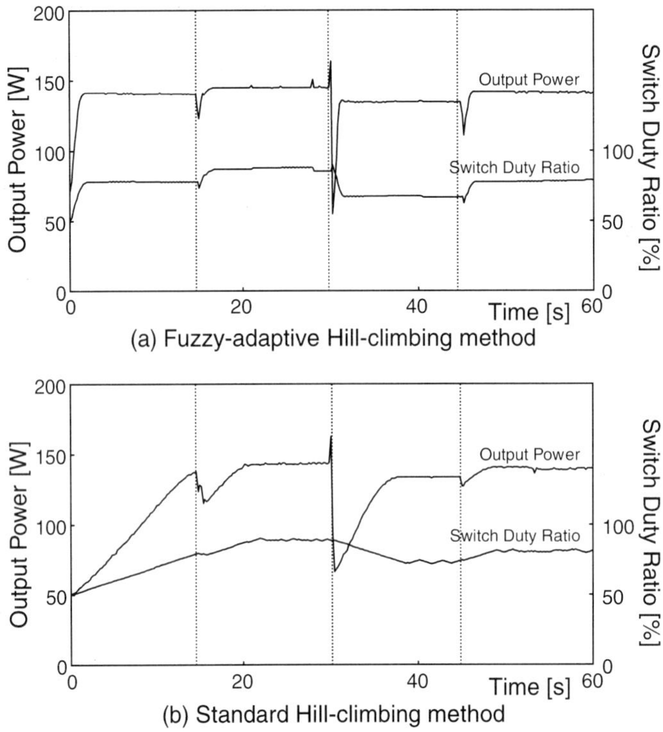

Fig. 9. Responses under conditions of sudden load changes (Average solar radiation: 100.2 [$\mathrm{mW/cm^2}$], average temperature: 7.2 [°C]).

fuzzy-adaptive hill-climbing method are shown in (a) and the results of a standard hill-climbing method with 0.4% as the change step of the switch duty ratio in (b) of each figure. The dotted lines in these figures show the points when solar radiation was changed.

4.2. *Response under conditions of sudden load changes*

Load changes are also considered as one of environmental changes. Hence, the performance of the proposed system was tested under rapid load change. The load was a variable resistance. The value of the resistance was set as 5.1 Ω at the start of the experiment. It was changed to 10.4 Ω 15 seconds later, to 2.5 Ω 30 seconds later and at 5.1 Ω 45 seconds later. Figure 9 illustrates the experimental results. The proposed control method can track the changing maximum power point in a short time, corresponding to rapid load changes.

4.3. *Response in the case that the constitution of solar modules differs*

In order to confirm that the proposed method does not require detailed information about the system, its performances were tested in the case when the set parameter P_{ref} is different from the actual constitution of the solar modules. The value of P_{ref} was set at 140 W as in the above experiments, in which five solar modules were used. Here, three solar modules were used, which were connected in parallel and the experimental environments were changed by changing the loads ($5.1\ \Omega \rightarrow 10.4\ \Omega \rightarrow 2.5\ \Omega \rightarrow 5.1\ \Omega$). Figure 10 illustrates the experimental results. Figure 10(c) shows the control results using the fixed fuzzy membership functions, the parameter p_{co} of which is not regulated ($p_{co} = 1.0$).

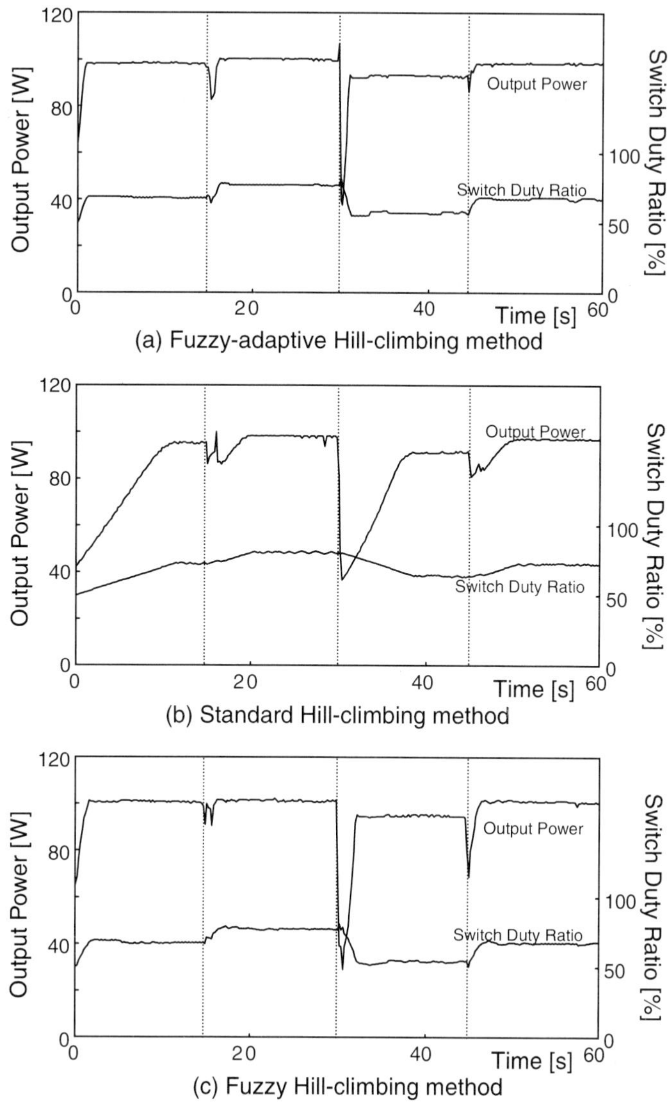

(a) Fuzzy-adaptive Hill-climbing method

(b) Standard Hill-climbing method

(c) Fuzzy Hill-climbing method

Fig. 10. Control results using three solar modules (Average solar radiation: 98.1 $[\mathrm{mW/cm^2}]$, average temperature: 6.8 $[^\circ\mathrm{C}]$).

The proposed method is considered to be able to reach the maximum power points earlier than these other two methods. Furthermore, high control performance can be maintained, because its membership functions are adaptively regulated, even if the wrong system information is given as parameter P_{ref}.

5. Conclusion

The fuzzy-adaptive hill-climbing method was proposed to maintain the output power of the PV system at its maximum power levels. It was applied to the experiments under various conditions and confirmed to be useful to obtain maximum power, coping with the characteristic changes corresponding to environmental changes in radiation, load and so on. The following two useful concepts are considered to make the proposed system work well.

(1) The estimation of the gradient values of the characteristic curves of the PV system by an adaptive algorithm: Adaptive algorithms can regulate the system parameters on the basis of detection of change in the characteristics and the environment of the controlled object. Hence, the non-linearity of the controlled object and its characteristic changes can be regarded as deviations of the parameters of the mathematical models [8]. That is, to estimate the two gradient values of the characteristic curves of the PV system is to seize the changing characteristics of the PV system.

(2) The effect of the parameter $p_{co}(k)$, which changes the membership functions: The width of the membership functions are changed corresponding to the shape of the characteristic curves, based on the measured output power. The parameter $p_{co}(k)$ let the change width of the switch duty ratio enlarged, when the output power is small. On the other hand, $p_{co}(k)$ allows the change in width of the switch duty ratio to be reduced, when the output is large. Therefore, this proposed system demonstrates fine control performance, because it is equipped with the regulation mechanism of the membership functions. Hence, the proposed system is applicable to a **PV** system, the characteristics of which cannot be completely recognized.

References

[1] B.K. Bose, P.M. Szczesny and R.L. Steigerwald, Microcomputer control of a residential photovoltaic power condition system, *IEEE Trans. Ind. Appl.* **IA-21** (1985), 1182–1191.

[2] J. Itsumi, M. Iwahori, T. Takahashi, S. Furukawa and Y. Ikeda, *Study on converter for photovoltaic power generator and its control* **SPC-84-23** (1984), 39–48.

[3] Chung-Yuen Won, Duk-Heon Kim, Sei-Chan Kim, Won-Sam Kim and Hack-Sung Kim, A new maximum power point tracker of photovoltaic arrays using fuzzy controller, *IEEE Power Electron. Spec. Conf.* **1994** (1994), 396–403.

[4] K. Takahara, Y. Yamanouchi and T. Matsuda, A MPPT system of photovoltaic power generation coping with environmental changes, *Trans. IEE Japan* **119-D** (1999), 1529–1534.

[5] K. Takahara, Y. Yamanouchi and H. Kawaguchi, Maximum Power Control for a Photovoltaic Power Generation System by Adaptive Hill-climbing Method, *Trans. IEE Japan* **121-D** (2001), to appear.

[6] S. Shinnaka, Adaptive algorithm, Sangyo-tosho, Tokyo, 1990.

[7] M. Mizumoto, Intelligible fuzzy theory III -fuzzy reasoning and fuzzy control, *Computrol* **28** (1989), 32–45.

[8] H. Wakamatsu and T. Kitamori, *Design of control system on the basis of partial knowledge about nonlinear objects represented by Volterra series and its application to artificial control of respiration*, Proc. IFAC 9th World Congr, 1984, 3045–3050.

International Journal of Applied Electromagnetics and Mechanics 13 (2001/2002) 181–185
IOS Press

A simulation of MPC application for linear motor position control

Keiichi Hirano*, Yongsu Um and Yoshio Kano
Department of Bio-Applications and Systems Engineering Tokyo University of Agriculture and Technology 2-24-16 Nakamachi, Koganei, Tokyo 184-8588, Japan

Abstract. Nonlinear modeling ideas of a Linear Reluctance Motor (Linear RM) as an MIMO system are proposed here. Fast settling time and an interesting input trajectory are performed on Model Predictive Controlled DC motor position simulation. Prospects for the nonlinear MPC simulation and the advantages to apply the method for linear motor controller are discussed.

1. Introduction

Model Predictive Control (MPC) has proved to be a strong method to control large MIMO systems and has gathered a large amount interest in the industry, especially within the process and petrochemical fields. The main benefit is the ability to handle constraints on various variables in the plant. The process input and output constraints are included directly in the problem formulation. So any future constraint violations are anticipated and prevented [1].

In industry for supervisory optimizing control, MPC is often preferred over other controller design methods, such as PID, LQ, and H^∞. PID controller is also easily tuned but can only be applied to lower order SISO systems. LQ and H^∞ can easily be applied to MIMO systems, but cannot handle signal constraints in an adequate way [2]. So we propose a plan to regard a linear motor as an MIMO system and to apply the MPC method for the controller.

The aim of this simulation is to examine the advantages of using MPC method for motor position control in order to apply the method for linear motor controller.

2. Nonlinear modeling ideas of a linear RM

The cross- section of the general structure of the Linear RM we modeled is shown in Fig. 1, which consists of the primary core stator, stator windings and the secondary core mover. The mover drives within the both ends of the stator, horizontally. There are 12 windings, that are described as w1 to w_{12}, in the primary core stator. The coil currents flow vertically on this figure. The unique point of our method is that all the voltage supplies to the windings are controlled individually to drive the linear motor with less energy.

*Corresponding author: Keiichi Hirano, Tel.: +81 42 388 7434; E-mail: spicy@cc.tuat.ac.jp.

1383-5416/01/02/$8.00

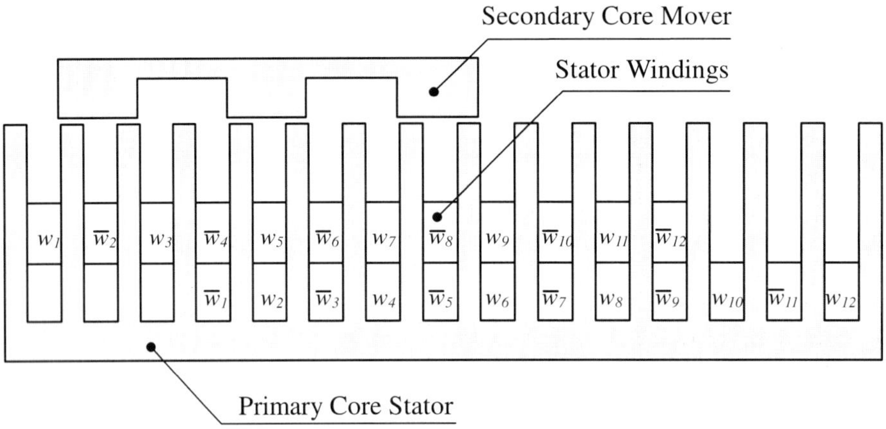

Fig. 1. Cross-section of the general structure of the Linear RM.

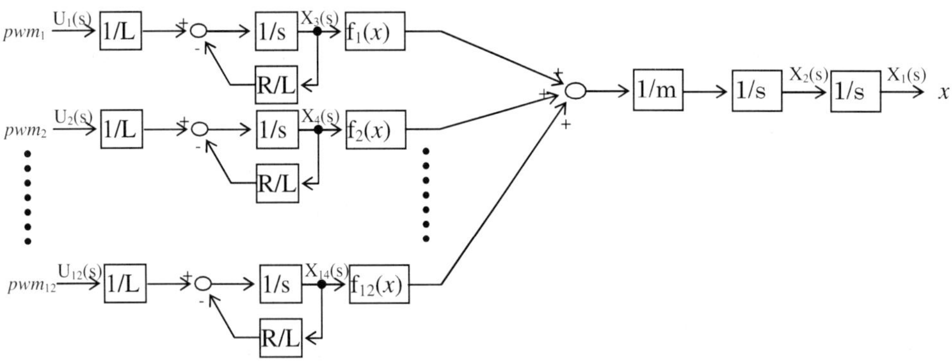

Fig. 2. Block diagram of the Linear RM.

The block diagram of the Linear RM is shown as Fig. 2. In this model, the inputs are the 12 individual voltage supplies, that are shown as pwm1 to pwm_{12} in the figure, to each coil and the output is the mover position, x. The motion force of the mover is not constant and it depends on its position even though the current through an individual coil is constant. So it can be said the relation between the mover position of Linear RM and each winding current is nonlinear. We described this feature as $f_1(x)$ to $f_{12}(x)$ in the block diagram, that are called nonlinear factors. So the state- space model is shown as (1) and (2). We should note here that the nonlinear model is still an experiment and it should also be considered that coil inductance changes according to secondary core position.

$$\frac{d}{dt}\begin{bmatrix} x_1 \\ x_2 \\ x_3 \\ x_4 \\ \vdots \\ x_{14} \end{bmatrix} = \begin{bmatrix} 0\,1 & 0 & 0 & \dots & \\ 0\,0 & f_1(x)/m & f_2(x)/m & \dots & f_{12}(x)/m \\ 0\,0 & -R/L & 0 & \dots & 0 \\ 0\,0 & 0 & -R/L & \dots & 0 \\ \vdots\,\vdots & \vdots & \vdots & \ddots & \vdots \\ 0\,0 & 0 & 0 & \dots & -R/L \end{bmatrix} \begin{bmatrix} x_1 \\ x_2 \\ x_3 \\ x_4 \\ \vdots \\ x_{14} \end{bmatrix} + \begin{bmatrix} 0 & 0 & \dots & 0 \\ 0 & 0 & \dots & 0 \\ 1/L & 0 & \dots & 0 \\ 0 & 1/L & \dots & 0 \\ \vdots & \vdots & \ddots & \vdots \\ 0 & 0 & \dots & 1/L \end{bmatrix} \begin{bmatrix} u_1 \\ u_2 \\ \vdots \\ u_{12} \end{bmatrix} \quad (1)$$

$$[y] = [x] = \begin{bmatrix} 1\,0\,0\,0 \ldots & 0 \\ 0\,0\,0\,0 \ldots & 0 \\ 0\,0\,0\,0 \ldots & 0 \\ 0\,0\,0\,0 \ldots & 0 \\ \vdots\,\vdots\,\vdots\,\vdots\,\ddots & \ddots \\ 0\,0\,0\,0 \ldots & 0 \end{bmatrix} \begin{bmatrix} x_1 \\ x_2 \\ x_3 \\ x_4 \\ \vdots \\ x_{14} \end{bmatrix} \quad (2)$$

where

$f_1(x), f_2(x), \cdots, f_{12}(x)$: nonlinear factors
$u, u_2, \cdots, u_{12}$: voltage supplies
R : electric resistance of a coil in the stator
L : electric inductance of a coil in the stator
m : mass of the mover

3. Simulation structures

DC Motor is modeled for the experimental simulation. The physical parameters of the DC motor are following:

- moment of inertia of the rotor $(J) = 3.2284 * 10^{-6}[\text{kg}^2/\text{s}^2]$
- damping ratio of the mechanical system $(b) = 3.5077 * 10^{-6}$ [Nms]
- electromotive force constant $(K = Ke = Kt) = 0.024$ [Nm/Amp]
- electric resistance $(R) = 4$ [ohm]
- electric inductance $(L) = 2.75 * 10^-6$ [H]

The outline of this simulation is to control the DC motor position from 0 to 1 [rad] with both MPC controller and PID controller to examine the generality that the MPC method settles quickly. This simulation is run on Simulink with MPC toolbox. The DC motor, the plant to be controlled is represented in Matlab S- function. In this simulation, motor controllers simply provide the voltage to the motor and receive the position information from the motor as a feedback.

For the PID simulation, we tuned the PID parameters to $K_p = 17$, $K_i = 600$, $K_d = 0.15$. These parameters are obtained experimentally [3].

MPC simulation block diagram is shown as Fig. 3. To run the MPC simulation, there are some parameters needed during its optimization process. So we set the input horizon to 2, the prediction horizon to 100 and the input constraint to $-10 \leqslant V \leqslant 10$ [V]. The way of the MPC parameter setting is quite different from PID. These are actual levels. Input horizon and prediction horizon are actual numbers of the definitions in the optimization calculation. Also the input constraint is the actual restriction in the calculation.

4. Simulation results and discussions

Input voltage trajectories to the DC motor are shown as (a) in Fig. 4. The MPC trajectory is quite different from the one controlled by the PID in the point of voltage transition speed. This is because the difference in the control method. PID control transition speed does not depend on its controllers processing speed but the MPC transition speed does, unless the MPC transition speed is restricted with

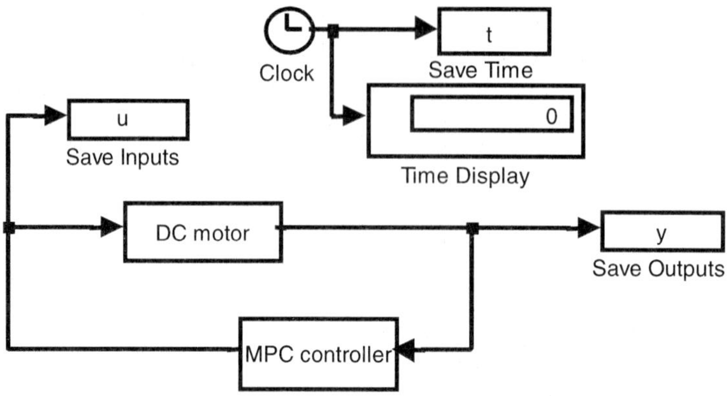

Fig. 3. MPC simulation block diagram.

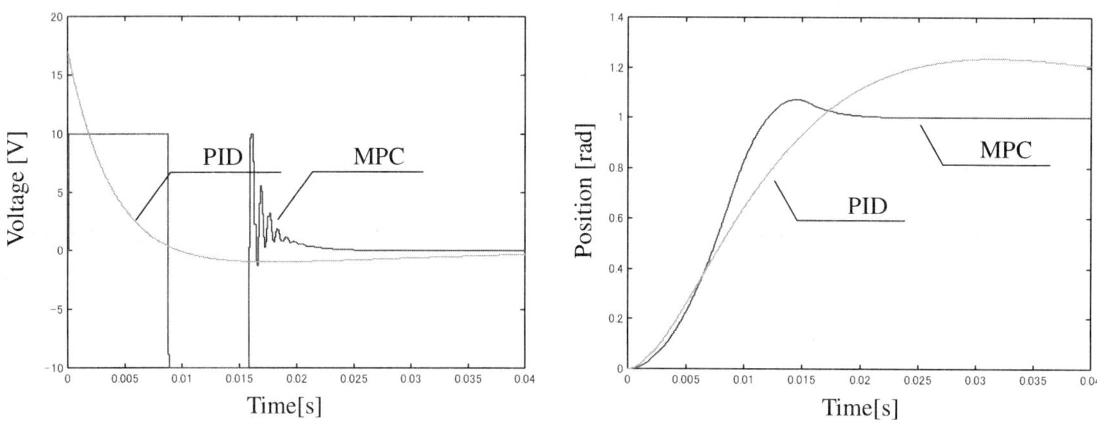

(a) Input voltages to the DC motor (b) Position of the DC motor

Fig. 4. Simulation results of a DC motor position control.

a transition speed constraint. So it can be said that the higher speed processor make the control more reliable and faster. Also we can see the MPC voltage trajectory varies exactly within the input constraint. PID controller can only restrict its proportional parameter as initial voltage input, but the restriction is not a proper way to restrict the input voltage.

Figure 4(b) shows the position trajectories of the DC Motor. The MPC trajectory has a settling time of roughly 20 [ms], and it has less than 10% overshoot. The difference is quite obvious. The PID trajectory has a settling time of roughly 80 [ms] and it has more than 20% overshoot. MPC settling time is quite fast in comparison with the one simulated with the PID controller.

These simulation results prove the high efficiency of applying MPC method to the DC motor controller to control the DC motor position faster. The proof is still only with SISO linear MPC control simulation. In the next section we consider about the prospects for the nonlinear MPC simulation of the Linear RM.

5. Prospects for the nonlinear MPC simulation

Algorithms for nonlinear MPC are often the extensions of linear MPC algorithms, though the number of report applications of nonlinear MPC is still very low [4].

A Linear RM position control using the nonlinear model and the nonlinear MPC method is simulated experimentally, with the way that the nonlinear factors are linearized every sampling period. As the simulation results, the position is actually settled and the multi- voltage- supply is performed, though the position trajectory is still hardly considered that optimized. The simulation program algorithm is still under our discussion.

6. Conclusions

A Linear RM position control using nonlinear MPC method is proposed. The idea consists of two independent studies. One is the nonlinear modeling of the control object, Linear RM. The nonlinear model and the modeling method are proposed. The other is the MPC study. A DC motor position control is simulated with the MPC and PID controller. The simulation results indicate the advantages of applying the MPC method to the SISO system. From the property of the MPC method, there is a possibility that the nonlinear MPC method satisfies the fast settling time demand of any motor control, and it gives greater benefit to linear motor control.

Acknowledgements

We would like to thank Dr. David Howe, Dr. Phil Churn, and Dr. Chris Bingham at the Department of Electronic and Electrical Engineering, University of Sheffield, U.K. for their helpful advice and the use of their simulation equipment.

We would also like to thank Dr. Kevin Greenslade for his assistance.

References

[1] S. Joe Qin and A. Thomas, Badgwell, An Overview of Industrial Model Predictive Control Technology, in: *Chemical Process Control- V*, Vol. 93, No. 316, IChe Symposium Series – American Institute of Chemical Engineers, 1997, pp. 232–256.

[2] J.J. Ton van den Boom, C.P.M. Ton Backx, *Model Predictive Control*, Lecture Notes for the MPC DISC Course, The Netherlands, winter, 1999.

[3] University of Michigan, PID Design Method for the DC Motor Position Control, Control Tutorials for Matlab, 1997.

[4] A. Helbig and W. Marquardt, *An Introduction to Model Predictive* Control, http://helio.diquima.upm.es˜mrod/docs/Mbm_revi.pdf

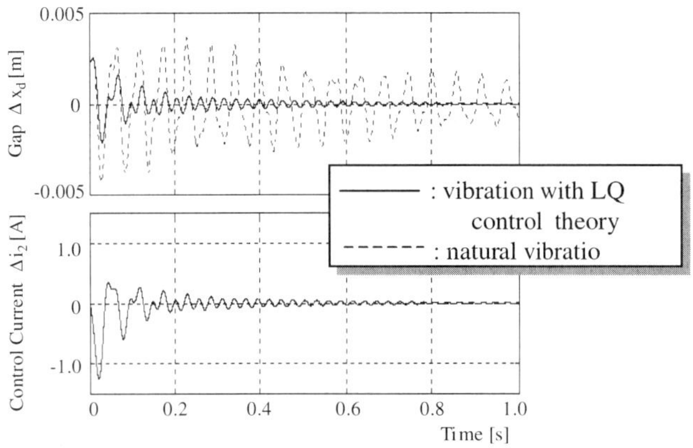

Fig. 6. Control effects for coil 2 (simulation).

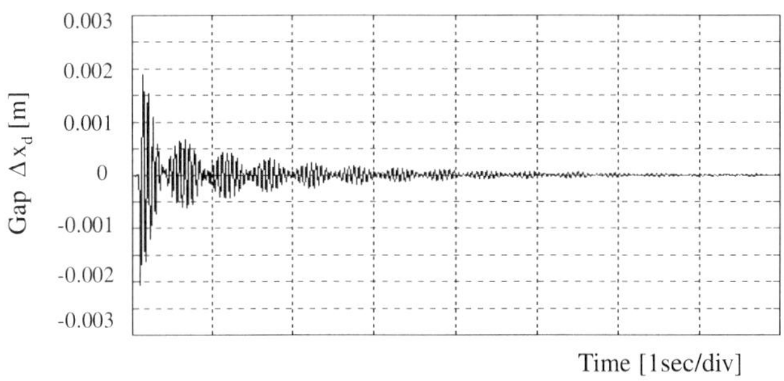

Fig. 7. Vibration of the strip withou control.

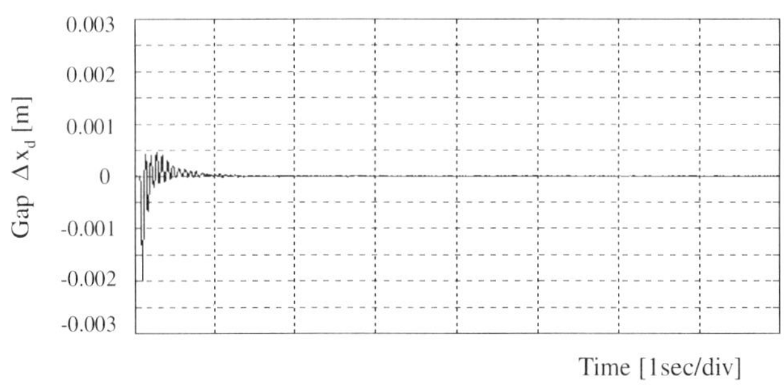

Fig. 8. Experimental result by using both the gap sensors and the current sensors.

4. Experimental device

The experimental device model in this study is shown in Fig. 4. In the strip, the width, length, thickness and the length between the strip and the electromagnet are 800 [mm], 1.2 [mm], 100 [mm] and 5 [mm] respectively. The tension is adjustable with the handle, and the natural frequency of the study can be set freely. The characteristics of the experimental device are shown in Table 1.

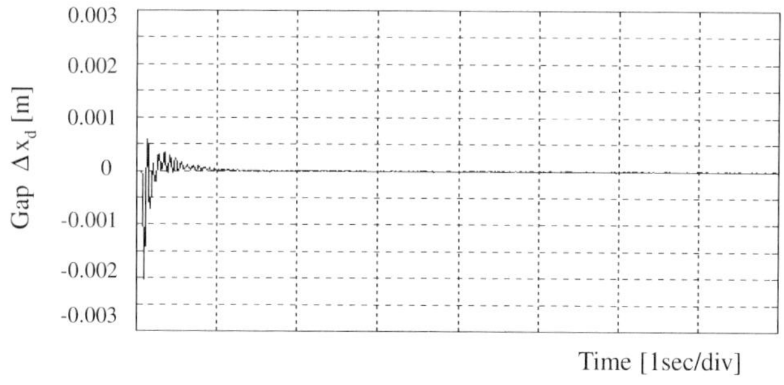

Fig. 9. Experimental result by using only the current sensors.

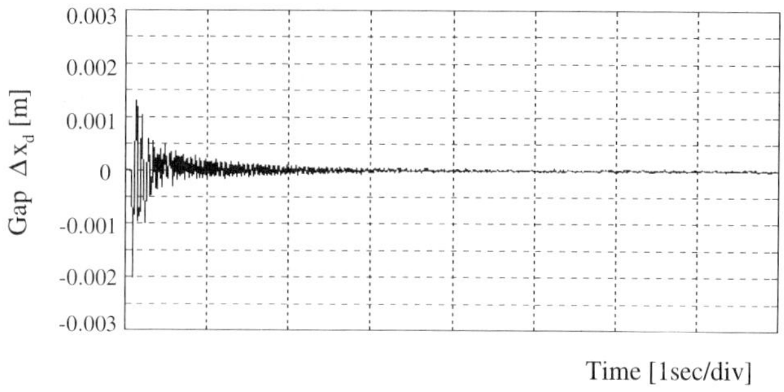

Fig. 10. Experimental result by using the observer sensors.

5. Simulation

The simulations are carried out for the initial displacement 5 [mm] at a point 160 [mm] from the upper end of the strip. The weighting matrixes Q_+, Q_- in Eqs (13) and (14) can be set as Eq. (19) from a viewpoint of control.

$$Q_+ = \begin{bmatrix} 0 & 0 & 0 & 0 \\ 0 & 0 & 0 & 0 \\ 0 & 0 & q_{+1} & 0 \\ 0 & 0 & 0 & q_{+2} \end{bmatrix}, \quad Q_- = \begin{bmatrix} 0 & 0 & 0 \\ 0 & q_{-1} & 0 \\ 0 & 0 & q_{-2} \end{bmatrix} \tag{19}$$

where, it is referred to as $q_{+1} = 100,\ q_{+2} = 100$ and $q_{-1} = 50,\ q_{-2} = 50$ respectively.

When the complete state feedback control is performed, the system is equipped fully with the gap sensors and the current sensors. The displacement of the strip and the control current not including the steady value 1.3 [A] is shown in the Figs 5 and 6.

From these results, it is demonstrated that the vibration of the strip is fully controlled when the feedback control with all the state variables is performed.

6. Experiments and the results

In Chapter 6, it is shown in the digital simulation that the vibration control effects are sufficient. This Chapter shows the vibration control effects by the vibration experiments using the experimental device. The initial condition is added by hammering at the point 160 [mm] from the upper end. The sensing point is the strip center. Figure 7 shows the experimental result of the strip without any control. On the other hand, Fig. 8 shows the result when the state feedback control is made with the gap sensors and the current sensors completely. Furthermore, Fig. 9 shows the result when the equivalent control to Fig. 8 is performed by replacing to the current variable by means of the circuit equations Eqs(3) $\sim$ (7), not using the expensive gap sensors but the cheap current sensors. Fig. 10 is the result when the equivalent control to Fig. 8 is performed by not using the gap sensors but using the current sensors by a observer.

From these results, it is shown that the strip amplitude becomes 1 [mm] peak to peak below and the damping time is also shortened by 1 sec. In the case of Fig. 9, the vibration control effects are the same as in the case of Fig. 8.

7. Conclusion

The following points became clear in this paper.

1. As the interaction among the electromagnets through the strip was theoretically cut by the appropriate mode transform, the simple control could be built.
2. Vibration control of the strip was accomplished by using the LQ control theory.
3. Sufficient vibration control was also obtained by using not the expensive gap sensors but only the cheap current sensors.

References

[1] S. Morii et al., *The development of the strip shape control of the process line,* Mitsubishi Heavy Industries, LTD. Technical report Vol. 32, March 1995.

[2] T. Nakagawa and T. Tanaka, *A study on vibration control and magnetic characteristic of the strip by using current sensors,* The 11th symposium on electromagnetics and dynamics October, 1999.

[3] Z. iwai, A. Inoue and S. kawaji, *Observer,* Corona Publishing co, LTD. October 1988.

International Journal of Applied Electromagnetics and Mechanics 13 (2001/2002) 195–202
IOS Press

Magnetic field standards – tracing to the maintained units

Kurt Weyand
Physikalisch Technische Bundesanstalt, Bundesallee 100, D-38116 Braunschweig, Germany
Tel.: +49 531 592 2240; Fax: +49 531 592 2154; E-mail: kurt.weyand@ptb.de

Abstract. The present state of realization, maintenance and dissemination the quantity "magnetic flux density" in SI units is discussed. Different measuring methods that are in use for the calibration of field coils and permanent or electromagnetic systems in the flux density range from the order of the Earth's magnetic field up to some Tesla are compared with respect to their uncertainties. Special attention is given to systems based on the gyromagnetic ratio of protons as a fundamental constant.

1. Introduction

For the realization and the maintenance of the flux density unit, field coils (with or without magnetic yoke) have to be regarded as the most accurate setups, if the field is controlled by Nuclear Magnetic Resonance. The application of the gyromagnetic ratio γ'_P of protons, as recommended by the Committee on Data for Science and Technology, CODATA [1], provides the highest accuracy. Nevertheless, careful checks are needed if the NMR probes differ from spherical container filled with pure water as specified together with the recommended γ'_P value.

A method of determining the coil constant of a particular arrangement of winding packages, which can also be connected to produce an inhomogeneous field, is to measure the field profile on its axis and to calculate its geometrical dimensions from the recorded profile. This method makes it possible to trace the coil constant directly to the length standard i.e., to a base quantity in the international system of units, the SI. The coil constant shall be considered as the quotient of the centre flux density over the coil current.

In cases of lower accuracy requirement the coil constant may be found by superposing the field of the coil under test to that of a NMR-based coil. Using this simple method the unknown coil constant is traced to the well known by measuring a current ratio.

Ac-field standards can be provided by means of various field coils, their dc-coil constant, too, being determined by means of NMR or being related to an NMR based coil by means of field superposition. Afterwards, the frequency dependence of the coil constant is checked by means of a simple but very reliable inductive measuring technique [2].

2. Role of fundamental constants

The gyromagnetic ratio of protons γ_P – which is the quotient of the proton magnetic moment μ_P over its angular momentum $1/2\hbar$–can be regarded as the earliest fundamental constant used as a quantum standard in the field of electromagnetic quantities. The value of γ'_P is determined by measuring the angular precession frequency $\omega_p = 2\pi f_p$ of the protons in aqueous samples in a magnetic field with the flux density B according to $\omega_p = \gamma_P B$. The prime indicates that the precession is observed using spherical samples of pure water i.e., the protons are diamagnetically shielded and the demagnetisation factor is well known. In general, there are two methods of determining the flux density B for a γ_P-experiment:

- B can be quantified by measuring the force F on a straight wire of length l carrying the current I according to $B = F/lI$ – B has to be then generated in the order of at least 0.5 T by means of an electro- or a permanent magnet to ensure a force, which is large enough to be measured with sufficient accuracy.
- B is calculated from the geometrical dimensions of an arrangement of circular turns carrying the current I according to $B = \mu_0 k_c I$, where k_c is the so-called coil constant of the turn arrangement with the dimension of m^{-1}. In this case, the flux density is limited to about 1 mT in order to avoid instability of k_c due to self-heating of the coil. If the high-field method is applied γ_P is linearly proportional to the current, whereas in the case of the low-field method γ_P is inversely proportional to I. The discrepancy between the γ_P values resulting from both methods gave for the first time rise to suspicion that the electrical base unit in the SI, the Ampere, might be erroneous.

This suspicion was confirmed to the same extent when experiments determining other fundamental constants, especially the Josephson constant and the von Klitzing constant, were increasingly becoming more accurate. It was finally proved within the 1986 adjustment of fundamental constants, that the unit of electrical current as maintained by the standard laboratories has to be enlarged by about 6 ppm. Figure 1 shows the γ_P – values and their relative standard uncertainties which in the past were evaluated by the various least square adjustments of fundamental constants. The gap between the dashed lines indicates the amount contributed by the error in the Ampere. Whereas for the1963 and 1968 adjustments [3,4] the direct γ_P determinations were dominant with respect to the finally adjusted value, this changed over the following years. The 1973 and 1986 adjustments [5,6] resulted in an adjusted value with an uncertainty of the same order as the most accurate direct determinations performed up to the respective date. It should be noted that after the last adjustment in 1998 [1] γ_P can be evaluated from other fundamental constants by at least a factor of 3 more accurate than it has ever been directly determined. This results from the fact that it is difficult to achieve sufficient precision when measuring the geometrical dimensions of current paths, which either are generating a magnetic field or which are affected by a magnetic field, what is needed to perform a direct determination as outlined above.

The Eqs (1) and (2) present two possible ways (among others) for evaluating γ_P from other fundamental constants. The meaning of the terms inside the braces is explained in Table 1, each term represents a quantity the uncertainty of which – column 4 in Table 1 – is considerably lower than estimated to be some parts in 10^{-7} for the most accurate direct γ_P determinations [7–9].

$$\gamma_p = \left\{\frac{h}{2m_e}\right\}\left\{\frac{\mu_p}{\mu_B}\right\}\left\{\frac{2e}{h}\right\} \tag{1}$$

$$\gamma_p = \frac{\mu_0^2 c_0^3}{16R_\infty}\left\{\frac{e^2}{h}\right\}^2\left\{\frac{\mu_p}{\mu_B}\right\}\left\{\frac{2e}{h}\right\} \tag{2}$$

The adjusted value of γ'_P as recommended by CODATA amounts to

$$\frac{\gamma'_p}{2\pi} = \left\{\frac{2\mu'_p}{h}\right\} = 42\,576\,388.8\,(1 \pm 4.2 \cdot 10^{-8})\mathrm{Hz/T} \tag{3}$$

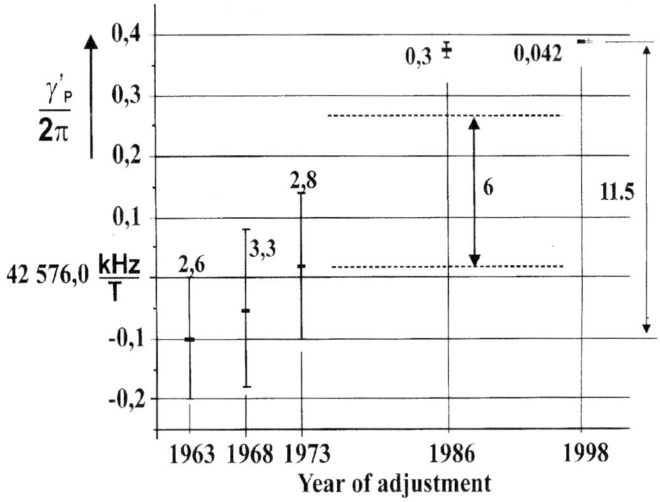

Fig. 1. Proton gyromagnetic ratio after the least square adjustments of fundamental constants in the past, protons in a spherical H_2O sample, relative standard uncertainties in ppm.

Table 1
List of the fundamental constants, from which γ'_P may be evaluated, and their relative standard uncertainties according to the 1998 least square adjustment

Symbol	Quantity	Units	Relative standard uncertainty
$\{h/2m_e\}$	quantum of circulation	m^2/s	$< 10^{-8}$
$\{\mu_P/\mu_B\}$	proton magnetic moment in Bohr magnetons		10^{-8}
{h/2e}	magnetic flux quantum	Vs	$4 \cdot 10^{-8}$
$\{h/e^2\}$	quantized Hall resistance	Ω	$< 10^{-8}$
μ_0	magnetic constant	Vs/Am	exact
c_0	speed of light	m/s	exact
R_∞	Rydberg constant	1/m	$< 10^{-11}$

where the prime indicates that this value holds for protons in a spherical sample of H_2O having a temperature of 25°C [1]. As it can be taken from the table, its uncertainty is dominated by that of the flux quantum. Without any doubt, tracing magnetic units to the SI base units is at present provided on the lowest level of uncertainty by applying proton magnetic resonance and using the recommended sample.

3. NMR based field standards

In order to apply NMR magnetometry the most common method is to observe the effect of energy absorption caused by the NMR resonance by measuring the damping of the tuned circuit of a marginal oscillator. If a small alternating field is superposed on the field to be measured, a quasicontinuous signal can be obtained. However, this very sensitive method can not be applied using samples containing pure water, because of its narrow line width which is in the order of 0.1 Hz. Modified sample substances are therefore in use, for instance natural gum or water which is doped by a small amount of salt containing paramagnetic ions. In this way not only the spin-spin relaxation time is reduced resulting in a widening of the absorption line width but also a shift in precession frequency is observed compared with that in pure H_2O. Other sources which possibly produce a frequency shift are a sample shape differing from the

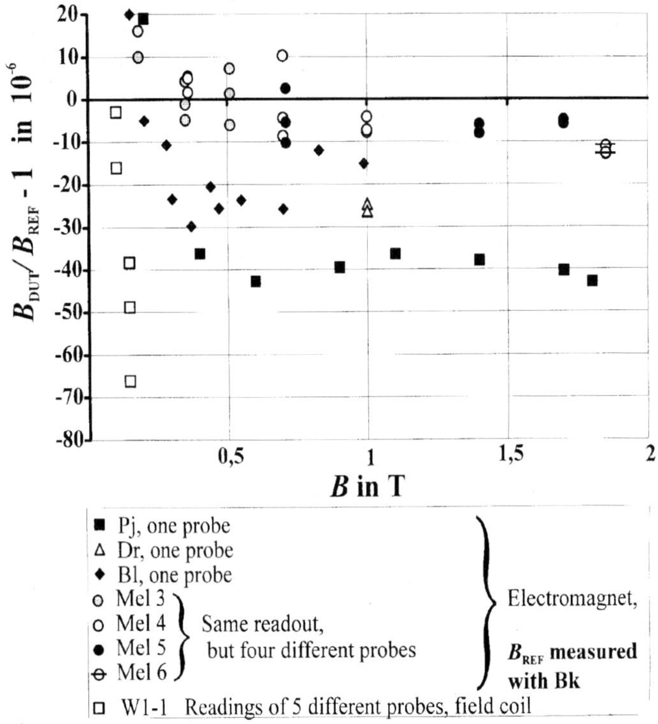

Fig. 2. Readings of various NMR magnetometers under test, BDUT, compared with the reading of an instrument declared as the reference at fields with the same flux density, BREF.

spherical resulting in a different demagnetisation factor and probe materials which are not magnetically neutral. The latter might be regarded as the most severe problem: the measurements with four different spherical samples containing pure H_2O which were fabricated with high accuracy by three standard laboratories have yielded relative differences of the precession frequency up to 5 parts in 10^{-7} when they were placed in the same field coil with constant flux density [7]. The problematic phenomenon ascribed to the probes became even more evident after a comparison of five commercially available magnetometers, all of them having been specified to be accurate within some parts in 10^{-6}. As can be seen from the measurement results in Fig. 2 – where one instrument was declared as the reference–the readings taken with the various readouts and probes differ up to $7 \cdot 10^{-5}$. The uncertainty of these measurements was estimated to be less than one part in 10^{-5}.

Figure 3 illustrates the situation prevailing at present at our laboratory with respect to the maintenance and dissemination of the flux density unit as a dc quantity. NMR-based calibration can be provided over a field range from 10 μT to 2.5 T corresponding to resonance frequencies f_P from about 400 Hz to 100 MHz if proton samples are used as indicated by the second axis of abscissa.

The top horizontal band shows the method by which NMR is observed in the particular field range. The next band refers to the availability of the instruments in use, and the last band indicates in which way the magnetic field is generated. For the range up to 200 mT, various field coils have been built. The lowest relative uncertainty is achieved at fields in the order of 1 to 2 mT, because in this range a γ'_P experiment was performed [10]. For fields down to 10 μT the free precession method with prepolarization of an H_2O sample has to be applied in order to increase the NMR signal level. The uncertainty increases with decreasing flux density for two major reasons: the contribution of fluctuating extraneous fields becomes

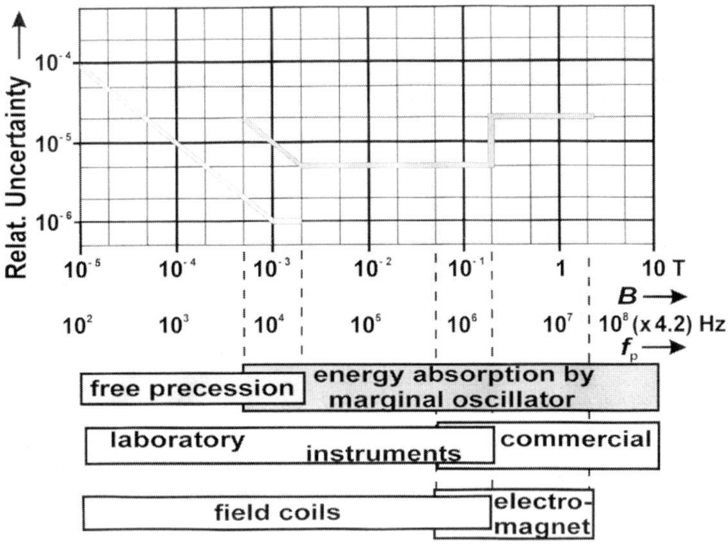

Fig. 3. Field range, uncertainties and other parameters of the calibration by means of NMR methods at PTB.

more significant and the resolution in the frequency measurement cannot be increased by the same order as the precession frequency decreases. Besides, the field across the sample has to be as homogeneous as necessary to provide an observation time of at least 1 s for the resonance frequency measurement.

The method of observing NMR resonance by damping the tuned circuit of a marginal oscillator is more convenient. Therefore, marginal oscillators have been further developed to increase their resolution with the result of a minimum detectable field at present in the order of 0.5 mT [11]. The gap between the uncertainty lines of the two NMR methods in Fig. 3 is due to the uncertainty in relating the samples doped with $CuSO_4$ to a pure water sample. There is another step in uncertainty at 0.2 T which marks the transition where the field must be generated by an electromagnet and where commercially available magnetometers are used.

In conclusion, it should be observed that the uncertainty in disseminating the quantity of magnetic flux density by means of proton resonance depends more on imperfection of the probes in use than on the state of the knowledge of the proton gyromagnetic ratio.

4. Tracing the flux density to current and length

The coil constant k_c of a field standard can be directly traced to the length standard by measuring the field profile on its axis and calculating its geometrical dimensions from the recorded profile. To apply this technique a coil arrangement producing a rapidly changing field profile is of advantage, because the portions of the field distribution which include the points of inflection are most informative with respect to the dimensions of the current paths generating the field [12]. This method is therefore well suited for coils made of winding packages. The principle of the method can be explained by a simple example. A winding package with N turns carrying the current I and distributed over a cross section, the edges of which are small compared with its mean diameter $2r$, generates the flux density B_x along the x-axis related to its value B_0 at $x = 0$ according to $B_x/B_0 = \{1 + (x/r)^2\}^{-3/2}$ as follows from Biot-Savart's law. The coil constant $k_C = \mu_0 N/(2r)$ can be found by measuring the fields in the centre plane and at

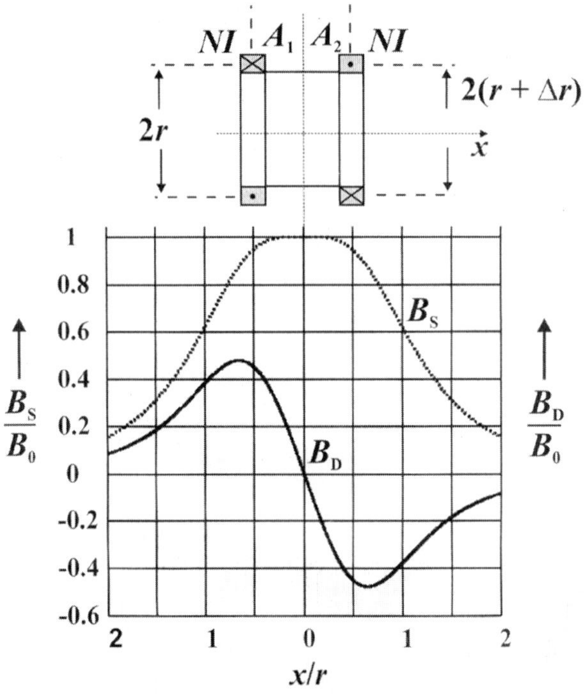

Fig. 4. Field profiles generated by symmetrically arranged winding packages, B_S windings series connected, B_D windings counterconnected related to the central field B_0 with the windings series connected.

an additional position x on the axis according to

$$k_c = \frac{\mu_0}{2x} N \left\{ \left(\frac{B_0}{B_x} \right)^{2/3} - 1 \right\}^{1/2} \tag{4}$$

provided the number of turns is known. Very good results are obtained for symmetrically arranged packages such as Helmholtz coils. In this case the profile is measured with the packages counterconnected and related to the centre field with the packages series-connected, which means the centre region of the field function becomes most informative as shown by Fig. 4.

Again, provided that the winding cross sections are small, the coil constant depends mainly on four geometrical parameters: the mean radii $r, r + \Delta r$, and the mean distances A_1, A_2 from the field profile zero crossing point. The parameters are to be determined by least square fitting a model function to the measured profile. Provided a magnetometer sensor well aligned to the coil axis, the ratio $\hat{B}_{\text{noise}}/(2B_0)$ has been found to be an important quality factor of this method, where B_{noise} represents fluctuations of extraneous fields but also includes the instability of the current through the coil and any instability of position. Using a fluxgate magnetometer with a sensor of 5 mm length and an optical position measuring system providing $1\mu\text{m}$ resolution, k_c of a Helmholtz coil of 30 to 40 cm in diameter can be determined with a relative uncertainty of less than 10^{-4}. Applying a SQUID magnetometer for the field and a laser interferometer for the position measurement a relative uncertainty of $5 \cdot 10^{-6}$ can be a chieved.

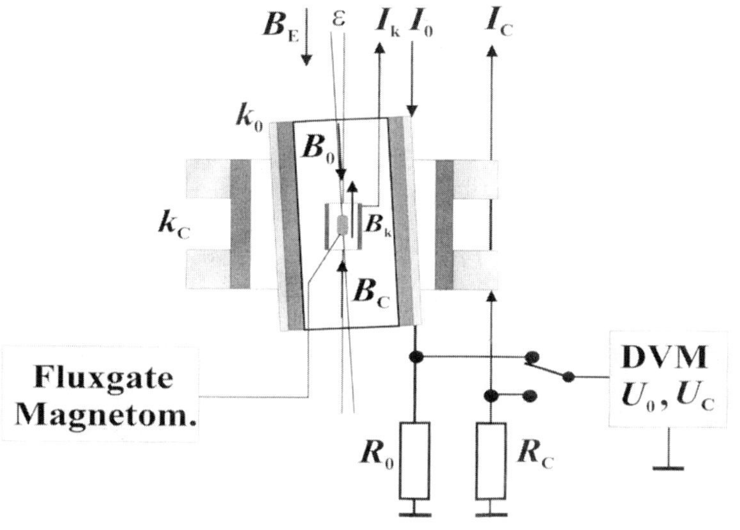

Fig. 5. Relating an unknown coil constant k_C to k_0 of a standard coil by means of superimposing their central fields B_C and B_0 in opposite directions.

5. Calibration by field superposition

For the sake of completeness, in case of lower accuracy requirement the dissemination of the flux density unit by means of field superposition is to be mentioned. The coil constant k_C of a field coil under test may be found by superimposing its field on that of a NMR-based coil with the constant k_0. With an arrangement of the coils as outlined in Fig. 5, where the fields B_C and B_0 generated in both coils in opposite direction by the currents I_C and I_0 and with a fluxgate magnetometer as a zero field indicator, the unknown coil constant is determined by means of a current ratio according to $k_c I_c = k_0 I_0 \cos\varepsilon$. If extraneous fields are compensated by means of B_k produced by a small coil fixed to the magnetometer sensor and if the currents I_C and I_0 are alternately measured by the same instruments, then, the angle ε between the coil axes is to be regarded as the main error source. Depending on field range and coil design, k_C can be determined with a relative uncertainty of 10^{-3} or even less; if k_C is close to k_0 the uncertainty of comparison may be in the order of 10^{-4}.

6. Conclusions

The value of the gyromagnetic ratio of the proton as recommended after least square adjustment of fundamental constants has proved to be more accurate than it had ever been directly determined in a magnetic field which was evaluated either by measuring a force or by measuring the geometrical dimensions of a current path. For that reason, the application of proton magnetic resonance is to be regarded as the most accurate method of maintaining the unit of magnetic flux density and of relating it to the base units of length and electrical current within the SI. Nevertheless, the possible high accuracy is only achieved if the proton resonance is observed using pure water contained in a spherical sample shape and if the magnetic properties of the probe material are known. Applying probes differing from the recommended one leads to a loss of accuracy by about one to two orders. Roughly the same level of accuracy can be achieved using particular field coils, if their geometrical dimensions are determined from their axial field profile. However, a relatively sophisticated technical equipment is needed to perform these measurements. An accuracy that might be sufficient in many cases is attained merely by relating the constant of a field coil under test to a standard coil by means of superimposing the fields generated in the centre of both coils. These statements were confirmed by an international intercomparison within EUROMET where the coil constant of a standard coil was determined by eight European standard institutes, whereby quite different measuring methods were applied [13].

References

[1] P.J. Mohr and B.N. Taylor, CODATA Recommended Values of the Fundamental Physical Constants 1998, *J. Phys. Chem. Ref. Data* **26** (1999), 1213–1852.

[2] K. Weyand, Maintenance and Dissemination of the Magnetic Field Unit at PTB, *IEEE Trans. Instrum. Meas.* **50** (April, 2001).

[3] E.R. Cohen and J.W.M. DuMond, *Rev. Mod. Phys.* **37** (1965), 537.

[4] B.N. Taylor, W.H. Parker and D.N. Langenberg, *Rev. Mod. Phys* **41** (1969), 375.

[5] CODATA Bulletin No.11, *Recommended consistent values of the physical constants, 1973*, Committee on Data for Science and Technology, Paris, 1973.

[6] E.R. Cohen and B.N. Taylor, *The 1986 Adjustment of the Fundamental Physical Constants*, CODATA Bulletin, No. 63, Pergamon Press, 1986.

[7] E.R. Williams, G.R. Jones, S. Ye, R. Liu, H. Sasaki. P.T. Olsen, W.D. Phillips and H.P. Layer, A Low Field Determination of the Proton Gyromagnetic Ratio in Water, *IEEE Trans. Instrum. Meas.* **38** (April, 1989), 233–237.

[8] B.P. Kibble, G.J. Hunt, A measurement of the gyromagneticratio of the proton in a strong magnetic field, *Metrologia* **15** (1979), 5–30.

[9] R. Liu et al., A Recent Determination for the SI Value of γ_P and $2e/h$ at NIM, *Acta Metrol. Sin* **16** (1995), 161–168.

[10] K. Weyand, First Results of the γ_P Experiment at the PTB, *IEEE Trans. Instrum. Meas.* **34** (June, 1985), 167–173.

[11] K. Weyand, Magnetometer Calibration Setup Controlled by Nuclear Magnetic Resonance, *IEEE Trans. Instrum. Meas.* **48** (April 1999), 668–671.

[12] K. Weyand and G.-D. Willenberg, Improved Determination of the Coil Constant in the γ_p' Experiment at the PTB, *IEEE Trans. Instrum. Meas.* **36** (1987), 275–279.

[13] K. Weyand et al., Intercomparison of Magnetic Flux Density by means of Field Coil Transfer Standard, *Metrologia* **38**(3) (2001), 187–191.

International Journal of Applied Electromagnetics and Mechanics 13 (2001/2002) 203–208
IOS Press

Magnetic hysteresis modelling for vector magnetic property

Masato Enokizono*
Faculty of Engineering, Oita University,700 Dannoharu, Oita 870-1192, Japan

Abstract. This paper presents a numerical modeling of vector (two-dimensional) magnetic properties considering hysteresis of silicon steel sheets. We define the magnetic reluctivity tensor newly and calculate it from the measurement data obtained by two-dimensional magnetic measurement. Our new modelings, which are "E&S Model" and "E&S2 Model", can be introduced into a finite element formulation for magnetic field analysis and can get the distribution of the magnetic field strength H-vector and the magnetic flux density B-vector.

1. Introduction

We can classify the magnetic measurement methods from the following two points of view, as far as magnetic properties of magnetic materials are considered. One is the relative evaluation of magnetic materials such as the Epstein testing and the Single Sheet Tester (SST). The other one is the general behavior for using these materials. The former requires the stability of the measured value, i.e. high reproducibility, as the IEC standard. The latter is located in the application side, and accurately measured values under various conditions are required. These form the magnetic property of arbitrary direction; rotational magnetic property and DC biased magnetic property. As a natural consequence, the general behaviors measuring method is different from the former measuring method.

We should call the measurement method refering to the general behavior "Two-Dimensional Magnetic Measurement" or "Vector Magnetic Measurement", and the conventional measurement method "One-Dimensional Magnetic Measurement" or "Scalar Magnetic Measurement". Especially, the latter measurement data are necessary for recent magnetic field numerical analysis such as the finite element method. Generally, it was found that the magnetic characteristics express the relationship between the B vector and the H vector, and do not only refer to the relationship of the magnitude but also to the nonlinear these quantitive phase angle [1]. Though we showed this relation as the tensor form, it was insufficient without taking into account the hysteresis. In this paper, we show two kinds of modelings as a vector magnetic hysteresis model, that is, "Rotational Flux Model [2]" and "E&S (Enokizono & Soda) Model [3]".

*Corresponding author: Masato Enokizono, Tel.: +81 975 54 7821; Fax: +81 975 54 7822; E-mail: menoki@cc.oita-u.ac.jp.

1383-5416/01/02/$8.00

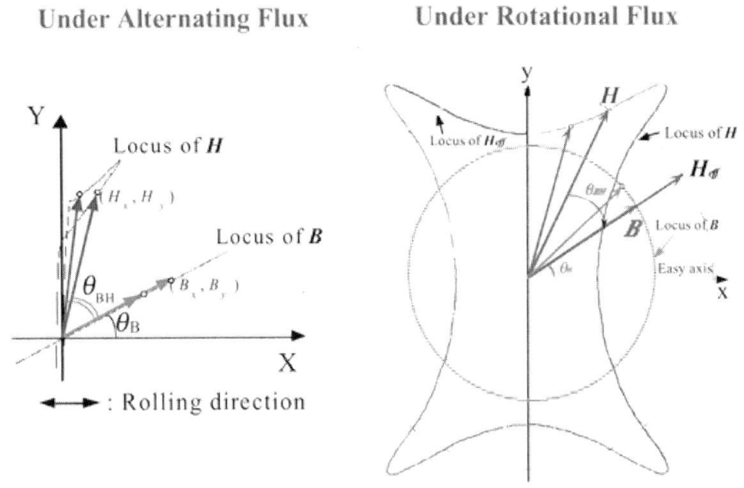

Fig. 1. Locus of H-vector and B-vector.

2. E&S and E&S^2 model

A. E&S model and fundamental equation

The relationship among the maximum magnetic flux density B_m, the inclination angle θ_B and the axis ratio $\alpha(= B_{\min}/B_{\max})$ under the alternating and the rotating flux condition were shown in Fig. 1.

From this vector magnetic property, we define the relationship between the B-vector and the H-vector considering both alternating and rotating hysteresis and B_x and B_y are expressed as follows

$$\begin{cases} H_x = v_{xr}B_x + v_{xi}\dfrac{\partial B_x}{\partial t} \\ H_y = v_{xr}B_y + v_{yi}\dfrac{\partial B_x}{\partial t} \end{cases}, \quad v_k = f_k(B, \theta_B), \quad k = xr, xi, yr, yi$$
$$B_x = B_{xm}\cos(\omega t + \phi_x), \quad B_y = B_{ym}\cos(\omega t + \phi_y), \tag{1}$$

where v_{xr}, v_{xi}, v_{xr} and v_{xi} are obtained from the measurement data. In this case, the data must be measured under the control of complete sinusoidal flux condition by two-dimensional excitation. The, B_{xm}, B_{ym} is the each maximum value of B_x, B_y.

Substituting Eq. (1) into Maxwell's equations for a two-dimensional quasi-static magnetic field, we can obtain the following equation with the vector potential $A(= A_z)$:

$$\frac{\partial}{\partial x}\left(v_{xr}\frac{\partial A}{\partial x}\right) + \frac{\partial}{\partial y}\left(v_{xr}\frac{\partial A}{\partial y}\right) + \frac{\partial}{\partial t}\left\{\frac{\partial}{\partial x}\left(v_{yi}\frac{\partial A}{\partial x}\right) + \frac{\partial}{\partial y}\left(v_{xi}\frac{\partial A}{\partial y}\right)\right\} = -J_0, \tag{2}$$

where J_0 is the exciting current density.

B. E&S^2 model and fundamental equation

On the other hand, we can define the integral form for the hysteresis modeling as follow:

$$\begin{cases} H_x = v_{xr}B_x + v_{xi}\int B_x dt \\ H_y = v_{xr}B_y + v_{yi}\int B_y dt \end{cases}, \quad v_k = f_k(B, \theta_B), \quad k = xr, xi, yr, yi$$
$$B_x = B_{xm}\cos(\omega t + \phi_x), \quad B_y = B_{ym}\cos(\omega t + \phi_y), \tag{3}$$

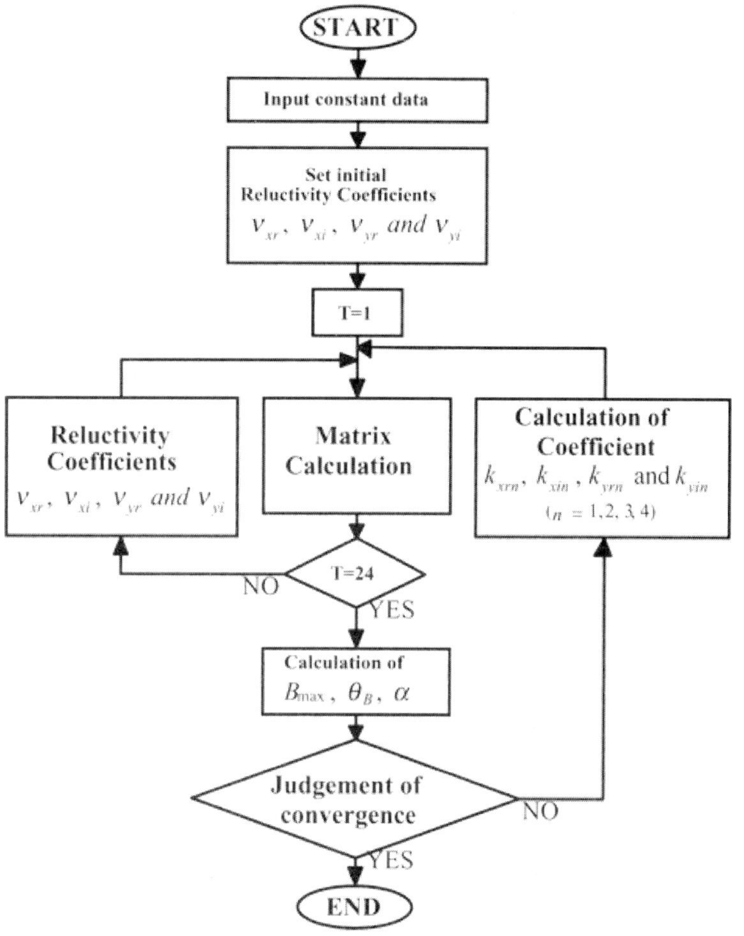

Fig. 2. Flow chart for magnetic field analysis.

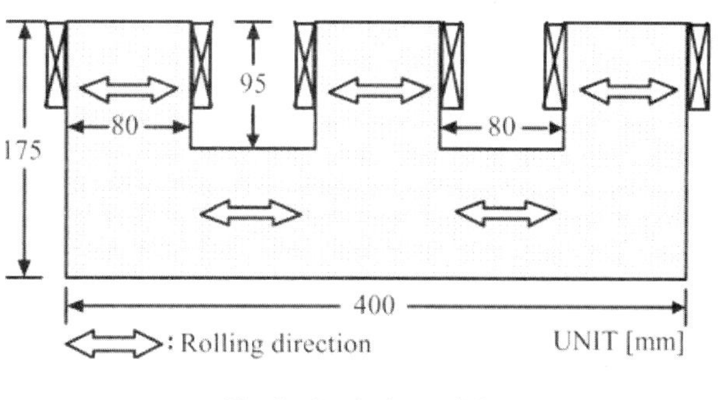

Fig. 3. Analysis model.

Then, the fundamental equation introduced Eq.(4) is expressed as follows:

$$\frac{\partial}{\partial x}\left(v_{xr}\frac{\partial A}{\partial x}\right)+\frac{\partial}{\partial y}\left(v_{xr}\frac{\partial A}{\partial y}\right)+\int\left\{\frac{\partial}{\partial x}\left(v_{yi}\frac{\partial A}{\partial x}\right)+\frac{\partial}{\partial y}\left(v_{xi}\frac{\partial A}{\partial y}\right)\right\}dt=-J_0, \tag{4}$$

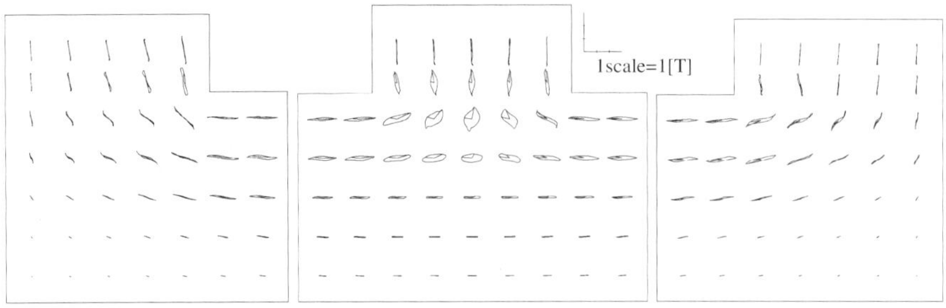

Fig. 4. Distribution of B-vector.

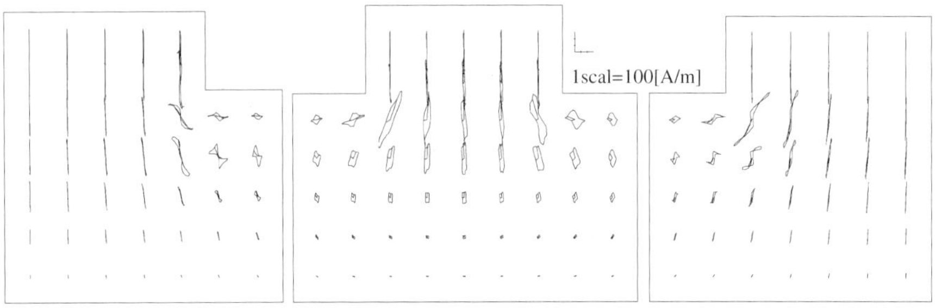

Fig. 5. Distribution of H-vector.

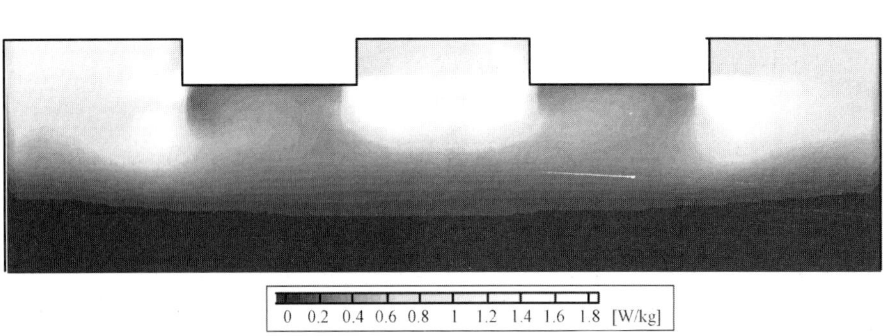

Fig. 6. Distribution of iron loss.

where the components of the reluctivity coefficients v_{xr}, v_{xi}, v_{yr} and v_{yi} have non-linearity as shown in Eqs (1) and (3). The coefficient v_{xr} and v_{yr} are named "magnetic reluctivity coefficient", v_{xi} and v_{yi} "magnetic hysteresis coefficient". Those coefficients are functions of B and θ_R, and are obtained by reference [4].We can carry out the non-linear magnetic field analysis considering both the alternating and the rotating hysteresis by using these models.

Next, we calculate the distribution of iron loss by using the distribution the H-vector and the B-vector in the iron cores. The iron loss can be calculated directly from the analyzed results by the following equation:

$$P_t = \frac{1}{\rho T} \int_0^T \left(H_x \frac{dB_x}{dt} + H_y \frac{dB_y}{dt} \right) dt \quad \text{[W/kg]}, \tag{5}$$

where T is one period of the exciting waveform ρ is the material density. The largest advantage of this

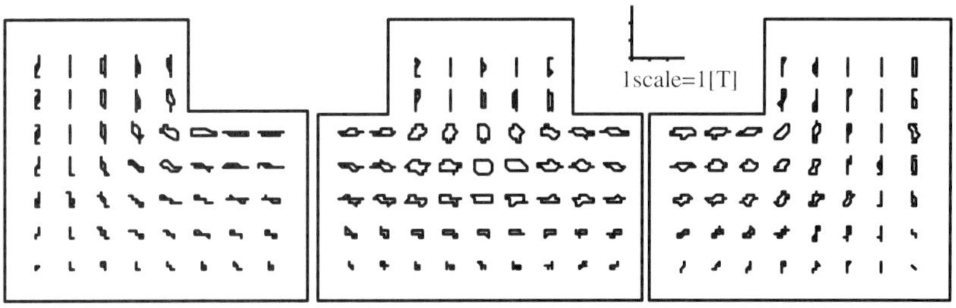

Fig. 7. Distribution of loci of B-vector.

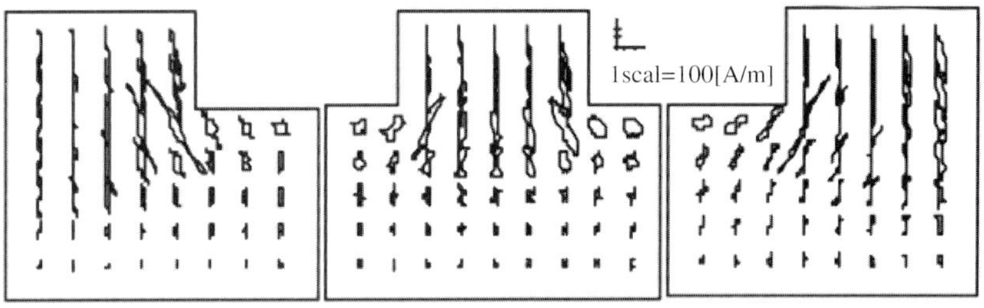

Fig. 8. Distribution of loci of H-vector.

magnetic field analysis method is to be able to calculate the loss directly. It was not possible to obtain the direct loss analysis results until now.

Figure 2 shows the flow chart for this analysis procedure.

3. Analytical results by improved finite element method with vector hysteresis model

We applied the expression presented here to a three-phase transformer core model, which is for the evaluation as shown in Fig. 3. Cutting off the unnecessary part made this core model from rectangular sheets. The specimen used in this study is a grain-oriented steel sheet (23ZDKH90 produced by NSC, 0.23 mm thickness). Since the core has symmetry, half of the region is analyzed by using the improved finite element method considering E&S Model.

Figures 4 and 5 show the distribution of loci of calculated B-vector and H-vector, respectively. It can be found that our analyzed results are unsymmetrical with respect to the symmetric axis at the central part in the core. These results are very different from the conventional one. Furthermore magnetic properties for the clockwise and anti-clockwise rotation of B-vector and H-vector were different due to the rotational hysteresis. Because the rotating direction of the rotating flux is clockwise in the left side in the central part and it's anti-clockwise in the right side, the both results become to be unsymmetrical. Furthermore, it is seen that the spatial phase angle exists between H-vector and B-vector. It is the most noticing point in this analytical result. It was not possible to obtain such characteristic behavior of H-vector and B-vector in the core by the conventional method. It is proved well that these results get a good agreement with the experimental result as shown in Figs 7 and 8. From these results the distribution of B-vector of Fig. 4 and H-vector of Fig. 5, and the distribution of iron loss is calculated by Eq. (5) are obtained. Figure 6 shows the distribution of iron loss and the comparison with measurement result as shown in Fig. 9.

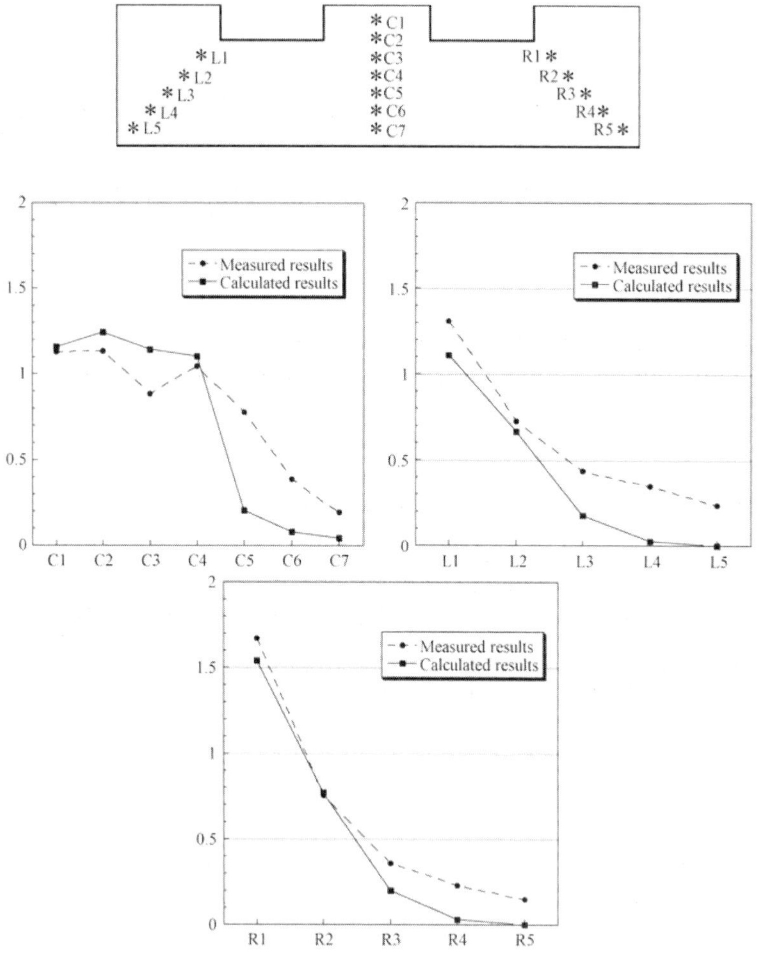

Fig. 9. Comparison of iron loss between analytican resul and measured result.

4. Conclusion

In this paper, two kinds of new modeling for vector magnetic properties for nonlinear iron loss analysis has been presented. We can analyze a three-phase transformer core model considering both alternating and rotating hysteresis. Moreover, it can be said that the presented method is very useful to analyze electrical machines and apparatus considering vector magnetic properties.

References

[1] M. Enokizono, T. Todaka, S. Kanao and J. Sievert, Two-dimensional Magnetic Properties of Silicon Steel Sheet Subjected to a Rotating Field, *IEEE Trans. on Magnetics* **29**(6) (1993), 3550–3552.

[2] M. Enokizono and N. Soda, Finite Element Analysis of Transformer Model Core with Measured Reluctivity Tensor, *IEEE Tran. on Magn.* **33**(5) (1997), 4110–4112.

[3] M. Enokizono and N. Soda, Direct Magnetic Loss Analysis by FEM Considering Vector Magnetic Properties, *IEEE Trans. on Magnetics* **34**(5) (1998), 188–195.

[4] N. Soda and M. Enokizono, E&S Hysteresis Model for Two-dimensional Magnetic Properties, *Journal of Magnetism and Magnetic Materials* **215–216** (2000), 626–628.

International Journal of Applied Electromagnetics and Mechanics 13 (2001/2002) 209–214
IOS Press

Dynamic magnetic hysteresis modelling using Elman recurrent neural network

Mahmoud Saghafifar, Anjan Kundu and Andrew Nafalski*
University of South Australia, School of Electrical and Information Engineering, Mawson Lakes, 5095, South Australia

Abstract. Dynamic magnetic hysteresis modelling is of crucial importance in determining the electromagnetic behaviour of magnetic cores. Among the different models of magnetic hysteresis, Preisach model is one of the more commonly applied. The dynamic Preisach model is a generalisation of the static Preisach model, and is obtained by adding the rate of change of the output variable, which makes the numerical implementation more complex.

Since dynamic hysteresis models require mapping of time dependent sequences, a feature available in some neural networks, in this paper the application of recurrent neural networks to model dynamic hysteresis was investigated. An Elman neural network was selected to test the concept. The network was trained with a set of measured dynamic data and was tested with another set of experimental data, showing acceptable accuracy.

1. Introduction

Dynamic modelling of a magnetic core is a challenging task due to the nonlinear characteristic of magnetic materials and complicated mechanism of core losses. In models of static magnetisation phenomenon, the output (magnetic flux density B) is only a function of the input (magnetic field strength H) and the past input extrema (input history). This means that static magnetisation modelling requires sequential mapping, which is independent from the rate of change of the output.

On the other hand, in models of dynamic magnetisation phenomenon, the output is a function of input and input history as well as of the rate of change of the output variable. This means that the dynamic magnetisation is rate dependent and needs mapping of time dependent sequences.

2. Dynamic preisach model

Dynamic Preisach model is a generalisation of the static model by introducing the dependence of μ-function on the speed of the output variations, df/dt. The μ-function is often known as the Preisach weighting function [3]. The model is represented by:

$$f(t) = \Gamma u(t) = \iint\limits_{\alpha \geqslant \beta} \mu(\alpha, \beta, df/dt)\gamma_{\alpha\beta} u(t)\, d\alpha\, d\beta \tag{1}$$

*Corresponding author: Andrew Nafalski, Tel.: +61 8 8302 3932; Fax: +61 8 8302 3384; E-mail: nafalski@unisa.edu.au.

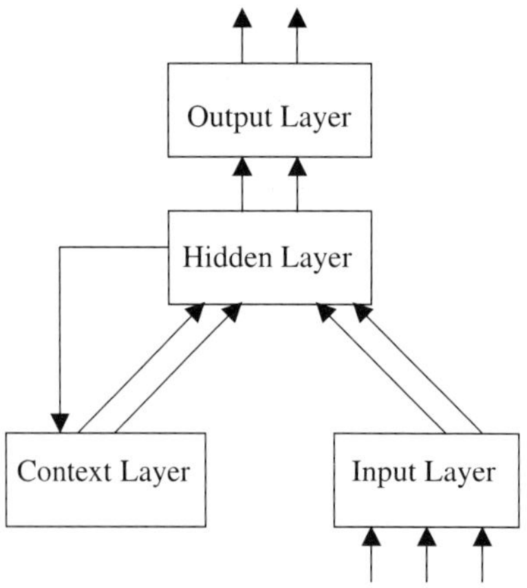

Fig. 1. A simple recurrent network.

where $u(t)$ is the input.

$f(t)$ is the output.

$\gamma_{\alpha\beta}$ is the rectangular hysteresis operator with $+1$ and -1 as the 'up' and 'down' switching output values.

α and β are the switching 'up' and 'down' input values of the operator.

$\mu(\alpha, \beta)$ is the weight corresponding to each pair of (α, β).

The numerical implementation becomes complex due to the dependency of the μ-functions on the unknown quantity df/dt. The complexity of the above model in the simulation of the magnetisation phenomena is well documented [4]. A recurrent neural network is seen as a potential alternative to the dynamic Preisach model in some applications.

3. Recurrent neural network

Recurrent neural network (RNN) is a generalised form of feedforward neural network, able to operate in a recursive mode, which has connections (feedback) from the output of one or more process elements (PEs) or neurons to the input of one or more PEs in the same or preceding layers or even self-feedbacked neuron connections. Incorporating feedback connections into multi-layer feedforward (MLFF) networks causes significant changes in its operation and learning process. The reward for the added complexity is the increased computational power of RNNs over the conventional MLFF networks. Figure 1 shows the architecture of an exemplary RNN. The hidden layer has a feedback onto itself through a buffer or "context" layer. Unlike conventional MLFF networks, RNNs exhibit dynamic behaviour i.e. they are capable of performing time and/or space dependent mapping. Therefore, they can perform more complex computations and train on context or time dependent sequences. For some dynamic recurrent applications MLFF networks may need a higher number of hidden layer nodes, compared with the corresponding RNNs, to perform with the same level of proficiency [3].

For a three-layer neural network the ith output y_i can be represented by:

$$y_i = a_o \left(bo_i + \sum_{q=1}^{n} w_{iq} h_q \right) \tag{2}$$

where a_o is the output activation function.

w_{iq} is the weight between the ith output and the qth hidden neuron. w is the m by n hidden to output weight matrix.

bo_i is the bias of the ith output neuron. bo is the m by 1 output layer bias matrix.

n is the number of hidden neurons.

m is the number of outputs.

h_q is the output of the qth hidden neuron.

In the case of having a conventional MLFF network h_q is:

$$h_q = a_h \left(bh_q + \sum_{j=1}^{k} v_{qj} x_j \right) \tag{3}$$

where a_h is the hidden layer activation function.

k is the number of inputs.

v_{qj} is the weight between the jth input and the qth hidden neuron. v is the n by k input to hidden layer weight matrix.

bh_q is the bias of the qth hidden neuron. bh is the n by k hidden layer bias matrix.

x_j is the jth input.

For the simple recurrent network in Fig. 1 the output of the context layer should be added to Eq. (3).

$$h_q = a_h \left(bh_q + \sum_{j=1}^{k} v_{qj} x_j + \sum_{p=1}^{n} u_{qp} c_p \right) \tag{4}$$

where u_{qp} is the weight between the pth context and the qth hidden neurons. u is the n by n context to hidden layer weight matrix.

c_p is the output of the pth context neuron.

In this case c_p is equal to the h_p of the previous time step.

$$c_p(t) = h_p(t-1) \tag{5}$$

Therefore, the context layer produces a delay of one sampling period to provide the system with memory [2].

4. The proposed model

Elman network is a three-layer RNN, with the middle layer recurrent. The middle layer consists of hidden and context neurons. Back propagation can be used to train the network because the connections to the context neurons are fixed which is a simplification over more complicated recurrent networks. The context neurons act as a delay in one sampling period. Therefore, the output is a function of the present state (supplied by the hidden neurons), the previous state (supplied by the context neurons with one sampling period delay), and the present input [5].

In the case of modelling scalar dynamic hysteresis, the input $x = H$ (magnetic field strength) and the output $y = B$ (magnetic flux density) are scalars which means that the input and output layers in the network need only one neuron each. Figure 2 shows the architecture of the Elman network used

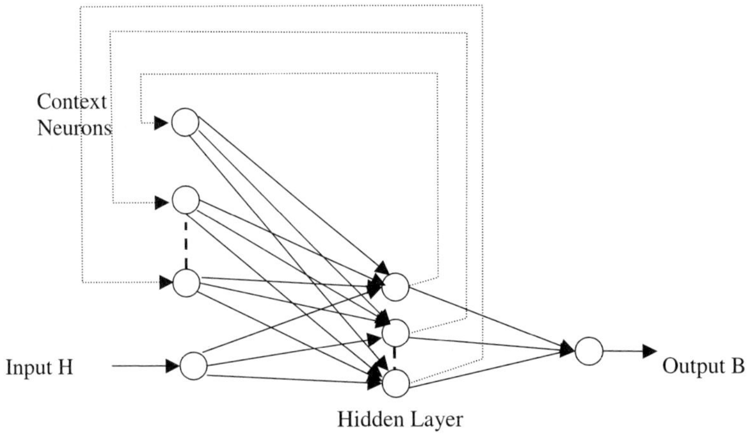

Fig. 2. Architecture of the Elman network used for dynamic hysteresis modelling.

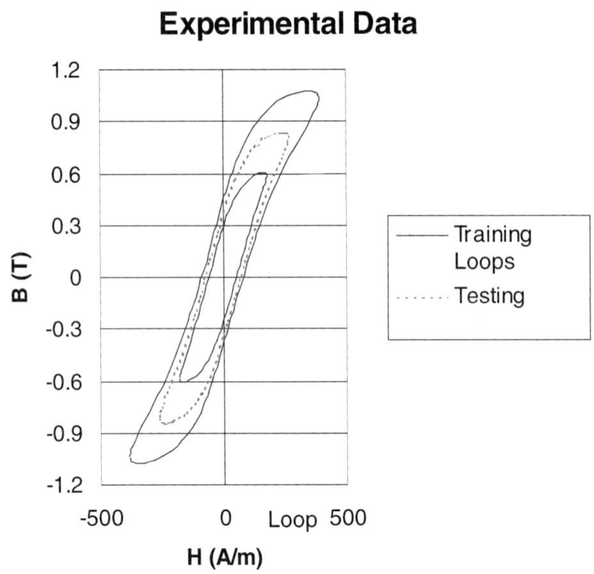

Fig. 3. Experimental data used for training and testing the model.

for the hysteresis modelling. By experimentations, it was found that the best performance in terms of accuracy and speed of convergence was achieved with 7 neurons each in the context and hidden layers with TANSIG (hyperbolic tangent sigmoid transfer function) and PURELIN (linear transfer function) as the hidden and output layer activation functions. Thus, Eqs (2) and (4) can be reduced to:

$$B = y = a_o \left(bo + \sum_{q=1}^{7} w_q h_q \right) \tag{6}$$

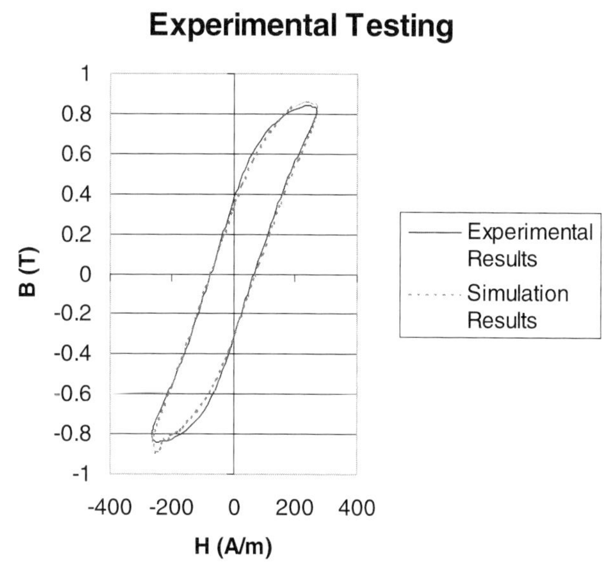

Fig. 4. Experimental testing of the proposed model.

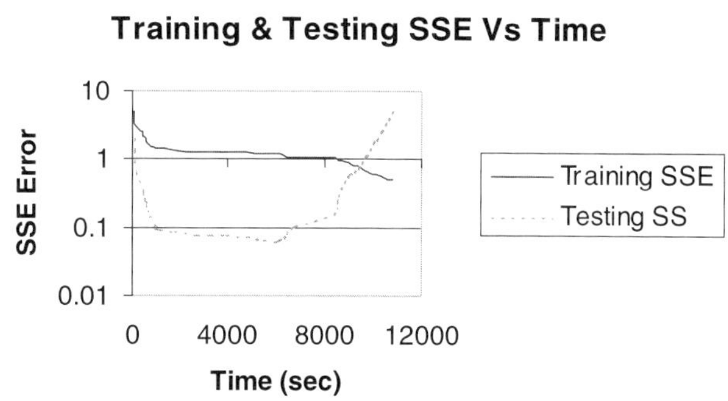

Fig. 5. Testing and training SSEs, demonstrating overtraining.

$$h_q = a_h \left(bh_q + v_q H + \sum_{p=1}^{7} u_{qp} c_p \right) \tag{7}$$

In this case the dimensions of the weight and bias matrixes are: 1 by 7 (w), 1 by 1 (bo), 7 by 1 (v), 7 by 1 (bh) and 7 by 7 (u).

The Matlab neural network toolbox was used to implement the network. The network was trained on two hysteresis loops and tested with another loop. The data used were obtained from [1]. Figure 3 shows the three loops with the middle one for testing the trained model. The simulation results agree with the experimental data to a high level of accuracy as shown in Fig. 4.

The sum squared error (SSE) is a measure of the accuracy of the simulation process. Training SSE is a measure of learning process from the available data while the testing SSE demonstrates the ability of the model to accurately predict the outcome. In the process of training, it was observed that with time the SSE for both training and testing reduced monotonically until a point was reached when further reduction of training SSE raised the testing SSE. This can be seen in Fig. 5. This is due to overtraining

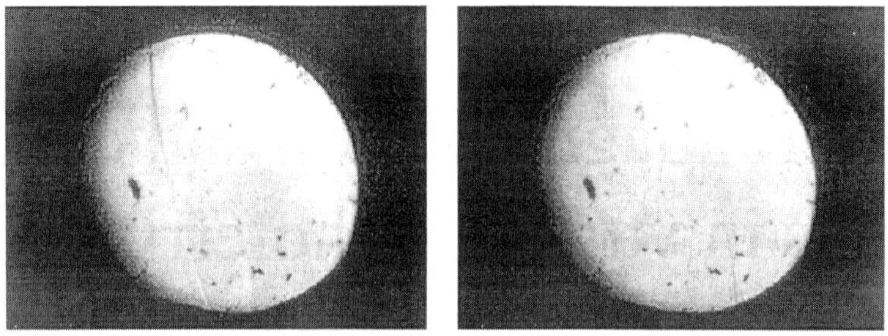

a) b)

Fig. 1. DW oscillations in the alternative magnetic field.

This information can be of special importance for practical applications. In orthoferrites the DW velocity reaches cosmic values – up to 20 km/s. This promises applications of orthoferrites in a wide variety of devices including wide bandwidth magnetooptical current transformers, sensors of mechanical quantities, spatial light modulators, optical switches, etc. [2,5]. All these applications need precise knowledge of the character of the DW motion. For a long time it was considered that at subsonic velocities, where the interaction of the DW with the phonon subsystem is negligible, the dependence of the DW motion on the driving magnetic field is linear. We found nonlinearities of the DW dynamics in this range [4]. We suppose that the lines are responsible for the nonlinearities in the sub-sound DW dynamics in orthoferrites which are not described by the existing models of the DW motion [12]. Further investigations of the lines can give answer to this problem and contribute in better understanding of the DW dynamics in weak ferromagnets.

2. Observations of DW inclinations

In the works [3,7] the observations of lines in weak ferromagnets were reported. Domain patterns in the orthoferrite plate were visualized by means of a dark field method. Several zones of sharp changes in the DW brightness were seen in the domain pattern images. DW section of different brightness were separated by Bloch lines. Changing the driving magnetic field resulted in BLs displacements.

The different brightness between parts of the DWs at different sides from the BLs was caused by the DW inclinations probably associated with the DW polarity. In this work the direct observation of the DW inclinations by means of the Faraday effect techniques is performed.

Figure 1 shows oscillations of a DW in a plate of yttrium orthoferrite ($YFeO_3$) under the influence of an alternating extorial magnetic field. The crystallographic a axis lies in the plane of the plate and is aligned horizontally in the image. Thickness of the plate was 90 μm, the plate was placed between crossed polarizers. As a light source an incandescent bulb with 1 W power was used. The driving magnetic field was produced by a wire coil with internal diameter of 1.5 mm attached to a surface of the orthoferrite plate. Figures 1a and 1b show DW in left and rightmost positions, respectively. The central part of the DW in the both cases has the minimum width and here the change of the DW brightness takes place. In the course of the DW motion the boundary between the bright and dark parts moves along the DW, and in the marginal positions the DW has the mutually opposite contrast.

The difference in brightness is associated with differences in the DW inclinations [7]. In this case the DW inclination is most significant at the top and bottom parts of the wall. The DW inclines in the direction of the crystallographic a axis. In earlier studies the DW inclination was observed in the direction of the

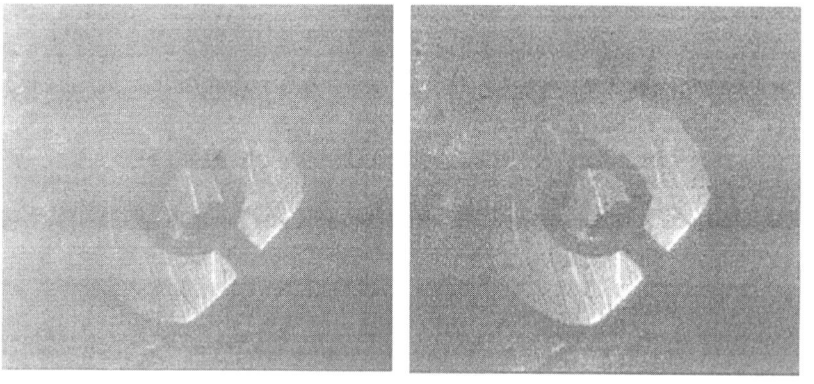

a) b)

Fig. 2. Domain pattern after magnetic field pulses.

easy axis [10]. DW inclinations in the direction of the a axis were observed in orthoferrites at rather high velocities, colse to the speed of sound (4.1 km/s). We were able to observe the DW inclinations at much smaller velocities as well as in statics.

Figure 2 shows photographs of a domain structure in a 120 μm thick orthoferrite plate. A shadow of the coil is seen. The coil produces single magnetic field pulses of 300 ns duration with an amplitude of 700 A/m. Figure 2a was obtained after a magnetic field pulse of 'positive' polarity. All 3 DWs inside the coil have bright images. On contrary, after application of a magnetic field pulse of 'negative' polarity, images of 2 DWs became dark. Zones of different brightness move along the DWs and remain in new positions after the termination of the pulse.

The sign of the DWs inclinations is definitely associated with the DW brightness and, we suppose, related to the DWs handedness. A good opportunity for the investigations of the inclinations provide thick plates of orthoferrites. Since thick plates are not transparent for visible light, the observations were performed with infrared light. A diode laser operating at the wavelength of $\lambda = 1.3\mu$m was used as a light source. Light transmitted through the analyzer was collected by an infrared viewer connected to a videocamera. The driving magnetic field H was produced by a wire coil, glued to the front side of the plate. The internal diameter of the coil was 1 mm, the amplitude of current in the coil equals 200 mA. When H = 0, the image of the DW is represented by a thin dark line separating two parts of the plate's image of equal intensity (see Fig. 3a). When H increases, the line transforms into a dark stripe (see Fig. 3b). One edge of the stripe coincides with the position of the line whereas the other edge moves in accordance with oscillations of H.

The origin of the DW image in this case differs from that for H = 0. When H = 0, the DW image is associated with a light diffraction at the phase edge discontinuity represented by the DW [8]. On the other hand the dark stripe is caused by transmission of light through two successive domains with the opposite senses of polarization rotation. Since the field of the coil is inhomogeneous in the thickness of the plate the force acting on the DW is maximum near the coil and strongly decreases at the opposite surface of the plate. This apparently causes inclinations of the DWs.

3. Conclusion

Observations of the lines and of the DW inclinations in the plates of yttrium orthoferrite are reported. The lines separate parts of the DWs of different brightness. Differences in brightness are associated with

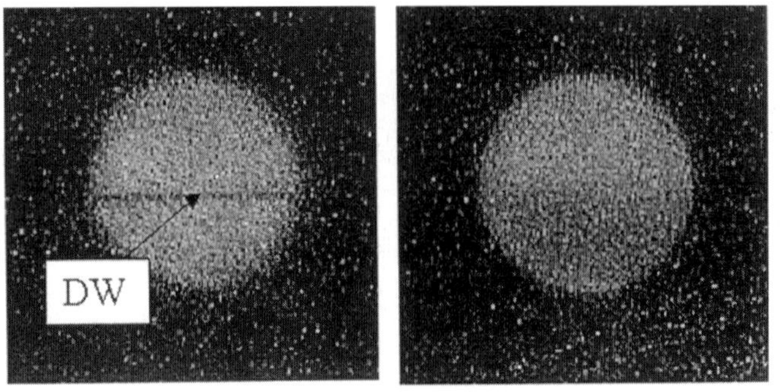

a) b)

Fig. 3. DW inclinations in a 1.2 mm thin plate of $YFeO_3$; a) $H = 0$, b) inclination at $H > 0$ (current $I = 200$ mA).

differences in the signs of DW inclinations with respect to the crystallographic a axis. The zones of different inclinations move along the DWs and remain in new positions after termination of the driving magnetic field pulses. DW inclinations in the direction of the a axis reach tens of degrees in rather weak magnetic fields. We believe that the study of DW inclinations in thick plates of orthoferrites for different kinds of driving magnetic fields can serve as an efficient tool for the study of the DWs and lines in weak ferromagnets.

Acknowledgement

We want to thank W. Toriser for technical assistance.

Financial support was provided by the Fonds zur Fõrderung der Wissenschaftlichen Forschung under Grant No. 13846-PHY.

References

[1] V.G. Bar'yakhtar, M.V. Chetkin , B.A. Ivanov and S.N. Gadetskii, *Dynamics of Topological Magnetic Solitons*, Springer-Verlag, New York, 1994.
[2] Y.S. Didosyan and H.Hauser, Magneto-Optical Sensors, in: *Magnetic Sensors and Magnetometers*, P. Ripka, ed., Artech House, Boston / London, 2001, pp. 243–266.
[3] Y.S. Didosyan and H. Hauser, Observation of Bloch lines in yttrium orthoferrite, *Phys. Lett. A* **238** (1998), 395–397.
[4] Y.S. Didosyan, H. Hauser, V.Y. Barash and P.L. Fulmek, Subsonic domain wall dynamics in yttrium orthoferrite, *J. Magn. Magn. Mater.* **177** (1998), 203.
[5] Y.S. Didosyan, H. Hauser, J. Nicolics and F. Haberl, Application of orthoferrites for light spot position measurement, *J. Appl. Phys.* **87** (2000), 7079–7081.
[6] Y.S. Didosyan, H. Hauser, J. Nicolics and I. Yavorsky, Light diffraction by a single domain wall, *J. Magn. Magn. Mater.* **203** (1999), 247–249.
[7] Y.S. Didosyan, G.A. Reider and H. Hauser, Bloch lines in domain walls of yttrium orthoferrite, *J. Appl. Phys.* **85** (1999), 5989.
[8] S. Huo et al., 3-D micromagnetic simulation of a Bloch line between C-sections of a 180° domain wall in a $\{100\}$ iron film, *J. Magn. Magn. Mater.* **218** (2000), 103–113.
[9] S. Konishi, A New Ultra-High-Density Solid State Memory: Bloch Line Memory, *IEEE Trans. Magn.* **19** (1983), 1838–1840.
[10] V. Laska and J. Miltat, Direct Observations of Neel Lines Propagation in Ferrimagnetic Garnet Epilayers with Canted Magnetization, *IEEE Trans. Magn.* **24** (1988), 1784–1786.
[11] A.P. Malozemoff and J.C. Slonczewski, *Magnetic Domain Walls in Bubble Materials*, Academic Press, New York, 1979.
[12] N. Papanicolaou, Dynamics of Domain Walls in Weak Ferromagnets, *Phys. Rev. B* **55** (1997), 12290–12308.

International Journal of Applied Electromagnetics and Mechanics 13 (2001/2002) 219–228
IOS Press

Differential motion restraint coupling using magnetorheological fluid

Kosuke Nagaya[a,*], Feng-Ying Tont[a], Iwanori Murakami[a] and Norio Tai[b]
[a]*Department of Mechanical Engineering, Gunma University, Kiryu Gunma 376-8515, Japan*
[b]*Viscodrive Japan CO. LTD, 1150 Daikouji-chou, Tochigi, 328, Japan*

Abstract. A differential motion restraint coupling is presented. In order to have strong torque, multi friction plates are installed in the friction unit of the device that involves magnetorheological fluid. Two permanent magnets lie on both sides of the friction unit. Hence, magnetic fluxes freeze the magnetorheological fluid. The friction discs have a number of slits in the radial direction, which generate torque due to shear forces between the frozen fluid and the slits. A design method has been presented for the device. To validate the present design formulae, experimental tests have been carried out. Calculated results are in good agreement with the experimental ones.

Keywords: Torque, differential motion, device, coupling, magnetorhrological fluid, clutch

1. Introduction

An automobile has a differential gear mechanism to run through curves smoothly. But, it transmits a driving torque to one of the wheels whose transmission torque is small. Hence, it is difficult to start the mobile on a snow road because of slip of the wheel with small friction. In order to improve this, the differential restraint device is stacked in the differential gear mechanism. There are two types of devices. One is the multi-friction disc type, and the other is the viscous coupling type [1,2]. Noise often occurs in the friction disc type device, so that the viscous coupling type is widely used. In the viscous coupling type, since the friction between discs and silicon oil generates transmission torques, the transmission torque is in proportion to the wheel velocity. This makes the mobile unstable when the mobile passes through curves. Hence, the coupling torque is designed to be an appropriate range. This implies that the transmission torque is not enough when the mobile starts on a snow road.

The object of this paper is to have a powerful viscous coupling whose torque has no dependence on the wheel velocity. MR fluid is used as the torque transmission fluid in our coupling, because the fluid freezes under magnetic fields [3–6]. The most important factor in the coupling is to have the strong transmission torque, but its size is limited. The size of our coupling is almost the same as the previous viscous coupling, and the torque is more than 100 $N \cdot m$ which covers the torques of viscous couplings, but has no dependence on the velocity.

*Corresponding author: Kosuke Nagaya, Tel.: +81 277 30 1563; Fax: +81 277 30 1599; E-mail: nagaya@me.gunma-u.ac.jp.

1383-5416/01/02/$8.00

2. Mechanisms for the present magnetorheological coupling

In order to have large torque in a small space, multi-disc type clutches are available. As for the multi-disc clutches, the friction between discs generates friction torques. In the magnetorheological fluid coupling, the friction torque between discs is not so large, but the shear torque between slits is large because the magnetorheorogical fluid is frozen by magnetic fields. Hence, the coupling made of multi-disc with slits is proposed in this paper.

2.1. Friction torque between discs

Consider a three layered sandwich structure of fixed disc, rotation disc and fixed disc which lie between N- and S-pole of the magnets. When the magnetorheorogical fluid is filled in the gaps, the friction torque due to a small radius dr at the radius r is $dT'_f = 2\pi r dr f_f r$. Hence, the torque T'_f is

$$T'_f = 2\pi f_f \int_a^R r^2 dr = \frac{2}{3}\pi f_f (R^3 - a^3) \tag{1}$$

where a is the inner radius of the disc, R is the outer radius, f_f is the friction stress between discs without slits, under the magnetic fields. Since there are two friction surfaces for a rotation disc, we have $T_f = 2T'_f$. Consider a reference disc whose inner radius is a_0, outer radius is R_0, friction stress is f_{f0}, and friction torque is T_{f0}. From Eq. (1), we have

$$f_{f0} = \frac{3T_{f0}}{4\pi(R_0^3 - a_0^3)}$$

Magnetic fluxes concentrate to the Fe-pate, and so the magnetic flux density is in proportion to the inverse of surface area of the Fe-plate when there is a Fe-plate between N- and S-poles of the magnets. Hence, the friction stress $f_f(\mathrm{N/m^2})$ between discs is a function of the area of the friction discs:

$$f_f(R) = \left(\frac{R_0^2 - a_0^2}{R^2 - a^2}\right) \frac{3T_{f0}}{4\pi(R_0^3 - a_0^3)} \tag{2}$$

Equation (2) illustrates the friction stress due to the magnetorheological fluid under magnetic fields.

The friction stress without magnetic field is

$$f_v = \frac{3T_{v0}}{4\pi(R_0^3 - a_0^3)} \tag{3}$$

where T_{v0} is the friction torque when there is no magnetic field. Equation (3) denotes the friction due to the viscosity of the MR fluid under non-magnetic field, so that it has constant value, which has no dependence on the radius of the disc. Hence, the friction stress due to the magnetic fields only $(f_m = f_f - f_v)$ for the disc with radii a and R is

$$f_m = \frac{3}{4\pi(R_0^3 - a_0^3)} \left\{ \left(\frac{R_0^2 - a_0^2}{R^2 - a^2}\right) T_{f_0} - T_{v_0} \right\} \tag{4}$$

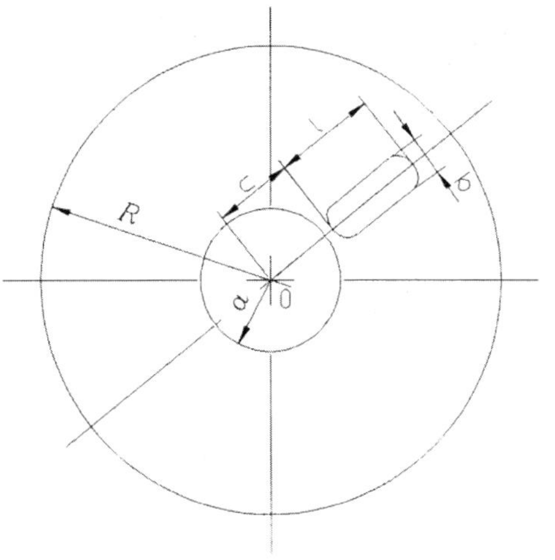

Fig. 1. Geometry of the friction disc with a slit.

2.2. Method to have large torques using slits

When there are slits in both fixed and rotation discs, the shear stress is large because the MR fluid in the slit is frozen under magnetic fields. If the slits lie in the radial direction as shown in Fig. 1, the torque due to the shear is $T_c = \int_c^{c+l} f_c r dr$, where $f_c(B)$ (N/m) is the shear force per unit length. Then the shear torque for the disc with n-slits is

$$T_c = \frac{\{(c+l)^2 - c^2\}n}{2} f_c \tag{5}$$

where c is the length from the center of the disc to the inner end of the slit, l is the slit length and n is the number of slits (see Fig. 1).

The friction area decreases when there are slits. The effects on the torque is

$$T_f'' = (c + l/2) bl f_f n$$

Then the total torque is $T_1 = 2T_f' - 2T_f'' + T_c$:

$$T_1 = 2(f_m + f_v)\{(2\pi/3)(R^3 - a^3) - (c + l/2)bln\} + \frac{\{(c+l)^2 - c^2\}n}{2}(f_{c0} + f_{cm}) \tag{6}$$

From which, we have

$$f_c = f_{c0} + f_{cm} = \frac{2[T_1 - 2f_f\{(2\pi/3)(R^3 - a^3) - (c + l/2)bln\}]}{\{(c+l)^2 - c^2\}n} \tag{7a}$$

Let the subscript 0 be the value for the reference disc. The shear force f_{c0} per a unit length for the reference disc, with parameters a_0, R_0, c_0 and l_0 under non-magnetic field $(B = 0)$, is obtained from Eq. 7(a):

$$f_{c0} = \frac{2[T_{10} - 2f_v\{(2\pi/3)(R_0^3 - a_0^3) - (c_0 + l_0/2)bl_0 n\}]}{\{(c_0 + l_0)^2 - c_0^2\}n} \tag{7b}$$

where f_{c0} is constant which has no dependence on the slit length.

The shear stress f_{cm0} per unit length along the slit for the reference disc under magnetic field $(B \neq 0)$ is

$$f_{cm0} = \frac{2[T_1 - 2(f_{m0} + f_v)\{(2\pi/3)(R_0^3 - a_0^3) - (c_0 + l_0/2)bl_0 n\}]}{\{(c_0 + l)^2 - c_0^2\}n} - f_{c0} \tag{7c}$$

Then the shear force for the disc with inner radius a and outer radius R due to the magnetic field is

$$f_{cm} = f_{cm0}\left(\frac{R_0^2 - a_0^2}{R^2 - a^2}\right) \tag{7d}$$

In the calculation of the torque, a fundamental experiment has to be performed for the reference discs. The torques for the reference disc with a_0 and R_0 without slits are first measured for zero-magnetic field and the reference magnetic flux density B_0. Substituting these values into Eqs (3) and (4) give the friction stress f_v and f_{m0}. In the next step, the torque T_{10} without magnetic field and that T_1 with reference magnetic flux density B_0 are measured for the reference discs with slits (c_0, l_0). Substituting the torques into Eqs (7b) and (7c), we have the shear force f_{c0} and f_{cm0}. Substituting f_{cm0} into Eq. (7d), the shear force f_{cm} due to the slits under the magnetic field is obtained for a disc with arbitrary sizes a and R.

The forces f_m and f_{cm} vary with the magnitude of the magnetic field. The attractive force between two magnetized Fe-plates is in proportion to the square of the magnetic flux density. Hence, the same assumption is used among Fe-particles in MR fluid:

$$\begin{aligned} f_{mB} &= f_m(B/B_0)^2 \\ f_{cmB} &= f_{cm}(B/B_0)^2 \end{aligned} \tag{8}$$

The total torque T_N for rotating discs is

$$T_N = N[2(f_{mB} + f_v)\{(2\pi/3)(R^3 - a^3) - (c + l/2)bln\} + \frac{\{(c+l)^2 - c^2\}n}{2}(f_{c0} + f_{cmB})] \tag{9}$$

where N is the number of the rotating discs.

3. Decision of design parameters

3.1. Magnetic flux density

Permanent magnets are used in our coupling, whose outer diameter is 75 mm, inner diameter is 30 mm, thickness is 20 mm, and magnetization strength is 0.875 T. The direction of magnetization is in the axial direction. The S-pole of the magnet faces N-pole of the other. The magnetic flux densities are measured by using Gauss meter in two cases. One is the case with a Fe-disc (made of SS-41 steel) between two permanent magnets, and the other is the case without the disc. Figure 2 depicts the magnetic flux density at the point ($r = 23$ mm from the magnet center, and d from the magnet surface) versus air gap $2d$. The magnetic flux density with Fe-disc is about 1.4 times of that without Fe-disc in this experiment:

$$B_{fe} = 1.4B_{\text{air}} \tag{10}$$

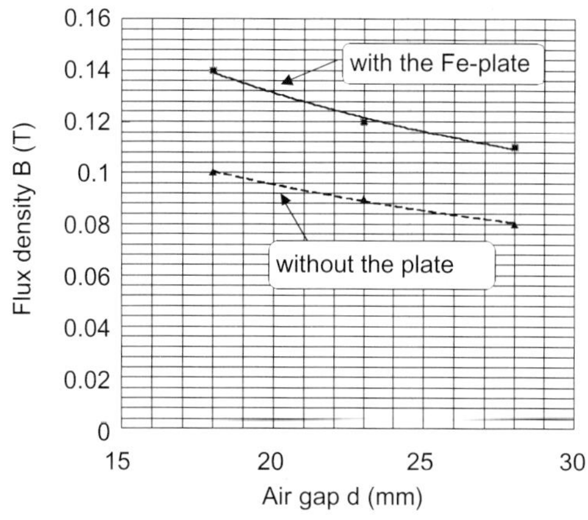

Fig. 2. Magnetic flux densities on the middle point between two permanent magnets.

Table 1
Dimesions of the rectangular discs used in the fundamental experiment(see Fig. 1)

No.	Length×width (mm)	Slit length I (mm)	Slit width b (mm)	Length c (mm)
1	74 × 74	25	10	8
2	70 × 70	22	10	8
3	64 × 64	19	10	8
4	60 × 60	18	8	8
5	54 × 54	15	8	8

3.2. Measurement of the fundamental torque

In order to have fundamental parameters, a thin clear acryl case was made. Two square iron plates with thickness 0.5 mm are connected to both sides of the case. A circular disc lies between two iron plates. The MR fluid is filled in the case. The circular disc has a shaft, and a torque meter is connected to the shaft. Hence, the torque between the fixed plate and the rotating disc can be measured when the shaft of the torque meter is rotated. Two permanent magnets (Inner radius = 75 mm, outer radius =30 mm and thickness = 20 mm) are attached to both sides of the case.

The MR fluid used in this experiment is as follows: Density=2.772, Fe = 86% (Weight percent), B = 0.1%, P = 0.3%, Si = 0.2% and K = 0.4%.

The friction torque T_{f0} was measured first using a circular disc without slits whose diameter was 72 mm and thickness was 0.5 mm, and two square plates without slits. The obtained results were $T_{f0} = 1.4$ N·m (for $B = 0.1$ T), $T_{f0} = 0.2$N·m (for $B = 0$). The magnetic flux density was $B_{\rm air} = B_0 = 0.1\ T$. The value was taken as a reference magnetic flux density. Substituting these into Eqs (3) and (4) give $f_v = 1023$Pa, $f_{m0} = 6140$Pa.

The shear force per unit length due to slits was obtained in the second step. Table 1 shows the sizes of the square plates used in the experiment (In order to have the same effect for magnetic flux density, the ratio between the area of the square plate to that of circular disc was taken to be constant). Table 2 shows the sizes for the discs. The sizes of the slits were the same in both fixed plates (square plates) and the rotating disc. The disc had six slits with the same circular pitch whose sizes were shown in Table 2. The shaft made of brass had 8 mm in diameter.

The experiments were performed for the combination of the same number of fixed plates and discs. Results were depicted by the black dots in Fig. 3. Taking the torque $T_{10} = 0.2\text{N}\cdot\text{m}$ for $B_0 = 0$ and

Table 2
Dimension of the circular discs used in the fundamental experiment (see Fig. 1)

No.	Outer Radius(mm)	Slit length I(mm)	Slit width b(mm)	Length c (mm)
1	36	25	10	8
2	33	22	10	8
3	31	19	10	8
4	28	18	8	8
5	26	15	8	8

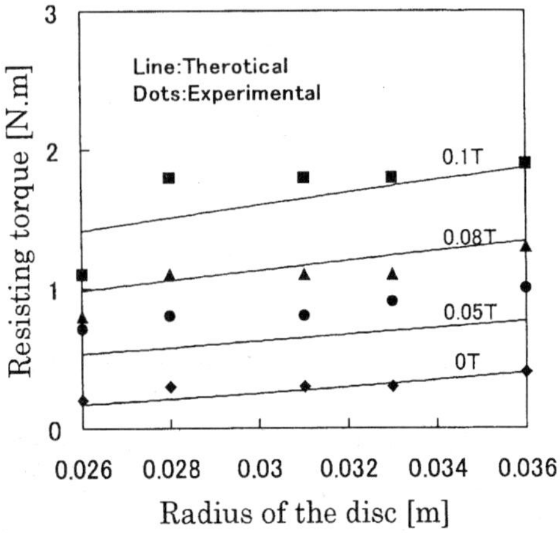

Fig. 3. Transmission torque versus radius of the disc.

$T = T_1 = 1.9\text{N} \cdot \text{m}$ for $B_0 = 0.1\text{T}$ to be the reference values, and substituting these into Eqs (7b) and (7c) yield $f_{c0} = 118$ N/m and $f_c m0 = 269$ N/m. Using these values, we had the torques for the discs shown in Table 2. The solid line in Fig. 3 shows the simulation results.

The torque increases with the radius of the disc (see Fig. 3) of course, but the increasing rate is not so large, because the magnetic flux density decreases with increasing values of the disc. Although some discrepancies are found between the theoretical and experimental results, both are in good agreement from the design-engineering point of view.

It is difficult to have the exact transmission torque for this coupling, so that analyses have not been reported. This paper presents the approximate analysis. The result can be used for designing the viscous coupling presented in this paper.

4. Multi-disc viscous coupling

4.1. Coupling using Fe-discs

Usually, a viscous coupling is attached to one of the wheel shafts to restrain differential motion. But, when the restraint torque is too large, the differential motion is restrained, and it will be dangerous. Hence, the transmission torque is taken in appropriate range ($50 \sim 130\text{N} \cdot \text{m}$ according to the kinds of automobiles). In this article, a coupling, which covers the torques, is designed. Figure 4 depicts the cross section and Fig. 5 shows the photos of the parts. The parts except the friction discs are made of stainless steel (SUS-304). Three torque generation units were made whose thickness is $D = 15$ mm, 20 mm and

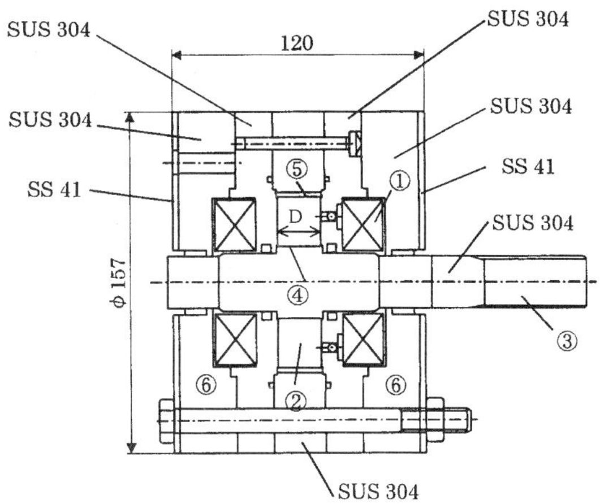

Fig. 4. Geometry of the multi-discs magnetic fluid coupling ①: Magnet, ②: Magnetic fluid unit, ③: Shaft, ④: Sprain, ⑤: Sprain, ⑥: Magnet holder).

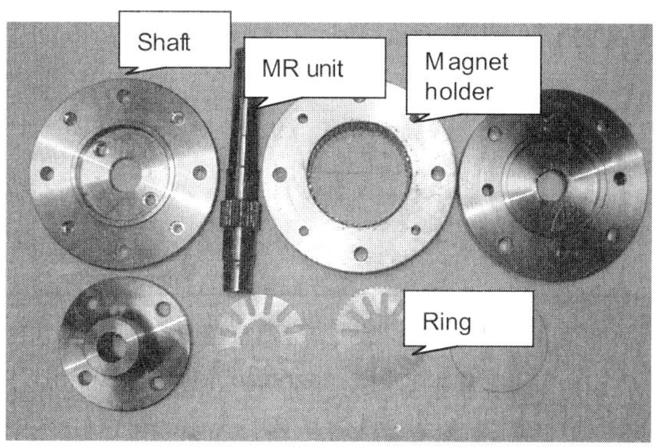

Fig. 5. Parts of MR fluid coupling.

Table 3
Number of coupling plates

Length of the magnetic fluid unit	Number of follower discs	Number of driving discs
15 mm	17	16
20 mm	22	21
30 mm	33	32

30 mm. The thickness of the disc is 0.4 mm. The photos are shown in Fig. 5, where the thickness of the rotating disc is 0.4 mm, outer radius is $R = 38$ mm, inner radius is $a = 17$ mm, slit length $l = 17$ mm, length from the center to the slit is $c = 21$ mm, and the number of slits is $n = 12$. The left side photo depicts the driver disc whose spline grooves are installed into spline grooves (4) of the shaft (3). The right side disc shows the follower disc whose outer spline grooves are installed into spline (5) of the MR fluid unit. The driver disc and follower disc are sandwiched with each other. The size of the driver discs is somewhat smaller than that of the follower discs. Spacer rings with 0.6 mm diameter lay outsides the driver discs. Hence, there is 0.1 mm gap between the driver disc and the follower disc. The frictions and shear forces at the slits of the driver discs are transmitted to the follower discs. The number of the discs, which are installed in the MR fluid unit, is shown in Table 3.

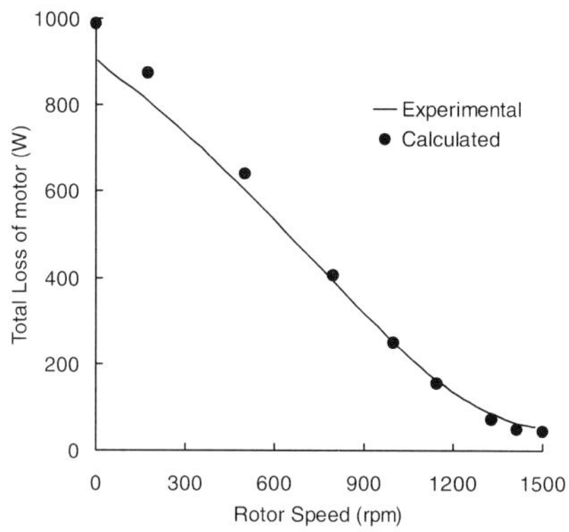

Fig. 3. Experimental and calculated total losses of motor.

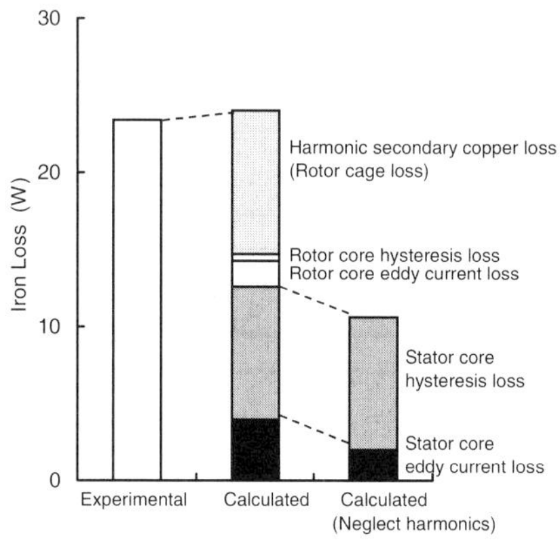

Fig. 4. Experimental and calculated iron losses (slip = 0).

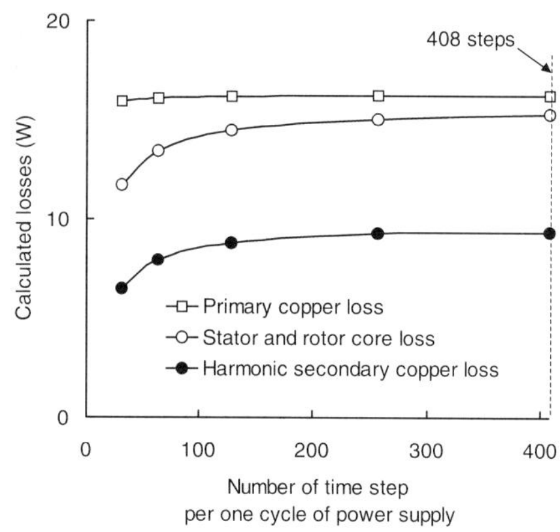

Fig. 5. Variation of calculated losses due to time-step.

Due to these assumptions, we can calculate the eddy current and hysteresis losses of the core W_{ce} and W_{ch} as

$$W_{ce} = \frac{K_e D}{2\pi^2} \int_{\text{iron}} \frac{1}{N} \sum_{k=1}^{N} \left\{ \left(\frac{B_r^{k+1} - B_r^k}{\Delta t} \right)^2 + \left(\frac{B_\theta^{k+1} - B_\theta^k}{\Delta t} \right)^2 \right\} dv \tag{7}$$

$$W_{ch} = \frac{K_h D}{T} \sum_{i=1}^{NE} \frac{\Delta V_i}{2} \left(\sum_{j=1}^{N_{pr}{}^i} (B_{mr}^{ij})^2 + \sum_{j=1}^{N_{P\theta}{}^i} (B_{m\theta}^{ij})^2 \right) \tag{8}$$

where D is the density of the core, K_e, K_h are the experimental constant by Epstein frame, N is the number of time step per one time period, Δt is the time interval, ΔV_i is the volume of i-th finite element, B_{mr}^{ij}, $B_{m\theta}^{ij}$ are the amplitude of each hysteresis loop.

3. Results

The proposed procedure is applied to the IEEJ induction motor K model [3], which is the 200 W class aluminum cage induction motor driven by a sinusoidal 100 V power supply. The number of the

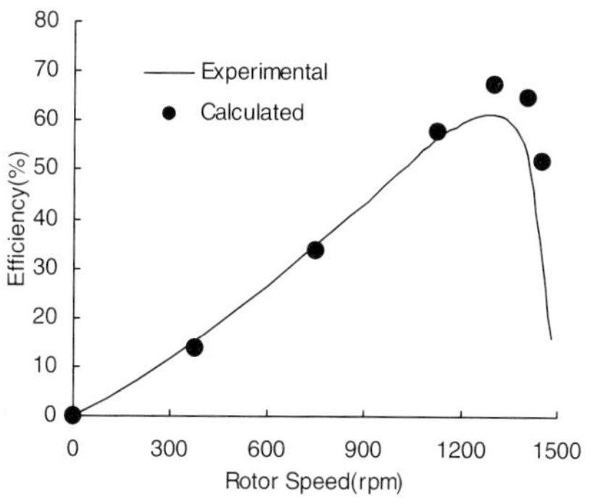

Fig. 6. Experimental and calculated efficiency of motor.

finite elements is 74,400 tetrahedron edge-elements for the 3D linear phasor analysis and 19,678 triangle elements in the 2D nonlinear time-stepping analysis, respectively. The number of time step is 256 per one cycle of power supply.

Figure 2 shows results of the flux distribution, time-variation of the flux density and the loss distribution at no load condition (1500 rpm, slip $= 0$). The magnetic fields at the stator yoke and stator teeth (a), (b) are nearly alternating fields of peripheral and radial direction, respectively. The waveforms of B_r and B_θ involve much time-harmonics near the air gap due to the movement of rotor slots. As a result, the locus of B_r and B_θ at top of the stator slot tooth (c) shows complex variation. The field at the top of the rotor slot tooth (d) also involves much time-harmonics due to the stator slot ripple. In this case, the locus of B_r and B_θ shows high frequency rotational magnetic field. As a result, the core loss concentrates at the rotor surface in this case. On the other hand, the magnetic fields at the rotor tooth and yoke (e), (f) are nearly DC fields of radial and peripheral direction, respectively because of slip $= 0$. The loss of the bar, that is, the secondary copper loss also concentrates at the surface due to the skin effect of the harmonic secondary currents.

Figure 3 shows the experimental and calculated total losses of the motor due to the rotor speed. The results agree well. It can be considered that the accuracy of the analysis is enough.

Figure 4 shows the experimental and calculated iron losses at no load condition. In the experiment, the iron loss W_i is determined due to the classical electric machine theory as follows

$$W_i = W_{\mathrm{noload}} - W_1 - W_m \tag{9}$$

where W_{noload} is the measured no load loss that is shown in the experimental result of Fig. 3 at 1500 rpm, W_1 is the measured primary copper loss and the W_m is the mechanical loss. In this classical definition, W_i includes not only core loss W_c but also harmonic copper loss W_2. Figure 4 indicate that the experimental iron loss W_i agrees well with the sum of the calculated harmonic secondary copper loss and the core losses $W_i + W_2$.

Figure 4 also shows the calculated result without the consideration of the harmonics caused by the rotor movement. When we neglect the harmonics, the secondary copper loss and the rotor core losses are assumed as zero at slip $= 0$. In addition, the calculated stator core losses are underestimated due to the neglect of the harmonics of the magnetic field. As a result, only less than half of the iron loss is estimated in this case. It is clarified that the consideration of the rotor movement is indispensable to estimate the losses of the motor and that the harmonic secondary copper loss and the rotor core losses at slip $= 0$ are not negligible.

Figure 5 shows the variation of the calculated core loss, primary and secondary copper losses at no load condition due to the number of time-steps of the time-stepping analysis. Figure 5 indicates that much time-steps are required for the accurate estimation of the core loss and the secondary copper loss (rotor cage eddy current loss) because much time-harmonics is included in the electromagnetic fields in the motor. On the other hand, the calculated primary copper loss, which is obtained from the primary coil currents, is not so much influenced by the number of the time-steps because the harmonics included in the primary coil currents are relatively small.

The results of 408 steps at one cycle of power supply in Figure 5 are obtained without the change of the finite element mesh at each time-step because the number of elements at one-pole pair along by the air-gap is 408. On the other hand, other results are obtained with the mesh changes. Figure 5 indicates that the effects of the mesh distortion can be considered as negligible because the results at 408 and 256 steps are almost identical. The calculation time is about 5 hours using PENTIUM 933 MHz PC at one cycle of power supply with 256 steps.

Figure 6 shows the calculated and experimental efficiency. The results agree well. The reasons of the slight overestimation of the calculated result can be considered as follows.

(a) Neglect of the effects of the eddy currents and hysteresis phenomena of the stator and rotor cores in the finite element analysis, which is the first step of the calculation procedure shown in Fig. 1.

(b) Neglect of the two-dimensional magnetic properties.

4. Conclusions

The losses and the efficiency of the induction motor are analyzed with the consideration of stray losses caused by the stator and rotor slot ripple. It is clarified that the losses concentrate at the surface of the stator and the rotor core and that the consideration of the time-harmonics caused by the rotor movement is indispensable to estimate the losses and the efficiency of the motor. The combined 3D-2D finite element formulation is useful from the point of the calculation time because much time steps are required for the correct estimation of the harmonic secondary copper loss and the core loss. Further work is required to consider the core loss and the 2D magnetic properties directly by the electromagnetic analysis with the estimation of the time-harmonics, which requires vast calculation time.

References

[1] B. Slates, P. Van Roy and R. Belmans, Energy Efficiency of Induction Machines, *Proc. ICEM2000* **3** (2000), 1503.

[2] K. Yamazaki, Induction Motor Analysis Considering both Harmonics and End Effects Using Combination of 2D and 3D Finite Element Method, *IEEE Trans. Energy Conversion* **14**(3) (1999), 698.

[3] K. Yamazaki and S. Shinfuku, Combined 3D-2D Finite Element Analysis Considering Variation of Neutral Point Potential in Star Connection, *IEEE Trans. Magn* (in press).

[4] M. Enokizono, T. Todaka and K. Okamoto, A Numerical Simulation of Induction Motor By Using Two-Dimensional Magnetic Properties, *Proc. CEFC2000* (2000), 182.

[5] J.D. Lavers, P.P. Bringer and H. Hollitscher, A simple method of estimating the minor loop hysteresis loss in thin lamination, *IEEE Trans. Magn.* **14**(5) (1978), 386.

[6] S.L. Ho, W.N. Fu and H.C. Wong, Estimation of Stray Losses of Skewed Rotor Induction Motors Using Coupled 2-D and 3-D Time Stepping Finite Element Methods, *IEEE Trans. Magn* **34**(5) (1998), 3102.

[7] N. Sadowski, J.P.A. Bastos, M.V. Ferreira Da Luz, P. Kuo-Peng and M. Lajoie-Mazenc, The Rain-Flow Method for Evaluation of Iron Losses in Electrical Machines, *IEEE Trans. Magn.* **36**(4) (2000), 1926.

International Journal of Applied Electromagnetics and Mechanics 13 (2001/2002) 235–239
IOS Press

High-coercivity $Nd_{90-x}Fe_xAl_{10}$ bulk amorphous alloys

H. Chiriac[a,*], N. Lupu[a], A. Takeuchi[b] and A. Inoue[b]
[a]*National Institute of Research and Development for Technical Physics 47 Mangeron Boulevard, R – 6600 Iasi, Romania*
[b]*Institute for Materials Research, Tohoku University 2-1-1 Katahira, Aoba-Ku, Sendai 980-8577, Japan*

Abstract. Extensive magnetic studies on $Nd_{90-x}Fe_xAl_{10}$ ($x = 35 \sim 50$) amorphous alloys in the shape of both melt-spun ribbons with thicknesses between 25 and 150 μm and rods with diameters no larger than 3 mm prepared by mould casting and suction casting indicate a complex magnetic behaviour depending on the Fe content. The microstructure and magnetic properties are strongly dependent on the cooling rate and the preparation technique. The high values of the coercive field of $Nd_{90-x}Fe_xAl_{10}$ amorphous samples in the as-cast state as well as their increase by the decrease of the cooling rate and temperature result from the formation of very small metastable or non-equilibrium clusters, whose size vary as a function of the quench rate. The presence of these magnetic clusters is evidenced by multiphase magnetic transitions observed from the ac-susceptibility measurements at low temperatures.

1. Introduction

The magnetism of RE-Fe glassy alloys produced by rapid quenching from the melt predicts a great stimulus for research in their preparation and behaviour. These alloys form a variety of magnetic structures ranging from ferromagnetic to sperromagnetic or sperrimagnetic, depending on the relative magnitude of the random magnetic anisotropy (RMA) with respect to the exchange energy [1,2]. The magnetic properties of these alloys are also of technological interest because of their wide range of applications as permanent magnets, magnetostrictive materials, and magneto-optical recording media.

Recently, it has been reported that Nd-Fe-(Al,Si) ternary amorphous alloys are formed in a wide range of compositions by melt spinning and mould casting technique and exhibit large coercivities at room temperature [3,4]. These results are in contradiction to those found in conventional ternary amorphous alloys, especially $Nd_2Fe_{14}B$, in which the amorphous microstructure determines soft magnetic characteristics with negligible coercivities [5], but they are in agreement with the high coercivities obtained in the past for Nd-Fe binary amorphous alloys [6,7].

To clarify the relationship between microstructure and coercivity in these bulk amorphous alloys, investigations on the glass formability, thermal stability and magnetic properties were performed on $Nd_{90-x}Fe_xAl_{10}$ melt-spun amorphous ribbons with thicknesses between 25 and 150 μm and cast amorphous rods up to 3 mm in diameter produced by mould casting and suction casting.

*Corresponding author: Prof. Dr. Horia Chiriac, National Institute of Research and Development for Technical Physics, 47 Mangeron Blvd., R-6600 Iasi, Romania. Tel.: +40 32 130680; Fax: +40 32 231132; E-mail: hchiriac@phys-iasi.ro.

International Journal of Applied Electromagnetics and Mechanics 13 (2001/2002) 241–244
IOS Press

Modeling the complex permeability spectra of Ni-Zn ferrite

Ovidiu Florin Caltun[a,*], Leonard Spinu[b], Alexandru Stancu[a] and Anca Sava[a]
[a] *"Alexandru Ioan Cuza" University, Faculty of Physics, Bd. Copou, 11, Iasi, 6600, Romania*
[b] *University of New Orleans, AMRI, New Orleans, LA 70148 USA*

Abstract. Ferrites properties depend very strongly on the chemical composition and on the microstructure. The complex permeability spectra of Cu substituted Ni-Zn ferrites produced by conventional ceramic process for high frequency were studied. The experimental and theoretical curves of real and complex permeability were compared. Is discussed the contribution of the domain wall and of the spin rotational motion on the complex permeability in the MHz region. A small amount of Cu ions can provide high performances to the Ni-Zn ferrite.

1. Introduction

Microstructure and magnetic properties of NI-Zn ferrite are highly sensitive to the preparation methodology, sintering conditions and the amount of constituent metal oxides, impurities or doping levels. Efforts are being made to develop a low-power loss material operating in the MHz region in accordance with the miniaturization of cores. The studies on microstructure and composition related magnetic properties have been reported for Ni-Zn ferrites by several researchers in this field [1–3]. Nakamura [4] has reported the low-temperature sintering of Ni-Zn ferrite at 900°C resulting in magnetic permeability over 200 at 1 MHz and density greater than 4.5 gcm^{-3}. The high permeability has been attributed to the domain wall contribution, which is controlled not only by the higher post sintering density but also by increased grain. The addition increases the density and decreases the sintering temperature and increases the resistivity of the material.

This paper focuses its discussion on dependence of permeability spectra on the frequency for three different samples of Ni-Zn ferrite added with copper oxide in sintering process. The model proposed by Nakamura was used to describe the permeability spectra. The experimental and theoretical curves are compared. The behavior was explained on the basis of their microstructural features. The contribution of the domain wall motion and of the spin rotation to the complex permeability is discussed.

2. Experimental results

Ni-Zn ferrite ceramics doped with CuO were syntesized using the usual ceramic technique [5]. The ring cores with outer radius 12 mm, inner radius 8 mm and height 6 mm were pressed at 150 MPa and

*Correspondence address: Dr. Ovidiu Florin Caltun, "Alexandru Ioan Cuza" University, Faculty of Physics, 11 Blvd. Copou, 6600 Iasi, Romania. Tel.: + 032 201176; Fax: + 032 201205; E-mail: caltun@uaic.ro.

Fig. 1. SEM micrograph of R1 ($x = 0.1$).

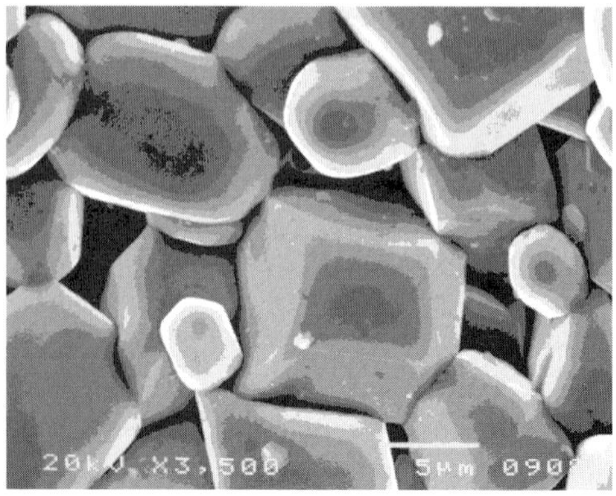

Fig. 2. SEM micrograph of R2 ($x = 0.2$).

Fig. 3. SEM micrograph of R3 ($x = 0.3$).

sintered at 1100°C/4 h in air. Three categories of sample with general formula $(NiZn)_{0.8-x}$ Cux $Zn_{0.2}$ Fe_2O_4 with x ranging in the from 0.1 to 0.3 in steps of 0.1 denoted by 1, 2 and 3 were obtained. X-ray diffractograms were recorded of the as dried, calcined and the sintered ferrites samples using Philips PW 1730 X-ray diffractometer with CuK(radiation. The characterization of the samples was completed by the EDAX analysis that confirmed the increasing concentration of the Cu ions. Microstructural features of the sintered samples were obtained using Jeol 5410 scanning electron microscope. Increasing the copper substitution the average grain size increases and the microstructure becomes more uniform without pores as is featured by Figs 1–3.

Spectra of the complex initial permeability at room temperature were measured by using the conventional technique based on the determination of the complex impedance of a circuit loaded with a toroid shaped sample (Hewlett Packard 4194A Impedance Analyzer).

The complex permeability of sintered ferrite is related to two different magnetizing mechanisms: the spin rotational magnetization and the domain wall motion. In order to make some detailed discussions, Nakamura decomposed the permeability spectra of the sintered ferrite into the spin rotational component, $\chi_{sp}(\varpi)$, and the domain wall component, $\chi_{dw}(\varpi)$. The spin rotational component is of relaxation type and its dispersion is inversely proportional to the frequency. The domain wall component is of resonance type and depends on the square of frequency. Finally he obtained the following equations:

$$\mu(\varpi) = 1 + \chi_{sp}(\varpi) + \chi_{dw}(\varpi) \quad (1)$$

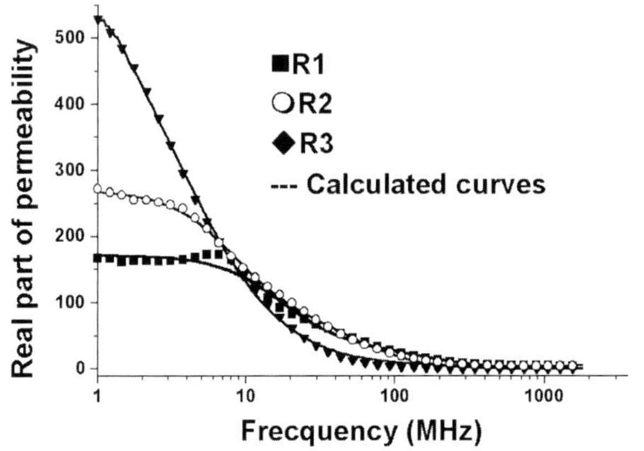

Fig. 4. Experimental and calculated real part μ' of the complex permeability at room temperature $B < 0.5$ mT for all the samples.

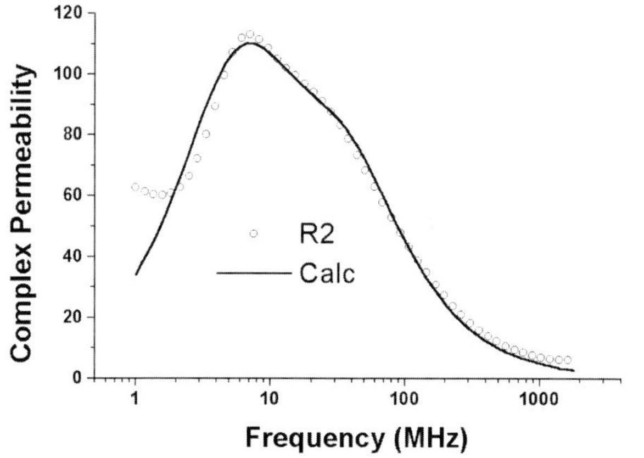

Fig. 5. Experimental and calculated imaginary part μ'' of the complex permeability at room temperature $B < 0.5$ mT for all sample R2.

Table 1

Permeability	K_{sp}	K_{dw}	ϖ_{sp} (MHz)	ϖ_{dw} (MHz)	β (MHz)
μ'R1	311 ± 40	1483 ± 44	171 ± 25	1.11 ± 0.23	0.71 ± 0.2
μ''R1	336 ± 52	1413 ± 58	129 ± 23	0.87 ± 0.19	4.7 ± 0.5
μ'R2	777 ± 35	1992 ± 44	57 ± 2.3	2.1 ± 0.2	5.5 ± 1.1
μ''R2	855 ± 74	1859 ± 81	53 ± 2.5	1.1 ± 0.6	6.2 ± 1.2
μ'R3	1463 ± 105	4249 ± 147	15.5 ± 0.7	2.2 ± 0.5	6.6 ± 1.4
μ''R3	1585 ± 113	4325 ± 165	14.4 ± 0.6	1.5 ± 0.8	6.9 ± 1.5

and each component can be written as follows:

$$\chi_{sp}(\varpi) = K_{sp}/\lfloor 1 + j(\varpi/\varpi_{sp})\rfloor, \tag{2}$$

$$\chi_{dw}(\varpi) = K_{dw}\varpi_{dw}^2/\lfloor \varpi_{dw}^2 - \varpi^2 + j\beta\varpi\rfloor, \tag{3}$$

with ϖ the magnetic field frequency, K_{sp} the static spin susceptibility, ϖ_{sp} the spin resonance frequency, K_{dw} the static susceptibility of the domain wall motion, ϖ_{dw} the domain wall motion frequency and β the damping factor of the domain wall motion.

$K_{sp}, \varpi_{sp}, K_{dw}, \varpi_{dw}$ and β can by determined by numerical fitting of the spectra. In Fig. 4 are compared the experimental and calculated curves. A very good agreement was observed for the real parts of the permeability spectra for all samples. Table 1 shows these dispersion parameters for the samples characterized by different microstructures.

It was found that the permeability in the 100 MHz region is determined mainly by the spin rotation magnetizing mechanism, but the contribution of the domain wall motion is not negligible in the 10MHz region. Spin component is dominated by the chemical composition, while the domain component is affected by both the composition and the microstructure [6]. Using the parameters deduced by the real permeability curves were calculated the imaginary permeability and the parameters are summarized in Table 1. In Fig. 5 are shown the experimental and calculated imaginary permeability for R2.

The results confirm a relation between the relaxation of the permeability and the changing in shape of hysteresis loop [7]. The real parts of the permeability are about 250, 350 and 750 for sample R1, R2 and

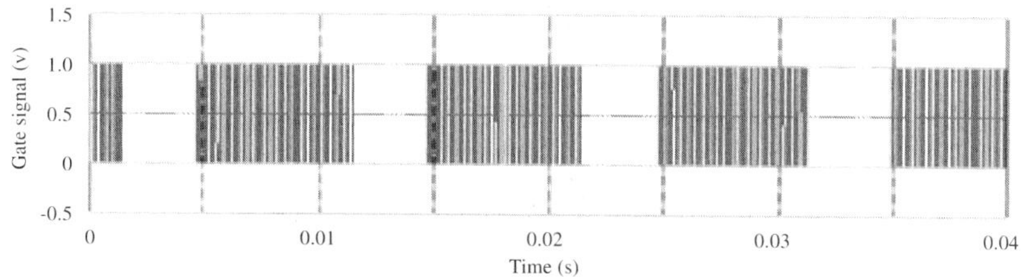

Fig. 6. Gate signal.

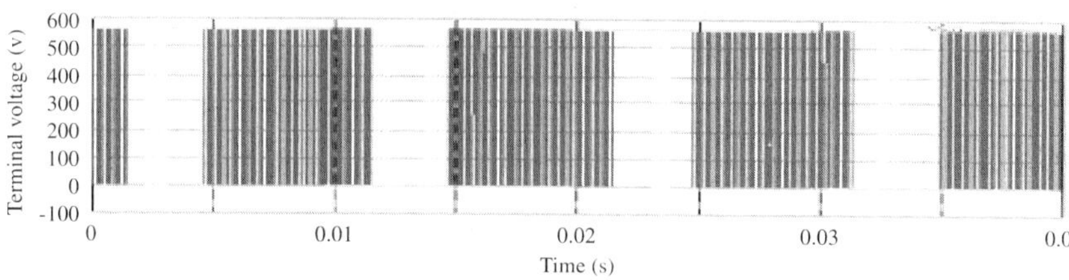

Fig. 7. Voltage waveform.

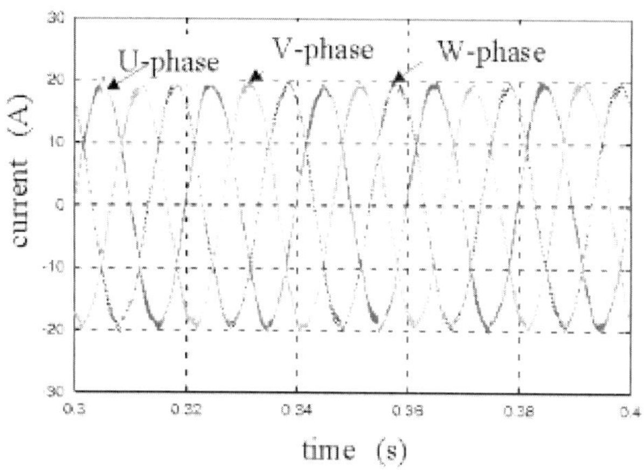

Fig. 8. Current waveform.

4. Analysis result

Figure 5 shows a control diagram for Interior Permanent Magnet motor (IPM motor). The control diagram is based on the characteristic equation of the d-q axis.

In this simulation, sampling time of current control is set to 250 μs. Sampling time of speed control is set to 2 ms. The torque constant of the IPM motor is set to 2.86 Nm/Arms.

Gate signal and terminal voltage are shown in Figs 6 and 7. Figure 8 shows the calculation result of current waveform. Figure 9 shows the FFT result of the current when the PWM frequency is set to 4 kHz, 8 kHz and 12 kHz. The main frequency components are 50 Hz and the PWM frequency. The 50 Hz on the scale is fundamental frequency of the current.

Using the FFT result of current waveform shown in Fig. 7, the iron loss is calculated by the two-dimensional finite element method. The rotor can be considered locked at the calculation in Pc because the current ripple frequency is so high. So Pf is calculated from a rotating rotor and Pc is calculated from

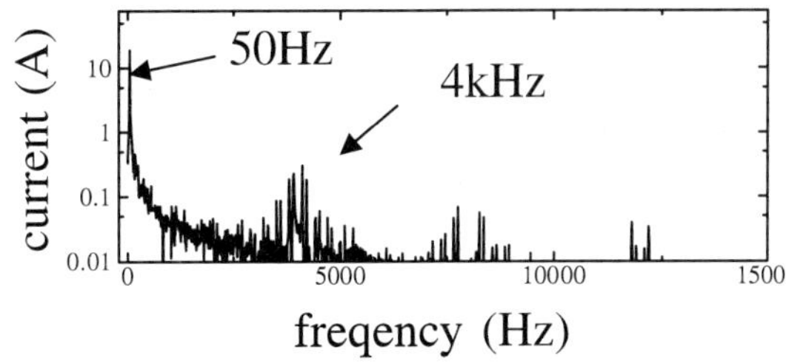

(a) Carrier frequency 4k

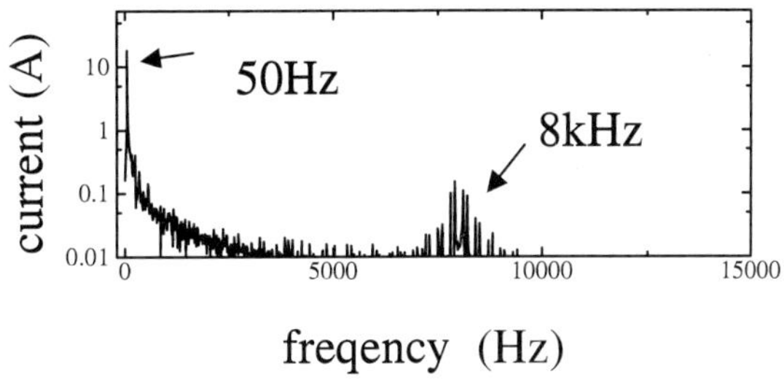

(b) Carrier frequency 4kHz.

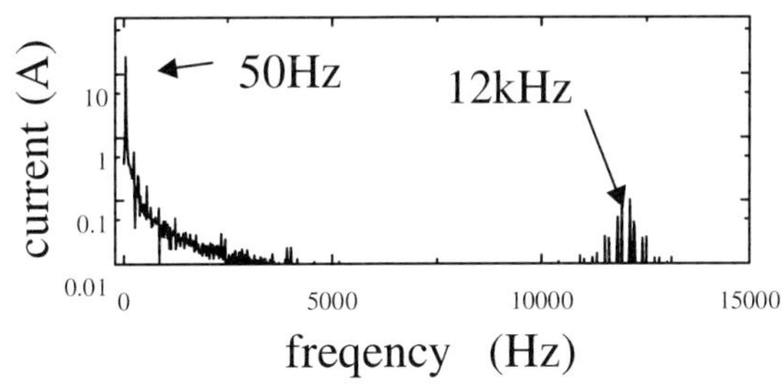

(c) Carrier frequecy 12kHz

Fig. 9. FFT of current waveform.

a locked rotor. Pf and Pc include the eddy current loss and hysteresis loss according to the analysis flow shown in Fig. 1. Figure 10 shows a magnetic flux distribution caused by fundamental current. Figure 11 shows a distribution of total iron loss density. The total iron loss is as follows;

$$\text{Ptotal} = Pf + Pc \tag{1}$$

Ptotal: total iron loss
Pf: iron loss by fundamental current component
Pc: iron loss by carrier frequency current ripple component

The iron loss mainly occurs in stator teeth. Iron loss appears on the surface of the rotor which is usually called surface loss. Figure 12 shows the influence of PWM frequency. The ratio of Pf/Pc is almost 20% at the 4 kHz PWM frequency. As the PWM frequency become higher, Pf/Pc becomes small.

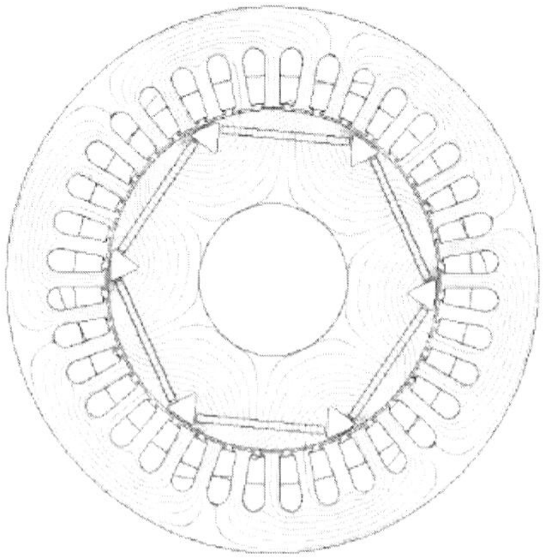

Fig. 10. Magentic flux ditribution caused by fundamental current.

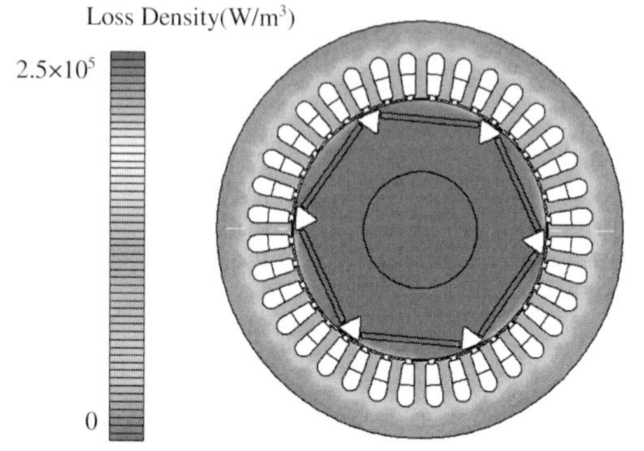

Fig. 11. Distribution of total iron loss.

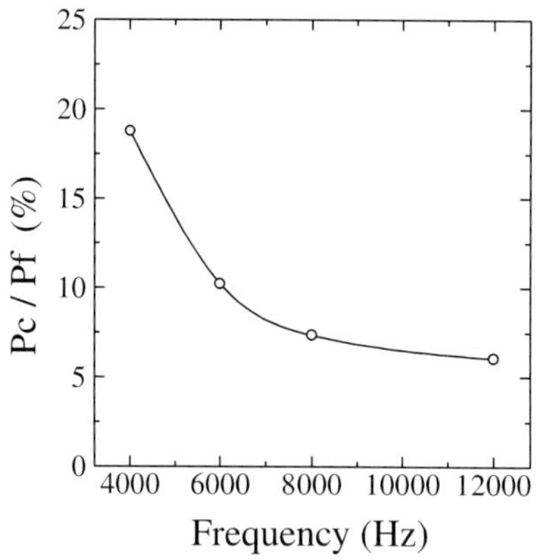

Fig. 12. Influence of carrier frequency.

5. Conclusion

A procedure has been developed to simulate iron loss under inverter drive. The iron loss is affected by the drive system. To calculate current waveform considering drive system, it is possible to predict iron loss precisely. The method was applied to an IPM motor and the previously simulated results such as iron loss distribution and the influence of PWM frequency were validated. This method can also be applied to other types of motors and drive systems.

References

[1] K. Atallah, A.Q.Zhu and D. Howe, An Improved Method for Predicting Iron Losses in Brushless Permanent Magnet DC Drives, *IEEE Trans. on Mag.* **28**(5) (September 1992).

[2] H. Kometani and S. Sakabe, Dynamic Electro-magnetic Analysis of a Permanent Magnet Synchronous Motor with the Support Ring Combined with the Inverter Circuit Equations, *Trans. IEEJ* **155-D**(12) (1995).

[3] S. Kitamura, Y. Ishihara, T. Todaka and Y. Inoue, Magnetic Field Analysis of DC Brushless Motor considered Volt-Ampere Characteristic of Feedback Diodes, *Trans. IEEJ* **116-D**(12) (Dec 1996).

International Journal of Applied Electromagnetics and Mechanics 13 (2001/2002) 251–256
IOS Press

Monte Carlo simulations of interaction effects on magnetic viscosity in ferromagnetic granular media

C.G. Verdes[a,*], B.R. Diaz[d], R.W. Chantrell[b], S. Thompson[d] and Al. Stancu[c]
[a]*Magnetic Materials Research Group, Center for Materials Science, University of Central Lancashire, Preston, PR1 2HE, UK*
[b]*Department of Physics, University of Durham, Durham, DH1 3LE, UK*
[c]*"Al. I. Cuza" University, Iasi, Romania*
[d]*Department of Physics, University of York, York, YO10 5DD, UK*

Abstract. A Monte Carlo time and temperature dependent model was developed to simulate magnetization processes in fine ferromagnetic particle media. The model uses a combination between a time dependent Stoner-Wohlfarth model and a standard Metropolis algorithm to find the magnetization of single-domain particle systems. Both magnetostatic and exchange interactions are introduced explicitly.

Variation of magnetic viscosity with interactions and temperature is studied. Results show that viscosity exhibits a maximum around coercive field. The value of this maximum is closely related to the strength and the origin of interactions as well as to temperature. Dipolar interactions decrease the viscosity and broaden the viscosity vs. field curve, while exchange interactions lead to an increase in viscosity around coercive field. An increase in temperature up to the critical temperature for the superparamagnetic behavior leads to an increase in viscosity. Further increase of temperature produces a decrease in viscosity.

1. Introduction

Magnetic recording media are used to store information for long periods of time, typically years, therefore the need for stable storage media so that information can be accurately read when necessary. The need for high densities requires that written bits become smaller, susceptible to thermal fluctuations that can affect the quality of the record. In ferromagnetic materials, in general, the value of the magnetization is time dependent. This dependence generally increases with temperature and decreases with particle size. To characterize the time dependence of magnetization, one uses the "magnetic viscosity", $S = \frac{dM}{d(\ln(t))}$.

Here, a computational model for single domain particle systems has been developed, a time and temperature dependent model capable to predict magnetic behavior of granular systems. The model uses a time-dependent Stoner-Wohlfarth approach technique to find the magnetization of a sample depending on the applied field, temperature, measuring time and previous state of the system. The model also

*Corresponding author: C.G. Verdes, Fax: 0191 3743749; E-mail: claudiu-georgel.verdes@durham.ac.uk.

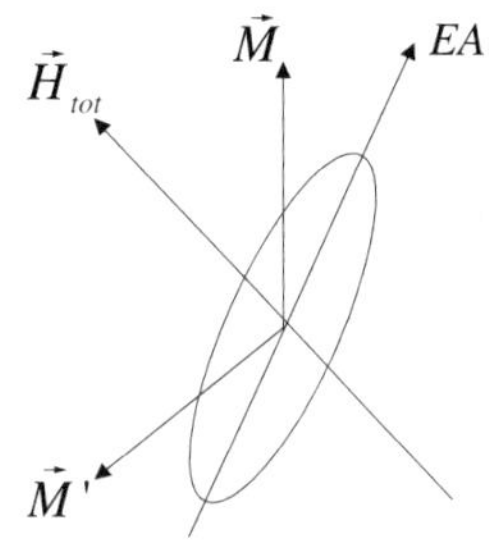

Fig. 1. The two possible equilibrium orientations for the magnetic moment.

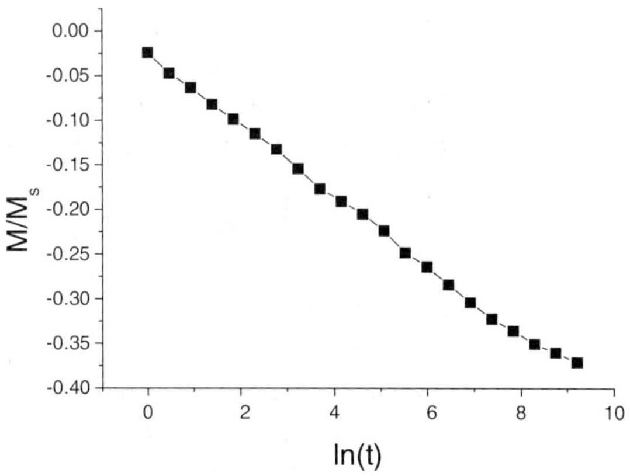

Fig. 2. Variation of M vs. $\ln(t)$ for a noninteracting system at coercivity.

includes the effect of thermal activation by means of Monte-Carlo simulations that is important in allowing the study of time dependent magnetization processes in fine particle systems.

Interaction effects are explicitly taken into account via the total interaction field (including exchange contributions) arising from other particles. The variation of interparticle interactions within the systems is obtained by using different packing fractions for the considered systems.

2. Description of the model

The model [1,2,5] consists of a cubic cell of interacting single domain particles. The particle sizes and the anisotropy fields are generated according to a lognormal distribution function. The anisotropy easy axes have a random three-dimensional distribution to make the system comparable to experimental granular systems.

The magnetic field that acts on each particle is the vector sum between the applied field $\vec{H}_{app}$ and the interaction field, which has two components, $\vec{H}_{int}$ and $\vec{H}_{ex}$ the dipolar field and the exchange-coupling field produced by the neighbor particles. The interaction field produced by the jth particle on the ith particle is:

$$\vec{H}_{ij} = \frac{3\left(\vec{\mu}_j \cdot \vec{r}_{ij}\right) \cdot \vec{r}_{ij}}{r_{ij}^5} - \frac{\vec{\mu}_j}{r_{ij}^3} + C_{ij} H_K \vec{\mu}_j \tag{1}$$

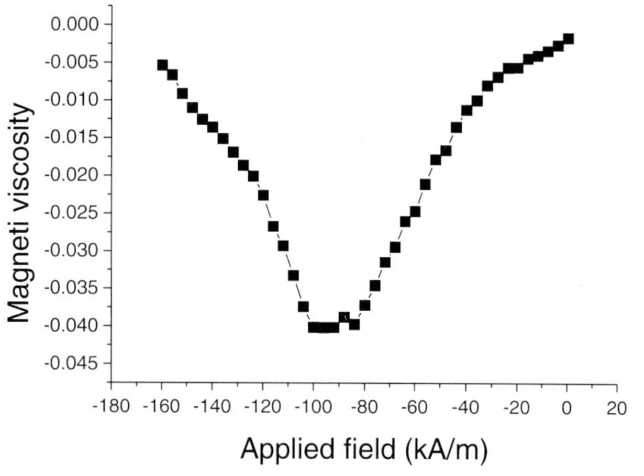

Fig. 3. Viscosity vs. applied field for a noninteracting system.

where $C_{ij} = \begin{cases} C^*, d_{ij} \leqslant d_{ex} \\ 0, \quad d_{ij} > d_{ex} \end{cases}$ with C^* the exchange coupling constant, d_{ij} the minimum distance between the surfaces of grains i and j and d_{ex} the maximum distance between the surfaces of two grains that allows exchange coupling interactions.

For every particle, the equilibrium position of the moment in the local field is calculated using the SW [4] model. However, if the SW model gives two equilibrium positions (this happens if the local field is smaller than the critical field for that orientation), the moment is allowed to jump between the two positions (see Fig. 1) with a probability p_r that depends on the temperature T, volume V, anisotropy constant K of the particle and the measuring time t:

$$P_r = 1 - e^{-t_m/\tau} \tag{2}$$

where t_m is the measuring time and τ is the relaxation time that has the expression

$$\tau^{-1} = f_0 \exp(-\Delta E/k_B T) \tag{3}$$

with ΔE the height of the total energy barrier for reversal, $f_0 = 9 \times 10^9$ Hz the frequency factor and k_B the Boltzmann constant

If transitions are not allowed, the magnetic moment goes to the equilibrium position that is closer to the actual orientation of the magnetic moment. If transitions are allowed the probability for the moment to go to one of the two positions is:

$$p_i = \frac{\exp(-E_i/k_B T)}{\exp(-E_1/k_B T) + \exp(-E_2/k_B T)} \tag{4}$$

where $i = 1, 2$, and E_i is the particle energy in the ith state.

After finding the equilibrium position, thermal equilibrium is achieved via a standard Metropolis [3] algorithm. Small random orientation changes are allowed for the magnetic moment and the energy E' for the new orientation and the energy change $\Delta E = E' - E$ with respect to the old orientation are calculated. The magnetic moment is allowed to stay in the new orientation with the probability:

$$p = \min\left(1, e^{-\Delta E/k_B T}\right) \tag{5}$$

so that the system reaches the thermal equilibrium state within the local energy minimum.

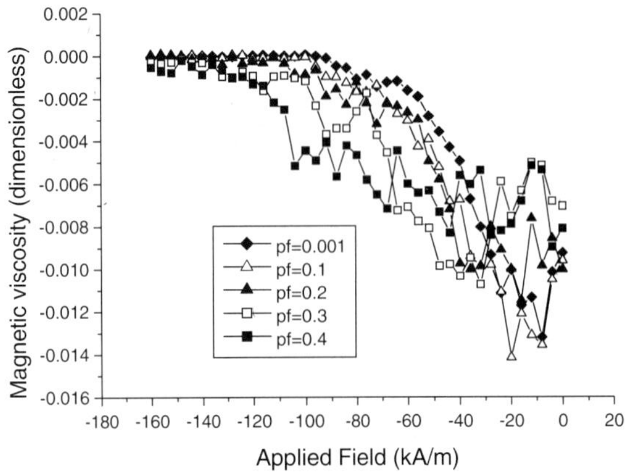

Fig. 4. Magnetic viscosity for different systems at $T = 300$ K.

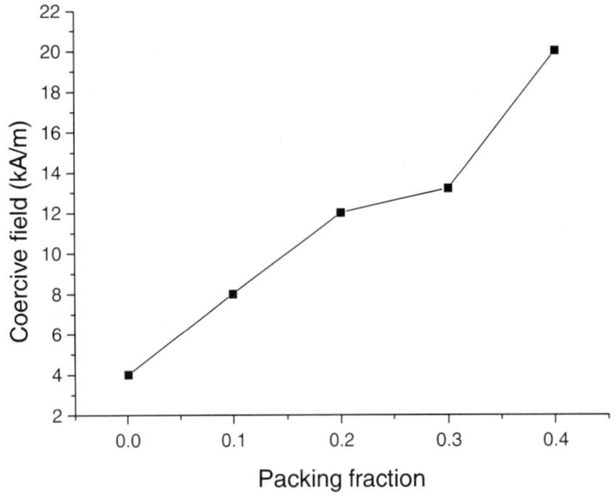

Fig. 5. Variation of coercivity with packing fraction at $T = 300$ K.

3. Results and discussions

The simulations were made using a primary cell usually containing 1000 particles with the anisotropy constant of $K = 2 \times 10^5$ J/m^3 and a saturation magnetization $M_s = 7 \times 10^5$ A/m. The average particle diameter was of 9 nm with a standard deviation of $\ln(D) = 0.1$.

In this paper we study the time dependence of the magnetization characterized by the "Magnetic viscosity coefficient" $S = dM/d(\ln(t))$. To measure the viscosity coefficient, the system is saturated with a high magnetic field. Then, the saturation field is removed and an opposite magnetic field, H, is applied. The magnetization of the sample is measured after different periods of time, and the viscosity is being defined as the slope of the M vs. $\ln(t)$ curve (Fig. 2).

Conventionally, the magnetic viscosity is represented as a function of the applied field H. Figure 3 shows such a curve for a system without interactions, at a temperature of 100 K.

The magnetic viscosity exhibits a maximum around coercivity, which is to be expected since at that point the energy barriers are the smallest and it is easy for the magnetic moment to switch from an equilibrium position to another. Figure 4 shows the variation of viscosity with packing fraction for systems at 300 K while Fig. 5 shows the variation of coercivity for the same systems.

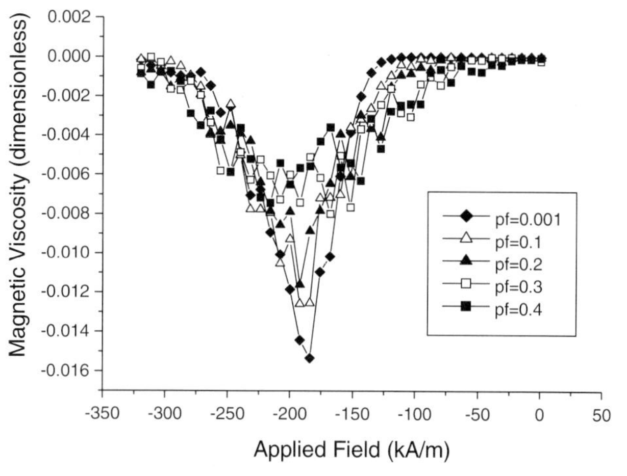

Fig. 6. Magnetic viscosity for different systems at $T = 20$ K.

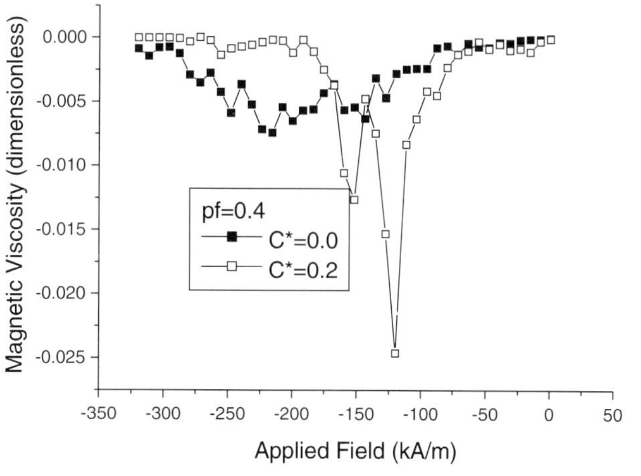

Fig. 7. Effect of exchange coupling on magnetic viscosity.

These results are for systems with dipolar interactions only and the KV/k_BT value is low enough so that for the system with low packing density most of the particles are superparamagnetic. In the high-density systems, interactions lead to an increase in the energy barrier so that some of the particles become stable.

The increase in coercivity with packing fraction, i.e. with increasing interactions, is due to occurrence of cooperative effects leading to a higher "effective volume" for thermal reversal and has been explained elsewhere [5].

Another effect of magnetostatic interactions is the decrease in the viscosity, as well as broadening the curve. This is also true for systems at low temperatures (20 K) where all particles are blocked (Fig. 6).

However, the exchange interactions, which occur mostly in high-density systems, produce a different behavior. Figure 7 shows that exchange coupling produces a substantial increase in viscosity at coercivity while it reduces its value at other fields.

This is consistent with the fact that for such systems, the slope of the hysteresis loop is quite high in this region, due to the formation of "domains" of particles that switch simultaneously [5]. For the

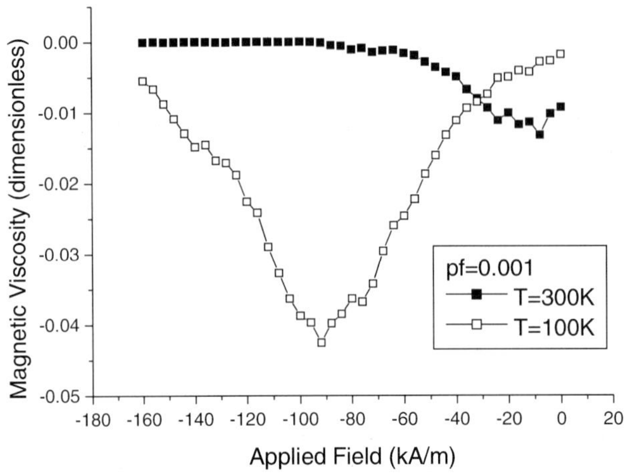

Fig. 8. Variation of viscosity with temperature.

relaxation mechanism a similar behavior is present, in the sense that the whole cluster will relax at once, producing big variations in the magnetic moment, i.e. high values for the magnetic viscosity.

Although the values for the maximum magnetic viscosity are about the same for superparamagnetic and blocked particles, there is a thermal dependence of this. Figure 8 shows that, for an intermediate temperature of 100 K, at the limit between blocked and superparamagnetic behavior for most of the particles, the viscosity dramatically increases. Presumably, this is due to the fact that the thermal fluctuation energy increases not enough to overcome the energy barriers over short time periods but sufficiently to produce reversals to the more stabile state on the long term, for most of the particles.

4. Conclusions

A micromagnetic model has been used to study the interaction and thermal effects on magnetic viscosity of granular systems. It has been shown that viscosity exhibits a maximum in the range of fields where the slope of the hysteresis loop has a maximum value, which usually is around coercivity. Magnetostatic interactions have been found to reduce the value of viscosity contributing to a greater stability of information while exchange coupling interactions further decrease the viscosity except around the coercive field. Highest values for viscosity are found when the temperature is about the value of the critical temperature for the superparamagnetic behavior.

References

[1] R.W. Chantrell, A. Lyberatos, M. El-Hilo and K. O'Grady, Models of slow relaxation in particulate and thin film materials, *J. Appl. Phys.* **76** (1994), 6497.

[2] M. El-Hilo, R.W. Chantrell and K. O'Grady, A model of interaction effects in fine particle systems, *J. Appl. Phys.* **84**(9) (1998), 5114.

[3] N. Metropolis, A.W. Rosenbluth, M.N. Rosenbluth, A.H. Teller and E. Teller, Equation of state calculations by fast computing machines, *J. Chem. Phys.* **21** (1953), 1087.

[4] E.C. Stoner and E.P. Wohlfarth, A mechanism of magnetic hysteresis in heterogeneous alloys, *Phyl. Trans. Roy. Soc. London Ser.* **A240** (1948), 599.

[5] C.G. Verdes, B. Ruiz-Diaz, S.M. Thompson, R.W. Chantrell and Al. Stancu, A computational model of the magnetic properties of interacting fine particles, *Phys. Rew. B.* (submitted to 2001).

International Journal of Applied Electromagnetics and Mechanics 13 (2001/2002) 257–262
IOS Press

A micromagnetic analysis of the Preisach model hypotheses on magnetic patterned media

Alexandru Stancu, Laurentiu Stoleriu and Mihai Cerchez*
"Alexandru Ioan Cuza" University Faculty of Physics, 11 Blvd. Carol, 6600, Iasi, Romania

Abstract. The interactions in particulate ferromagnetic media are generally described by magnetostatic interactions and exchange interactions. The Preisach type models describe the interactions by the use of an interaction distribution and a mean interaction field. It is usually considered that the mean interaction field is determined by the exchange interactions in the system, while the statistical interactions are determined by the magnetostatic interactions. This paper presents a micromagnetic analysis of the hypotheses used in the Preisach models to describe the magnetic particulate system.

1. Introduction

The Preisach type models describe the interactions in the system by a distribution of interactions which, basically, have the role of shifting the hysteresis loops of the individual particles and a mean interaction field which is proportional to the value of the magnetic moment of the sample. The Classical Preisach Model (CPM), is essentially a description of the system as a distribution of hysteresis loops distributed as a function of coercivity (width of the loops) and interaction fields (shift of the loops). To clarify the physical grounds of the hypotheses used in the Preisach model we have used a physical micromagnetic model.

2. The model

We have generated 3D and 2D samples with different geometries and up to four thousand Stoner-Wohlfarth particles. The system is created using a lognormal distribution of volumes and a Gauss distribution of the orientation of the easy axes of the particles.

The magnetic field applied to each particle is calculated as the sum of the external applied field and the interaction field given by its neighbors. The orientation of the magnetic moment of each particle is computed by solving the system of Landau-Lifshitz equations which describe the motion of the coupled magnetic moments. Boundary free conditions have been used in order to avoid edge effects.

For each value of the applied field one calculates the interaction field as the vector sum of the magnetic fields coming from the rest of the particles of the system and its projection and the coercive field of the particle are used to obtain the critical fields of the particles to represent the particle in the Preisach plane.

*Corresponding author: Mihai Cerchez; Tel.: +40 32 201175; Fax: +40 32 201205; E-mail: CERCHEZ@UAIC.RO.

1383-5416/01/02/$8.00

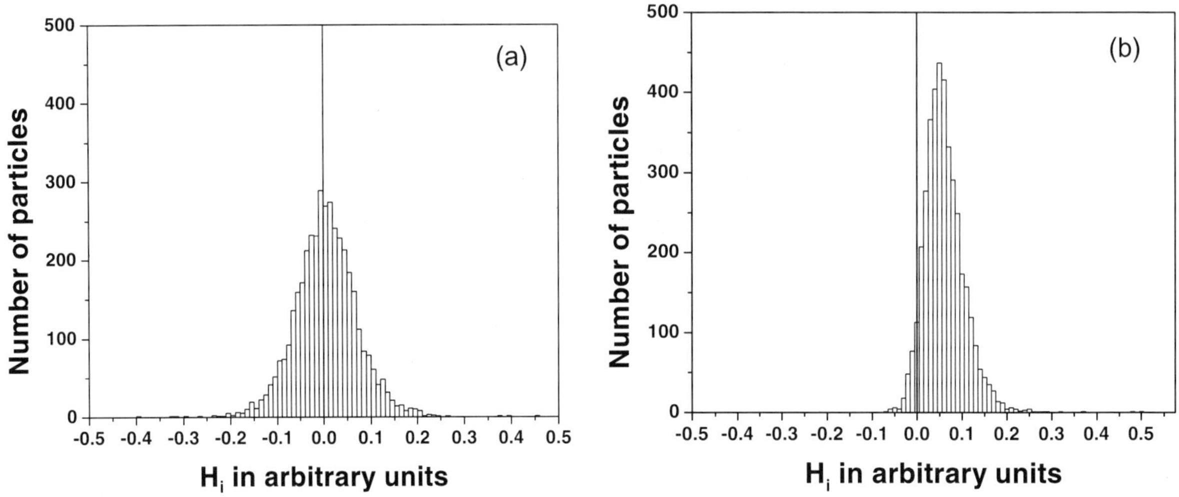

Fig. 1. The interaction field distribution in two magnetic states: (a) demagnetised state, (b) saturated state.

3. Statistical interactions

The assumptions of the statistical interaction fields is due to the distribution of the volumes of the particles and implicitly of the magnetic moments of the particles, as well as the variable distance from the particle to particle which lead to magnetic interactions which are random in nature. In the classical Preisach model [1] the mean interaction field is zero. In the Moving Preisach models [2], as a first approximation to describe the changes in the interaction field distribution during the magnetisation processes, a mean interaction field is considered and the interaction field distribution is shifted with the value of the mean interaction field which is proportional to the magnetic moment of the sample. This shift is called moving effect and the proportionality constant between the mean interaction field and the magnetic moment of the sample is called the moving parameter.

Figures 1 (a) and (b) presents the interaction field distribution for a 2D system in two magnetisation states: the DC demagnetised state (Fig. 1(a)) and the magnetic saturated state (Fig. 1(b)). The interaction field distribution is represented as the number of particles which are subjected to the same interaction field from their neighbours. A clear shift of the distribution is seen in the magnetic saturate state but one may also observe a slight asymmetry of the distribution. This asymmetry is observed for all geometrical configurations of the system, including a 2D system where all particles are positioned on a uniform grid with magnetic moments oriented along the saturation field direction. Therefore, one may assume that this asymmetry is due to the volume distribution considered which determines the magnetic moments magnitude distribution. The interaction field distribution for the same sample in a demagnetised state is symmetric and has a standard deviation significantly larger than that calculated for the saturation state [3]. Calculating the interaction field as the vector sum of the fields coming from the neighbour particles and the coercive field for each particle in the applied field direction, one may represent (Fig. 2) each particle in the Preisach plane with the coordinates given by the switching fields of the particle (H_α and H_β). For a simple magnetisation process one may notice a quite good concordance with the hypotheses of the Preisach model.

If one calculates the trajectory of the particles in the Preisach (Fig. 3) plane during the magnetisation process, one observes that the particles are changing positions with a random behaviour, but the distribution evolves in a predictable manner. However the initial and final points on this trajectory, which

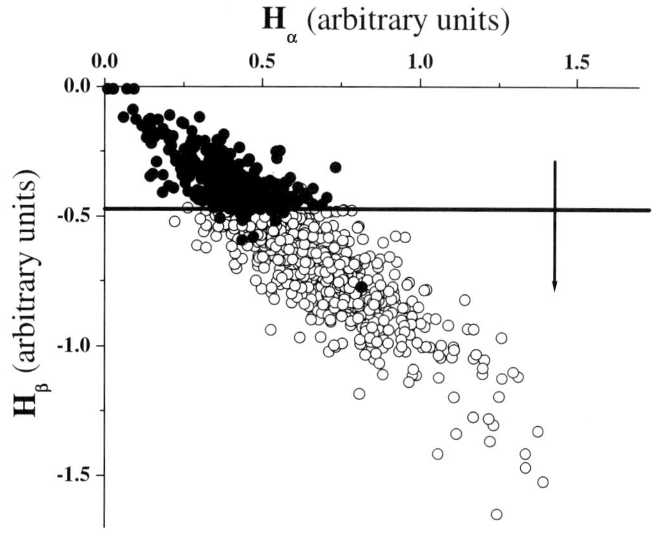

Fig. 2. The Preisach plane during the MHL process.

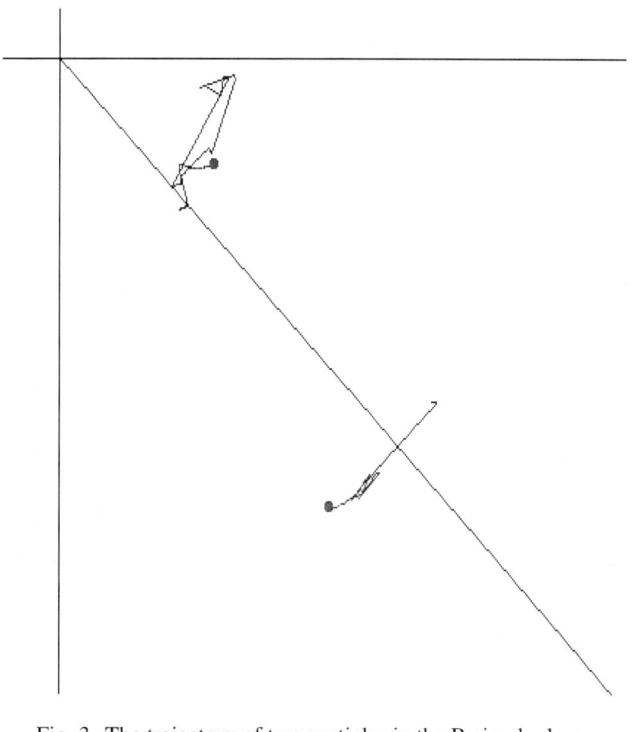

Fig. 3. The trajectory of two particles in the Preisach plane.

correspond to the positive and negative saturation states are quite symmetrical with respect to the bisector of the forth quadrant of the Preisach plane. The movement of the particles in the Preisach plane affects the memory zones. Thus, one may notice particles that are already reversed even though the field is lower than the theoretical field required to reverse the particle in Preisach models (Fig. 2). The statistical stability [4] reefers to the fact that a particle may change place during the magnetisation process but

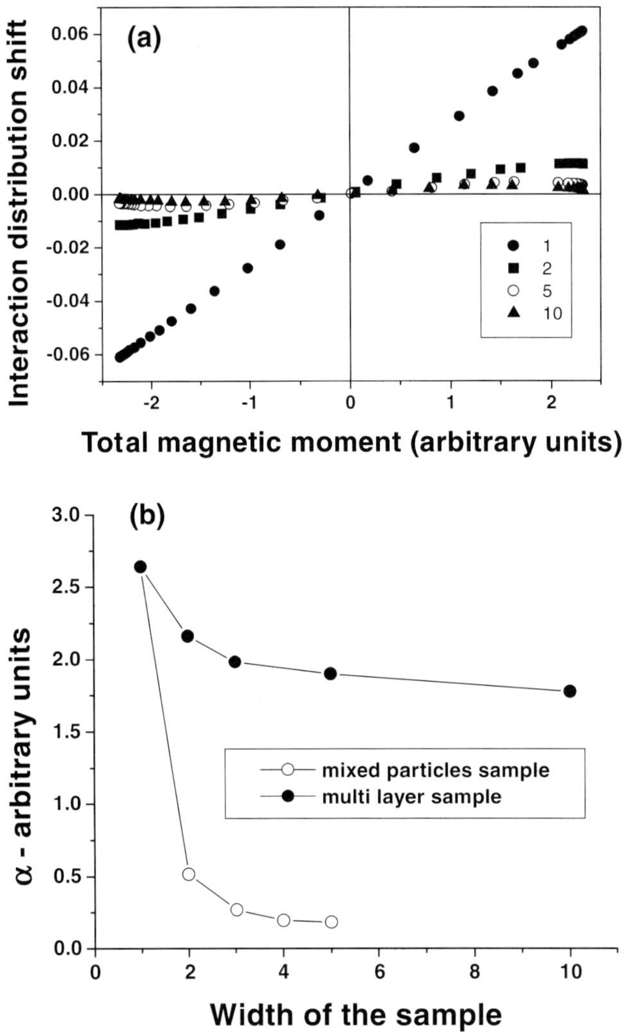

Fig. 4. (a) – Moving of the most probable value of the interaction field distribution for four mixed particle samples (with width from 1-2D sample, to 10 times the maximum diameter of the particles contained in the 2D sample); (b) – Moving constant (α) as a function of the sample width for the mixed particle sample and the multi-layer sample.

there is always another particle which takes its place. However, this is not observed in the system as the distribution is moving towards higher interaction fields and its variance depends on the magnetic moment of the system.

4. Mean field interactions

We have also analysed the effect of the sample width on the moving constant (the shift of the interaction distribution as a function of the magnetic moment of the sample). We have made simulations on samples with increasing width, starting from a 2D-like sample where all the particle centres were taken in the same plane (3D sample of one particle width). We also have made simulations on multi-layer systems

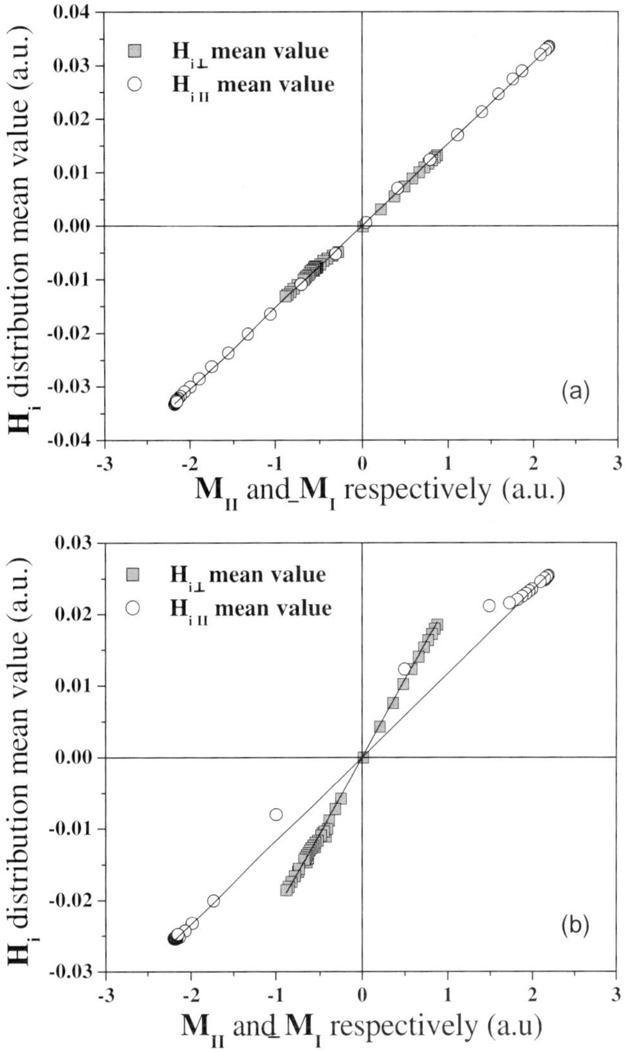

Fig. 5. Moving of the interaction field distribution for two directions (longitudinal and transversal) and for two samples (a) – random arranged particles and (b) – particles arranged in columns.

made from a number of parallel, interacting, 2D samples. The first sample-type was named "mixed particle" sample while the second one was named "multi-layer" sample.

In Fig. 4(a) one presents the evolution of the interaction distribution shift as a function of the magnetic moment of the sample on the major hysteresis loop. There are presented the results for four mixed-particle samples with the width increasing from the 2D sample (denoted by 1) to a width equal to 10 times the maximum diameter of the particles in the 2D sample (noted with 10). One observes that for all samples a linear moving term is a good approximation. However, we have obtained different slopes (proportional to the moving constant) depending on the sample width. The 2D sample has the maximum slope and as the width is increasing the slope is decreasing to zero. For the multi-layer samples the slope is not decreasing to zero, but to a positive finite value. These results are presented in Fig. 4(b). The results show a strong influence of the sample geometry on the value of the mean field parameter in samples where the exchange interactions could be neglected. Some authors (e.g. [5]) consider different moving

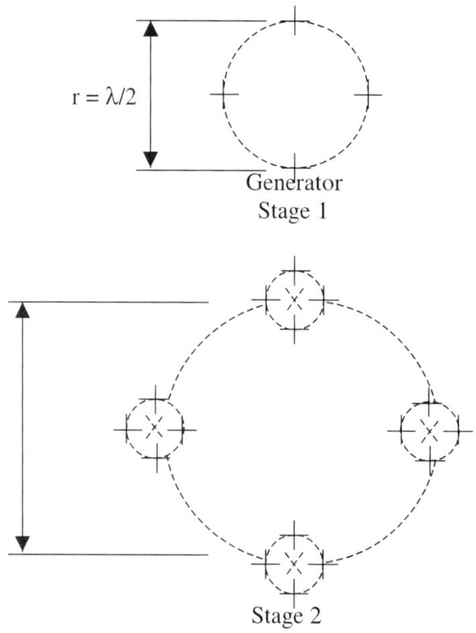

Fig. 1. Illustration of the procedure used in the design of a fractal array based on a two element sub-array.

2. The fractal approach

By definition [2] a fractal is a recursively generated object, shape or volume having a fractional dimension. There are two main terms used to describe a fractal, either it is random or deterministic. An example of a random near fractal structure could be a coastline or a bolt of lightning, while a deterministic fractal structure uses a known 'generator' to deduce successive size scales, where each size scale is referred to mathematically as iteration. An example of a deterministic structure is the concentric circular ring which will have the form shown in Fig. 1.

The generator has diameter of the half wavelength and is then scaled using the scaling factor δ which is the relationship between the desired frequencies $\delta = f_2/f_1$, where $f_2 > f_1$. The quarter wavelength monopoles situated at each appropriate X (Fig.1) constitute the design. The distance between two monopoles $\delta r = \delta \cdot r$, where r is the half wavelength (λ/2).

3. Generator design

Experiences with the design of the tracking antennas in [1], suggested that in order to achieve circular polarisation of the radiation pattern, a cross-Yagi design needed to be used. A similar approach was adopted with the design of the fractal antennas.

The commercial antenna simulation software NEC4WIN95 was used for simulations. The first stage was to investigate a single half wavelength (λ/2). The classical doughnut field shape in free space was obtained (Fig. 2), with gain in forward direction of 2.2 dBi and the half of the gain lost in the backward direction.

The next stage was to simulate a cross dipole structure where the second element was placed a quarter of the wavelength (λ/4) in free space above the first, or the feed to the second element was delayed by and extra quarter of the wavelength (λ/4) to produce the necessary phase shift between the two

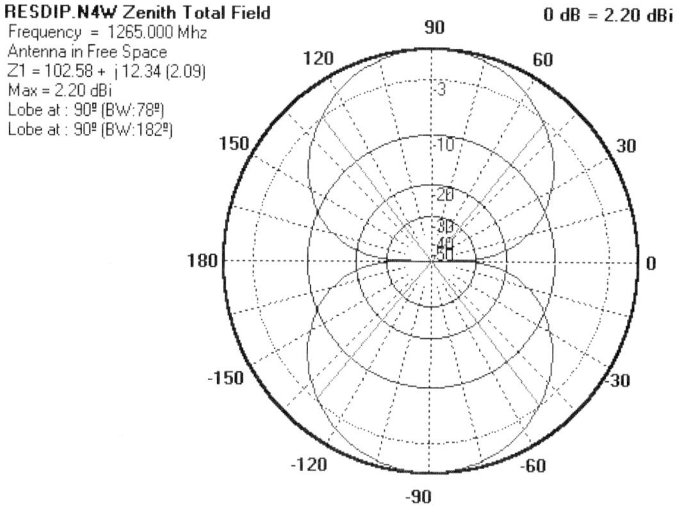

Fig. 2. Radiation pattern of $\delta/2$ dipole.

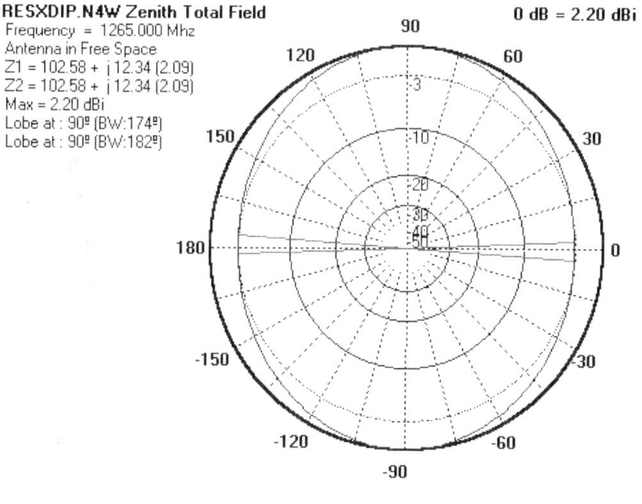

Fig. 3. Radiation pattern of cross-dipole structure.

elements. The doughnut effect was eliminated but the gain was still 2.2 dBi and half of the power was in the backward direction (Fig. 3). Placing the structure 0.01 m above the ground plane eliminated any backward radiation resulting in gain of 8.04 dBi as well as the circular polarisation (Fig. 4). The 3 dB or half power beamwidth was now 90°. This structure was used as the generator for the fractal antenna design.

4. Up-link design and simulation

For the FedSat up-link design the following parameters were used: $f_1 = 313.55$ MHz, for the Merlion up-link $f_2 = 1265$ MHz. The resulting fractal antenna coefficients were $\delta = 4.0344$, $r = \delta/2 = 0.118$ m

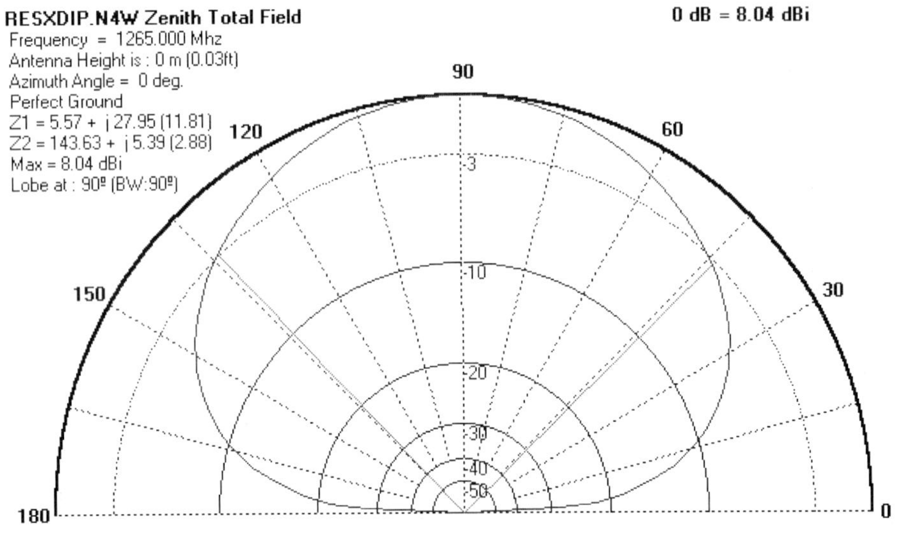

Fig. 4. Radiation pattern of cross dipole placed above ground plane.

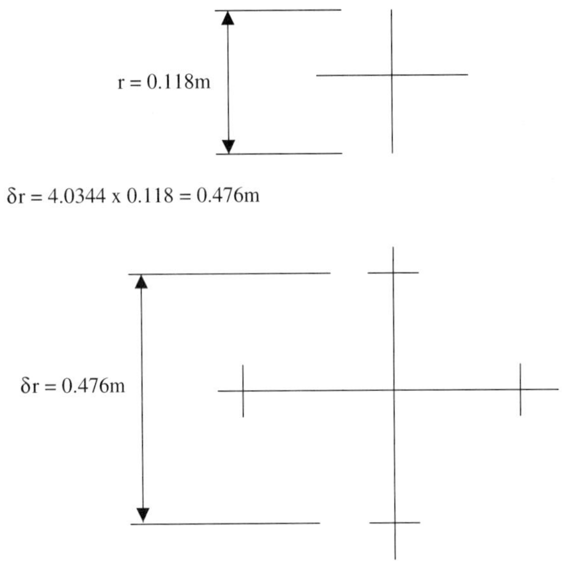

Fig. 5. Two stages of the up-link fractal satellite antenna design with circular polarisation.

and $\delta r = 0.476$ m. The two stages of the design, i.e. for the generator and for the final two element sub-array, are illustrated in Fig. 5. The radiation patterns obtained using NEC4WIN95 at frequencies f_1 and f_2 are shown in Figs 6 and 7, respectively. A plot of the gain achieved for this antenna for frequencies between 100 MHz and 2000 MHz is shown in Fig. 8. Results obtained from the design satisfy the requirements for the antenna at both specified frequencies. Analogous design and simulation process was repeated for the down-link antenna. Design specifications and simulation results obtained from designs for the up- and down-link fractal antennas are shown in Table 1.

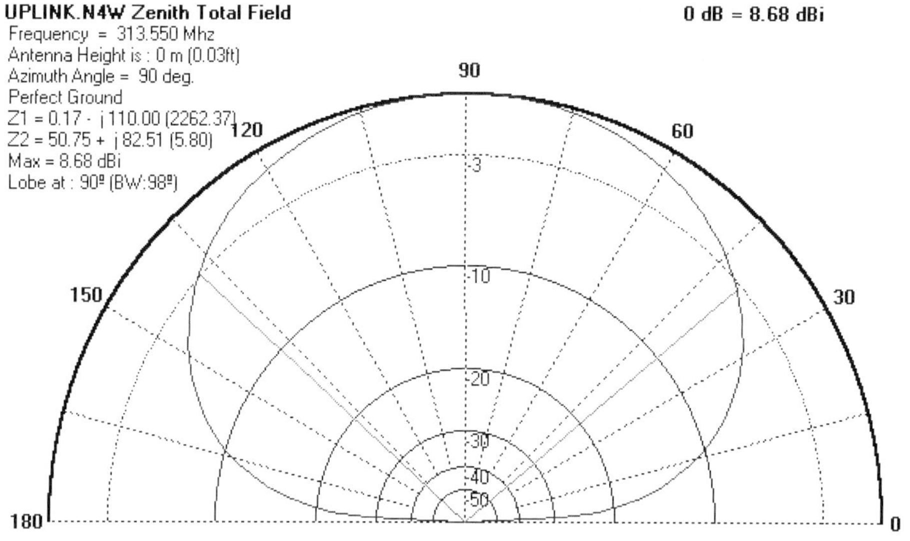

Fig. 6. Radiation pattern of up-link antenna at 313.55 MHz.

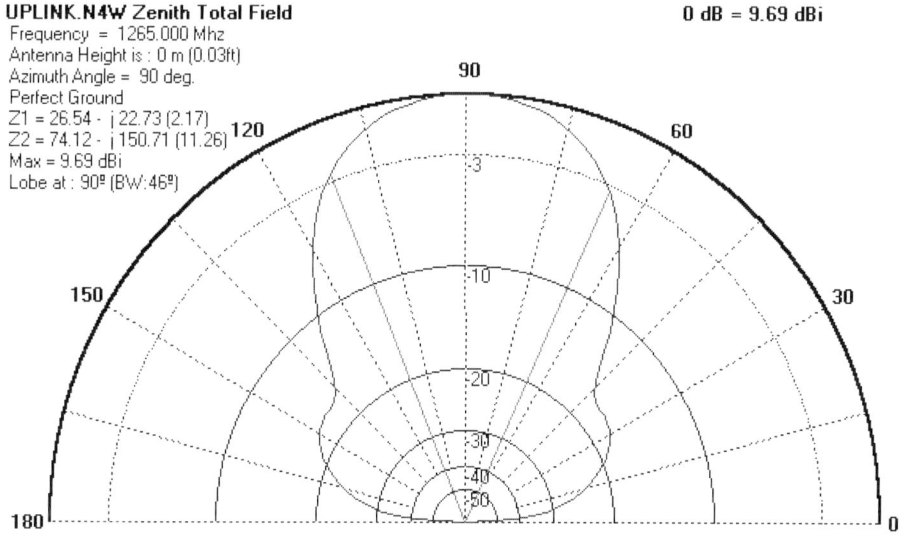

Fig. 7. Radiation pattern of up-link antenna at 1265 MHz.

5. Conclusions

Using design techniques associated with fractal antenna arrays, two antennas have been designed and simulated to act as either the up-link or down-link to the FedSat and Merlion satellites. The theory was to use the fractal antenna characteristic of multi-band, frequency independence to reduce the number of antennas necessary for the satellite communication links from four to two. Simulation results show us that the design specifications for the gain of each antenna have been met.

The project undertaken that led to this investigation was done so as a final year project for the Bachelor

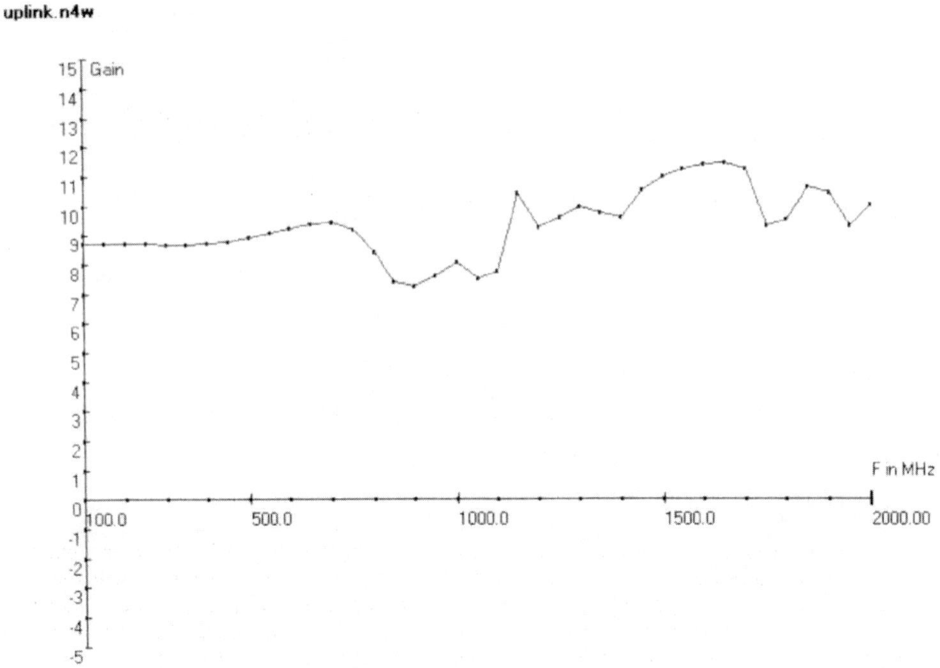

Fig. 8. Gain of up-link antenna over the frequency range from 100 MHz to 2000 MHz.

Table 1
Summary of design specifications and simulated results of the two designs of fractal antennas

Up-link Frequencies MHz	Required Gain dB	Simulated Gain dB
313.55	7	8.68
1265	7	9.69
Down-link Frequencies MHz	**Required Gain dB**	**Simulated Gain dB**
400.4	7	8.83
2425	7	7.78

of Engineering (Telecommunications), School of Electrical and Information Engineering, University of South Australia. As such it is believed that the University may take this investigation further through the manufacture and testing of such antenna. This may take the form of either post-graduate study or a project for current final year students in this field. The results of any further study should then be used to more accurately assess the suitability of the fractal antenna for this application.

References

[1] J. Verringer and B. Tovirac, Tracking antennas for FedSat and Merlion, Final year project, University of South Australia, School of Electrical and Information Engineering, 2000.
[2] T. Vernon, Fractal antennas offer benefits, *Radio World* (1 September, 1999).

International Journal of Applied Electromagnetics and Mechanics 13 (2001/2002) 277–283
IOS Press

Temperature characteristics of a new magneto-optical current transformer

Yuri S. Didosyan, Hans Hauser* and Walter Toriser
Institute of Industrial Electronics and Material Science, Vienna University of Technology, Gusshausstrasse 25-29/366, A-1040 Vienna, Austria

Abstract. In contrast to existing current transformers utilizing the Faraday rotation of light polarization, the principle of operation of the new sensor is based on the direct registration of the domain wall positions in an orthoferrite, a transparent ferromagnet. The results of the measurements are not affected directly by temperature changes of Faraday rotation, birefringence, fluctuations of the incident light beam, polarization distortions in fibers, etc. The amplitude of domain wall motion remains constant in a wide range of driving magnetic field frequencies from dc to hundreds kHz. With the further growth of the frequency the amplitude decreases in two times at 5 MHz. In the temperature range from 20°C to 130°C the domain wall mobility and the bandwidth decrease in about three times.

1. Introduction

Based on the Faraday effect, magnetooptical current transformers (MOCTs) provide numerous advantages as compared with conventional measuring transformers [7]. The measurements are contactless and the measured signal is transmitted by a light beam propagating in air or in optical fibers. This eliminates costly problems of potential separation. Moreover, MOCTs provide a wide frequency bandwidth and can measure even dc in an extremely high dynamic range. In the vast majority of the MOCTs developed to date the main parameter is the angle of polarization rotation. The specific rotations of the polarization plane in paramagnets and diamagnets are very low and therefore – to provide substantially large angles – the lengths of the sensors must be rather large (up to many meters). This imposes problems of birefringence which depends on temperature and on mechanical factors and seriously affects the measurement results.

Transparent ferromagnets, especially the ones with a strong uniaxial magnetic anisotropy, have much higher values of Faraday rotation [9]. But the values of Faraday rotation in the domains strongly depend on the temperature of the crystal and this leads to considerable uncertainties of measurements. Moreover, in all existing MOCTs additional efforts are made to reduce the influence of variations of both power and wavelength of the light source, of polarization distortions of the optical fibers connecting different parts of the sensor, of variations in the sensitivity of photodetectors, for example, on the measurement results.

*Corresponding author: Tel.: +43 1 58801 36614; Fax: +43 1 58801 36695; E-mail: Hans.Hauser@TUwien.ac.at.

1383-5416/01/02/$8.00

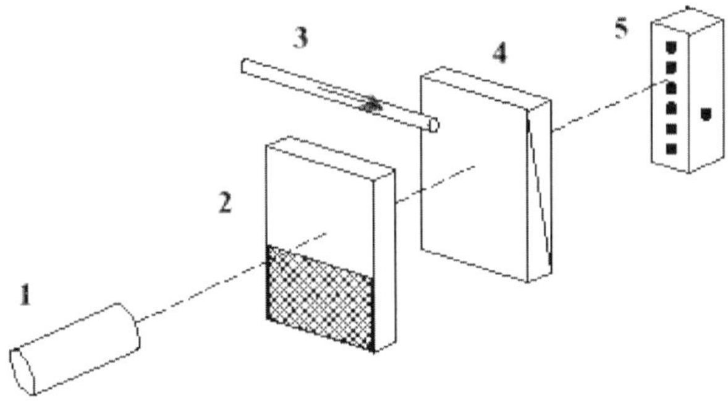

Fig. 1. Scheme of a MOCT based on DRDWP.

2. MOCT based on domain wall motion

A new sensor based on orthoferrite crystals was described recently [4,6,5]. The new sensor provides opportunity to get rid of the sources of uncertainties mentioned above. The principle of operation of the new sensor is based on the direct registration of the domain wall (DW) positions (DRDWP). Figure 1 shows the scheme of such a MOCT. A polarized light beam from the source (1) illuminates the sensor material, an orthoferrite plate (2). By means of two small magnets (not shown in the figure) a two-domain structure is formed in the plate. The plate is located in the vicinity of the conductor (3) carrying the current to be measured. The light emergent from the sensor propagates through the analyzer (4) and is incident on the position sensitive photoreceiver (5). The only function of the photoreceiver (5) is to determine the location of the boundary between the two parts of different brightness, which is the DW position. It depends on the applied field H produced by the current. The measurement results are not affected by temperature changes of the Faraday rotation, birefringence, fluctuations of the incident light beam, polarization distortions in fibers, and so on. These factors can reduce the contrast of the domain structure and thus the signal-to-noise ratio of the measurements but they cannot affect the main parameter – the position of the DW.

3. Measurements

In this work we investigate the temperature dependence of the DW mobility of yttrium orthoferrite.

Orthoferrites are magnetically uniaxial crystals. The action of a magnetic field on these crystals does not result in a magnetization rotation in the domains but rather in the motion of the DWs. Respectively the speed of this motion is a very important parameter responsible for the dynamic properties of the DRDWP sensors. In a wide range of the external field H – and of the velocities, from zero to the speed of sound (4.1 km/s) – the DW velocity v is proportional to H:

$$v = \mu H, \tag{1}$$

where μ is the DW mobility. Investigations of the DW mobility in weak ferrromagnets are of importance both for basic research [2,8] and for applications. There are different ways of measurements of the mobility μ. One way consists in measurements of the dependence of the velocity v on the applied magnetic field H. They can be performed by means of double exposure high speed photography [1] or by means of the dark field method [3]. Another way consists in measuring the amplitude of the DW oscillations under action of the magnetic field of a periodic electric current.

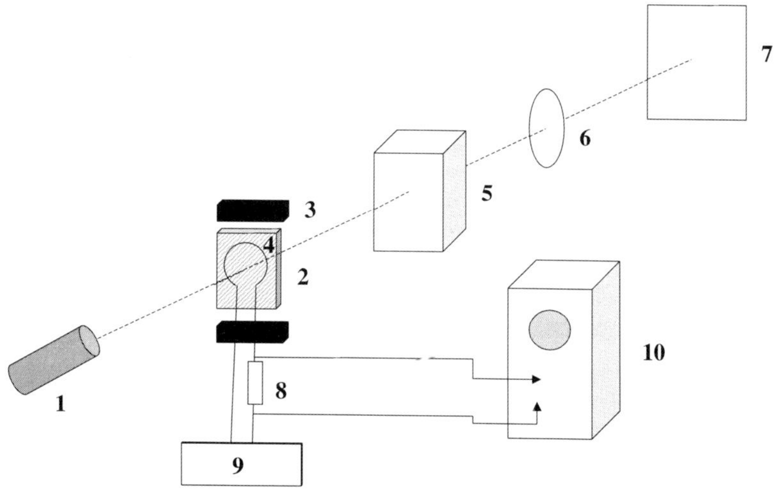

Fig. 2. Setup for DW mobility measurements.

The magnetic field H acting on the DW consists of two parts. One part is an inhomogeneous permanent magnetic field

$$H_k(x) = kx \tag{2}$$

with the gradient k along the x-axis producing and stabilizing a single DW in the plate. The second part consists of a homogeneous magnetic field $H_h(t)$ of the amplitude h:

$$H_h(t) = h \cdot f(t) \tag{3}$$

Thus Eq. (1) reads

$$\frac{dx}{dt} = \mu[kx + h \cdot f(t)] \tag{4}$$

For the case of the step-like time dependence ($H_h(t) = \pm h$) the solution of Eq. (4) is

$$x(t) = -\frac{h}{k}\left(1 - \frac{2\exp(-\mu kt)}{1 - \exp(-\mu kt/2)}\right) \tag{5}$$

One can see that the amplitude A of the DW oscillations can be written in the form

$$A = \frac{h}{k}\frac{3^{(\nu_o/\nu)} - 1}{3^{(\nu_o/\nu)} + 1}, \tag{6}$$

where ν_o is the frequency at which the amplitude of the domain wall oscillations decreases in two times as compared with the shift of the DW in a permanent magnetic field:

$$\nu_o = \frac{\mu k}{2\ln 3} \tag{7}$$

Thus by measuring ν_o the DW mobility μ can be determined.

Fig. 3. DW oscillations in an yttrium orthoferrite plate cut normally to the optical axis.

3.1. Frequency dependence

Measurements of the DW mobility were performed on the setup shown in Fig. 2. Permanent magnets (3) produce the inhomogeneous magnetic field with a gradient $k = 5.6 \times 10^6$ A/m^2 along the crystallographic a axis of the orthoferrite plate (2). Helmholtz coils (4) are attached to both sides of the plate (2) producing the pulsed magnetic field driving the DW. The source (9) drives the current in the coils (4) which is measured by a shunt (8) and an oscilloscope (10). The plate is cut normally to the optical axis of the crystal (valid at the wavelength of $\lambda = 0.63\mu$m). The laser (1) illuminates the plate. Analyzer (5) and objective (6) form an image of the domain structure in the plane of the screen (7).

Figure 3 shows a photo of the domain structure subjected to the action of the magnetic field pulses with a step-like time dependence. The zone of intermediary brightness represents the zone of the DW oscillations. By measuring the width of this zone the amplitude of the DW oscillations is determined. As compared with measurements performed by means of a photoreceiver they have the advantage of being free from the frequency dependence of the photoreceiver' response, from variations of the Faraday rotation, of the laser output power, and so on.

Figure 4 shows the dependence of the DW oscillations amplitude on the frequency of the current in the coils. One can see that the amplitude decreases in two times at $\nu_o = 5$ MHz. By substituting this value in Eq. (6) the calculated curve was obtained. This curve is in a good agreement with the measured data with the exception of the high frequency margin of the plot. The latter divergence is probably caused by distortions of the current in the coils.

3.2. Temperature dependence

Equation (7) gives the value of the DW mobility as $\mu = 2.1$ m^2/sA at room temperature.

To study the temperature dependence of the DW mobility the amplitudes of the DW oscillations at two frequencies (5 kHz and 5 MHz) were measured in the temperature range 20°C–130°C. Figure 5 shows the dependence on temperature of the amplitude of the DW oscillations at 5 MHz. At 5 kHz the ratio

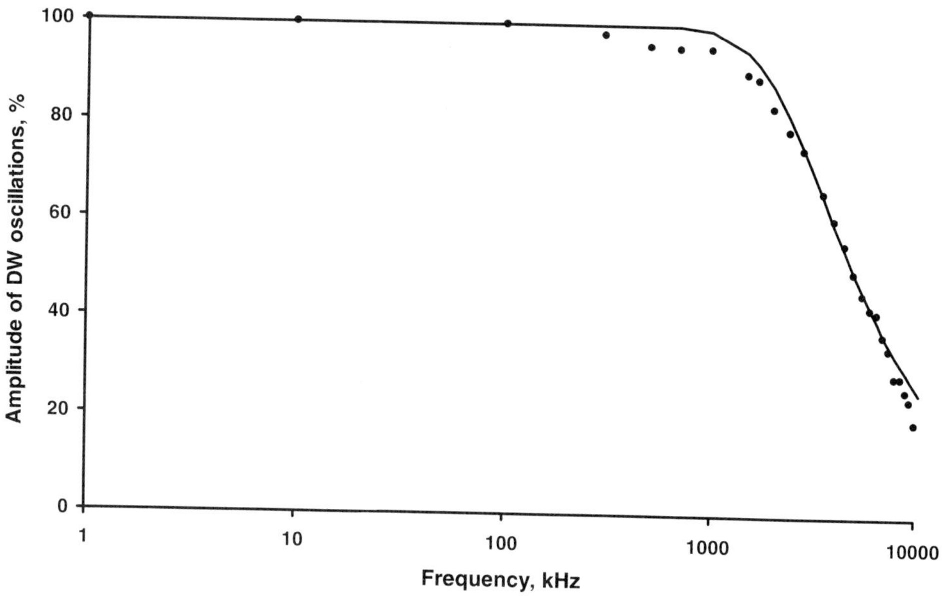

Fig. 4. Dependence of amplitude of DW oscillations on frequency.

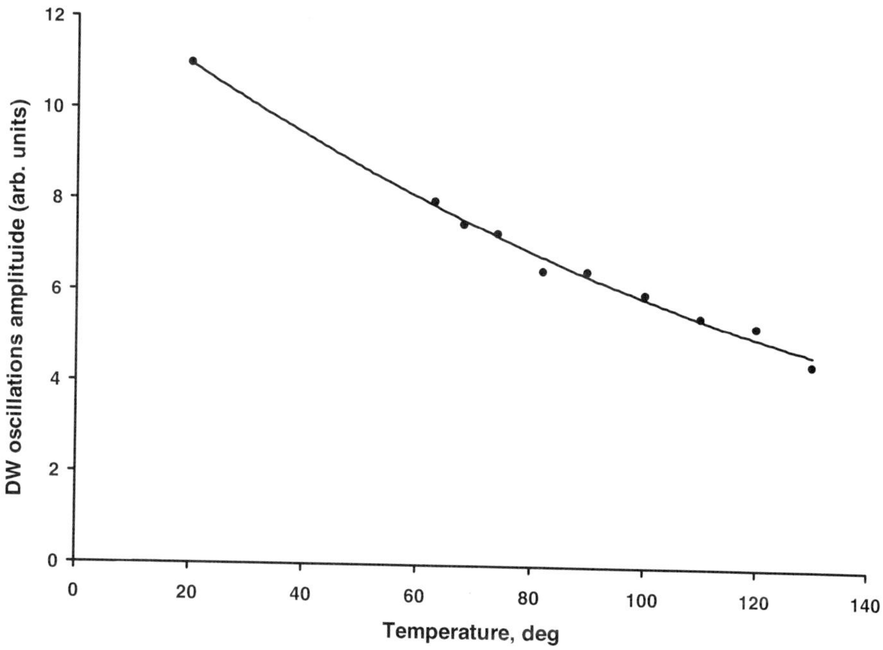

Fig. 5. Amplitude of the DW oscillations versus temperature. Frequency: 5 MHz.

of the DW oscillation amplitude A_T at the temperature T to that at the room temperature A_{20} gave the decrease of the field gradient k:

$$\frac{k_{20}}{k_T} = \frac{A_T}{A_{20}} \tag{8}$$

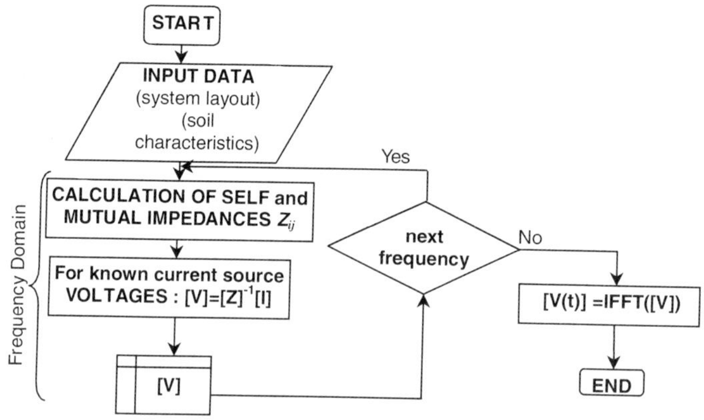

Fig. 1. Flowchart of grounding system analysis program.

where $I_0 = 32$ kA, $a = -9091$, $\beta = -727300$.

2.2. *Calculations using EMTP*

EMTP (ElectroMagnetic Transients Program) [1] has been used for calculation of grounding system response. In literature EMTP-based and other methods are summarized in two categories:

1. Pi-circuits methods which involve division of the electrode in elementary segments and its analysis as ladder network with lumped parameters [2–5]
2. Methods based in modeling grounding electrodes as lossy transmission lines [6–10]. More particularly the J.Marti model for transmission lines calculations has been successfully applied [7–10].

In this paper the second methodology has been applied, as it is valid for a wide frequency range. Thus the method is suitable for calculations involving high frequencies as lightning.

2.3. *Impulse response calculation using a suitable code*

The frequency behavior of grounding systems elements has been calculated for frequencies up to 10 MHz using the algorithm shown in Fig. 1. The method of moments has been used for the solution of the potentials problems [14–17]. Inverse Fourier Transformation is used to translate frequency domain to time domain results.

3. Application results

3.1. *Contribution of points located far from the energization point – long horizontal conductor*

Effective dimensioning of extended grounding systems usually involves calculation of the effective length of horizontal grounding electrodes. Effective length is the length above which, no considerable reduction of the high frequency impedance of the electrode is observed. It depends on the frequency and can be determined using a diagram of $|Z|$ vs. length or $|Z|$ vs frequency. Grounding electrode is considered equivalent to an open-ended transmission line, thus having input impedance $z = z_c coth(\gamma\ell)$.

In Fig. 2, $|Z|$ of 10, 20 and 30 m long electrodes vs. frequency have been plotted. For the 20 m long it remains to a low value until 100kHz where it begins to increase. Plot of $|Z|$ of a longer electrode

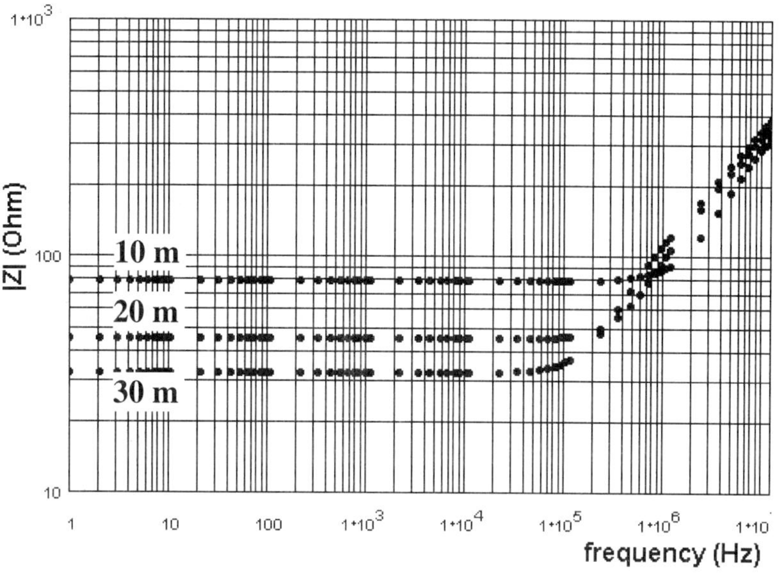

Fig. 2. Impedance of horizontal conductor in 500 Ωm soil vs. frequency.

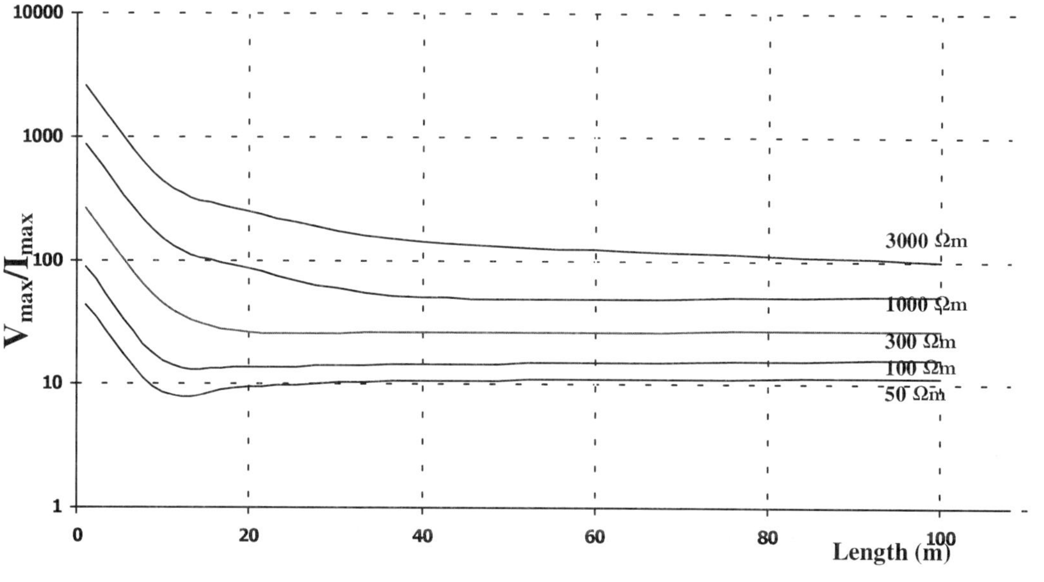

Fig. 3. $V_{\max}/I_{\max}$ of horizontal conductor under fast impulse.

would show the expected increase at a lower frequency value, as it would be more reactive. However the effective length of any longer electrode above a certain high frequency f is almost equal.

The ratio of peak voltage to peak current has been plotted when a fast 1/4 μs impulse current has been injected to show the influence of effective length (Fig. 3).

The faster the injection current rises to its maximum value, the shorter the effective length value. It can be generally observed that lengths which can be used to notably reduce raised potentials in all cases do not exceed 25–30 m.

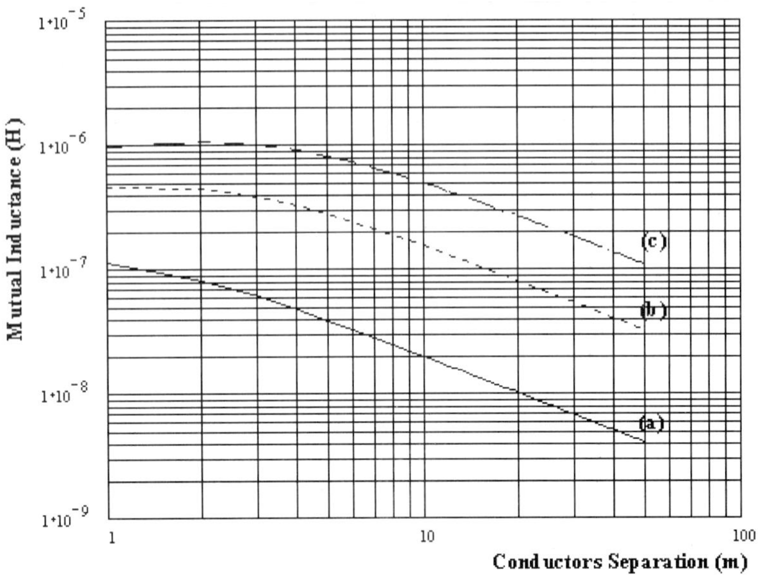

Fig. 4. Mutual Inductance vs. conductor separation for Conductors length a) 1 m, b) 2 m, c) 3 m.

3.2. Contribution of conductors located far from the energization point

The effect of components of the grounding system located far from the energization point is examined. Two horizontal conductors have been placed in parallel as shown below.

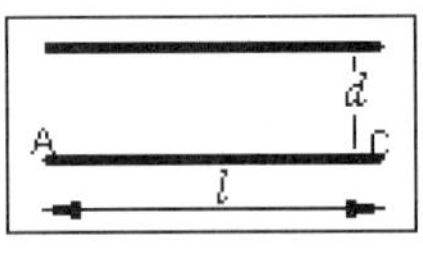

d=1 to 50 m
l=1 to 3 m
Conductor radius=0.005m
Burial Depth=0.5 m
Soil Resistivity =500 Ω m

Mutual conductance is calculated according to [18]. Mutual inductance is given by Grover's formula [19].

Mutual inductance is not dependent on soil resistivity. It depends on the length of conductors and their separation distance. It can be observed in Fig. 4 that mutual inductance values decrease considerably when electrodes separation exceeds 10 m.

Mutual conductance is dependent on soil characteristics and conductors separation. In Fig. 5 it can be observed that only the electrodes placed in a distance < 15 m from energization seriously contribute to the grounding resistance decrease. A similar conclusion can be drawn in case of non-parallel grounding segments.

3.3. Interconnected grounding arrangements

The case of interconnected grounding arrangements is examined next with particular reference to windfarms. A practical windfarm consisting of 12 windturbines as shown in Fig. 6 has been used as study case. Windturbines are connected in series with horizontal electrodes. Single windturbine arrangement has a grounding resistance of 12 Ω, in 500 Ωm soil. Impedance "seen" in windturbine 9 is lower than impedance in WT 1 but increases rapidly in high frequencies. The effective length of

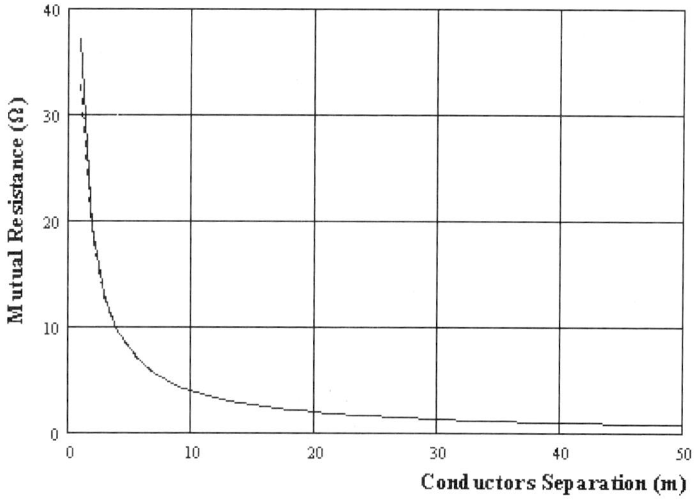

Fig. 5. Mutual Resistance vs. conductor separation for Conductors length a) 1 m, b) 2 m.

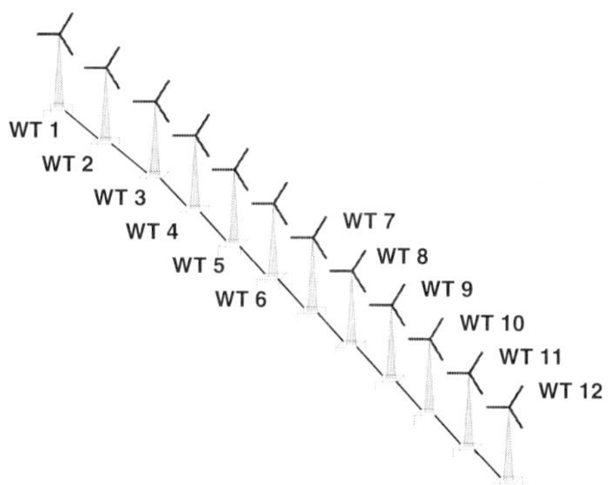

Fig. 6. Schematic representation of the windfarm used as study case.

interconnection electrodes results in slightly lower high frequency values of the impedance seen from WT 1. In Fig. 7 the voltages at WT1, WT2 and WT3 are plotted. Lightning is supposed to strike WT1. Transferred potentials to wind turbines situated at a distance greater than 500 m from WT1, are less than 5% of the max GPR (Ground Potential Rise).

In Fig. 8 the voltages at WT9, WT10 and WT11 are plotted. Lightning strikes WT9. Transferred potentials to other wind turbines except from WT10 and WT8 are negligible. It is generally observed that:

- Interconnection electrodes reduce the max GPR values compared to a single WT arrangement. It is also lower in case the WT arrangement is connected from both sizes to other WTs
- Percentage of Transferred GPRs to neighboring WTs is lower when lightning hits WT9 rather than if it hits WT1.
- Transient impedance is maximum when lightning hits the first or last WT. It can be observed that it reduces in case lightning strikes a middle WT.

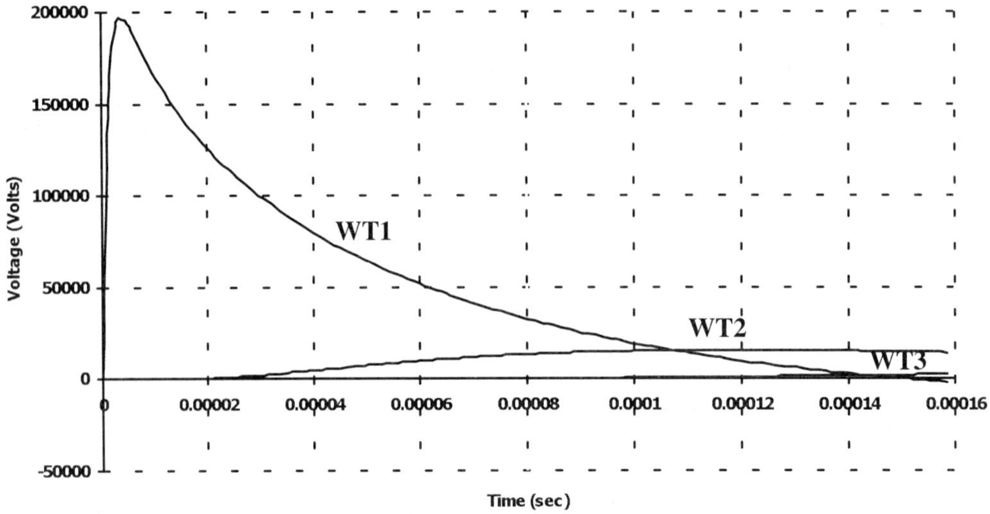

Fig. 7. Potential at WT1 and transferred potentials to WT2 and WT3 when lightning hits WT1.

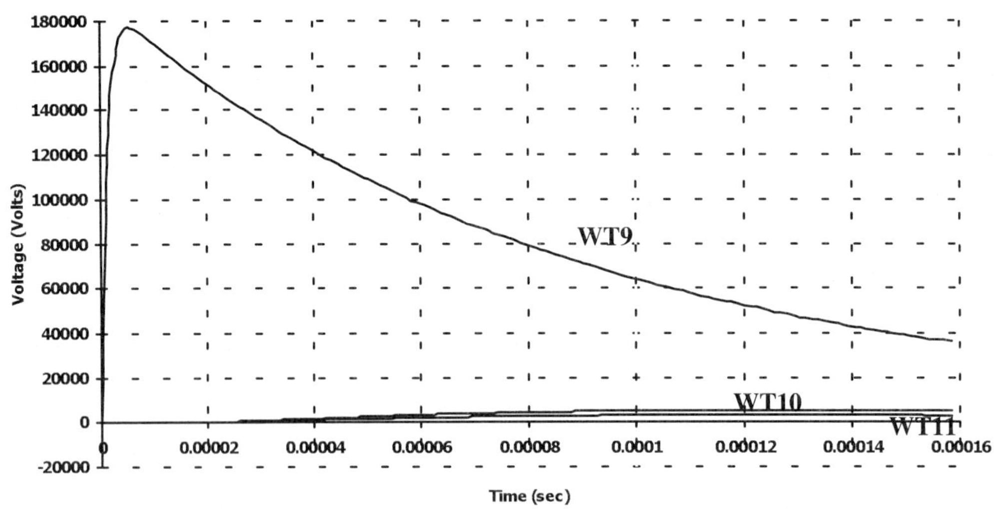

Fig. 8. Potential at WT9 and transferred potentials to WT10 and WT11 when lightning hits WT9.

4. Conclusions

Grounding systems serve the purpose of effectively dispersing fault or lightning currents into the ground. Dimensioning of grounding systems for lightning protection involves calculation of their transient response. In this paper, impulse response of extended grounding systems is calculated and the results are commented. The impulse impedance of an extended system is higher than their low frequency grounding resistance resulting in higher values of raised potentials. The following examples of extended grounding systems have been analyzed:

- Long horizontal electrodes
- Interconnected grounding arrangements resulting in a large grounding system with particular reference to windfarms.

Effective length values are given for most practical cases of grounding electrodes, while effective area of grounding grids is limited to few square meters around the energization point. In case of a windfarm, interconnection electrodes, lower low frequency grounding resistance but result in transferred potentials to the neighboring windturbines. In addition not all the length of interconnections contributes to the reduction of transient potential rise.

References

[1] W. Dommel, *EMTP Theory Book,* BPA, Oregon, 1986.
[2] A.P. Meliopoulos and M.G. Moharam: Transient Analysis of Grounding Systems, *IEEE Trans. on Power Apparatus and Systems* **PAS-102**(2) (Feb. 1983).
[3] A.D. Papalexopoulos and A.P. Meliopoulos, Frequency Dependent Characteristics of Grounding Systems, *IEEE Trans. on PWRD* **4** (Oct. 1987).
[4] N.D. Hatziargyriou and M. Lorentzou, Grounding Systems Design Using EMTP, *23rd European EMTP Users Group Meeting,* Barcelona, 9–11 November 1997.
[5] M. Lorentzou and N.D. Hatziargyriou, *EMTP Modelling of Grounding Electrodes 32nd UPEC Conference,* Manchester, 10–12 September 1997.
[6] R. Velazquez and D. Mukhedhar, Analytical Modeling of Grounding Electrodes Transient Behavior, *IEEE Trans. on Power Apparatus and Systems* **PAS-103**(6) (June 1984).
[7] F. Menter, Accurate Modeling of conductors imbedded in Earth with Frequency Dependent Distributed Parameter Lines, *Proceeding of the 1st EMTP User Group Meeting,* 1992.
[8] F. Menter, EMTP Based Model for Grounding System Analysis, *IEEE Trans on PWRD* **9**(4) (October 1994).
[9] F. Menter, Accurate Modeling of conductors imbedded in Earth with Frequency Dependent Distributed Parameter Lines, *Proceeding of the 1st EMTP User Group Meeting,* 1992.
[10] F. Menter and L.D. Grcev, EMTP Based Model for Grounding System analysis, *IEEE Trans. on PWRD* **9**(4) (Oct. 1994).
[11] M.A. Uman, *The Lightning Discharge,* Academic Press, 1977.
[12] M. Lorentzou and N. Hatziargyriou, Modeling of Long Grounding Conductors Using EMTP, *IPST 99 – International Conference on Power Systems Transients,* Budapest, 20–24 June 1999.
[13] E.D. SundeEarth, *Conduction Effects in Transmission Systems,* Dover Publ. N. York, 1949.
[14] L. Grcev, Improved Earthing System Design Practices for Reduction of Transient Voltages paper 36-302 CIGRE Session, 1998.
[15] L. Grcev and F. Dawalibi, An electromagnetic model for transients in grounding systems, *IEEE Trans. on Power Delivery* **5** (1990), 1773–1781.
[16] F.P. Dawalibi et al., Transient Performance of Substation Structures and Associated Grounding Systems, *IEEE Trans. on Industry Applications* **31**(3) (May/June 1995).
[17] L.D. Grcev, Computer Analysis of Transient Voltages in Large Grounding Systems, *IEEE Trans. on PWRD* **11**(2) (April 1996).
[18] Meliopoulos, *Power Systems Grounding and Transients,* Marcel Dekker, N. York, 1988.
[19] F.W. Grover, Inductance Calcuations, D. Van Nostrand Company Inc., New York, 1949.

International Journal of Applied Electromagnetics and Mechanics 13 (2001/2002) 299–306
IOS Press

Magnetic levitation of a metal orb

Zsuzsanna Matlák* and Tamás Barbarics
Department of Electromagnetic Theory, Budapest University of Technology and Economics, H-1521, Budapest, Hungary

Abstract. This paper presents the analysis of the force determination that exerts an influence in the levitation-melting method model. The goal of the present study is a numerical investigation of the problem results concerning the alternation of the magnetic field modified by the sphere, which melts during the levitation-melting method. The numerical investigation is realised by global variational method, completed with the R-functions.

1. Introduction

In the practice the melting technique becomes important in improving the quality of metal and in developing new materials. This method can also be adopted in melting metal without contamination, because the molten metal is levitated by electromagnetic force in space.

In this experiment the metal crucible, comprising 8 segments and supplied with internal cooling, is placed into a high-frequency induction coil. The ingot applied in course of the experiment melts on the effect of induction heating. The position of the molten metal is stabilised since there is a repulsing magnetic force induced between the molten metal and the cold crucible [3].

This research concentrates on the definition of an optimal distance, where the position of the sphere becomes stationary. The calculation is also necessary since rapid rotational movement makes scattering of the molten metal by the centrifugal force. It is therefore indispensable to control and stabilise the rotation before melting. The rotational movement of the metal must be defined taking into account the improvement of the technique from the point of view of uniform heating [2].

In course of these calculations a simplified experimental set is taken into account. Consequently these calculations can be represented as a 2D case. At this time the crucible has been neglected. The six-turn, double winding exciting coil has been modelled with 6 circle conductors. The ingot is considered as a perfect sphere.

Figure 1 shows the 2D model. The parameters of the circle conductors and the sphere are shown in the picture. The value of the exciting current assumed in the magnetic field intensity analysis is selected 1 kA. In previous study the magnetic field intensity has been specified to different conditions and number of circle conductors since the conductors can be excited independently of each other, with the same value of exciting current [5]. In case of further settings the magnetic field of more than one circle conductors with the superposition of the field's components can be determined. The results of these calculations are considered in course of present research.

*Corresponding author: Zsuzsanna Matlák, Tel.: +36 1 463 3190; Fax: +36 1 463 3189; E-mail: mano@evtsz1.evt.bme.hu.

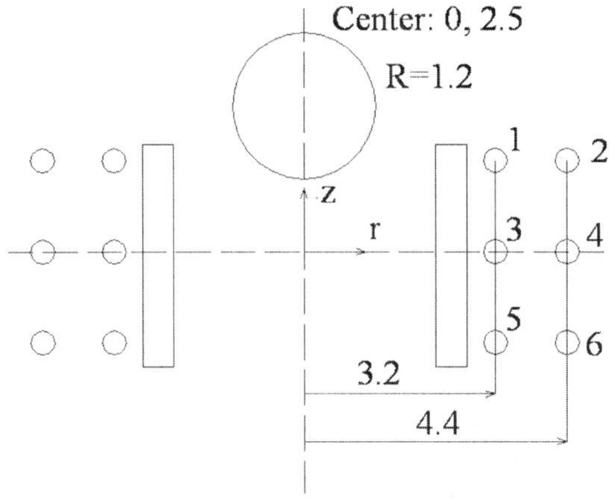

Fig. 1. The geometrical model for analysis.

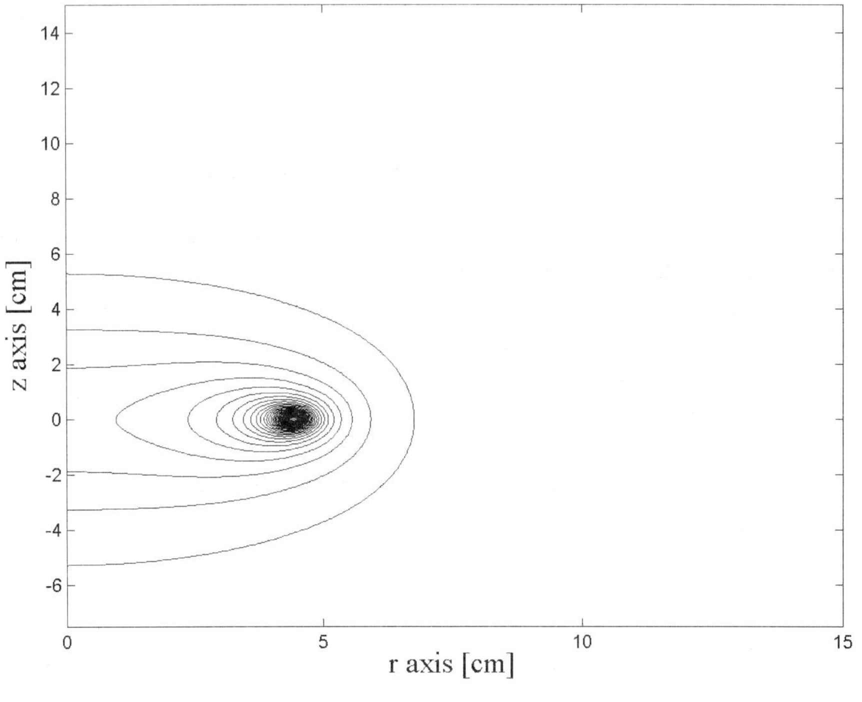

Fig. 2. Magnetic force lines with 1 conductor.

2. Governing equations

The results can be determined on the basis of the magnetic field of the eddy currents as

$$\nabla \times \boldsymbol{H} = \sigma \boldsymbol{E} + \boldsymbol{J}_s, \quad \nabla \times \boldsymbol{B} = 0, \quad \nabla \times \boldsymbol{E} = -\frac{\partial \boldsymbol{B}}{\partial t} = -\mu_0 \frac{\partial \boldsymbol{H}}{\partial t}\left(1 + \frac{\partial \boldsymbol{M}}{\partial \boldsymbol{H}}\right). \tag{1}$$

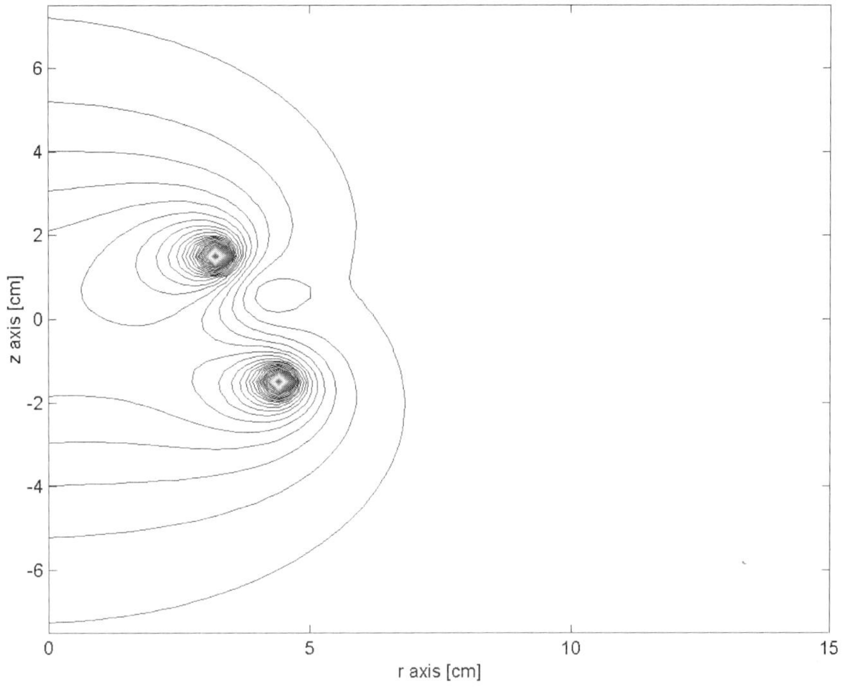

Fig. 3. Magnetic force lines with 2 conductors.

where $\boldsymbol{J}_s$ is the exciting current density and $\boldsymbol{M}$ is the magnetisation vector. The value of the $\frac{\partial \boldsymbol{M}}{\partial \boldsymbol{H}}$ can be calculated from the magnetic hysteresis characteristics [1].

Taking into account Eq. (1) the result can be expressed from the following equation

$$\nabla \times (\nabla \times \boldsymbol{H}) + \mu_0 \sigma \left(\mathbf{1} + \frac{\partial \boldsymbol{M}}{\partial \boldsymbol{H}}\right) \frac{\partial \boldsymbol{H}}{\partial t} = \nabla \times \boldsymbol{J}_s. \tag{2}$$

For modelling the problem the magnetic field is supposed to be composed of a source component $\boldsymbol{H}_s$, generated by the current of the circle conductors and its modification $\boldsymbol{H}_m$ due to the eddy current arising in the sphere. Taking into account Eq. (2) the magnetic field intensity can be also formulated with these terms as

$$\nabla \times \boldsymbol{H}_s = \boldsymbol{J}_s, \quad \boldsymbol{H}_m = \boldsymbol{H}_{m_r}(r, z, t) + \boldsymbol{H}_{m_z}(r, z, t). \tag{3}$$

and the total field yields the diffusion equation for the primary selected magnetic field intensity

$$\nabla \times (\nabla \times \boldsymbol{H}_m) + \mu_0 \sigma \left(\mathbf{1} + \frac{\partial \boldsymbol{M}}{\partial \boldsymbol{H}}\right) \frac{\partial \boldsymbol{H}_m}{\partial t} = -\mu_0 \sigma \left(\mathbf{1} + \frac{\partial \boldsymbol{M}}{\partial \boldsymbol{H}}\right) \frac{\partial \boldsymbol{H}_s}{\partial t} \tag{4}$$

During the determination of the field quantities the solution of Eq. (4) can be evaluated by reducing the problem to a minimum value problem of the energy related functional

$$W(\boldsymbol{H}) = \int_\Omega \left[\nabla \times \boldsymbol{H}^* \cdot \nabla \times \boldsymbol{H}_m + \mu_0 \sigma \boldsymbol{H}^* \left(1 + \frac{\partial \boldsymbol{M}}{\partial \boldsymbol{H}}\right) \frac{\partial \boldsymbol{H}_m}{\partial t} + \mu_0 \sigma \boldsymbol{H}^* \left(1 + \frac{\partial \boldsymbol{M}}{\partial \boldsymbol{H}}\right) \frac{\partial \boldsymbol{H}_s}{\partial t}\right] d\Omega. \tag{5}$$

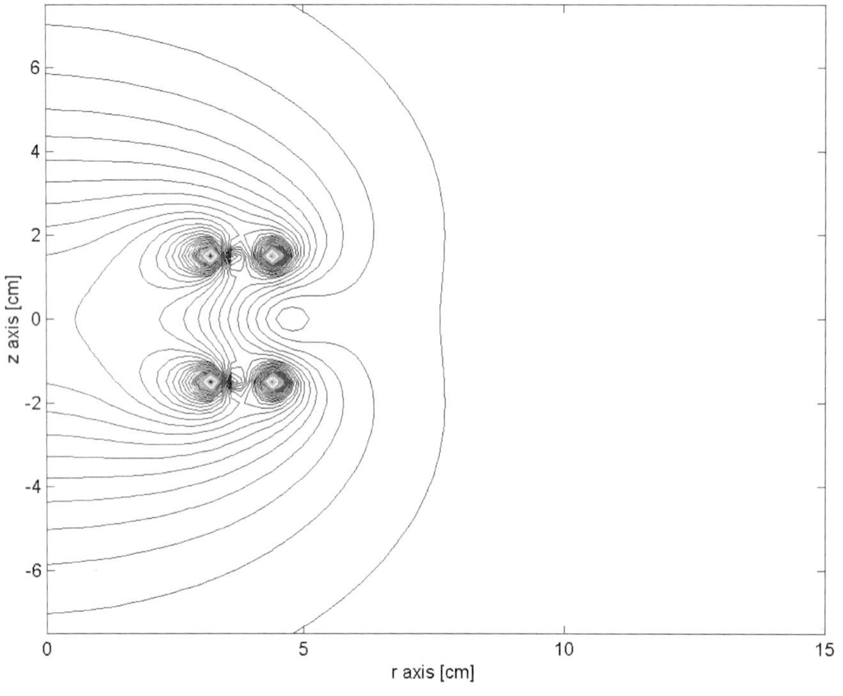

Fig. 4. Magnetic force lines with 4 conductors.

where Ω is the examined region.

For every time step the extreme value of the functional Eq. (5) yields the solution of the ordinary differential equation [4]

$$\mathbf{K}_1\mathbf{a} + \mu_0\sigma\mathbf{K}_2\left(\mathbf{1} + \frac{\partial \boldsymbol{M}}{\partial \boldsymbol{H}}\right)\frac{\partial \mathbf{a}}{\partial t} = -\mu_0\sigma\mathbf{G}, \tag{6}$$

where the elements of the matrixes can be constructed as:

$$\mathbf{K}_1[k,q] = \int \nabla \times F_k \cdot \nabla \times F_q d\Omega, \quad \mathbf{K}_2[k,q] = \int F_k \cdot F_q d\Omega, \quad \mathbf{G}[k] = \int F_q \cdot \dot{\boldsymbol{H}}_s d\Omega. \tag{7}$$

3. Force calculation

Maxwell's force tensor is one of the most efficient general methods for the calculation of forces acting on bodies under the influence of magnetic field. The force can be determined as

$$\boldsymbol{F} = \int_\Omega \boldsymbol{J} \times \boldsymbol{B} d\Omega, \tag{8}$$

which after some algebraic manipulation yields

$$\frac{d\boldsymbol{F}}{dS} = \mathbf{e}_r\mu_0 H_r H_z + \mathbf{e}_z\frac{\mu_0}{2}\left(H_z^2 - H_r^2\right), \tag{9}$$

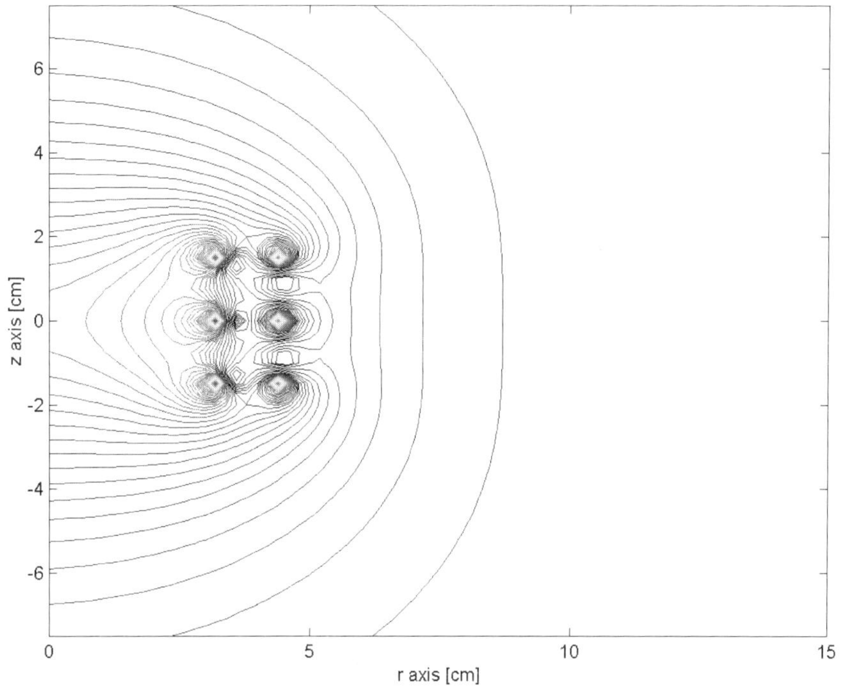

Fig. 5. Magnetic force lines with 6 conductors.

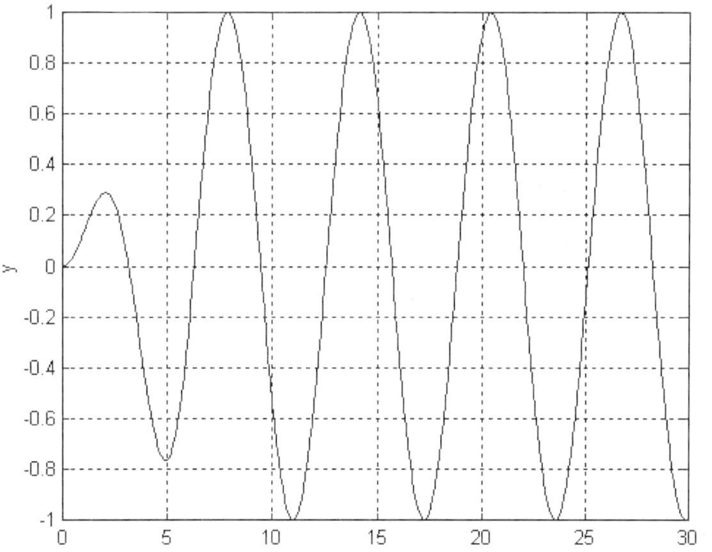

Fig. 6. The applied excitation.

where $S(\Omega)$ is the surface of the body. Upon simplification for the magnitude of $d\boldsymbol{F}$ for an elementary ds surface becomes

$$d\boldsymbol{F} = \frac{1}{2}\mu_0 \left(H_z^2 + H_r^2\right) ds. \tag{10}$$

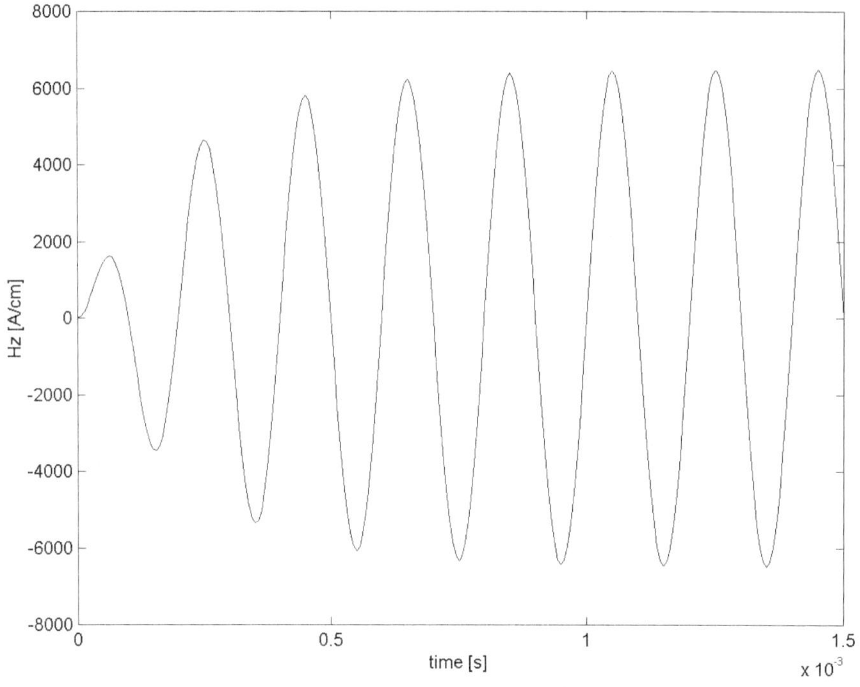

Fig. 7. The value of Hz at the bottom of the sphere.

The direction of $d\boldsymbol{F}$ is obtained from the tangent of the angle Θ, which $\boldsymbol{H}$ forms with the normal, and for the angle α, which $d\boldsymbol{F}$ forms with the normal is

$$\alpha = \tan^{-1}\left(\frac{2\tan\Theta}{1-\tan\Theta}\right) = 2\Theta. \tag{11}$$

It is obvious that, to evaluate the total force acting on the body, all differential forces dF_r and dF_z have transfer to the global system of coordinates.

4. Numerical results

As a numerical result, first the magnetic field intensity is determined without the sphere, taking into account 1, 2, 4 and 6 current conductor. The magnetic field intensity for these cases are plotted respectively in Figs 2, 3, 4 and 5.

Taking into account an exponentially increasing sinusoidal excitation, as it shows in Fig. 6, the value of the magnetic force at the bottom of the sphere in the z direction can be seen in Fig. 7. The force effected on the sphere is determined, and plotted in Fig. 8. The amplitude of the forces around the sphere can be seen in Fig. 9, where the lengths of the arrows are commensurable to the amplitude of the force. The solution is symmetric, so it means that the total force has only component in z direction, so the force wants to lift the sphere. The gravitation is not taken into account in the figure, so in practice the excitation has to increase up to the value, where the magnetic force should be greater than the gravitation to lift the sphere.

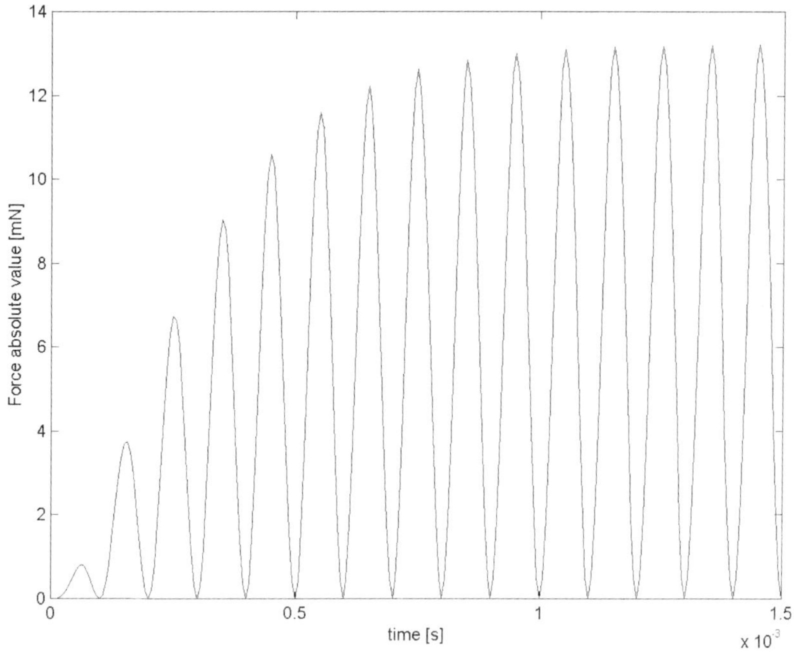

Fig. 8. The total force on the sphere.

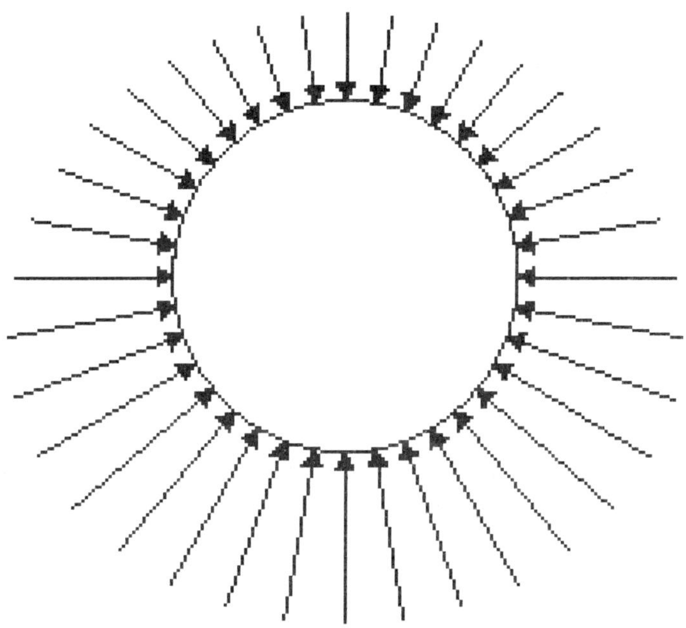

Fig. 9. The force around the sphere.

5. Conclusion

The 2 dimensional magnetic field is examined. The force effected on the sphere is determined. At this time just a 2D model is applied so the effect of gyration is not taken into account. However it is planed to develop a 3D model to consider this effect.

References

[1] A. Iványi, *Hysteresis Models in Electromagnetic Computation,* Akadémiai Kiadé, Budapest, 1997.
[2] H. Tsubodi, M. Tanaka, F. Kobayashi and T. Misaki, Three-dimensional eddy current analyis of induction melting in cold crucible, *IEEE Trans. on Magn.* **29** (1993), 1574–1577.
[3] M. Enokizono, T. Todaka, K. Yokoji, I. Matsumoto, K. Kubo and Y. Wada, Experimet and analysis of levitation-melting method of induction furnace, *IEEE Trans. on Magn.* **31** (1995), 4205–4207.
[4] T. Barbarics, A. Kost, D. Lederer and P. Kis, Electromagnetic Field Calculation For Magnetic Shielding with Ferromagnetic Material, *IEEE Trans. on Magn.* **36**(4) (2000), 986–989.
[5] Zs. Matlák, T. Barbarics and A. Iványi, Field Calculation of Levitation-melting Method of Induction Furnace, *Proceedings of the 9th International IGTE Symposium on Numerical Field Calculation in Electrical Engineering,* 11–13 Sep 2000, pp. 395–400.

International Journal of Applied Electromagnetics and Mechanics 13 (2001/2002) 307–310
IOS Press

Stress effect on hysteretic magnetization curve of nickel

Takaaki Suzuki*, Hidehiro Baba and Eiji Matsumoto
Department of Energy Conversion Science, Kyoto University, Yoshida-Honmachi, Sakyo-Ku, Kyoto 606-8501, Japan

Abstract. In this paper, the magnetic constitutive equation of the magnetoelastic material with hysteresis is presented. The proposed equation exhibits both the hysteresis and the stress dependence of the magnetization curve. To verify the constitutive equation, the magnetization under the stress is measured. In this test pure nickel is used, which has especially large stress dependence of magnetic properties. Theoretical results are found to be in very good agreement with the experimental data.

1. Introduction

Magnetic properties of ferromagnetic materials are changed by applying the stress. This is known as the effect of the magnetoelastic coupling, cf. Matsumoto [1]. The effect is expected for application to sensors, actuators and nondestructive evaluation. On the other hand, the case is recently increasing that ferromagnetic structural materials are used under the magnetic field. Therefore, it is also important to predict the influence of the stress on the equipment using ferromagnetic materials.

In this paper, as a theoretical model on magnetoelastic effects, Jiles and Atherton's theory [2] neglecting the stress is extended to describe the stress dependence of the hysteretic magnetization curve. That is, the material parameters concerning the magnetic domain structure in their theory are assumed to depend on the stress. Thus, the modified magnetic constitutive equation takes a similar form to Jiles and Atherton's one with parameters having the same physical meaning, and it can also exhibit both the hysteresis and the stress dependence of the magnetization curve. Nickel is known to have especially large stress dependence of magnetic properties among ferromagnetic materials. Thus, to verify the proposed constitutive equation, magnetization tests on pure nickel (99.7%) under the stress are carried out.

2. Theoretical analysis

Jiles and Atherton described a model of the magnetization curve which gives rise to the hysteresis loops by considering impedances to domain wall motion caused by pinning sites, e.g., nonmagnetic inclusions

*Corresponding author: Takaaki Suzuki, Tel./Fax: +81 75 753 3565; E-mail: suzuki@system.energy.kyoto-u.ac.jp.

1383-5416/01/02/$8.00

and voids, inhomogeneous strain regions and so on. According to them, the differential equation for the hysteretic magnetization curve is

$$\frac{dM}{dH} = \frac{1}{(1+c)} \frac{1}{\delta k/\mu_0 - \alpha(M_{an} - M)}(M_{an} - M) + \frac{c}{(1+c)} \frac{dM_{an}}{dH}, \tag{1}$$

where c is experimentally determined by the ratio of the initial differential susceptibilities of the normal and the anhysteretic magnetization curves, α a mean field parameter representing interdomain coupling, μ_0 the permeability of the vacuum, and k the parameter characterizing the pinning energy. The parameter δ takes the value $+1$ when H increases in the positive direction and -1 when H increases in the negative direction, and hence this parameter generates the hysteresis in the magnetization curve. The function M_{an} is the anhysteretic magnetization, for which Langevin function $L(x)$ was employed in the above theory. Then we have

$$M_{an}(H, M) = M_s L(x) = M_s[\coth(x) - 1/x], \tag{2}$$

where M_s is the parameter corresponding to the saturation magnetization of M_{an}, a a parameter with the dimension of the magnetic field which characterizes the shape of the anhysteretic magnetization curve, and $x = (H + \alpha M)/a$.

In this study, the above differential Eq. (1) is extended to take into account the stress dependence of the magnetization curve. Since the deformation and the stress are small within the elastic limit, the parameters M_{st}, α_t, a_t are assumed to be the first order with respect to the stress:

$$M_{st} = M_s + M_s'T, \quad \alpha_t = \alpha + \alpha'T, \quad a_t = a + a'T, \tag{3}$$

where the prime denotes the coefficient of the stress T in the corresponding parameter. Here, it will be experimentally confirmed that each parameter depends linearly on the stress as in Eq. (3). On the other hand, it is assumed that the stress dose not influences two parameters c and k characterizing the pinning effects of the magnetic domain motion. Therefore, the values for free stress are employed for the tensile stress. Thus by replacing the tensile stress in Eq. (1) with the stress dependant parameters in Eq. (3), we have

$$\frac{dM_t}{dH} = \frac{1}{(1+c)} \frac{1}{\delta k/\mu_0 - \alpha_t(M_{ant} - M_t)}(M_{ant} - M_t) + \frac{c}{(1+c)} \frac{dM_{ant}}{dH}. \tag{4}$$

The hysteretic magnetization curve is obtained by solving Eq. (4).

3. Experimental method

The magnetization curve under the tensile stress is measured. The test piece is made from pure nickel. We used a tabular test piece for tensile test, width of the section 25 mm, thickness 10 mm, length of the parallel part 150 mm. From the tensile test, a linear relationship exists between the stress and the strain up to the stress 50 MPa. So the elastic limit is defined by 50 MPa in this paper, and in this experiment the stress is applied within the elastic limit.

First, the test piece is held by the tensile testing machine with a constant uniaxial stress. Next, by the electromagnet, the magnetic field parallel to the uniaxial stress is applied to the test piece. The magnetic field is measured by the H-coil of enamel wire wound around the thin acrylic plate laid on the test piece surface. The magnetization is measured by the B-coil of enamel wire wound around the test piece.

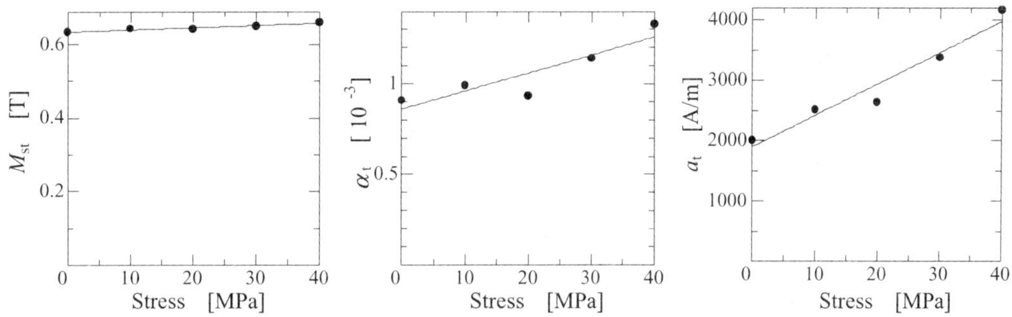

Fig. 1. Parameters with magnetization curve versus tensile stress.

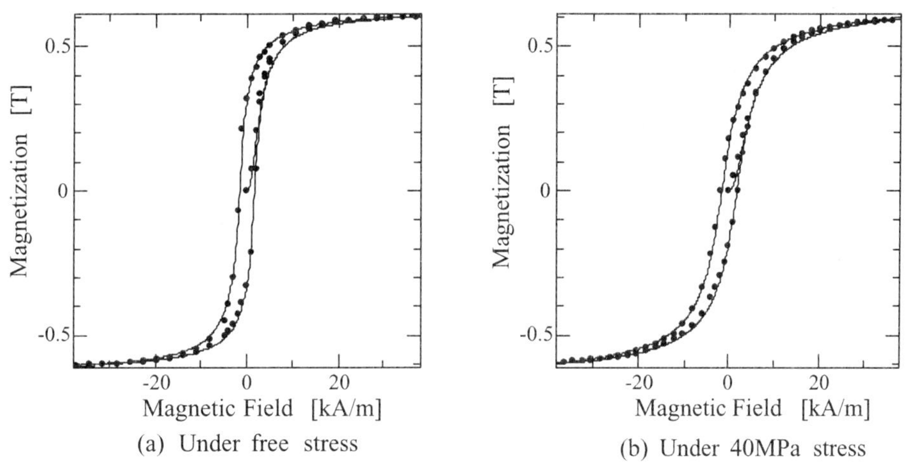

Fig. 2. Magnetization curve under tensile stress.

4. Identification of parameters

The values of the stress dependent parameters for each tensile stress M_{st}, α_t and a_t are determined from the experimental data by using the least square method. The calculated parameters for each stress are plotted by the dot in Fig. 1. From this figure, it can be seen that three parameters linearly depend on the tensile stress as assumed to be the first order in the stress. Therefore, the parameters are determined in the linear approximation by the least square method as shown Fig. 1.

5. Theoretical and experimental data

Figure 2 shows the hysteretic magnetization curve for the obtained parameters by numerically solving the proposed equation using Runge-Kutta method. In the figure, the black dot denotes the experimental data, and the solid line the theoretical ones. From the experimental data for each stress, we find that the magnetization is saturated and there is the hysteresis in the magnetization curve for the applied magnetic field.

Comparing the experimental data for 0 and 40 MPa, we find that the material is slight hardly magnetized under the tensile stress. That is, the susceptibility and the saturation magnetization decrease by applying

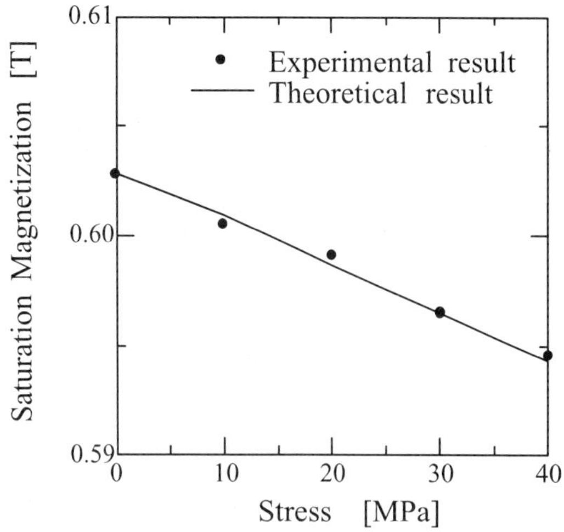

Fig. 3. Saturation magnetization versus tensile stress.

the tensile stress. More precisely, the saturation magnetization linearly depends on the tensile stress in both of the experimental and the theoretical result, as shown in Fig. 3. In both figures, it can be seen that the theoretical results are in very good agreement with the experimental data, so that the proposed constitutive equation can describe both the hysteresis and the stress dependence of the magnetization curve.

6. Conclusion

In this paper, the magnetic constitutive equation for nickel is presented to describe the stress dependence of the magnetization curve based on the effect of the magnetoelastic coupling. To verify the proposed constitutive equation, magnetization tests under the stress are carried out, and the following conclusions are obtained.

(1) The magnetic constitutive equation which can reproduce the stress dependence of the magnetization curve is derived by extending Jiles and Atherton's theory.
(2) It is experimentally confirmed that the magnetization of nickel decreases with increasing tensile stress and that the magnetization curve with hysteresis depends on the stress.
(3) Since the theoretical results are in very good agreement with the experimental data, the proposed magnetic constitutive equation is applicable to describe the stress effect on the hysteretic magnetization curve.

References

[1] E. Matsumoto, Solid mechanics under magnetic field, *Mater. Sci. Res. Int. (Japan)* **5** (1999), 1–12.
[2] D.C. Jiles and D.L. Atherton, Theory of ferromagnetic hysteresis, *J. Magn. Magn. Mater.* **61** (1986), 48–60.

International Journal of Applied Electromagnetics and Mechanics 13 (2001/2002) 311–316
IOS Press

Stress-magnetization change in amorphous ribbon due to magnetic field heat treatment

Yoshio Iwami[a], Yasuo Okazaki[a] and Akihiko Saito[b]
[a]*Department of Electrical and Electronics, Faculty of Engineering, Gifu University, 1-1 Yanagido, Gifu, Japan*
Tel./Fax: +81 58 293 2696; E-mail: iwami@violin.ocn.ne.jp, okazaki@cc.gifu-u.ac.jp
[b]*Fukuoka Jo Gakuin University, 3-42-1 Osa, Minami-ku, Fukuoka, Japan*
Tel./Fax: +81 92 581 1492; E-mail: asaito@fukujo.ac.jp

Abstract. Stress-magnetization changes in Co-based amorphous ribbons annealed under magnetic field (magnetic field heat treatment) have been investigated. The stress-magnetization changes due to tensile stress were observed in positive increment of magnetization for the sample annealed under transversal magnetic field to its ribbon axis. However, a Co-based amorphous ribbon annealed under applied magnetic field parallel to the ribbon axis (longitudinal direction) did not show positive magnetization increment under applied tensile stress but showed stress insensitive region in magnetization change, which extended with increasing annealing temperature.

1. Introduction

Amorphous ribbon, well known as a soft magnetic material, has been utilized widely in power electricity and sensors [2], etc. It has very magnetic properties sensitive to applied stress, especially after heat treatment. However, this mechanism of magnetization change under stress in amorphous ribbon has not yet been clearly understood. Saito et al., has reported that a Co-based amorphous ribbon which has negative magnetostriction constant showed conversion of the magnetostriction constant under applying tensile stress due to heat treatment [1,7]. In addition to that, there was a tensile insensitive region in terms of magnetization change due to heat treatment, and that region increased at higher annealing temperature [8]. However, a mechanism that could interpret both the extension of tensile insensitive region and apparent conversion of magnetostriction constant due to magnetic heat treatment has not been identified clearly. Cho et al. [6] reported on investigation of an amorphous ribbon, METGLAS 2714A, that anisotropy of the sample was induced in the same direction of applied magnetic field during annealing but was induced in the transverse direction to the ribbon length when the sample was annealed without magnetic field. However, the direction of induced anisotropy would change from transverse to longitudinal to the ribbon axis, when the surface of the annealed sample was removed. Though the mechanism of this change was not clarified enough, the surface removal caused magnetic characteristics change of the sample. According to a saturation magnetostriction constant, Vazquez et al. [5], and J.M. Blanco et al. [3], reported that the annealing temperature varied the value of amorphous ribbon magnetostriction constant and even its sign. However, according to our investigation by means of Small Angle Magnetization Rotation (SAMR) [4], the results gave still negative value of -1×10^{-7} for the sample annealed even at 300°C temperature.

In the present work, the stress-magnetization change of the amorphous ribbon having slightly negative magnetostriction constant and its relation to the induced anisotropy have been studied.

1383-5416/01/02/$8.00

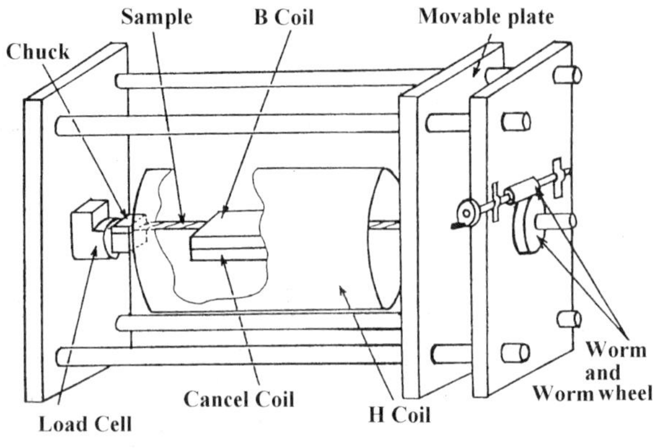

Fig. 1. A sketch of stress-magnetization measurement.

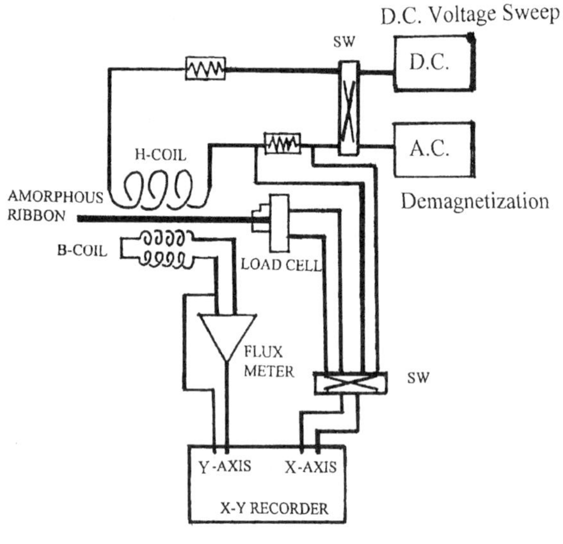

Fig. 2. A schematic diagram of the experiment.

2. Sample and experiments

The sample is an amorphous ribbon (METGLAS 2705M) composed of $Co_{69.0}$ $Fe_{4.0}$ $Ni_{1.0}$ $Mo_{2.0}$ $B_{12.0}$ $Si_{12.0}$ by ALLIED Ltd. The specification of the sample is as follows: saturation magnetization 0.78 (T), Curie temperature 365°C, crystallization temperature 550°C (nominal values by the supplier), and magnetostriction constant has small negative value of -0.2×10^{-6}. The width, thickness and length of the ribbon are 25 mm, 20 μm and 45 cm, respectively.

The sample was annealed for one hour at desired temperature with filling Argon gas for avoiding any oxidation during magnetic field (4400 A/m) heat treatment in the parallel direction to the sample axis (longitudinal), width (transversal) and thickness (normal to its surface) in each annealing case. Figure 1 shows a sketch of our experiment for measuring the sample hysteresis loci and its stress-magnetization change. The sample is tightly clipped at ends, one end to a load cell and the other end to a movable plate so that tension could be applied uniformly on it. Figure 2 shows a schematic diagram of the

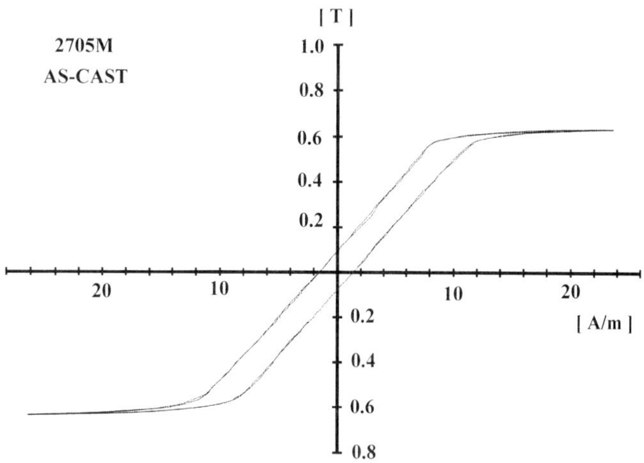

Fig. 3. A hysteresis loop for the as-cast sample.

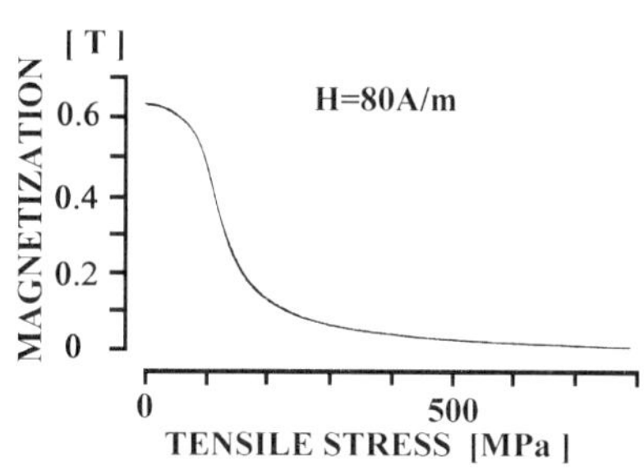

Fig. 4. The stress-magnetization loci of the as-cast sample.

entire measurement system. DC power is fed for sweeping magnetic field and AC power is utilized for demagnetization of the sample at each measurement.

3. Experimental results

Figure 3 shows a hysteresis curve of the as-cast sample. Large Barkhausen jumps and kinks for a rectangular shape can be seen on the hysteresis locus. Figure 4 shows the stress-magnetization change of the as-cast sample measured in 80 A/m external magnetic field. This locus shows reversible magnetization change of the sample. At the beginning of tensile stress application, magnetization started decreasing gradually. For further stress application, it decreased drastically. This would be an evidence that the sample has anisotropy field which slightly prevented change in magnetization from decreasing.

Figure 5 shows hysteresis loci of the samples annealed at 300°C with magnetic field in directions longitudinal (a), transverse (b), to its axis and normal (c) to its surface as shown in the figure. It is obviously understood that induced magnetic anisotropy is fabricated due to magnetic field application during annealing in each direction. Figure 6 shows the stress-magnetization loci of the samples annealed with magnetic field in longitudinal direction to its axis at various temperatures. For at 100°C annealing, the sample showed almost the same stress-magnetization characteristic as of as-cast. However, the stress-magnetization insensitive region was appearing and expanding with increasing annealing temperature.

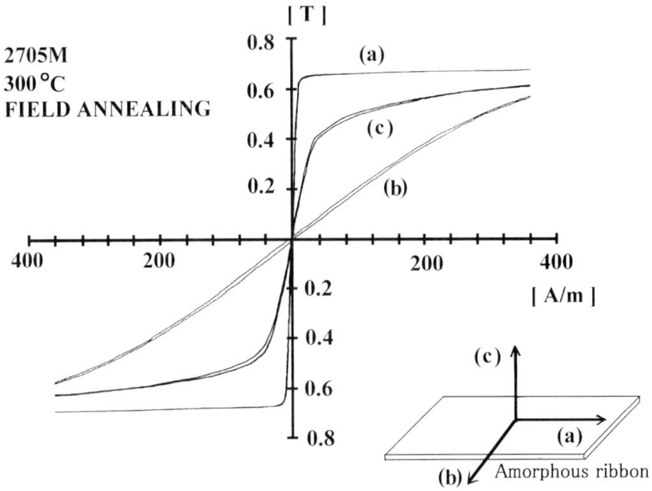

Fig. 5. Hysteresis loci of the samples with field annealed in direction of (a) longitudinal, (b) transverse to the axis, and (c) normal to its surface.

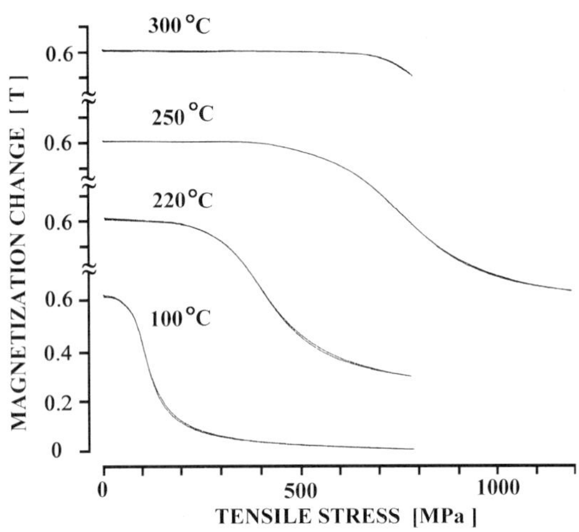

Fig. 6. The stress-magnetization loci of the sample annealed with field in direction of its axis at temperatures of 100° C, 220° C, 250° C and 300° C.

Figure 7 shows the stress-magnetization characteristic of the sample annealed at 300 °C with field in transverse direction to the axis. The numerical values shown at each magnetization locus are intensities of external magnetic field, in which the stress-magnetization change was measured. At around 300 A/m external magnetic field, the stress-magnetization change showed a unique behavior that the magnetization change was increasing up to 300 MPa stress application, and then decreasing for further stress application, though there were few changes in magnetization at lower external magnetic field and at higher field. Figure 8 shows the stress-magnetization characteristic of the sample annealed at 300 °C with field in normal direction to its surface. The numerical values shown at each magnetization locus has the same meaning of Fig. 7.

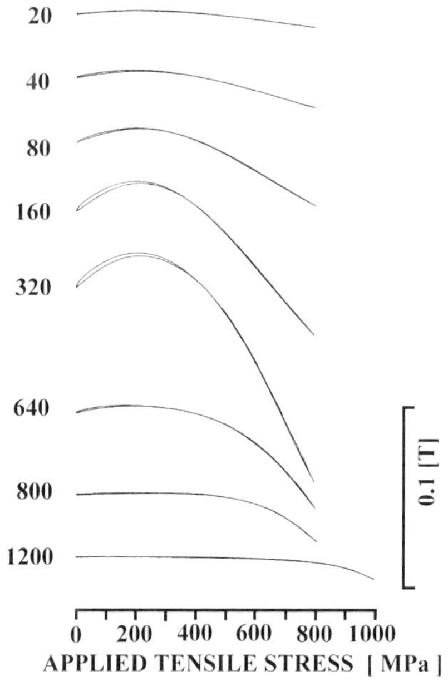

Fig. 7. The stress-magnetization loci of the sample annealed at 300°C with field in transverse direction to its axis.

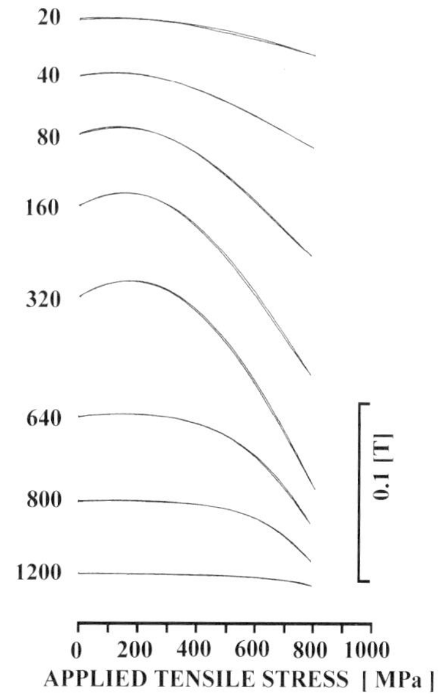

Fig. 8. The stress-magnetization loci of the sample annealed at 300°C with field in normal direction to its surface.

4. Discussions

Figure 9 shows magnetic domain configurations of the sample annealed at 300 °C with longitudinal magnetic field to its axis, (a) under remanence and (b) under 80 A/m exciting field conditions. The sample shows 180° domain walls parallel to the axis. A 180° domain wall indicated by an arrow in Fig. 9(a) moved parallel to the ribbon axis after field application of 80 A/m as indicated in Fig. 9(b). According to the small angle magnetization rotation (SAMR) [4] measurement, the sample annealed at 300 °C with longitudinal magnetic field to its axis had negative magnetostriction constant of mean value -1×10^{-7}. The magnetization change due to stress application can be explained by induced anisotropy field, though the domain observation under tensile stress application has not been done successfully. When the stress is applied to the sample, the magnetization change will occur from domain wall movement at first, then magnetization vector rotations toward the ribbon width will follow due to negative magnetostriction material. The expansion of stress-magnetization insensitive region, in Fig. 6, would be therefore related to a degree of difficulties in a movement of the domain walls, which was induced by heat treatment.

On the other hand, the sample annealed with transversal magnetic field to its axis, already had induced anisotropy field in direction to the ribbon width, as seen in Fig. 10(a) at a residual condition. Figure 10(b) shows a magnetic domain patterns observed under condition of 80 A/m external field applied in parallel to its ribbon axis. The magnetic domains were separated into small pieces and their walls were diverted in complicity. When a tensile stress would be applied to the sample of Fig. 10(b), the magnetic domain configuration would be reorganized toward previous domain configuration as of Fig. 10(a). During this reorganizing process, the magnetization vector would rotate to the sample axis in a transition stage whereby the external field was applied. Consequently, the magnetization vector of the sample would rotate to the width direction.

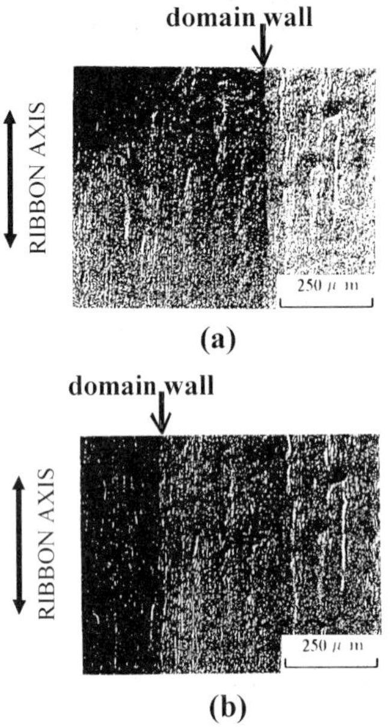

Fig. 9. The magnetic domains observed in the sample annealed at 300°C with field in direction of its axis, under (a) remanence and (b) 80 A/m applied field conditions.

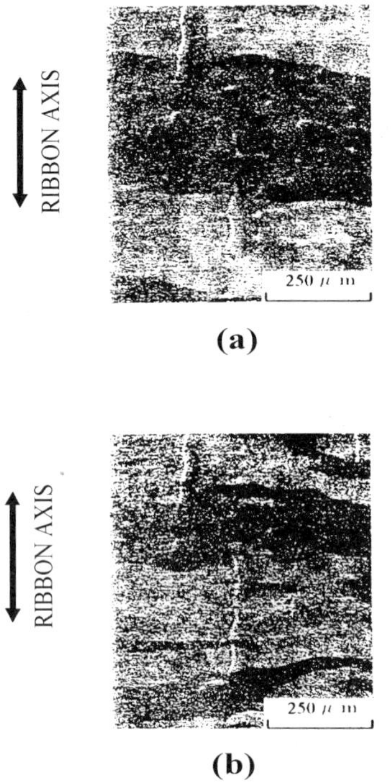

Fig. 10. The magnetic domains observed in the sample annealed at 300°C with field in transverse of its axis, under (a) remanence and (b) 80 A/m applied field conditions.

References

[1] A. Saito, T. Yamazaki, T. Sugahara and S. Murashige, Effect of Heat Treatment on the Stress-Magnetization Properties of an Amorphous Ribbon, *J. Magn. Jpn.* **20** (1996), 457–460.

[2] M. Takajo and J. Yamasaki, Magnetic Characteristics of Amorphous Ribbons with Induced Anisotropy in the Circumferential Direction, *J. Magn. Jpn.* **24** (2000), 663–666.

[3] J.M. Blanco, L. Dominguez, P. Aragonesez and J. Gonzalez, Large Barkhausen Jumps in relaxed Co-rich amorphous alloy ribbons, *IEEE Trans. Magn.* **33**(5) (1997), 3778–33780.

[4] K. Narita, J. Yamasaki and H. Fukunaga, Measurement of saturation magnetostriction of a thin amorphous ribbon by means of Small-Angle Magnetization Rotation, *IEEE Trans. Magn.* **16**(2) (1980), 435–439.

[5] M. Vazquez, J. Gonzalez and A. Hernando, Induced Magnetic anisotropy and change of the magnetostriction by current annealing in Co-based amorphous alloys, *J. Magn. Magn. Mater.* **53** (1986), 323–329.

[6] W.-S. Cho, H. Lee, S.-W. Lee and C.-O. Kim, The effect of Anisotropy in Co66Fe4Ni1B14Si15 Ribbon on the Transverse Incremental Pearmiability and Magnetoimpedance, *IEEE Trans. Magn.* **36**(5) (2000), 3442–3444.

[7] Y. Iwami, T. Yamasaki, T. Shimizu, M. Ohkita and A. Saito, *Effect of Heat Treatment on the Magnetization Changes of Amorphous Ribbon under the Stress,* (Vol. 13), IOS Press, Amsterdam, 1998, pp. 551–554.

[8] Y. Iwami, Y. Okazaki, T. Shimizu, T. Hirakawa and A. Saito, Effect of magnetic annealing on magnetization changes of amorphous ribbon under stress, *J. Magn. Magn. Mater.* **215–216** (2000), 443–445.

International Journal of Applied Electromagnetics and Mechanics 13 (2001/2002) 317–321
IOS Press

The influence of Mo addition and thermal treatment on the Hall effect and magnetoresistance of $Co_{68.25-x}Fe_{4.5}Si_{12.25}B_{15}Mo_x$ alloys

Horia Chiriac, Mihai Lozovan, Maria Neagu*, Cornelia Hison and Nicoleta Lupu
National Institute of R & D for Technical Physics, 47 Mangeron Blvd., 6600 Iasi 3, Romania

Abstract. Results concerning the influence of Mo addition and thermal treatment on the Hall effect and magnetoresistance of $Co_{68.25-x}Fe_{4.5}Si_{12.25}B_{15}Mo_x$ ($x = 1; 2$) ribbons in the amorphous, nanocrystallized and crystallized states are reported. The value of the Hall resistivity of the samples increases continuously with applied magnetic field up to about 7 kOe and 13 kOe for $Co_{67.25}Fe_{4.5}Si_{12.25}B_{15}Mo_1$ and $Co_{66.25}Fe_{4.5}Si_{12.25}B_{15}Mo_2$ alloys, respectively and then it approaches saturation. For $Co_{67.25}Fe_{4.5}Si_{12.25}B_{15}Mo_1$ ribbons the ferromagnetic anisotropic resistivity (FAR) values are 2.78×10^{-4}, 2.60×106^{-4} and 2.84×10^{-4} for amorphous, nanocrystalline and crystalline states, respectively. The FAR values of $Co_{66.25}Fe_{4.5}Si_{12.25}B_{15}Mo_2$ ribbons are 1.35×10^{-4}, 1.78×10^{-4} and 18×10^{-4} for amorphous, nanocrystalline and crystalline states, respectively.

1. Introduction

Amorphous and nanocrystalline materials with soft magnetic properties have been a subject of increasing attention for engineering applications as well as for fundamentals in nanostructures formation and magnetism [1–3]. For these materials, the coexistence of different magnetic phases, and the possibility to modify their relative volume fraction through annealing can be used to investigate interactions among constituent magnetic particles, the surface effects, transport properties etc. $Co_{68.25-x}Fe_{4.5}Si_{12.25}B_{15}Mo_x$ (x = 1; 2) amorphous ribbons are excellent soft magnetic materials having nearly zero magnetostriction and very low magnetic anisotropy induced during the fabrication process. The substitution of Co by Mo improves the glass-forming ability of the alloy and increases the thermal stability of the amorphous phase. Structural changes (nanocrystallization and crystallization) are induced in these materials after isothermal annealing up to 1 hour, at temperatures above 400°C. As a consequence, the magnetic, magnetoelastic and galvanomagnetic properties of the materials change [1–5].

The aim of this work is to analyse the influence of Mo addition and thermal treatments between 400°C and 650°C on the Hall resistivity and magnetoresistance of $Co_{68.25-x}Fe_{4.5}Si_{12.25}B_{15}$Mox ($x = 1; 2$) ribbons.

*Corresponding author: Dr. Maria Neagu, National Institute of Research and Development for Technical Physics, 47 Mangeron Blvd., 6600 Iasi 3, Romania. Tel.: +40 32 130680; Fax: +40 32 231132; E-mail: mneagu@phys-iasi.ro.

2. Theory

The curves of Hall resistivity, ρ_H, vs. magnetizing field are fitted by the formula (in e. m. u., as generally used in literature) [4,6]:

$$\rho_H = E_y/j_x = hV_H/i_x = R_0[H_a + 4\pi M_z(1 - N)] + 4\pi M_z R_s \tag{1}$$

where: j_x is the electric current density; E_y is the electric field; V_H is the Hall potential; h is the sample thickness; i_x is the current in the sample; H_a is the applied magnetic field; N is the demagnetizing factor; M_z is the magnetization; R_0 is ordinary Hall constant; R_s is the extraordinary (or spontaneous) Hall constant.

At saturation it follows that $H_a = 4\pi M_s$ and the Hall coefficients R_0 and R_s are determined directly from ρ_H or V_H curve [4,6].

The resistivity of ferromagnetic materials is sensitive to magnetic field and, according to the magnetic domain theory, it depends on the angle between magnetization and electric current in the sample. The magnetoresistance is characterized by relative change in resistivity

$$\Delta\rho/\rho = [\rho(H) - \rho(0)]/\rho(0) \tag{2}$$

where: $\rho(H)$ and $\rho(0)$ are the electrical resistivity measured at H and zero applied magnetic field, respectively.

The ferromagnetic anisotropic resistivity (FAR) is defined as $(\rho_{\|} - \rho_{\perp})/\rho(0)$ where: $\rho_{\|}$ and $\rho_{\perp}$ are the resistivities of the sample obtained in a saturating magnetic field parallel and perpendicular to the current direction, respectively.

3. Experimental details

The amorphous $Co_{68.25-x}Fe_{4.5}Si_{12.25}B_{15}Mo_x$ ($x = 1; 2$) ribbons were prepared by rapid solidification from the melt. The amorphous and nanocrystalline states of the samples were examined by X-ray diffraction, differential thermal analysis (DTA) and thermomagnetic measurements.

The Hall voltage measurements were carried out by the Van der Pauw method at room temperature, using external magnetic inductions up to 2 T, and dc sample biasing currents between 5 and 60 mA [4–6]. The source of magnetic field for the Hall voltage measurements was a Weiss electromagnet with a iron core connected to a dc power supplier.

The longitudinal and transversal magnetoresistance measurements were performed at room temperature using a four-points probe method [4,5]. The source of magnetic field for the magnetoresistance measurements was a solenoid connected to a dc power supplier. The curves $\Delta\rho_{\|}/\rho = [\rho_{\|}(H) - \rho(0)]/\rho(0)$ and $\Delta\rho_{\perp}/\rho = [\rho_{\perp}(H) - \rho(0)]/\rho(0)$ versus applied magnetic field are obtained experimentally by locating the ribbon sample with its long axis parallel and perpendicular respectively, to the magnetic field and with its plane parallel to the magnetic field. The relative changes in electrical resistivity are positive when monitored as function of the longitudinal magnetic field $\Delta\rho_{\|}/\rho$ and negative when detected as function of the transverse magnetic field $\Delta\rho_{\perp}/\rho$.

The nanostructured and crystallised states were obtained after isothermal annealing in vacuum for 0.5 hour, at 525°C and 650°C, respectively, for $x = 1$ and for 1 hour at 475°C and 600°C, respectively, for $x = 2$.

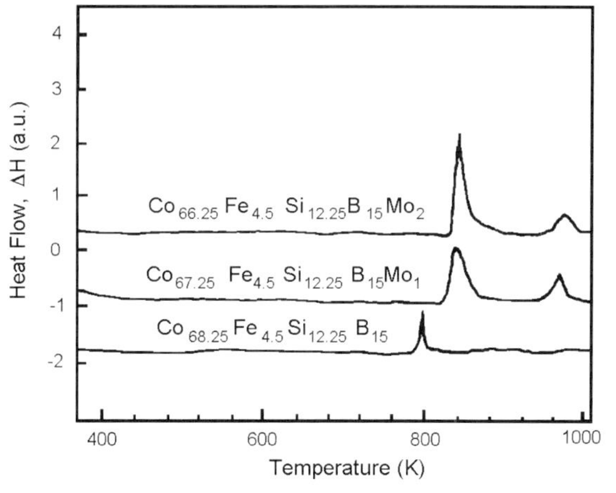

Fig. 1. DTA curves for $Co_{68.25}Fe_{4.5}Si_{12.25}B_{15}$; $Co_{67.25}Fe_{4.5}Si_{12.25}B_{15}Mo_1$ and $Co_{66.25}Fe_{4.5}Si_{12.25}B_{15}Mo_2$ ribbons.

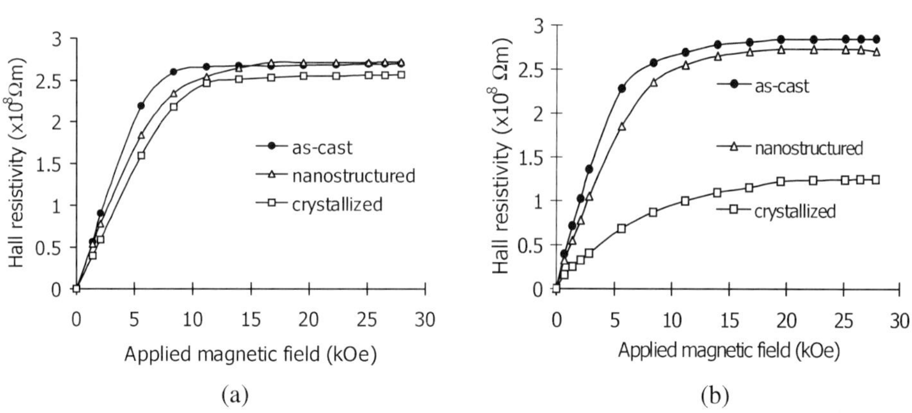

Fig. 2. The dependence of the Hall resistivity on the applied magnetic field for as-cast, nanostructured and crystallised: (a) $Co_{67.25}Fe_{4.5}Si_{12.25}B_{15}Mo_1$; (b) $Co_{66.25}Fe_{4.5}Si_{12.25}B_{15}Mo_2$ ribbons.

4. Results and discussion

Figure 1 shows the DTA curves for $Co_{68.25}Fe_{4.5}Si_{12.25}B_{15}$, $Co_{67.25}Fe_{4.5}Si_{12.25}B_{15}Mo_1$ and $Co_{66.25}Fe_{4.5}Si_{12.25}B_{15}Mo_2$ ribbons. As it can be seen, the substitution of Mo for Co in CoFeSiB alloys determines an increase of the crystallization temperature. For 1 and 2% Mo, two exothermic peaks appear indicating a two steps crystallization process.

Figures 2(a) and (b) presents the Hall resistivity dependence on the applied magnetic field for as-cast, nanostructured and crystallized $Co_{67.25}Fe_{4.5}Si_{12.25}B_{15}Mo_1$ and $Co_{66.25}Fe_{4.5}Si_{12.25}B_{15}Mo_2$ ribbons respectively. The value of the Hall resistivity of the samples increases continuously with applied magnetic field up to about 7 kOe and 13 kOe for $Co_{67.25}Fe_{4.5}Si_{12.25}B_{15}Mo_1$ and $Co_{66.25}Fe_{4.5}Si_{12.25}B_{15}Mo_2$ alloys and then it approaches saturation. The changes in the Hall resistivity of samples after annealing are due to the structural changes, in agreement with the observed magnetostrictive behaviour. In the as-cast state, the saturation magnetostriction value is quite small (about 10^{-7}) and negative and shifts to positive values with annealing due to changes in the relative volume fractions of the crystalline and amorphous phase [1,2,5].

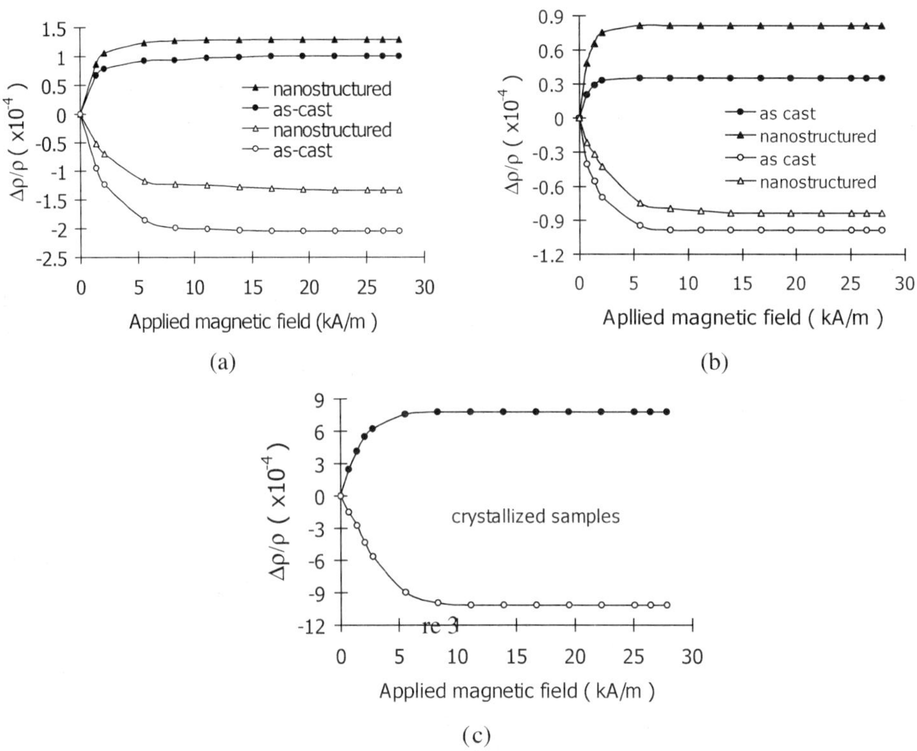

Fig. 3. The magnetoresistence dependence on the applied magnetic field for: (a) and (b) as-cast and nanostructured $Co_{67.25}Fe_{4.5}Si_{12.25}B_{15}Mo_1$ and $Co_{66.25}Fe_{4.5}Si_{12.25}B_{15}Mo_2$ ribbons, respectively; (c) crystallized $Co_{66.25}Fe_{4.5}Si_{12.25}B_{15}Mo_2$ ribbons.

Figures 3(a) and (b) presents the magnetoresistence dependence on the applied magnetic field for as-cast and nanostructured $Co_{67.25}Fe_{4.5}Si_{12.25}B_{15}Mo_1$ and $Co_{66.25}Fe_{4.5}Si_{12.25}B_{15}Mo_2$ ribbons. Figure 3c presents the magnetoresistence dependence on the applied magnetic field for crystallized $Co_{66.25}Fe_{4.5}Si_{12.25}B_{15}Mo_2$ ribbons. For $Co_{67.25}Fe_{4.5}Si_{12.25}B_{15}Mo_1$ ribbons, the FAR values are 2.78×10^{-4}, 2.60×10^{-4} and 2.84×10^{-4} for amorphous, nanocrystalline and crystalline state, respectively. The FAR values of $Co_{66.25}Fe_{4.5}Si_{12.25}B_{15}Mo_2$ samples are 1.35×10^{-4} and 1.78×10^{-4} for as-cast and nanocrystalline samples, respectively, increasing up to 18×10^{-4} during crystallization. For as-cast state there is a pronounced asymmetry of resistivity changes as direction of the applied magnetic field. The resistivity changes with the transverse field are higher than the corresponding changes for the longitudinal field. The asymmetry is due to the longitudinal anisotropy induced in the samples during fabrication process. After annealing, the asymmetry of the curves with respect to the field axis becomes very small, because the magnetic anisotropy along the longitudinal axis of the ribbons is almost eliminated and the axes of the nanocrystallites or crystallites are randomly oriented [3]. The galvanomagnetic behaviour as result of crystallisation can be explained on the basis of the evolution of relative volume fractions of the Co-based crystallites with positive FAR and the Fe-Si crystallites with negative FAR [1, 3].

The investigations were made in order to provide information about the transport properties in these alloys, with are important for technological application based on galvanomagnetic effects.

5. Conclusions

The influence of the thermal treatments on the Hall effect and magnetoresistance of $Co_{67.25}Fe_{4.5}Si_{12.25}B_{15}Mo_1$ and $Co_{66.25}Fe_{4.5}Si_{12.25}B_{15}Mo_2$ amorphous, nanocrystalline and crystalline ribbons was analysed.

The value of the Hall resistivity of the samples increases continuously with applied magnetic field (up to about 7 kOe and 13 kOe for $Co_{67.25}Fe_{4.5}Si_{12.25}B_{15}Mo_1$ and $Co_{67.25}Fe_{4.5}Si_{12.25}B_{15}Mo_2$ alloys) and then it approaches saturation.

For crystallised $Co_{67.25}Fe_{4.5}Si_{12.25}B_{15}Mo_1$ ribbons, the ferromagnetic anisotropic resistivity (FAR) value (2.84×10^{-4}) is about the same with that of the amorphous state (2.78×10^{-4}). For $Co_{67.25}Fe_{4.5}Si_{12.25}B_{15}Mo_2$ ribbons, the FAR value in the crystallyzed state is about 13 times higher than in the amorphous state.

References

[1] J. Gonzalez, N. Murillo, J.M. Blanco, P. Quintana, E. Amano and R. Valenzuela, Effects of nanocrystallization on the magnetostriction of Co-based amorphous alloys, *IEEE Trans. Mag.* **30** (1994), 4812–4814.

[2] H.K. Lachowicz, R. Zuberek, M. Kuzminski and A. Slawska-Waniewska, Magnetic hardening in gradually devitrified Co-based glassy alloys, *J. Magn. Magn. Mater.* **196–197** (1999), 151–153.

[3] G. Bordin, G. Buttino, A. Cecchetti and M. Poppi, Hall effect and magnetoresistance in Co-based amorphous and crystallized alloys, *J. Magn. Magn. Mater.* **172** (1997), 291–300.

[4] P. Quintana, E. Amano, R. Valenzuela and J.T.S. Irvine, Effects of nanocrystallization upon the soft magnetic properties of Co-based amorphous ribbons, *J. Appl. Phys.* **75–10** (1994), 6490–6498.

[5] H. Chiriac, M. Lozovan and M. Neagu, On Hall Effect of $Co_{68.25}Fe_{4.5}B_{15}Si_{12.25}$ Amorphous Ribbons, *Acta Physica Slovaca* **68** (1998), 695–698.

[6] C.M. Hurd, *Hall Effect in Metals and Alloys,* Plenum Press, New York, 1972.

International Journal of Applied Electromagnetics and Mechanics 13 (2001/2002) 323–328
IOS Press

Analysis of thermal critical curves in temperature and time dependent Preisach type models

I.D. Borcia[a,*], L. Spinu[b] and Al. Stancu[a]
[a]*Faculty of Physics, "Al. I. Cuza" University, Iasi, 6600, Romania*
[b]*AMRI, University of New Orleans, LA 70148, USA*

Abstract. The shape and the movement of the critical curves in the Preisach plane are investigated using the numerical solution of master equation with variable magnetic field and temperature. It is shown that the shape of the thermal critical curve is a curve line. The movement of the thermal boundaries is simulated in magnetization processes with constant or variable field. Using this model we can explain the dependence of the coercivity on the variation rate of the magnetic field.

1. Introduction

In the last time becomes important the study of ultrafine and nano-particles ensembles because of their applications on high density recording media. At this kind of systems the phenomena of relaxation and particle interactions are very important. In order to study these aspects many models containing the combined effect of magnetic relaxation and interparticle interactions was proposed. A phenomenological theory based on the intrinsic energy barrier has been used by Chantrell et al. [1]. The Preisach model [2] was proved efficient in describing the interactions in the particulate systems. Time and temperature dependent Preisach models have been developed lately [3–6]. In order to simplify the calculations, linear energy barrier versus field were used in [3,4]. This gives straight critical lines, separating superparamagnetic and blocked particles. If the Stoner-Wohlfarth model is considered in the calculus of the energy barrier, nonlinear dependence is obtained. The calculus is more complicated since the critical curves become non-linear as well.

In this paper an analysis of the shape and movement of the thermal boundaries is performed using a model in which the dynamic of all subsystems in the Preisach plane (particles associated to a certain point in the Preisach plane) is described by the master equation. Some particular processes are studied using this model.

*Corresponding author: Ion Dan Borcia, Faculty of Physics, "Al. I. Cuza" University, 11 Blvd.Copou, IASI – 6600, Romania. Tel.: +40 32 201176; Fax: +40 32 201205; E-mail: iborcia@mail.uaic.ro.

1383-5416/01/02/$8.00

2. The model

We consider an ensemble of uniaxial single domain particles distributed in the Preisach plane, having their easy axis aligned, and with a field making an angle ψ with this direction. For a given particle one has two positions of stable equilibrium for the polarization in the Stoner-Wohlfarth model [7] corresponding to angles θ_+ and θ_- between the easy axis and the polarization separated by unstable equilibrium position θ_m. The corresponding values of the magnetic moment are:

$$m_\pm = VP_S\cos(\psi - \theta_\pm) \tag{1}$$

where V is the particle volume, P_S is the saturation polarization of the particle. In general the energy barrier is a non-analytical function. We have used in our simulations the approximation given by Pfeiffer [8]. The expressions of the energy barrier corresponding to the polarization vector in and out the field direction is:

$$\Delta E_\pm(H,\psi) = \frac{VP_SK_K}{2}\left[1 \pm \frac{H}{H_c(\psi)}\right]^{g(\psi)} \tag{2}$$

with

$$g(\psi) = 0.86 + 1.14\frac{H_c(\psi)}{H_k}. \tag{3}$$

To describe the time evolution of the number of particles in the two stabile equilibrium states is used the master equation:

$$\frac{dn_+}{dt} = \frac{n_-}{\tau_-} - \frac{n_+}{\tau_+}. \tag{4}$$

where n_+ and n_- are relative number of particles in and out the field direction, respectively. One operate for the relaxation times, τ_+ and $_-$ with the expression given by Néel [9]:

$$\tau_\pm = \tau_0 \exp\left(\frac{\Delta E\pm}{kT}\right). \tag{5}$$

where τ_0^{-1} is a microscopic attempt frequency, which is about 10^9 s^{-1}.

If the relaxation not depends on time one have analytical solution for the master equation:

$$n_\pm = \frac{\tau_\pm}{\tau_+ + \tau_-} \pm \left(n_{+0} - \frac{\tau_+}{\tau_+ + \tau_-}\right)\exp\left(-\frac{t}{\tau}\right). \tag{6}$$

where

$$\tau = \frac{\tau_+\tau_-}{\tau_+ + \tau_-}. \tag{7}$$

Is the relaxation time and n_{+0} is the initial relative number of particles in the positive equilibrium state at $t = 0$. The particles with $\tau \gg t$ are called blocked particles and the particles for which $\tau \ll t$ are named superparamagnetic. The firsts rest in the initial state and the seconds arrives to the thermodynamic equilibrium. The separation lines between the two sorts of particles are named thermal boundaries or critical curves.

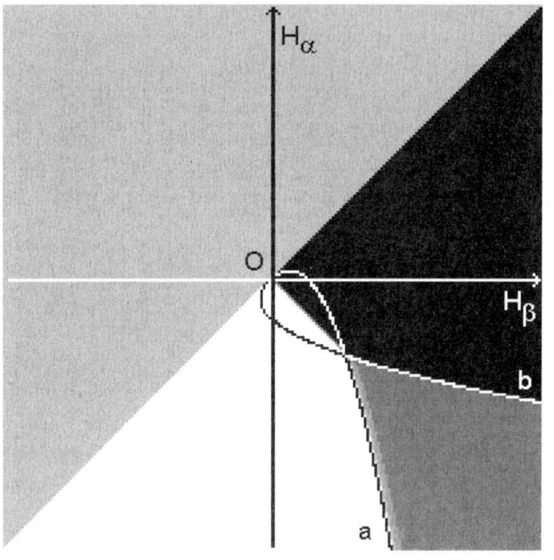

Fig. 1. The Preisach plane after 500 s in zero field, starting from a thermal demagnetized state; Oa and Ob are the calculated thermal critical curves. The white zones correspond to $n+ = 1$, the black zones- $n+ = 0$.

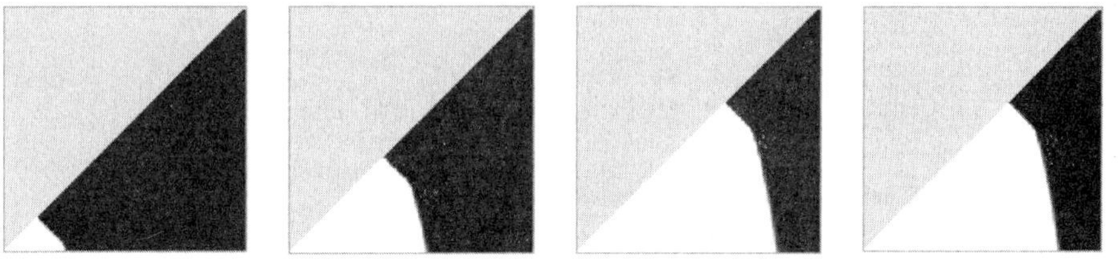

Fig. 2. The Preisach plane in the time of a process with constant temperature and variable field with the rate $R = 8$ kA/m/s: (a) – initially thermal demagnetized; (b) – after 27; (c) – after 53 s; (d) – after 80 s.

3. Simulations

In this paper we simulate the behavior of an ultra fine particle system with parallel easy axis distributed in a Preisach plane ($H_\alpha OH_\beta$). For a lattice of 400x400 particles we have numerically solved the master equation for the values of the field where there are two stable equilibrium states. In Fig. 1 we have represented the final state calculated for the system initially thermally demagnetized after 500 s of standing in zero field parallel with the easy axis. The white zones corresponds to $n_+ = 1$ $(n_- = 0)$, the black zones to $n_+ = 0$ $(n_- = 1)$ and the gray ones to intermediate values proportional with the color intensity. Comparatively one represents the position of the calculated thermal boundaries Oa and Ob using nonlinear barrier of energy (see [3,4]). One can observe that the calculated boundary separate the zones with blocked and superparamagnetic particles from our simulation.

Using the previous model [3,4] we can calculate the position of the thermal boundaries only for processes with constant relaxation times. To know the behavior of the thermal boundaries while the field or the temperature change we have extended our studies for processes with variable field or temperature. Processes with constant temperature and variable field were simulated using the exact solution of the master equation. In [6] it was considered that the field is maintained constant a certain time and it varies

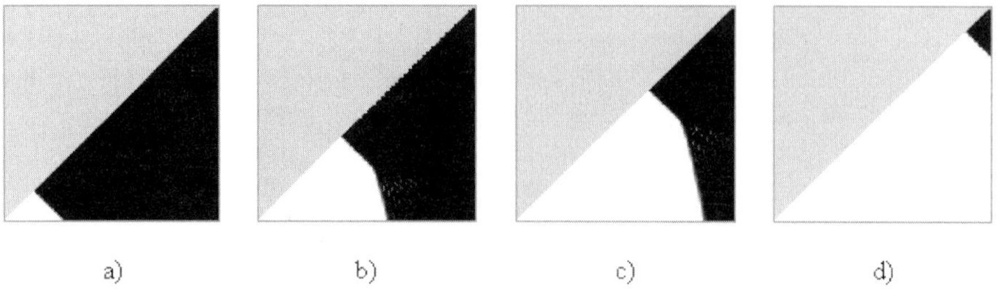

Fig. 3. The Preisach plane in the time of a process with constant temperature and variable field with the rate 8 A/m/s: (a) – initially thermal demagnetized; (b) – after 7.5 h, (c) – after 15 h; (d) – after 22 h.

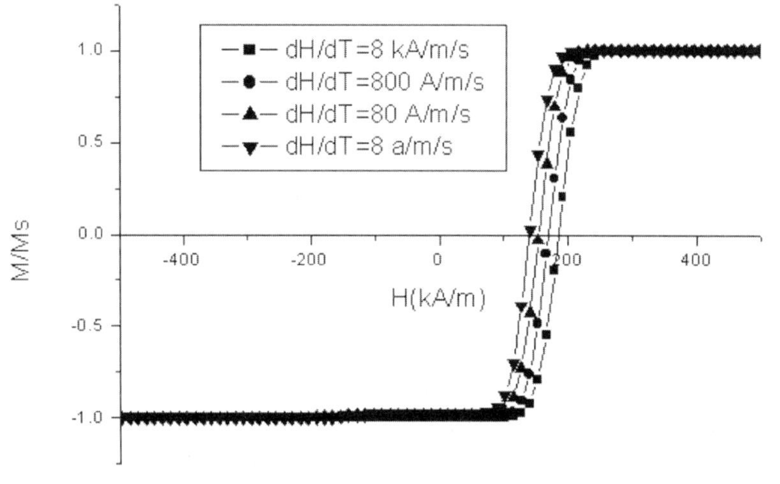

Fig. 4. The magnetization curves starting from negative saturation for different field rates; for the field rate of 8 A/m/s $H_c = 875$ A/m, for 80 A/m/s $H_c = 965$ A/m, for 800 A/m/s $H_c = 1060$ A/m and for 8 kA/m/s $H_c = 1160$.

instantaneously to a neighbor value. Using the actual model we can consider continuous variations of the field. We have obtained similar effect of the temperature on the major hysteresis loop like in [6]: the decrease of the coercive field with the increase of the temperature. The explication of this effect resides in the shift of the thermal boundary towards higher coercive fields in Preisach plane when the temperature increases. The contrary effect produces the increases of the field rate (the decrease of the coercive field). Figure 2 presents a temporal succession of sequences corresponding to the process with a constant field rate $R = 8$ kA/m/s. The first image is the initial thermal demagnetized state. Images b, c, d correspond to the movement of the thermal boundary over the Preisach distribution. In the Fig. 3 we present the same process with a field rate of 8 A/m/s. One superposes thermal boundaries calculated for the instantaneous field, the temperature of the processes and different times of applied field for each image in Figs 2 and 3. For a constant field rate the equivalent time rests the same independently of the moment we consider. It depends only on the field rate not on the total time of the processes. The magnetic moment versus the applied time in the processes with the field rate varying between 8 A/m/s and 8 kA/m/s vas represented in Fig. 4. As we can observe from Fig. 5, the coercive field increases almost logarithmical with the field rate. This can be explained considering that the critical curves moves toward higher coercive fields in the Preisach plane like the particles are maintained in constant field times inverse proportionally with the field rate.

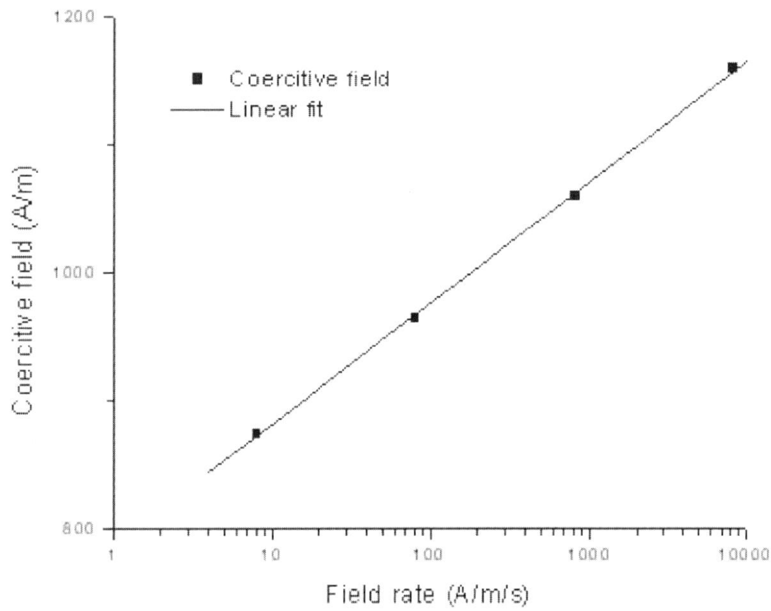

Fig. 5. The variation of the coercive field versus the field rate $R = \partial H/\partial t$.

4. Conclusions

The thermal boundary behavior in various time and temperature magnetization processes was studied. The shape of the thermal boundary is a curve line for nonlinear energy barrier. The consideration of non-linear energy barrier enables us to find connections between the identification parameters and the real properties of the particle system starting from physical characteristics. This method is more flexible permitting to consider the orientation of the particles and distributions of orientations. It was shown that the movement of the thermal boundaries and implicit the final moment depend not only on field and temperature but also on the magnetic field rate. The position of the thermal boundaries in processes with variable relaxation times deduced using our model could be integrated in more complex models that considered until now only constant field and temperature [5,6]. In such processes the results using the two models coincide.

Acknowledgements

This work was partially supported by Romanian CNCSIS under the Grants D/42 and A. The authors would like to acknowledge the support from ISEM'2001 organizers.

References

[1] R.W. Chantrell, A. Lyberatos, M. El-Hilo and K. O'Grady, *J. Appl. Phys.* **76** (1994), 6407–6412.
[2] F. Preisach, *Z. Phys.* **94** (1935), 277–302.
[3] P.D. Mitchler, E. Dan Dahlberg, E.E. Wesseling and R.M. Roshko, *IEEE Trans. on Magn.* **32** (1996), 3185–3194.
[4] T. Song and R.M. Roshko, *IEEE Trans. on Magn.* **36** (2000), 223–230.
[5] L. Spinu and A. Stancu, *J. Mag. Mag. Mater.* **189** (1998), 106–114.

[6] A. Stancu and L. Spinu, *IEEE Trans. on Magn.* **34** (1998), 3867–3875.
[7] E.C. Stoner and E.P. Wohlfarth, *Philos. Trans. R. Soc.* **A240** (1948), 599–642.
[8] H. Pfeiffer, *Phys. Stat. Sol.* (*a*) **118** (1990), 295–306.
[9] L. Néel, *Annales Géophysique* **5** (1949), 99–136.

International Journal of Applied Electromagnetics and Mechanics 13 (2001/2002) 329–334
IOS Press

Extraction of B-H relation based on the inverse magnetostatic problem

Daniel Ioan* and Mihai Rebican
"POLITEHNICA" University of Bucharest, Electrical Engineering Department, Numerical Methods Laboratory, Splaiul Independentei 313, 77-206 Bucharest, Romania

Abstract. The paper describes a magnetic field inverse problem and an efficient numerical method to solve it. Starting from the experimental flux-current $\Phi - I$ data obtained using a non-uniform magnetic field device with known geometry, the B-H nonlinear characteristic of the homogeneous magnetic core is extracted. Both functions B-H and $\Phi - I$ are piecewise linear approximated with the same number of breakpoints. The correct formulation (solution uniqueness) of this particular inverse problem is discussed. The method was successfully applied to characterize the ribbon core material of a FLUXSET magnetic field sensor used in Eddy Current Testing. In this case, the hysteresis loop and the equivalent lumped magnetic nonlinear circuit were extracted. Comparison with experimental results validates the proposed method.

1. Introduction

The classical methods to measure the B-H relations of magnetic materials preferably use devices with uniform magnetic field, such as: coils with toroidal core or Epstein frames. In this case the B-H characteristic is obtained by appropriate scaling of the flux-current $\Phi - I$ relation. Due to technological restrictions (e.g. parasitic air gaps), not even these devices can ensure a perfectly uniform magnetic field in the ferromagnetic material sample.

If the sample has small dimensions or an irregular shape, it can be tested only in an open magnetic circuit. In these particular cases the magnetic field is strongly nonuniform and the measured data are influenced by on the shape-dependent demagnetization factor. To eliminate this drawback it is necessary to use a method able to extract the B-H relation, even if the magnetic field is a non-uniform one. An original method to solve this problem was first presented in [1]. The solution uniqueness and the control of the error is discussed for the linear core in [2].

An extension of the algorithm for the media with hysteresis was presented in [3]. Due to its originality, the proposed approach generated a lot of debates regarding its correctness. In the present paper are discussed theoretical aspects related to the correct formulation of the nonlinear discrete inverse problem.

2. Problem formulation

Let us consider a homogeneous nonlinear ferromagnetic body $\mathcal{D}_m$ (having $\mathbf{B}||\mathbf{H}$ and a monotonic, bijective B-H relation) surrounded by air, and a non-magnetic coil $\mathcal{C}$ with the magnetic flux φ generated by the current i.

*Corresponding author: Daniel Ioan, Tel.: +40 1 411 1190; Fax: +40 1 411 1190; E-mail: daniel@lmn.pub.ro.

1383-5416/01/02/$8.00

The magnetostatic field equations are:

$$\begin{aligned} &\operatorname{div}\mathbf{B} = 0 \quad \text{in } \mathbb{R}^3; \\ &\operatorname{curl}\mathbf{H} = \mathbf{J} \quad \text{with } \mathbf{J} = 0 \text{ in} \mathbb{R}^3 \backslash \mathcal{C}; \\ &\mathbf{B} = f(H)\cdot \mathbf{H}/H \quad \text{in } \mathcal{D}_m; \\ &\mathbf{B} = \mu_0 \mathbf{H} \quad \text{in } \mathbb{R}^3 \backslash \mathcal{D}_m. \end{aligned} \tag{1}$$

Considering a given geometrical configuration and knowing function $g : \mathbb{R} \to \mathbb{R}$ with $\varphi = g(i)$, the problem is to find the function $f : \mathbb{R} \to \mathbb{R}$, which is the magnetic characteristic $B = f(H)$ of the ferromagnetic material in the domain $\mathcal{D}_m$, assuming that f is not dependent on the point $P \in \mathcal{D}_m$, and $f(0) = 0$.

The current density $\mathbf{J}$ depends linearly on the total current i, and for the simplicity we assume that the coil wire is relatively thin. Therefore the current is distributed uniformly across its section.

The $formulatedproblemisaninverseone$, because it is not the field but the material characteristic that must be found, starting from the "external" data. This continues problem has a unique solution [2].

3. The uniqueness theorem for the inverse magnetostatic discrete problem

In the linear case, $\mathbf{B} = \mu\mathbf{H} = \mu_0(1+\chi)\mathbf{H}$. The integral equations method for discrete magnetisation $m \in \mathbb{R}^n$ gives the following $\varphi - i$ relation:

$$\varphi = \lambda_0 i + l^{\mathrm{T}}(I + \chi^{-1}A)^{-1} l i,$$

where: λ_0 is the inductance without core, $l \in \mathbb{R}^n$ has as components the fluxes generated by a unit magnetisation, $A \in \mathbb{R}^{n\times n}$ is the mutual influence matrix (a_{ij} being the average field in the subdomain i produced by a unit magnetisation in domain j). Due to the field reciprocity theorem and energy positiveness, A is a symmetric and positive definite matrix. Therefore $A = U^{\mathrm{T}}\Lambda U$, when U is a unitary matrix and Λ a diagonal matrix with positive elements (eigen values of A).

We shall prove that for a given inductance $L = \varphi/i$ and a given geometry, with linear and homogeneous core, there is only one permeability μ of the core. The following lemma supports this statement.

For any given $L \in [L_0 = \lambda_0, L_\infty = l^2]$ *the equation* $\lambda_0 + l^{\mathrm{T}}(I + \chi^{-1}A)^{-1} l = L$ *has only one positive solution* χ.

Considering $\mathbf{B} = \mu_0(\mathbf{H} + \chi\mathbf{H} + \mathbf{M}_p) = \mu\mathbf{H} + \mu_0\mathbf{M}_p$, and using Ampèrian current model (equivalent permanent magnetisation current) and Maxwell equation for inductances, the $\varphi - i$ relation is also an affine one and has the form:

$$\varphi = \lambda_0 i + l^{\mathrm{T}}(I + \chi^{-1}A)^{-1} l i + l^{\mathrm{T}} m_p,$$

where $m_p \in \mathbb{R}^n$ is the permanent magnetisation vector. If the $\varphi - i$ relation is the same for two materials with parameters (χ_1, m_{p_1}) and (χ_2, m_{p_2}), then $\chi_1 = \chi_2$ and $l^{\mathrm{T}}(m_{p_1} - m_{p_2}) = 0$. Considering $m_p = [\eta, \eta, \ldots, \eta] = \eta u$, $(\eta_1 - \eta_2) l^{\mathrm{T}} u = 0$ with $l^{\mathrm{T}} u > 0$, and therefore $\eta_1 = \eta_2$ and $m_{p_1} = m_{p_2}$. Considering the limit $n \to \infty$ then conclusions in both cases are valid also for the continuous case.

For the nonlinear case, let us consider a particular nonlinear B-H relation with three parameters (μ_i, H_s, μ_f) as in Fig. 1.

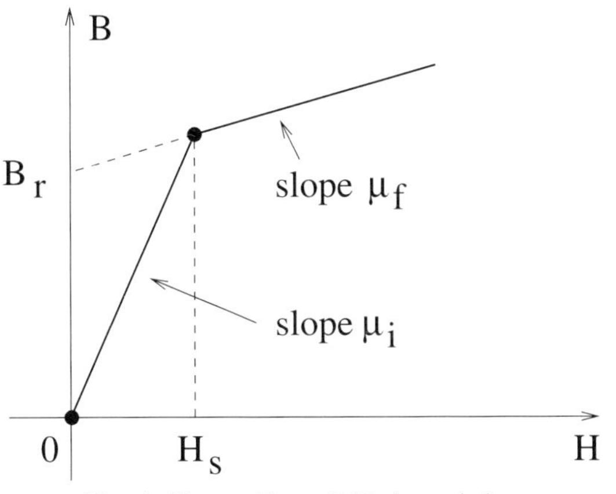

Fig. 1. The nonlinear B-H chacteristic.

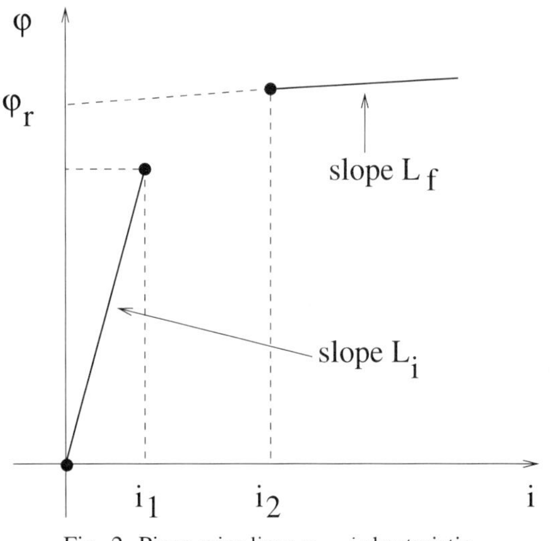

Fig. 2. Piece-wise liner $\varphi - i$ chacteristic.

The corresponding $\varphi - i$ relation is presented in Fig. 2, when currents i_1, i_2 are chosen so that

$$\max_{P \in \mathcal{D}} H(P, i_1)\Big|_{B=\mu_i H} = H_s \qquad \min_{P \in \mathcal{D}} H(P, i_2)\Big|_{B=\mu_f H + B_r} = H_s.$$

Based on previous results, there are bijective relations between L_i and μ_i, φ_r and B_r, L_f and μ_f, therefore there are no two different three-parameter B-H relations in an equivalence class (generating the same $\varphi - i$ relation). By complete induction one can be prove the same property for piecewise linear B-H relation with m breakpoints. At the limit $m \to \infty$, the property is valid for any B-H continuous monotonic relation, with $B(0) = 0$.

4. B-H Characteristic Extraction Algorithm

The algorithm we propose to solve the problem Eq. (1) has the following steps:

- **Generate a piece-wise linear approximation** of the $\varphi - i$ characteristic with an imposed error, by the selection of appropriate m points (φ_k, i_k) from the experimental data (Fig. 2).
- Consider a linear magnetostatic field problem with $B = \mu H$ in D_m, and find $\mu = \mu_1$ such that the magnetic flux due to the current i_1 is φ_1. This is accomplished **by iteratively solving the nonlinear equation**: $\varphi(i_1) = \varphi_1$, where φ is the numerical solution of the forward magnetostatic field problem Eq. (1), linear at low field.
- Compute the maximum magnetic field in D_m (H_{max}, $B_{max} = \mu_1 H_{max}$) and set the first two points on the extracted B-H characteristic: $H_0 = 0$, $B_0 = 0$ and **the first breakpoint** $H_1 = H_{max}$, $B_1 = B_{max}$.
- **Iteratively solve the forward magnetostatic field** problem Eq. (1) with a non-linear B-H characteristic with $k = 1$ breakpoints (as in Fig. 1) and **find** μ_2 such that the magnetic flux due to the current i_2 is: $\varphi(i_2) = \varphi_2$.
- Each point of $\mathcal{D}_m$ has different static permeability. The maximum magnetic field H_{max}, B_{max} due to i_2 represents **the second breakpoint** of the B-H characteristic.

Table 1
Numerical results

m	5	10	23	36
B_s [T]	0.7043	0.7042	0.7057	0.7068
μ_{ri}	40235	46485	48047	45703
CPU Time [s]	120	260	840	1520

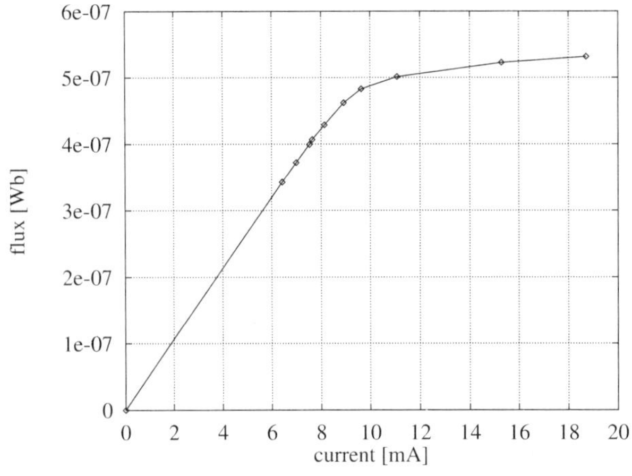

Fig. 3. Piece-wise linear $\varphi - i$ characteristic.

– **Repeat iteratively the last two steps** considering $k = 2, 3, \ldots$, until the breakpoint number m of the $\varphi - i$ characteristic is reached.

The computing effort to solve the inverse problem is directly linked to the number m of breakpoints on the $\varphi - i$ characteristic. This number is related to the approximation error generated by linearization on each segment. If the imposed error ε is known (usually of the order of measurement error), the number of breakpoints can be reduced by the selection step 1 of the algorithm. Starting from 418 experimental data points, we obtained $\varepsilon = 1\%$ with $m = 5$ breakpoints. Roughly speaking, the halving of the approximation error leads the doubling of the number of breakpoints.

5. Field computation and results

An important decision to be taken in order to apply the proposed algorithm concerns the numerical method used to compute the magnetic field.

In the case of the FLUXSET Sensor, the nonlinear magnetostatic field problem was solved with an integral equations method [4].

Table 1 presents the main numerical results (saturation flux density B_s, the relative initial permeability μ_{ri} and CPU time) for different numbers of breakpoints.

The B_s value is not sensitive to the number of breakpoints (the modification is less than 0.3%) but the initial permeability has values within $\pm 8\%$ range.

Less than 11 iterations (field computations) were enough for each k to obtain the solution with a numerical relative error less than the experimental one. Starting from the $\varphi - i$ dependence with $m = 10$ breakpoints represented in Fig. 3, the B-H characteristic of the core's material presented in Fig. 4 was extracted.

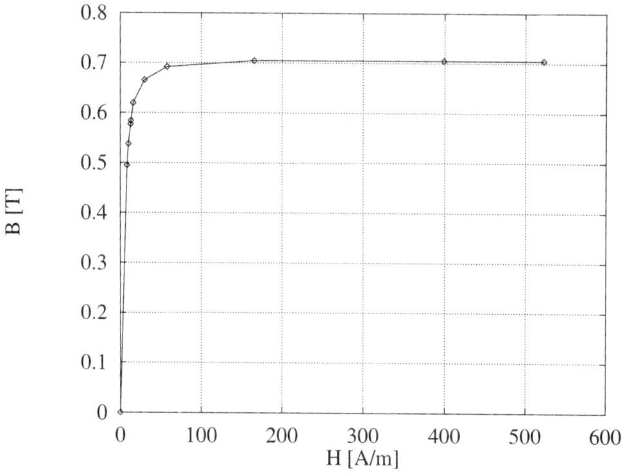

Fig. 4. The extracted B-H characteristic.

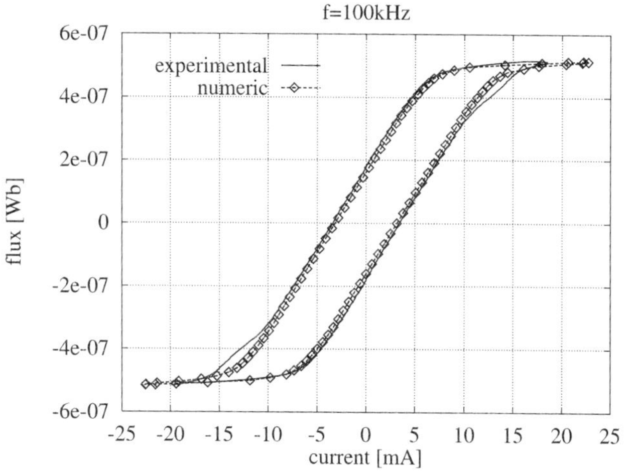

Fig. 5. The hysteresis loop at 100 kHz.

Figure 5 shows the numerical and experimental $\varphi - i$ hysteresis loops. Numerical data are obtained by SPICE simulation of the lumped circuit (Fig. 6) built using the B-H loop of the core material of a FLUXSET sensor, extracted from field solution.

6. Conclusions

The paper presents a uniqueness theorem for the solution of the inverse magnetostatic nonlinear problem and an algorithm able to solve this problem and to extract the B-H characteristic of the tested ferromagnetic samples from the $\varphi - i$ dependence, regardless the sample's shape and field uniformity.

The non-uniform magnetic field in the nonlinear core was numerically computed using the integral equations method. An adapted version of Katzenelson method was successfully used to solve the nonlinear system of equations obtained by discretisation. This approach of forward problem solving

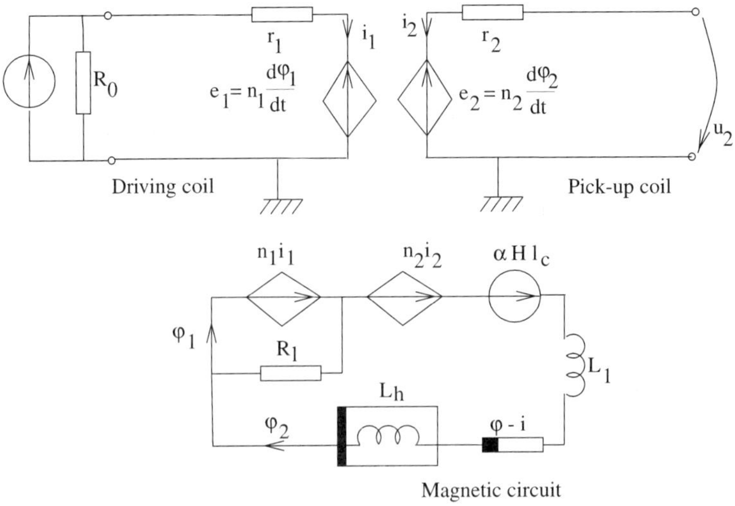

Fig. 6. The SPICE limped circuit.

proved to be more efficient in our case than other method such as: FEM or hybrid numerical-analytical techniques.

Using SPICE simulation of the extracted lumped circuit, good agreement with the experimental data was obtained. The proposed method is able to determine in a very effective manner the sensibility and frequency bandwidth of a newly designed FLUXSET sensor.

References

[1] D. Ioan, M. Rebican and A. Gasparics, B-H characteristic extraction using devices with non-uniform field, *COMPEL, The Int. J. for Comp. and Math. in Electr. and Electron. Eng.* **18**(3) (1999), 469–481.

[2] R. Albanese, A. Bossavit, R. Fresa, D. Ioan, G. Rubinacci, A. Tamburrino and F. Villone, Identification of B–H curve from external measurements using complementary formulations, *Physica B (Condensed Matter)* **275** (2000), 228–232.

[3] Daniel Ioan and Mihai Rebican, B-H Hysteresis Loop Extraction using Devices with Non-uniform Magnetic Field, in: *Conf. Record of the 12th Conf. on the Comp. of Electromagn. Fields COMPUMAG'99*, Sapporo, Japan, October, 1999, pp. 800–801.

[4] R. Albanese, A. Formisano, G. Rubinacci and R. Fresa, Assessment of performance for FLUXSET sensor core, in: *Electromagnetic Nondestructive Evaluation*, vol. II. R. Albanese et al., eds, IOS Press, 1998, pp. 180–187.

International Journal of Applied Electromagnetics and Mechanics 13 (2001/2002) 335–338
IOS Press

Modeling the temperature dependence of magnetization processes in soft ferrite cores

Ovidiu Florin Caltun[a,*], Petru Andrei[b] and Alexandru Stancu[a]
[a] *"Alexandru Ioan Cuza" University, Faculty of Physics, Bd. Copou, 11, Iasi, 6600, Romania*
[b] *University of Maryland, Department of Physics, 7002 Dartmouth Ave., College Park, MD 20740, USA*

Abstract. The shape of the hysteresis loop of MnZn ferrites depends on temperature. By increasing the temperature, the squareness of the loop increases. Consequently, the magnetic properties change the power losses and the performances of the ferrites. In this paper, a comparison is made between the calculated hysteresis loops in Jiles-Atherton model and the experimental loops. The influence of temperature and of the microstructure (size of the grains, pores and segregation) on the model's parameters is discussed.

1. Introduction

The dependence of saturation magnetization on temperature is due to the antiferromagnetic properties of MnZn ferrites. The total magnetization of saturation depends on the saturation magnetization of each sub-lattice and it is the difference between them. The convexity of the magnetization vs. temperature curve for ferrite is lower than that of ferromagnetic materials. The purpose of this work is to study the influence of temperature on the hysteresis loop and on the magnetic properties of some Mn-Zn ferrite cores. The experimental results are compared with the theoretical ones obtained in Jiles-Atherton model.

2. Experimental results

The samples were prepared with a conventional ceramic process [1]. The microstructure of the samples was studied using a Philips scanning microscope (SEM). To study the influence of temperature on the hysteresis loop, the experimental set-up described in [2] was used. This paper is restricted to the magnetization processes at high magnetic field strength of 100 A/m and 10 kHz excitation frequency. The temperature was varied between room temperature and 500 K, which is greater than the Curie temperature established by initial permeability measurement [1]. The hysteresis shape and the magnetic properties depend on temperature. The change in shape of the normalized hysteresis loop for the sample due to the temperature is represented in Fig. 1.

One observes that the shape of the hysteresis loops of the samples depends on temperature and doping level. By increasing the temperature, the squareness of hysteresis loops becomes greater (see

*Corresponding author: Dr. Ovidiu Florin Caltun, "Alexandru Ioan Cuza" University, Faculty of Physics, 11 Blvd. Copou, 6600 Iasi, Romania. Tel.: +032 201176; Fax: +032 201205; E-mail: caltun@uaic.ro.

1383-5416/01/02/$8.00

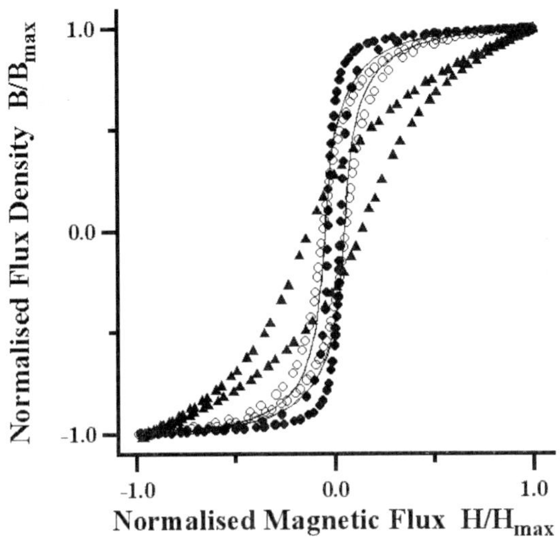

Fig. 1. Normalized hysteresis loops vs. temperature Sample 1 (▲) $T_C + 20°C$; (○) T_C; (−) $T_C - 20°C$; (•) 25°C.

Fig. 1). The dependence of saturation flux density versus temperature, characteristic for Mn-Zn ferrites. The convexity of the curves depends on the sample (doping level, sintering condition etc.). These measurements agree with those reported by Kumar [3] for Ni-Zn ferrites. The coercive magnetic field strength decreases when the temperature increases, taking a minimum value at temperatures close to the Curie temperature. Beyond this temperature, the coercive magnetic field strength increases.

3. Theoretical results in Jiles-Atherton model

In the Jiles-Atherton model, the magnetic susceptibility is a function of the magnetization M and the applied magnetic field H [4,5]:

$$\frac{dM(H)}{dH} = (1-c)\delta_M \frac{M_{an}(H_e) - M(H)}{k(1-c)sgn(\dot{H}) - \alpha[M_{an}(H_e) - M(H)]} + c\frac{dM_{an}(H_e)}{dH_e} \tag{1}$$

where H_e is the effective field given by:

$$H_e = H + \alpha M, \tag{2}$$

and M_{an} is the anhysteretic magnetization which is considered a Langevin function:

$$M_{an}(H_e) = M_s L(H_e/a) = M_s[coth(H_e/a) - a/H_e]. \tag{3}$$

δ_M is defined as follows:

$$\delta_M = [\begin{cases} 0, \text{ if } \dot{H} < 0 \text{ and } M_{an}(H_e) - M(H) \geqslant 0; \\ 0, \text{ if } \dot{H} > 0 \text{ and } M_{an}(H_e) - M(H) \leqslant 0; \\ 1, \text{ otherwise.} \end{cases} \tag{4}$$

In the above equations, a is the form factor for the anhysteretic curve, c is approximate to the ratio of the initial susceptibility (on the first magnetisation curve) to the initial anhysteretic differential susceptibility,

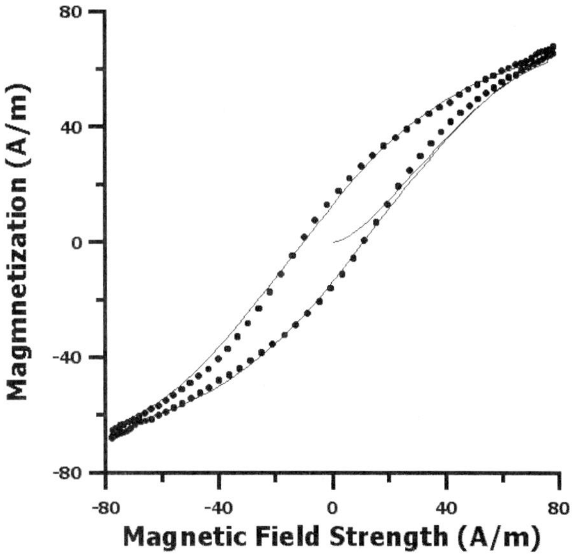

Fig. 2. Hysteresis loop for Sample 1 at 100°C (•) experimental and (−) calculated.

α is a parameter representing the coupling between domains, k is the pinning constant (it gives a measure of the width of the hysteresis loop) and M_s is the saturation magnetisation. All these parameters can be calculated [2] from experimental values of the coercive field (H_c), remanent magnetisation (M_r), saturation magnetisation (M_s), initial anhysteretic susceptibility (χ_{an}), initial susceptibility measured on the first magnetisation curve (χ_{in}), maximum differential susceptibility (χ_c) and differential susceptibility at remanence (χ_r).

In Fig. 2 is presented the experimental and calculated hysteresis loops obtained at 100 °C for the Sample 1 for the excitation frequency $\nu = 10$ kHz. The identification has been performed in each case for all the samples. A few facts are obvious from the above results. Firstly, the saturation magnetization decreases when the temperature increases. Secondly, the parameter c also decreases when T increases. In the original Jiles-Atherton model the parameter a is considered inversely proportional to the temperature and depends on the average magnetic moment of the particles. However, because of the simplicity of the model the calculated hysteresis loops are in good agreement with the experimental data only if the parameters of the model are correlated with the variation of the temperature and changed.

4. Conclusions

The experimental study has demonstrated how the shape of the hysteresis loops depends on temperature. At temperatures close to the Curie temperature, the squareness of the hysteresis increases. The dependence of the magnetic properties, like saturation flux density and coercive field, was taken into account when finding the parameters of the Jiles-Atherton model. Even if the experimental data agrees with the calculated data, the parameters of the model change almost randomly. The parameter α increases monotonously when the temperature increases. An extended study and comparison of the values of the Jiles-Atherton parameters for different samples and ferrites will be the subject of future work.

Acknowledgements

This work was partially supported by Romanian CNCSIS under the Grants D/42 and A. The authors would like to acknowledge the support from ISEM'2001 organizers.

References

[1] P. Andrei, O. Caltun, C. Papusoi, Al. Stancu and M. Feder, Losses and magnetic properties of Bi2O3 doped MnZn ferrites, *Journal of Magnetism and magnetic Materials* **196–197** (1998), 362–364.
[2] Al. Stancu, O. Caltun and P. Andrei, Model of hysteresis in magnetic cores, *Journal de Physique IV France 7, Colloque C1, Supplement au journal de Physique III de mars* **C1** (1997), 209–210.
[3] P.S. Anil Kumar, S.R. Sainkar, J.J. Shrotri, S.D. Kulkami, C.E. Deshpande and S.K. Date, Particle size dependence of rotational responses in Ni-Zn ferrite, *Journal of Applied Physics* **83**(11) (1998), 6864–6866.
[4] D.C. Jiles, The frequency dependence of hysteresis curves in "non-conducting" magnetic materials, *IEEE Transaction on Magnetics* **29**(6) (1993), 3990–3997.
[5] J.H.B. Deane, Modeling the dynamic of nonlinear inductor circuits, *IEEE Transaction on Magnetics* **30**(5) (1994), 2795–2802.

International Journal of Applied Electromagnetics and Mechanics 13 (2001/2002) 339–342
IOS Press

Some effects of grain boundaries on the field distribution on the surface of grain oriented electrical steels

A.J. Moses* and S.N. Konadu
Wolfson Centre for Magnetics Technology, School of Engineering, Cardiff University, P.O. Box 925, Cardiff, CF24 0YF, UK

Abstract. A non-destructive system using a Hall effect I.C. sensor has been used to measure the distribution of the vertical component of stray magnetic field on the opposite surfaces of a magnetised specimen of high permeability, grain-oriented electrical steel. An optical microscope was used to establish positions of grain boundaries in order to explain differences in field profiles outside the systematic accuracy of measurements. It was found that unlike previously assumed, grain boundaries may not necessarily be perpendicular to the sheet surface so different field distribution may occur. This should be considered when evaluating local power loss from surface sensor movements. Future studies are also necessary to quantify the effect of this phenomenon in the interpretation of surface or bulk properties of electrical steels using surface field sensors.

1. Introduction

Electrical steels are used in many transformer and motor core applications where a.c. magnetic permeability or losses are important. They are usually used in sheet form typically 0.23 mm to 0.65 mm thick. Grain oriented 3.25% Silicon Steel has a strong [001](110) texture with grains of up to 10 mm in diameter. Its surfaces are usually coated with complex non-magnetic layers to impose beneficial tensile stress in the steel as well as providing the necessary interlaminar or interturn insulation when built into cores. The energy loss under a.c. magnetisation is proportional to the integral of the product of the instantaneous tangential component of surface field and the instantaneous spacial averaged rate of change of flux density.

In grain oriented electrical steel the bulk of the material in the demagnetised state comprises anti-parallel bar domains with 180° walls. The domain, structure which depends on the grain size and orientation as well as several other factors is changed by application of an external field and during the process magnetic losses occur. The surface magnetic field in the magnetised or demagnetised state can give an indication of material structure which in turn has a large influence on the losses and permeability.

The surface magnetic field can be measured in several ways. Hall sensors have been used for the measurement of the tangential component of surface field of the coated material at scanning steps of around 0.3 mm, sensor length being typically 3 mm [1,2]. More recently smaller magneto resistive sensors are used with greater flexibility [3]. Close relationships are found between grain structure orientation and tangential field [4].

*Corresponding author: A.J. Moses, Tel.: +44 029 20876729; Fax: +44 029 20876729; E-mail: MosesAJ@cf.ac.uk.

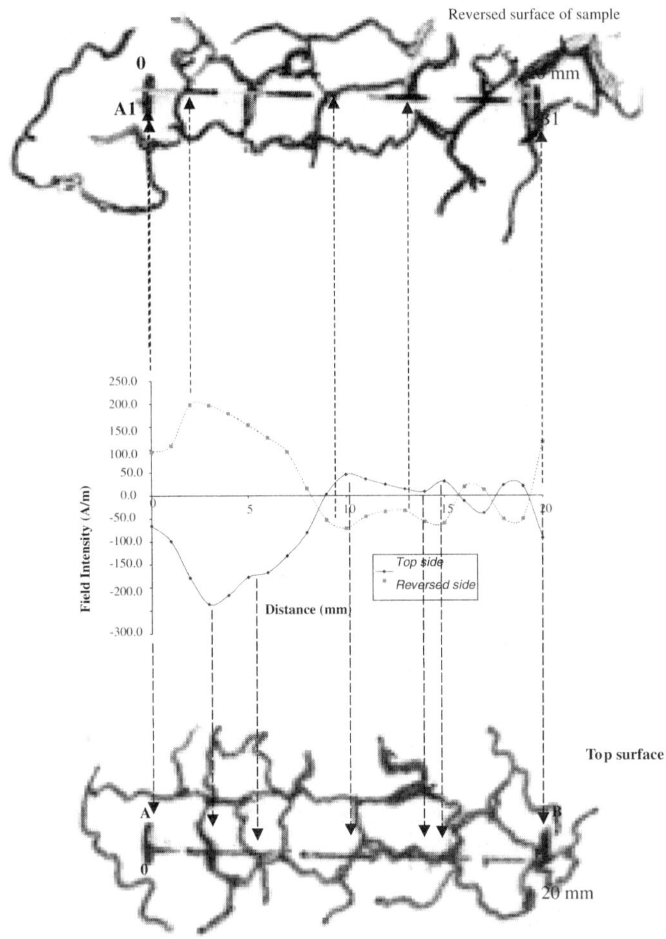

Fig. 1. Illustration of the mean results of the scans carried out on either surface of the sample related the observed grain structure on either side of the sample.

Previously surface magnetic domain structures have been visualised from variation of the vertical component of surface field [5]. In this paper some effects of the grain structure on the distribution of the vertical (normal) component of the magnetic field are measured. The vertical component of field is made up of three parts: H_b, a stray field generated at grain boundaries, H_s a stray field generated at 180° domain walls and H_β, a stray field generated at domain on the surface of the grains with misorientation angle not equal to zero [6]. Generally $H_\beta > H_s$ and if a high d.c. external, in-plane field is applied then the domain walls will mostly be annihilated leaving two possible vertical field components originating at grain boundaries and on the saturated grain surfaces where $\beta > 0$.

2. Experimental procedure and results:

A rectangular sheet of high permeability grain oriented electrical steel 116 mm (rolling direction) by 78 mm by 0.23 mm thick was chemically decoated to reveal a grain structure with average size of around 60 mm^2. A Linear Hall effect I. C. sensor with an active length of 1 mm was calibrated in a uniform

Table 1
Measurement of positions of grain boundaries along arbitrary paths parallel and perpendicular to the rolling direction

Boundary	Line PQ (mm) (Ref. Point P)	Line P_1Q_1 (mm) (Ref. Point P_1)	Line RS (mm) (Ref. Point R)	Line R_1S_1 (mm) (Ref. Point R_1)
1	3.8	7.0*	0.1	2.1*
2	4.9	8.5	1.7*	10.7*
3	7.3*	8.9	10.6*	18.6*
4	9.4*	9.1*	11.0	25.6*
5	10.0*	10.5*	18.8*	28.5*
6	16.3*	11.0	25.7*	32.5*
7	23.7*	12.2	28.3*	
8	27.7*	14.5	32.8*	
9	30.5	17.3*		
10	32.8	21.3*		
11	34.0	28.7*		
12	34.5*	35.0*		
13	35.9*	35.6*		
14	36.1	37.7*		
15	36.5	40.1		
16	38.0*	41.2		

solenoid carrying a d.c. current and its sensitivity was found to be 0.03 mV/A/m. The sensor was moved in 1 mm steps along the surface of the sheet at a height of approximately 0.25 mm above the surface. The sensor was energised by a 3.5 mA current and its output voltage was measured on a digital voltmeter with an accuracy of ± 5 mV. The peak value of the surface field was measured in this way to the accuracy of ± 0.41 A/m

The sheet was magnetised along its rolling direction by a d.c. external field of 1 kA/m making use of a C-core magnetising yoke. The sensor was scanned along a 20 mm path close to the centre of the lamination and along its rolling direction. The vertical components of field was plotted along this path on one side of the specimen which was then inverted and the process repeated on the corresponding position on the other surface

Figure 1 shows the average results of 6 scans carried out on either side of the sample along the same path AB. The positive and negative values indicate field directed out of or into the surface respectively. There is a difference between the distributions on either surface although they broadly follow the same pattern, the maximum field of about 250 A/m occurs at around 2–3 mm from the reference position directed into the sample at the top surface. Figure 1 shows also the approximate positions of grains observed on either side of the sample path AB using a travelling microscope. It is probable that several subgrains are present but could not be easily observed in this way.

In order to help attempt to explain the difference of the field patterns, the position of grain boundaries through the thickness of the material was investigated. It is normally assumed that the boundaries lie at 90° to the sheet surfaces and are therefore in identical relative positions on either surface. A piece of material 10 mm by 10 mm, of same sort that was scanned was cut from the sheet and edges PQ and RS were ground, polished with diamond paste and etched in 3% nitric acid to reveal grain boundaries by use of a X100 optical microscope. Boundaries in the typical examples shown in Fig. 2 are inclined at angles from 89° to 26° degrees to the sheet surface. Figure 2(d) shows the case of a grain visible from only one surface. Table 1 indicates the position of all visible grain boundaries on either side of the sheet lines PQ and P_1Q_1 corresponding to positions along the rolling direction and lines RS/R_1S_1 positions along the transverse direction. The double arrowed lines indicate what might be positions of individual

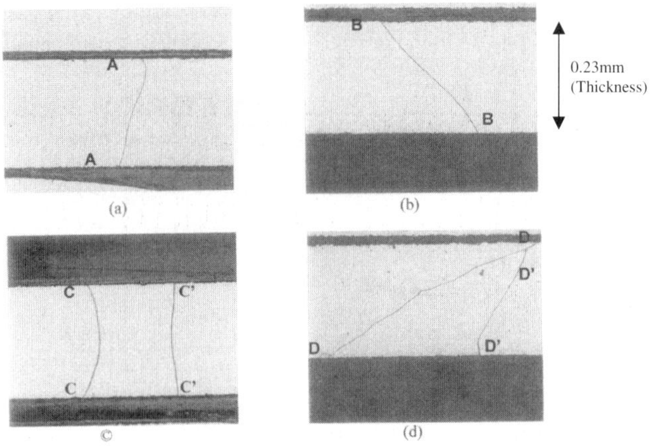

Fig. 2. Micrographs showing typical grain boundary profiles through the thickness of a sheet of grain oriented steel.

grain boundaries on either surface. Obviously if large angles of inclination are present, a given grain boundary could emerge at the opposite surfaces displayed by 1 mm or more so the links in Table 1 are only postulations. It is interesting however that along the lines RS/R_1S_1 the inclination angles appear to be all close to 90° to the sheet plane.

3. Discussion and conclusion

The high d.c. field eliminates most of the surface domains so it is possible that the field distribution shown in Fig. 1 mainly originate due to the angle of dip, of the surface grains. The field H_b should have a minor effect since H_β appears to be far larger. However, distances between the 0 mm point and the other grain boundary positions (arrowed in Fig. 1), compared with the distances on the magnetic field profile show that the turning points on the profile occur where grain boundaries are visible [2,7]. The magnitude of the surface field varies a great deal and does not appear to be directly related to individual grains. The very high field region extends over a region with grains of different size and orientation.

In conclusion the field distribution, although repeatable, is not the same on either surface. If surface field measurements are used to predict local losses or domain structures then care is needed in the interpretations of the results obtained from one side of a sheet only because small changes in tangential field will also occur on either side. Normally it is assumed that any specific bar domain will be identical at either surface but this is expected not to be the case in practice because of the grain boundary profiles. Future studies are necessary to quantify the effect of this phenomenon in the interpretation of surface or bulk properties of electrical steels using surface field sensors.

References

[1] H. Pfutzner, *J. Magn. Mag. Mat.* **19** (1980), 27–30.
[2] K. Mohri and T. Fujimoto, *Mem. Kyushu Inst. Technology Eng.* **7** (1977), 33–37.
[3] B.B. Mohd Ali and A.J. Moses, *J. Fiz. Mat.* **9**(75) (1988), 77–80.
[4] S. Tumanski and M. Stabrauski, *J. Magn. Mag. Mat* **29**(6) (1993), 2998–3000.
[5] M.H. So et al., *IEEE Trans Mag.* **31**(6) (1995), 3370–3372.
[6] H. Pfutzner et al, *Jpn. J. Appl. Phys* **22** (1983), 361–364.
[7] P. Beckley, *Electrical Steels, A handbook for Producers and Users*, European Electrical Steels, ORB Works, 2000, pp. 47.

International Journal of Applied Electromagnetics and Mechanics 13 (2001/2002) 343–348
IOS Press

Effects of magnetism on the electronic structure and pairing mechanism in oxide superconductors

V.A. Gavrichkov[a], E.V. Kuz'min[b], S.G. Ovchinnikov[c,*] and A.G. Mamalis[c]
[a]*L.V. Kirensky Institute of Physics, Krasnoyarsk, 660036, Russia*
[b]*Krasnoyarsk State University, Krasnoyarsk, 660062, Russia*
[c]*National Technical University of Athens, Athens, 10682, Greece*

Abstract. Generalized tight-binding method is developed to strongly correlated electron systems. Evolution of the band structure of CuO_2 layer with hole doping from the undoped charge transfer antiferromagnetic insulator to the optimally doped paramagnetic metal is obtained. Effective low energy Hamiltonian is given by a t-J-I model that allows to compare magnetic mechanism and symmetry of pairing in cuprates and ruthenates on equal footing.

1. Introduction

Up to now there is no widely accepted theory for the electronic structure of the undoped antiferromagnetic insulator, it's concentration evolution with doping to metal state and for the pairing mechanism in cuprates of p- and n-type. A spin-fluctuation mechanism of pairing is probably the best candidate to describe HTSC in copper oxides [1,2]. One of the unsolved problems is the underlying electronic structure and how the specific electronic properties of a given material provide the rich variety of different properties in cuprates of p- and n-type, ruthenates and layered $RuSr_2GdCu_2O_8$. We discuss this problem in the framework of the multiband p-d model of transition metal oxides.

2. Generalized tight binding approach

The necessary ingredient for discussing possible mechanisms of HTSC is the band structure of the fermion-like quasiparticles. However, it is a difficult subject for *ab initio* calculations due to the strong electron correlations. For this reason we here use a model approach which is based on a possibility to rewrite the exact full Hamiltonian of the system in terms of larger-scale variables. Described briefly, the essence of the approach is as follows. Suppose we have a set of Wannier functions belonging to a chosen elementary cell, which is generated by some single-electron problem for the crystal symmetry of interest. Then the full electron field operator can be written in the form of an expansion:

$$\therefore\varsigma\,(r) =_{jL} \Phi_{jL}(r) a_{jL}.$$

*Corresponding author: S.G. Ovchinnikov, Fax: +7 3912 438923; E-mail: sgo@iph.krasn.ru.

1383-5416/01/02/$8.00

Here the index j R_j gives the coordinate of the chosen elementary cell of the lattice, L is composite index, $L = \{C, \zeta\}$, Olabels bands and ζ/stands for projector of spin. Due to orthogonality of the Wannier functions we have $\{a_{jL}, a^{+}_{j'L'}\} = \Gamma_{jL,j'L'}$. Full non-relativistic Hamiltonian can be written in the form of a sum of intra- (H_c) and inter-cell (H_{cc}) terms:

$$
\begin{aligned}
H &= H_c + H_{cc} =_i H_i^c +_{ij} H_{ij}^{cc}, \\
H_i^c &= {}_{LL'} h_i^{LL'} a^{+}_{iL} a_{iL'} + \frac{1}{2}{}_{LL'} \Theta_{iL,iL',iL'',iL'''} a^{+}_{iL} a^{+}_{iL'} a_{iL''} a_{iL'''}, \\
H_{ij}^{cc} &= {}_{LL'} h_{ij}^{LL'} a^{+}_{iL} a_{jL'} + \frac{1}{2}{}_{LL'} \Theta_{iL,jL',jL'',jL'''} a^{+}_{iL} a^{+}_{jL'} a_{jL''} a_{iL'''} \\
&\quad - \frac{1}{2}{}_{LL'} \Theta_{iL,jL',iL'',jL'''} a^{+}_{iL} a^{+}_{jL'} a_{iL''} a_{jL'''}.
\end{aligned} \tag{1}
$$

Suppose now that we are able to find the many-electron states for the intra-cell Hamiltonian in an approximation better than Hartree-Fock, so, that it takes into account (at least, partly) strong intra-atomic and intra-cluster interactions. Then

$$
H_i^c |i, \Gamma_n\rangle = E^{(0)}_{n\Gamma} |i, \Gamma_n\rangle \tag{2}
$$

and the full Hamiltonian can be rewritten in terms of the Hubbard's operators

$$
X_i^m \; X_i^{\Gamma_{n-1},\Gamma_n} \; |i, \Gamma_{n-1}\rangle\langle i, \Gamma_n|, \quad Z_i^{[} \; Z_i^{\Gamma_n,\Gamma'_n} \; |i, \Gamma_n\rangle\langle i, \Gamma'_n|. \tag{3}
$$

The Hubbard model is often used to study electronic structure of strongly correlated electron systems (SCES). To take into account the chemistry of metal oxides it is generalized to the p-d model, a simplest version of such a model has been proposed by Emery [3] and Varma et al. [4]. In this 3-band p-d model only $d_{x^2-y^2}$ Cu and p_σ O orbitals are considered. There are many indications on the importance of the d_{z^2} Cu orbital (see the review in [5]). The multiband p-d model with both $d_{x^2-y^2}$ and d_{z^2} states has been proposed by Loktev et al. [6]. To calculate electronic structure in SCES the generalized tight-binding (GTB) method combining the exact diagonalization of the Hamiltonian for small clusters (intracell part of the Hamiltonian) with cluster perturbation theory for the intercell part of the Hamiltonian has been proposed by Ovchinnikov and Sandalov [7]. Recently the electronic structure of the undoped antiferromagnetic (AF) insulator like $Sr_2CuO_2Cl_2$ and its evolution with doping have been studied by Gavrichkov et al. [8]. The dispersion equation in GTB method for two sublattice AF state is given by

$$
\left\| \frac{(E - \Omega^G_m)}{F^G_\zeta(m)} \Gamma_{mn} - 2_{OO} \partial^*_{O\zeta}(m) T^{PG}_{OO}\left(\vec{k}\right) \partial_{O\zeta}(n) \right\| = 0, \tag{4}
$$

where the zero frequency of the transition $\Omega^{(0)}_{i,m} = E^{(0)}_{n+1,\Gamma} - E^{(0)}_{n,\Gamma}$ is renormalized by the Coulomb interactions within the analogue of the Hubbard-I approximation as follows:

$$
\begin{aligned}
\Gamma_{mm_1} \Omega_{i,m} = \Gamma_{mm_1} \Omega^{(0)}_{i,m} +_{j(i)} \Big(& B_{ij}^{\Gamma_1\Gamma_2,\Gamma_3\Gamma_4} H_{m_1}^{m,[\Gamma_1,\Gamma_2]} \left\langle Z_j^{\Gamma_3,\Gamma_4} \right\rangle \\
& - B_{ji}^{\Gamma_1\Gamma_2,\Gamma_3\Gamma_4} H_{m_1}^{m,[\Gamma_3,\Gamma_4]} \left\langle Z_j^{\Gamma_1,\Gamma_2} \right\rangle \Big),
\end{aligned} \tag{5}
$$

where the matrix elements $B_{ij}^{\Gamma_1\Gamma_2,\Gamma_3\Gamma_4}$ are:

$$
B_{ij}^{\Gamma_1\Gamma_2,\Gamma_3\Gamma_4} = \Theta_{iL,jL',jL'',iL'''} \left[a^{+}_{iL} a_{iL'''} \right]^{\Gamma_1\Gamma_2} \left[a^{+}_{jL} a_{jL'''} \right]^{\Gamma_3\Gamma_4}. \tag{6}
$$

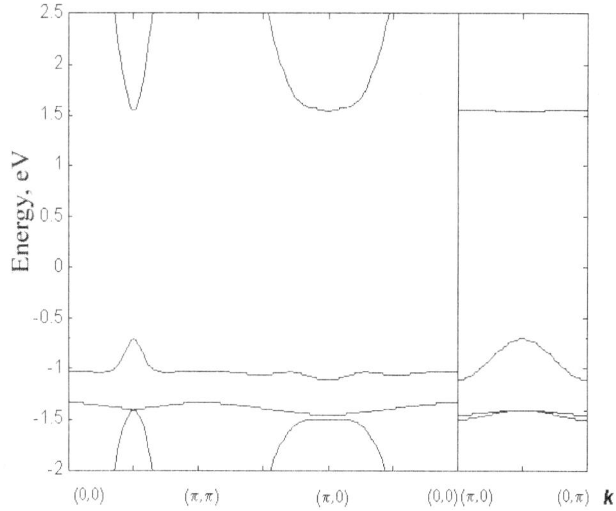

Fig. 1. Band structure of the undoped CuO_2 layer.

Here the transitions $[\Gamma_1, \Gamma_2], [\Gamma_3, \Gamma_4]$ are Bose-like ones, and the splitting in fermion-like transitions originates, obviously, from the crystal-field splitting of the states Γ by Coulomb interactions with the neighbouring clusters. On the background of $\Omega^{(0)}_{i,m}$, which is of the order of Hubbard U of the cluster, these effects are small. Below we consider the antiferromagnetic state and the lattice index i runs over the sublattices P, G.

Here $m \leftrightarrow (p, q)$ numerate a quasiparticle described by the Hubbard operator X^{pq}, and $\vartheta_{O\zeta}(m)$ is a parameter of X-operator representation for the single-electron annihilation operator with orbital λ and spin σ, $a_{fO\zeta} =_m \vartheta_{O\zeta}(m)X^m_f$. Equation (4) has the same structure as the usual tight-binding equation of the single electron approach but differs in the following: i) the local energy Ω_m are given by the multielectron resonance $\Omega_m = E_{n+1}(p) - E_n(q)$, between n- and $(n+1)$-electron terms of the cell, ii) filling factors $F(m) = \langle X^{pp}\rangle + \langle X^{qq}\rangle$ result in concentration and temperature dependence of the band structure.

The electronic structure for the undoped case with hole concentration $n_h = 1$ is shown in Fig. 1. It was calculated for the following set of the model parameters (in eV):

$$t_{pd} = 1, \quad H(d_{x^2-y^2}) = 0, \quad H(d_{z^2}) = 0, \quad H(p_x) = 1.6, \quad H(p_z) = 0.5,$$
$$t_{pp} = 0.46, \quad t'_{pp} = 0.42, \quad U_d = 9, \quad U_p = 4, \quad V_{pd} = 1.5, \quad J_d = 1.$$

The wide charge transfer gap $E_{ct} \approx 2$ eV separate the empty conductivity band mainly of $d_{x^2-y^2}$ character and the filled valence band formed by a complex mixture of different copper and oxygen orbitals.

The dispersion of the top of the valence band is shown in the Fig. 2 together with the ARPES data [9]. It should be noted that the valence band is formed by hole excitations from the initial single-hole state $|1, \zeta\rangle$ to different final two-hole states, both singlet $|S\rangle$ and triplet $|T, M\rangle, M = -1, 0, +1$ are essential. One more unusual result of GTB calculations [8] is the dispersionless virtual level shown by dotted line in Fig. 2. It has zero weight in the undoped case but acquires the spectral weight $\sim x$ in the hole doped

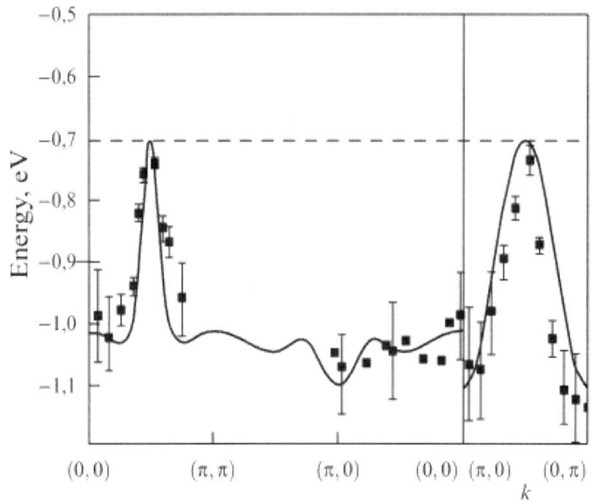

Fig. 2. The dispersion of the top part of the valence band in comparison with ARPES data on $Sr_2CuO_2Cl_2$.

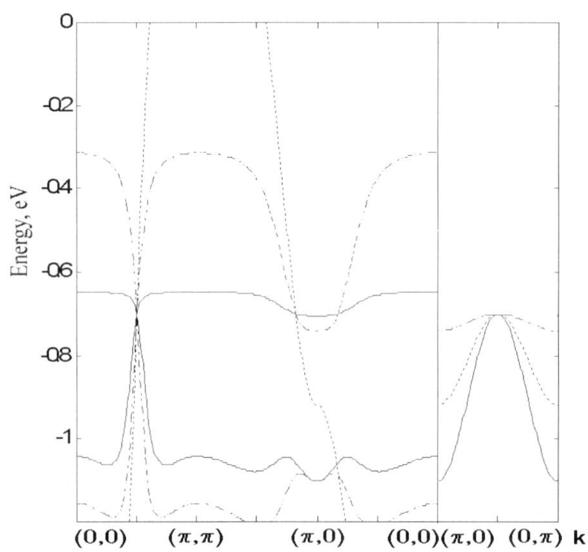

Fig. 3. Concentration dependence of the band structure in the antiferromagnetic ($x = 0.01$, solid line; $x = 0.1$, dot-dashed line) and paramagnetic ($x = 0.2$, dashed line) phases.

system with $n_h = 1 + x$, providing formation of the in-gap state at the top of the valence band. The dispersion of the in-gap state and its concentration evolution is shown in the Fig. 3.

The calculation [8] was not self-consistent, magnetic order is given *ad hoc*. Thus the curve for $x = 0.1$ in Fig. 3 is the interpolation, it's justification is large correlation length of the short antiferromagnetic order.

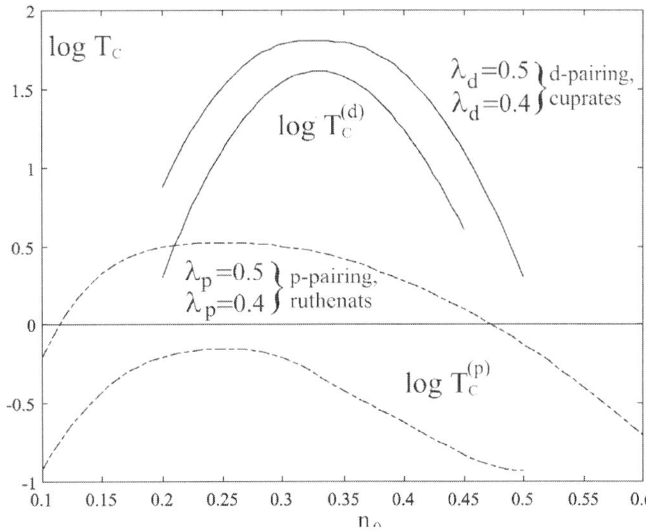

Fig. 4. Concentration dependence of $T_c, O_p = I/t, O_d = J/t$. Here $t = 0.1$ eV.

3. Comparison of magnetic mechanisms of pairing for cupratesand and ruthenates

The different electronic structure formed by $(t_{2g} - p) - \Sigma$ bonding in ruthenates in comparison with $(d_{x^2-y^2} - p) - \zeta$ bonding can result in a ferromagnetic coupling, I, between nearest Ru^{4+}-ions. To compare superconductivity in cuprates and ruthenates the t-J-I model has been proposed. Its mean-field solution [10,11] has shown that for $J \gg I$ (cuprates) there is the $d_{x^2-y^2}$ – gap while for $J \ll I$ (ruthenates) the triplet p-symmetry gap occurs. The two order of magnitude difference in T_c values for cuprates and ruthenates results from different symmetry of pairing (Fig. 4). The equation for T_c is similar for both cases, nevertheless the different gap symmetry results in two orders of magnitude difference of T_c values.

The van-Hove Singularity gives its contribution to the effective density of states for $d(x^2 - y^2)$-gap $\Delta_d = \Delta_0(\cos k_x + \cos k_y)$ than increase T_c in cuprates. For ruthenate with p-symmetry gap $\Delta_p = \Delta_0(\sin k_x + \sin k_y)$ the van Hove contribution is cancelled, thus T_c is low.

The coexistence of ferromagnetism and superconductivity in $RuSr_2GdCu_2O_8$ [12] can also be studied in the framework of the t-J-I model. Superconductivity in CuO_2 layer ($J \gg I$) has – symmetry while for RuO_2 layer ($J \ll I$) there is a competition of the ferromagnetic and triplet superconducting order. According to the mean-field phase diagram [13] small decrease of hole concentration in RuO_2 layer in comparison to Sr_2RuO_2 results in the magnetic phase.

References

[1] Yu.A. Izyumov, Spin-fluctuation mechanism of high T_c superconductivity and symmetry of order parameter, *Uspekhi. Fiz. Nauk.* **169** (1999), 225.

[2] T. Moria and K. Ueda, Spin fluctuation mechanism of high-T_c superconductivity, *Adv. Phys.* **49** (2000), 555.

[3] V.J. Emery, Theory of high-Tc superconductivity in oxides, *Phys. Rev. Lett.* **58** (1987), 2794.

[4] C.M. Varma, S. Schmitt-Rink and E. Abrahams, Charge transfer excitations and superconductivity inionic metals, *Solid. St. Commun.* **62** (1987), 681.

[5] S.G. Ovchinnikov, Quasiparticles in strongly correlated electronic systems in cuprate oxides, *Physics-Uspekhi.* **40** (1997), 993.

[6] Yu.B. Gaididei and V.M. Loktev, On the theory of electronic spectrum and magnetic properties of high-T_c superconductors, *Phys. Stat. Sol.* **B147** (1988), 307.
[7] S.G. Ovchinnikov and I.S. Sandalov, The band structure of strongly-correlated electrons in $La_{2-x}Sr_xCuO_4$ and Yba_2CuO_{7-y}, *Physica* **C161**(3) (1989), 607–617, *Physica* **C198** (1989), 607.
[8] V.A. Gavrichkov, S.G. Ovchinnikov, A.A. Borisov and E.G. Goryachev, Evolution of the quasiparticle band structure with doping in copper oxides in the framework of the generalised tight-binding method, *JETP* **91** (2000), 369.
[9] B.O. Wells, Z.X. Shen, A. Matsuura, D.M. King, M.A. Kastner, M. Greven and R.J. Birgeneau, E versus k relations and many body effects in the model insulating copper oxide $Si_2CuO_2Cl_2$, *Phys. Rev. Lett.* **74**(6) (1995), 964–967, *Phys. Rev. Lett.* **741** (1995), 964.
[10] E.V. Kuz'min, S.G. Ovchinnikov and I.O. Baklanov, Superconductivity of strongly correlated electrons in copper and ruthenium oxides within the t-J-I model, *JETP* **89** (1999), 349.
[11] E.V. Kuz'min, S.G. Ovchinnikov and I.O. Baklanov, Comparison of superconductivity in Sr_2RuO_4 and copper oxides, *Phys. Rev.* **B61** (2000), 15392.
[12] C. Bernhard and J.L. Tallon, Coexistence of ferromagnetism and superconductivity in the hybrid ruthenate-cuprate compound $RuSr_2GdCu_2O_8$ studied by muon spin rotation and dc magnetization Ch, *Niedermayer et at. Phys. Rev.* **B59** (1999), 14099.
[13] E.V. Kuz'min, S.G. Ovchinnikov, I.O. Baklanov and E.G. Goryachev, Microscopic model of the coexistence of superconductivity and ferromagnetism in the hybrid ruthenate-cuprate oxide $RuSr_2GdCu_2O_8$, *JETP* **91** (2000), 353.

International Journal of Applied Electromagnetics and Mechanics 13 (2001/2002) 349–357
IOS Press

Design analysis on the magnetic field control system in compact ITER

Ikuo Senda*, Teruaki Shoji, Masanori Araki, ITER Japan Home Team and ITER Joint Central Team
Japan Atomic Energy Research Institute, Naka Fusion Research Establishment, 801-1 Mukoyama, Naka-machi, Naka-gun, Ibaraki, 311-0193 Japan

Abstract. The system analyses of the magnetic field control system in the International Thermonuclear Experimental Reactor (ITER) are described. The analyses are performed by using a model of the closed loop system, which consists of the data analysis of diagnostics, the controller, the power supply of coils and the tokamak plant. The characteristics of the magnetic field control system in ITER are discussed and it is shown that the control system designed during the ITER engineering design activities satisfies its requirements. The important roles of eddy currents in optimizing the magnetic field control system in ITER are also discussed.

1. Introduction

The demonstration of the fusion energy as a possible choice of energy resources is an important milestone in the fusion research. In the engineering design activities (EDA) of the international thermonuclear experimental reactor (ITER), a design of the tokamak device to accomplish this objective was established. This paper describes the design activity of the magnetic field control system in ITER.

One of the objectives of the compact ITER is to achieve burning plasmas with the ratio of fusion power to auxiliary heating power at least 10 ($Q > 10$) [1]. The control of burning plasmas requires new aspects, which have not been experienced in existing machines. In the kinetic control, accurate identification of burning conditions and suitable burn controls by actuators, such as the fueling, pumping, auxiliary heating and external current drives, are needed. For the magnetic field control system, important requirements are the strong plasma shaping to maintain plasma confinement quality and the avoidance of plasma-wall contacts to protect plasma-facing components. These kinetic and field controls have to be combined into a unified system to achieve good performance of the device.

The structure of the control system in ITER is the same as ordinary industrial systems. Figure 1 shows the overview of the ITER control loop, which consists of subsystems such as the measurements of plasmas, the data analysis to identify burning conditions, the generation of control commands and the control of actuators. In design of a large plant, such as ITER, the system integration is an important issue. The hardware has to be designed with a close contact with the system analyses of the plant. The interactions between the hardware design and the system analyses are repeated until the required

*Corresponding author: Ikuo Senda, Fax: +81 29 270 7388; E-mail: sendai@fusion.naka.jaeri.go.jp.

1383-5416/01/02/$8.00

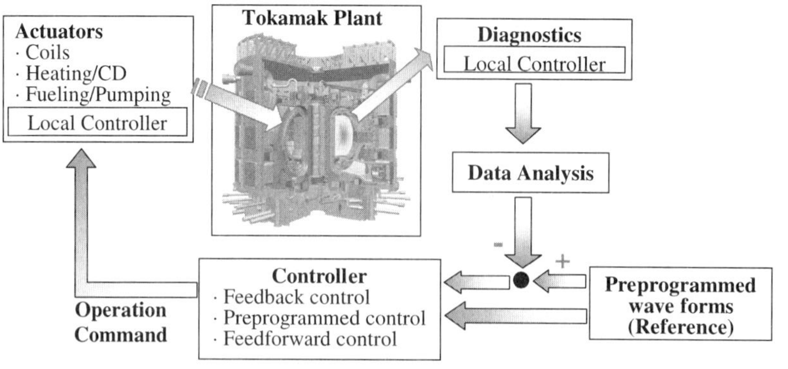

Fig. 1. The illustration of the plasma control system in ITER.

performance of the plant is achieved. To illustrate the system analyses in ITER EDA, the model of the magnetic field control system and the optimization the system by simulation analyses are discussed in this paper.

In Section 2, the magnetic field control system in ITER and its characteristics are described. The modeling of the magnetic field control system is explained in Section 3. The optimization and simulation results are presented in Section 4. Section 5 is used for conclusion.

2. Magnetic field control system in ITER

The magnetic field control system in ITER is, in principle, an extension of systems in existing tokamaks. Combinations of magnetic and non-magnetic measurements are used to identify plasma parameters in equilibrium configurations. Figure 2 shows the actuators in the field control. The poloidal field coils are used to control toroidally symmetric modes and the correction coils installed outside the vacuum vessel control non-toroidally symmetric modes, such as MHD activities and error fields [1,2].

The equilibrium field configurations are controlled both statically and dynamically. Many characteristics of the field control in the ITER originate from the design of the vacuum vessel with the low toroidal loop resistance ($\approx 7.4\ \mu\Omega$). There are two important reasons for this design choice. The first is the design concept of the reactor, namely this design is chosen to protect superconducting magnets from plasma disturbances. Since ITER is an experimental device, frequent occurrences of plasma disturbances are expected. The magnetic energy released in disturbances are dissipated mainly in the vacuum vessel due to its low resistance, thus the quench of magnets is avoided. The second reason is related to the strategy to obtain high plasma performance. The main plasma is designed on the basis of the plasma energy confinement scaling law [2]. Since the scaling law has a dependence on the plasma elongation as $\tau_E \sim \kappa_a^{0.78}$, where κ_a is the elongation at the plasma surface. Therefore, high elongation has an advantage to obtain better confinement. In the nominal operation of ITER, the plasma configuration with $\kappa_a = 1.85$ is considered. However, elongated plasmas inherit the positional instability [2,3]. The eddy current induced in the vacuum vessel is effective to stabilize this instability. The eddy current has to be sufficiently large to stabilize the plasma, and then the low resistance of the vacuum vessel is needed. On the other hand, eddy currents hinder magnetic measurements and the penetration of control fields produced by the external coils. Therefore, measurements and control of plasmas have to be assessed in combination with the effect of eddy currents in the vessel and invessel components.

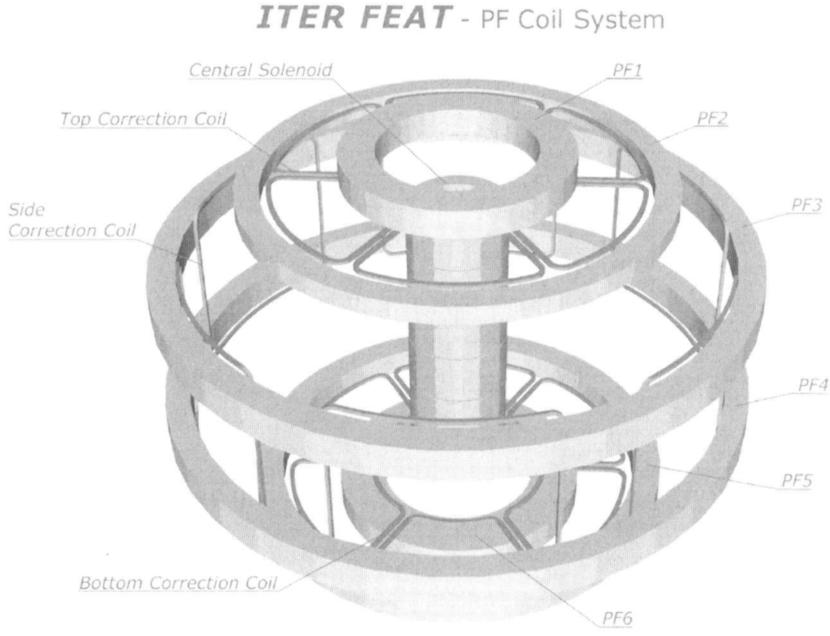

Fig. 2. The coil system in ITER. The toroidally symmetric PF coils are used for the control of the equilibrium configuration. The correction coils are used for the control of MHD activities and correction of the error field.

It is useful to compare typical decay times of eddy currents in exiting tokamaks and ITER. The principal mode of eddy currents in the compact ITER has a decay time about 0.6 s. The decay time of the eddy currents in existing large tokamaks are 0.03 s and 0.005 s for JT-60U and JET, respectively. Therefore, one can expect that the eddy currents in ITER have large influence on the field control system.

3. Model of magnetic field control system in ITER

The design of the magnetic field control system requires an integrated analysis of the closed loop, which is illustrated in Fig. 1. In this section, the modeling of the closed loop system is described.

The model of the tokamak plant consists of the vessel and invessel structures, coils, probes, power supplies and the core plasma. To close the control loop, the data analysis and the controller are added. One of the key items is the modeling of plasmas. In ITER EDA, the participating home teams use their own models of plasmas and the overall consistency between results from different models has been checked. In Japan Home Team, the eigenmodes of eddy currents in structures and their couplings with coils, probes and plasmas are computed by EDDYCAL code [4]. To simulate deformable evolutions of plasmas, ACCORD3 code is used [3]. Figure 3 is the model of the vacuum vessel and the blanket modules, which is used to analyze eddy currents. The model in Fig. 3 is 10-degree part of the structures in the toroidal direction and 36-fold toroidal symmetry is assumed. It is an advantage of EDDYCAL code that the eddy currents in a realistic geometry of structures are analyzed. Using the results of eddy current analysis, ACCORD3 code produces the state space description of the tokamak plant.

The output of the controller is the voltage commands to the power supply. The response of the power supply is an important element to determine the characteristics of the closed loop system. In ITER EDA, the response characteristics of the power supply are analyzed by using a numerical code such as

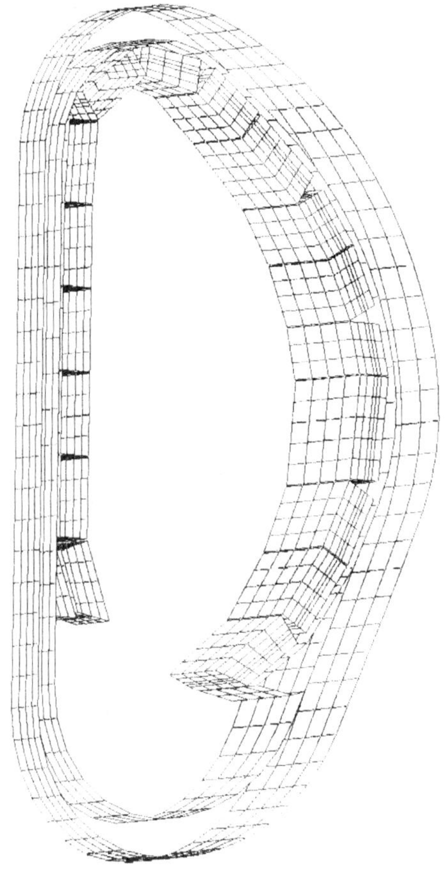

Fig. 3. The model of the vessel and invessel structures to analyze eddy currents in ITER. The model includes the double-wall vacuum vessel and the blanket modules.

EMTDC [5]. The model of converter modules is produced consistently with the design of the power supply. It is possible to combine the model of the power supply with the model of the tokamak, however it is convenient to describe the responses of the converters by the transfer function,

$$F_{\text{Convertor}} = e^{-\tau_D \cdot s} \frac{1}{1 + s \cdot \tau_{1-lag}}, \tag{1}$$

where the dead time τ_D and 1st-order time lag τ_{1-lag} are determined consistently with the numerical analysis of the converter modules. For the converters with fast responses, $\tau_D/\tau_{1-lag} = 2.5$ ms/7.5ms is used and $\tau_D = \tau_{1-lag} = 15$ ms is used for the converters with slow responses.

The reconstruction of the plasma parameters by analyzing probe signals is an important part of the control loop. Since principles of data analyses have already been demonstrated in existing devices, the data analyses in ITER can be a natural extension of the exiting techniques, except that the influence of eddy currents in structures is significantly large compared with the existing tokamaks. There are two kinds of data analyses, namely one is the physics-based method and the other is the physics-independent method. In the physics-based methods, the field configuration is reconstructed by analyses such as solving the Grad-Shafranov equation, the filament method and using the exact solutions of the Maxwell

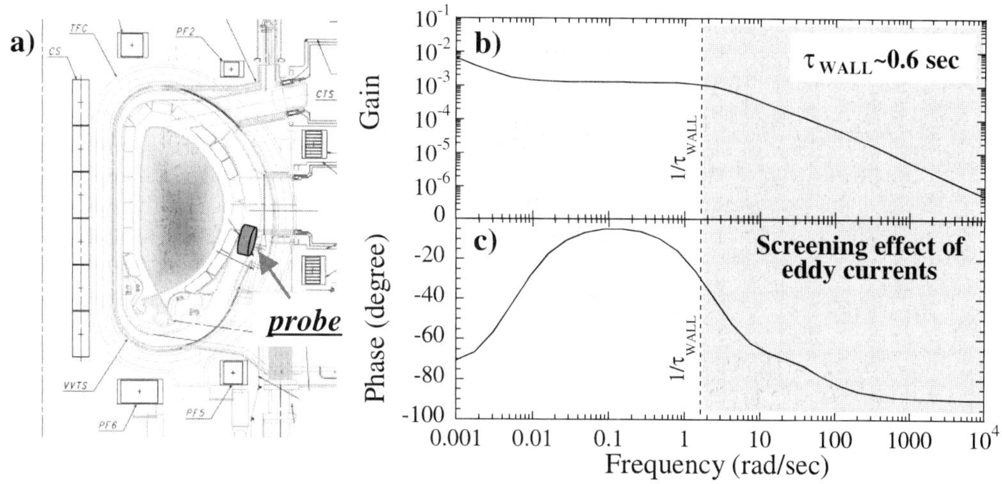

Fig. 4. The frequency response of the magnetic probe to the plasma fluctuation. a) shows the position of the probe, b) and c) are the bode diagram of the normal field probe.

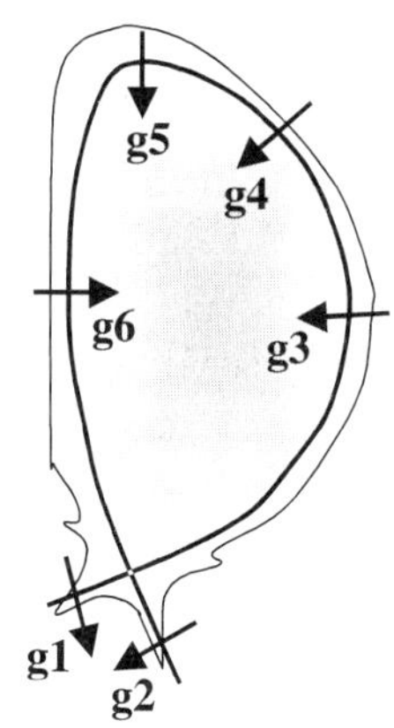

Fig. 5. Six locations on the separatrix are used as marker points to control the shape of the plasma during plasma disturbances. The arrows describe the positive motions of the separatrix.

equation. In these methods, the physical pictures of events are clear, however enough accuracy may not be obtained in large plasma disturbances. The physics-independent methods are analyses such as, the regression, the neural network and the data mining. These analyses do not provide clear physical pictures, however they have advantage that the applicable parameter domain is extended when a suitable database is used to train the reconstruction parameters. To illustrate the dependence of the control system on the data analysis, the regression method is used in the analyses of next section.

Regression is a fitting of a database by using the least square method. For example, let us assume that a deviation of a plasma parameter, δX, is given by a linear combination of probe signals,

$$\delta X = \sum_{i;\mathrm{Probe}} a_i \cdot S_i,$$

where S_i is an integration of i-th probe signal. The coefficients a_i are obtained by minimizing the

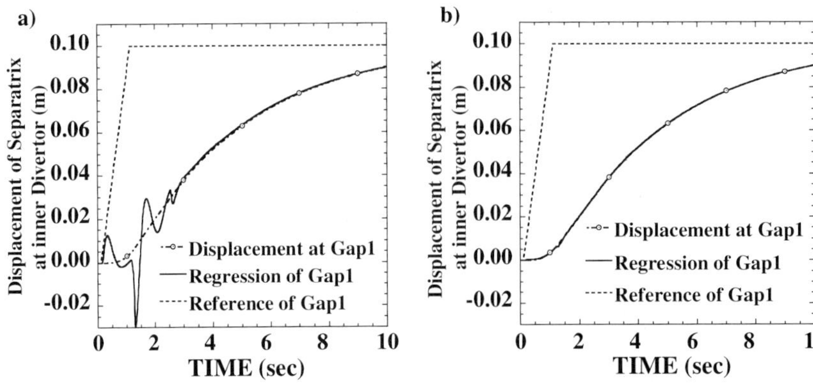

Fig. 6. The comparison of the plasma control simulations when the plasma position and shape are identified by using the regression coefficients derived by the static database (a), and by the database which includes both the static and transient data (b). The simulation is to move the inner divertor leg (gap1 in Fig. 5) by 10 cm.

function Q defined by

$$Q = \sum_{k \in \{\text{database}\}} w_k \cdot \left(\delta X_k - \sum_{i;\text{Probe}} a_i \cdot S_{i,k} \right)^2 , \tag{2}$$

where $\{w_k\}$ are weights of data. The quality of the data analysis depends on the database used to derive coefficients $\{a_i\}$.

4. Simulation study of the magnetic field control system

In design of the magnetic field control system, it is important to understand the characteristics of the magnetic probes. In ITER, the magnetic probes are located behind the shield blankets to avoid intense neutron irradiations. This constraint is unfavorable because the probes become far from the plasma and probe signals include more influence of eddy currents. Figure 4 shows the frequency response of a normal field probe located inside the vacuum vessel. The location of the probe is shown in Fig. 4(a). Figures 4(b) and (c) are the bode diagram of the probe signal. In Fig. 4, it is seen that the shielding effect of eddy currents in structures becomes significant at high frequency region, $> 1/\tau_{\mathrm{WALL}}$, where $\tau_{\mathrm{WALL}} \approx 0.6$ s is the decay time of the principle eddy eigenmode. Therefore, the effect of eddy currents has to be considered in the data analysis to observe rapid motions of plasmas.

The important function of the field control system is to avoid plasma-wall contact in the occurrence of the plasma disturbances. Figure 5 shows the marker points on the plasma separatrix and the deviations of the separatrix at the marker points are controlled by the feedback control loop.

Results in Fig. 6 are the step responses of the magnetic field control system to the command to shift the inner divertor leg by 10cm toward the divertor dome. In the simulation of Fig. 6(a), the regression coefficients are derived by using the database of static plasma equilibriums. In the simulation of Fig. 6(b), the database including both static and transient data is used to derive the regression coefficients, thus the effect of eddy currents is taken into account in the reconstruction. The dotted lines with circles are actual displacements of the inner divertor leg. The divertor leg moves to the reference position with time delay, which is attributed to the limited capacity of the power supply. The solid lines are the outputs of

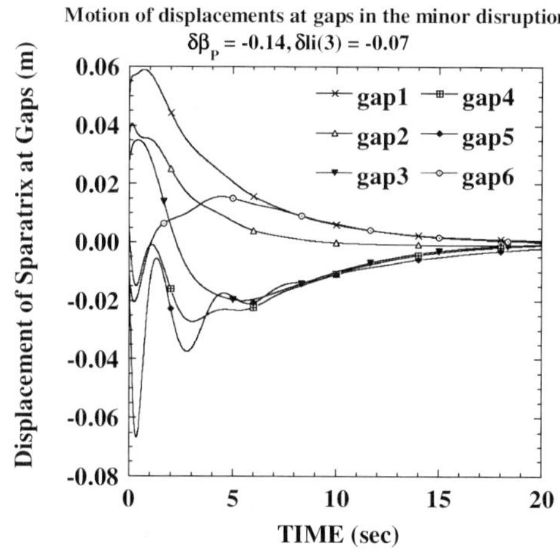

Fig. 7. The control simulation of the plasma shape in the minor disruption.

the data analyses using the regressions, which are used in feedback control. One sees that the regression produces large errors in Fig. 6(a) at the beginning of the simulation, where large eddy current is induced. However, in Fig. 6(b) a sufficient agreement is achieved. Therefore, it is essential to consider effects of eddy currents in the data analysis.

Various control algorithms have been investigated as controllers of the magnetic field control system in ITER. It was shown that a wide range of control algorithms could achieve the required performances. It is, of course, possible to improve the control quality by using sophisticated controller, such as robust controllers. An example of control simulations is shown in Fig. 7, which shows the displacements of separatrix at the marker points in the minor disruption. In this simulation, the controller is a simple decoupling controller of the plasma vertical position, shape and current, and the regression coefficients in the data analysis are derived by the database containing both static and transient data.

The plasma operation in ITER is defined in domains of plasma parameters. In general, the plasma control becomes demanding as the plasma is close to the operation boundary. Figure 8 shows the vertical displacements in the control simulation of the plasma with a low thermal energy ($\beta_P = 0.1$) and a peaked current profile ($\ell i(3) = 1.0$), which is the most demanding in the operation domain. The minor disruption with 20% loss of thermal energy is considered. In Fig. 8(a), the same regression coefficients as Fig. 7 are used. Since the positional instability of this plasma is large, the control waveform is oscillating. An interesting attempt to stabilize this oscillation is the application of the predictive regression algorithms. The advantage of physics-independent data analysis is that it is possible to predict future motion of plasmas without explicit knowledge of the plasma internal parameters [6]. In the predictive algorithm, the regression coefficients are derived by replacing $\delta X_k(t)$ in Eq. (2) by $\delta X_k(t + \delta t)$, where $\delta t = 10$ ms is used. Figure 8(b) is the simulation of the same plasma as Fig. 8(a) by using the predictive regression algorithm. The expansions of vertical motions from 0 s to 0.4 s are shown in the small frames. Figure 8(b) shows that the regression analysis successfully predicts plasma motion 10 msec later and the control is improved.

The data analysis by the regression and the neural networks is useful not only for the position/shape control, but also for the disruption avoidance. In existing tokamaks, several methods to avoid disruptions have been demonstrated by sensing precursors and/or keeping away from disruption boundaries [6,7]. These methods are useful for the effective operations of machines and a similar system is also considered in ITER.

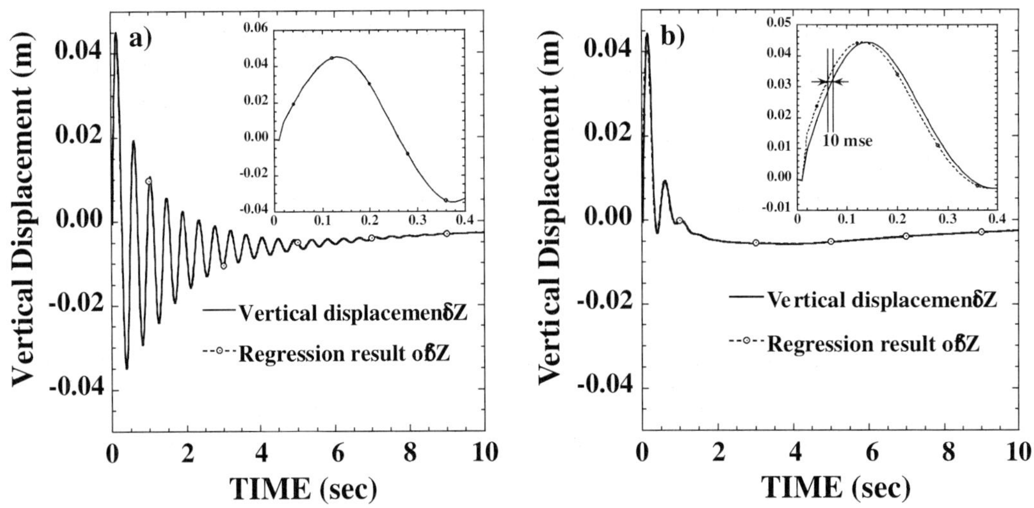

Fig. 8. The control simulation of the equilibrium near the operation boundary ($\beta_P = 0.1$, $li(3) = 1.0$), a) when the normal regression is used and b) when the predictive regression is used. In both simulations, the regression coefficients are derived by the database including static and transient data.

The simulation studies have been carried out in wide range of plasma parameters and in various plasma disturbances. The outputs of these simulation studies are the envelopes of the plasma motions, the control powers, current/voltage waveforms of converters. By these simulations, it is confirmed that the design of the magnetic field control system is sufficiently optimized to achieve its required performance. The magnetic field control is related to many aspects of the tokamak design. Some examples are listed below.

- The waveforms of active/reactive powers are used to design the power station, such as reactive power compensation and harmonic filter systems.
- The eddy current losses at the 4 K structures near the superconducting magnets during plasma control actions are evaluated. The results are used as design conditions of the cryogenic system.
- The distribution of magnetic probes is assessed based of control simulations.
- The locations of the plasma facing components are optimized by considering plasma motions in control actions.

5. Conclusion

In ITER EDA, the design of the tokamak plant is optimized based on the system analyses. As an example, the closed loop analyses of the magnetic field control system are presented. An important characteristic of the magnetic field control in ITER is that the eddy currents in the structures play essential roles in optimizing the system. It is also shown that the predictive data analysis is useful to improve performance of the control system.

In the construction and operation schedule of ITER, the construction is assumed to continue for about 10 years. Therefore, the detailed specifications of the data analysis system and the controllers are not needed at the present stage. Since the computer system in ITER is fully digitalized, it is possible to improve these softwares during the operation. An important issue in ITER EDA is to design the tokamak plant, which is able to achieve its required performance. The purpose of system analyses is to investigate the performance of the plant system and to point out how to improve the plant system if its performance is insufficient. As we have discussed in this paper, the closed loop analyses of the magnetic field control system in ITER have shown that the field control system is able to achieve its required performance. Though the levels of system analyses depend on individual systems, the activities to investigate the performance of systems have been carried out in various systems.

The outputs of these activities are important achievements of ITER EDA.

References

[1] Technical Basis for the ITER-FEAT Outline Design, IAEA 1999,. Y. Shimomura et al., *Nucl. Fusion* **41** (2001), 309.
[2] ITER Physics Basis, *Nucl. Fusion* **39** (1999), 2137.
[3] I. Senda, T. Shoji, T. Tsunematsu, T. Nishino and H. Fujieda, *Nucl. Fusion* **37** (1997), 1129.
[4] S. Nishio and T. Horie, *IEEE Trans. Magn. MAG* **26** (1990), 865, *A. Kameari, J. Comp. Phys.* **42** (1981), 124.
[5] Manitoba HVDC Research Center, http://www.hvdc.ca/.
[6] R. Yoshino, J.K. Koga and T. Takeda, *Fusion Tech.* **30** (1996), 237.
[7] E. Frederickson et al., *Phys. Plasmas* **2** (1995), 4216, D. Wroblewski, G.L. Jahns and J.A. Leuer, *Nucl. Fusion* **37** (1997), 725.

International Journal of Applied Electromagnetics and Mechanics 13 (2001/2002) 359–364
IOS Press

Evaluation on fracture toughness of low activation ferritic steel (JLF-1)

H.K. Yoon[a,*], S.P. Lee[a], C.S. Seok[b] and A. Kohyama[c]
[a]*Department of Mechanical Engineering, Dong-Eui University, 24, Kaya Dong, PusanJin-Ku, Pusan, 614-714, Korea*
[b]*Department of Mechanical Engineering, SungKyunKwan University, Suwon, 440-330, Korea*
[c]*Institute of Advanced Energy, Kyoto University, Gokasho, Uji, Kyoto, 611-0011, Japan*

Abstract. Low activation ferritic steels have been developed for a candidate of structural materials in nuclear fusion reactor applications. This paper dealt with the effects of rolling direction and side groove on the fracture toughness of low activation ferritic steel, JLF-1 Steel (9Cr-2W-V-Ta). The fracture toughness of JLF-1 steel showed the anisotropy in the rolling direction. In the case of J-R curve, J_Q and slope of the longitudinal JLF-1 steel (BML) for the rolling direction were a slight larger than those of transverse JLF-1 Steel (BMT) at room temperature. In order to obtain the valid fracture toughness of JLF-1 Steel through the UC method, the CT specimen with a side groove of 40% was recommended.

1. Introduction

As one of low activation ferritic steels for the IEA test program, the JLF-1 (9Cr-2W-V-Ta) steel has been developed and investigated by the active cooperation of Japanese Universities [1,3,4]. The low activation ferritic steels mounted in a high magnetic field such as fusion reactor components may be affected by the operating cycle and the mechanical vibration that is associated with the magnetic damping. For the sound design of fusion reactor, it is necessary to investigate the effect of magnetic damping on the mechanical properties of JLF-1 steel. Prior to its widespread, practical use, however, several fracture properties of the JLF-1 steel including tensile property, fatigue property and fracture toughness must be evaluated, preferentially. Especially, the investigation of fracture toughness and J-R curve is essential for practical applications of the JLF-1 steel into fusion areas, since it suffered from the elastic-plastic fracture behavior caused by the history of operation temperature. When the JLF-1 steel is used as the blanket structural material of fusion reactors, the experimental data for tensile properties and fracture toughness are very important to estimate safety life span of the blanket by critical crack length with the irradiation. The tensile properties and fracture toughness characteristics of unirradiated materials are required before the estimation of irradiated materials [2,5]. Finally, the correlativity between magnetic field and fracture properties of JLF-1 steel should be established for its active application. Due to the limited amount of materials, the unloading compliance method with single-specimen method (ASTM E813-89), which was the alternative of multi-specimen method, was recommended for the examination of fracture toughness [6,9].

*Corresponding author: H.K. Yoon, Department of Mechanical Engineering, Dong-Eui University, 24, Kaya Dong, Busanjin-Ku, Busan, 614-714, Korea; Tel.: +82 51 890 1642; Fax: +82 51 890 2232; E-mail: hkyoon@dongeui.ac.kr.

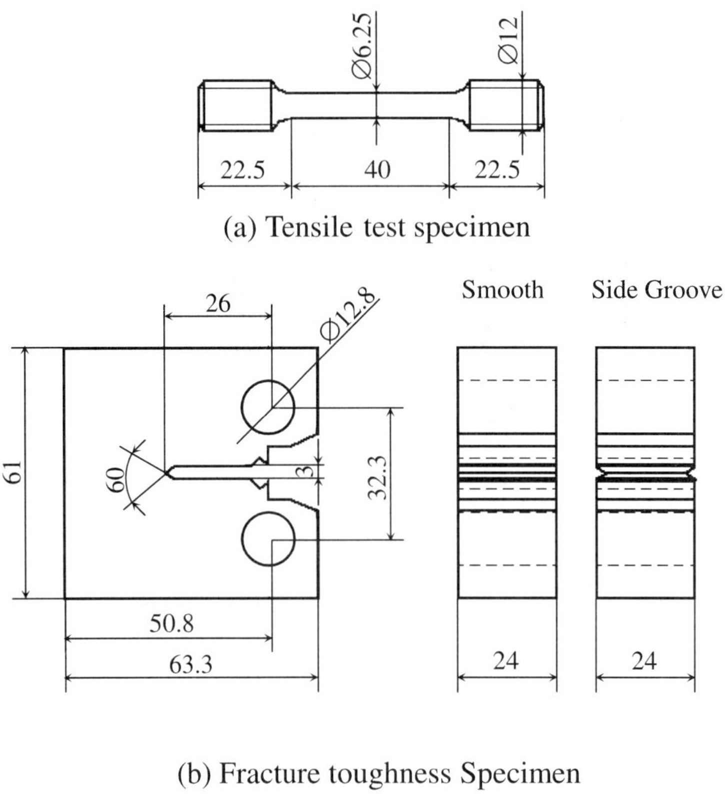

Fig. 1. Configuration of test specimen.

Table 1
Chemical Composition (wt%) of base metal (JLF-1)

	C	Si	Mn	P	S	Al	Cr	W	V	Ta	N	Ti	B
BM	0.10	0.05	0.45	0.003	0.002	0.003	8.85	1.99	0.2	0.08	0.0231	–	0.0002

In this study, tensile properties and fracture toughness of JLF-1 steel, which is associated with rolling and transverse directions, were investigated at room temperature. The effects of side grooves on fracture toughness of JLF-1 steel were also investigated.

2. Experimental procedures

The chemical composition of the JLF-1 steel plate (BM) is shown in Table 1. The plate with the thickness of 25 mm was normalized at 1323 K for 1 hr under the air cooling and then tempered at 1050 K for 1 hr. The geometry and the dimensions of test specimen for evaluating tensile properties and fracture toughness of the JLF-1 steel are shown in Fig. 1. The tensile test specimens have a diameter of 6.25 mm and a parallel part of 40 mm. The tensile test was performed with crosshead speeds of 0.1 mm/min and 1.0 mm/min. The fatigue precrack of CT specimens was induced to 3 mm for the overestimation of fracture toughness. The side groove depth was about 20% and 40% of the specimen thickness. The root radius was 0.1 mm. All the specimens were machined to the orientation of rolling direction. In

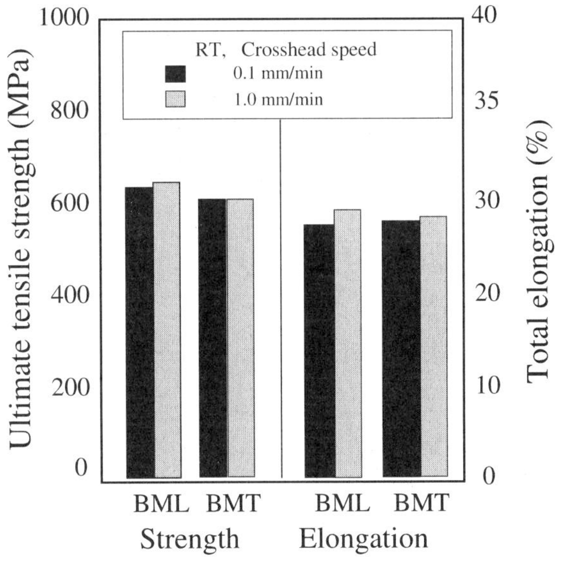

Fig. 2. Relation between mechanical properties and cross head speed.

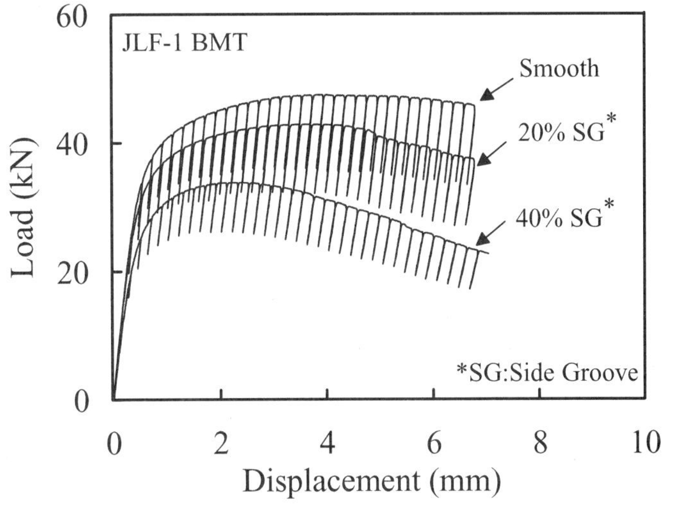

Fig. 3. Load and load–line displacement curves of JLF-1 steel with smooth.

this experiment, the test specimens were defined as the longitudinal direction (BML) and the transverse direction (BML) to the rolling direction. The fracture toughness tests were subjected to ASTM E813-89 by the unloading compliance method. All tests were performed at room temperature.

3. Test results and discussion

The tensile test results for the JLF-1 steel are shown in Fig. 2. The ultimate tensile strength of BML and BMT materials were 650 MPa and 610 MPa, respectively. The elongation of BML and BMT materials were also about 27%. The JLF-1 steel entirely showed the isotropy in tensile properties, even

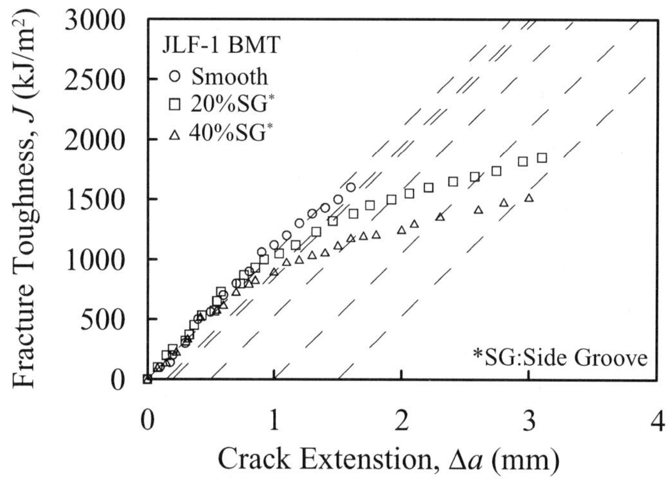

Fig. 4. J-R curves of BMT of JLF-1 steel with smooth, 20%, 40% side groove.

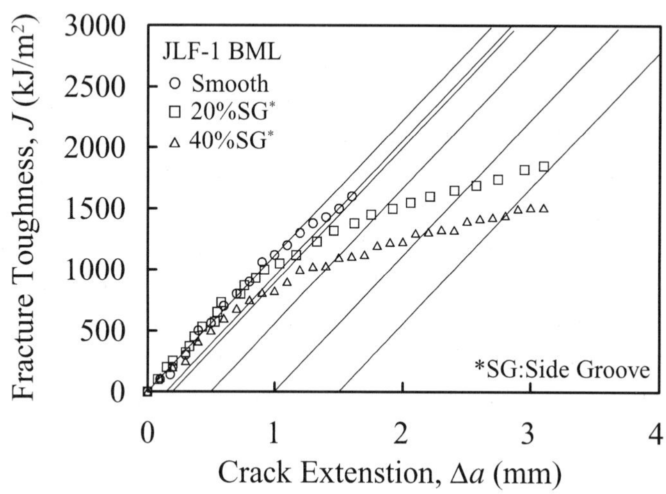

Fig. 5. J-R curves of BML of JLF-1 steel with smooth, 20%, 40% side groove.

if the ultimate tensile strength of longitudinal direction material (BML) was slight higher than that of transverse direction material (BMT). However, the tensile properties of each material associated with the rolling direction were not affected by the variation of crosshead speed. From these results, it was found that the JLF-1 steel possessed a considerable homogeneity within the acceptable range. Figure 3 shows the load-displacement curves of the BMT-CT specimens with side grooves of 20% and 40% used for the estimation of fracture toughness. As a comparison, the load-displacement curve of BMT-CT specimen without a side groove is also shown in this figure. It can be found that these curves were very smooth, regardless of the amount of side grooves. Unloading and reloading lines were also very clear, even if the load decreased with decreasing the depth of side groove.

The effects of side grooves on the load-displacement curves of BML and BMT specimens are also shown in Figs 4 and 5. It was found from Fig. 4 that the J_Q of BML specimen without the side groove did not satisfy the definition of ASTM standard, even if the results of smooth specimen were located

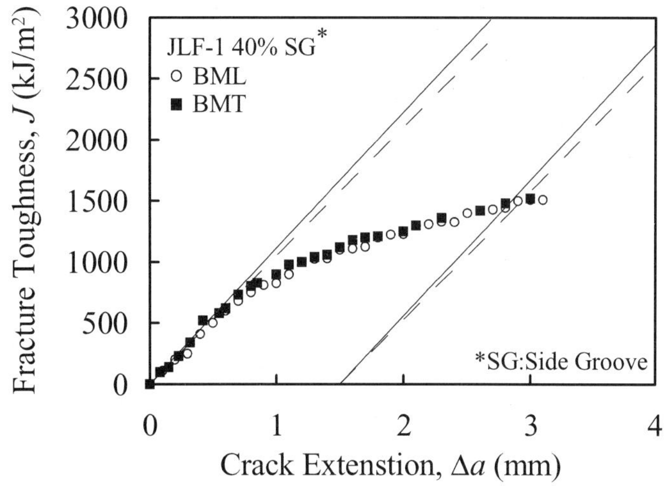

Fig. 6. *J-R* curves of JLF-1 steel with 40% side groove in the rolling direction of BMT and BML.

on the blunting line. However, the J_Q values of BMT specimen with side grooves of 20% and 40% were about 1000 KJ/m^2 and about 900 KJ/m^2, respectively. As shown in Fig. 5, the J_Q value of BML specimen also was similar to that of BMT specimen. The J_Q value and the slop of J-R curve of BMT and BML specimen with side groove of 40% was smaller than those with side groove of 20%. It can be found that the CT specimen with a side groove of 40% was recommended to obtain the valid fracture toughness of JLF-1 steel based on the UC method (ASTM E813-89). Figure 6 shows *J-R* curves of BMT and BML specimens depending on the rolling direction. The amount of side groove was 40%. The J_Q value and theslop of *J-R* curve of the BML specimen were similar to those of BMT specimen. This showed that the fracture toughness of JLF-1 steel had the isotropy in the rolling direction. Therefore, it can be recognized that the low activation ferritic JLF-1 steel possesses an excellent fracture toughness.

4. Conclusion

This paper dealt with the tensile properties and the fracture toughness of JLF-1 steel at room temperature. The results obtained in this study is as follows.

1. Tensile properties of JLF-1 steel at the room temperature showed considerable homogeneity, irrespective of the rolling direction and the variation of crosshead speed.
2. The JLF-1 steel had excellent room temperature fracture toughness higher than 450 kJ/m^2. However, in order to obtain the valid fracture toughness of JLF-1 steel based on the unloading compliance method (ASTM E813-89), the CT specimen with a side groove of 40% seemed to be valuable, even if this study is extremely limited.
3. The room temperature fracture toughness of JLF-1 steel showed the anisotropy to the rolling direction. A J_Q value and a slope of J-R curve of BML material were a slight large compared with those of BMT material.

Acknowledgements

The authors would like to thank the financial support of Japan society for the promotion of science (JSPS) and Brain Korea of Dong-eui University.

References

[1] A. Kohyama, A. Nishimura, D.S. Gelles, R.L. Klueh, W. Pietz and K. Ehrlich, Low-Activation Ferritic and Martensitic Steel for Fusion Application, *J. Nucl. Mater.* **233–237** (1996), 138–147.

[2] A. Kohyama, A. Hishimura, Y. Kohno, K. Shiba and A. Sagara, The Development of Low Activation Ferritic Steel for fusion Application, *SCI, Rep. Ritu* **A45** (1997), 137–141.

[3] A. Nishimura, A. Kohyama, R.L. Klueh, D.S. Gelles, D. Diez and K. Ehrlich, Current Status and Future R&D for Reduced-Activation Ferritic/Martensitic Steels, *J. Nucl. Mater.* **258–263** (1998), 1936–1204.

[4] A. Kohyama, Y. Kohno, M. Kuroda, A. Kimura and F. Wan, Production of Low Activation Steel (JLF-1), Large Heats-Current Status and Future plan, *J. Nucl. Mater.* **258–263** (1998), 1319–1323.

[5] A. Nishimura, N. Inoue and T. Muroga, Fracture Toughness of Low Activation Ferritic Steel (JLF-1) Weld Joint at Room Temperature, *J. Nucl. Mater.* **258–263** (1998), 1242–1247.

[6] Annual Book of ASTM Standards E 813-89, Standard Test Method for J_{IC}, A Measure of Fracture Toughness, 3.01, 1992, pp. 732–746.

[7] H.K. Yoon and J.S. Park, Evaluation on Elastic Fracture Toughness of SA533B-1 Steel by Load Ratio Analysis and Multiple Specimen Method, *KSME Journal* **A**(23) (1999), 1598–1605.

[8] H.K. Yoon, S.P. Lee, M.B. Lim and A. Kohyama, *Tensile Strength and Fracture Toughness Characteristics of Low Activation Ferritic (JLF-1) Steel*, KSME 2001 Conference, 2001, pp. 11–15.

[9] J.D. Landes and J.A. Begley, *Recent Developments in JIC Testing*, ASTM STP632, 1976, pp. 57–79.

International Journal of Applied Electromagnetics and Mechanics 13 (2001/2002) 365–372
IOS Press

Depressurization behavior of vacuum vessel pressure suppression systems in fusion reactors at multiple first wall pipe break events

K. Takase* and H. Akimoto
Japan Atomic Energy Research Institute (JAERI), Tokai-mura, Ibaraki-ken, 319-1195, Japan

Abstract. An integrated ICE (Ingress-of-Coolant Event) test facility was constructed to demonstrate that the ITER (International Thermonuclear Experimental Reactor) safety design approach and design parameters for the ICE are adequate. Major objectives of the integrated ICE test facility are to estimate the performance of an integrated pressure suppression system and obtain the validation data for safety analysis codes and clarify the two-phase flow behavior under vacuum. The integrated ICE test facility simulates the ITER components with a scaling factor of 1/1600. The modified TRAC-PF1 code is used to verify the integrated ICE experimental results. From the present study the effectiveness of the ITER pressure suppression system was verified experimentally and analytically, and then, it was clarified quantitatively that the prediction accuracy of the modified TRAC-PF1 code is very high.

1. Introduction

If cooling tubes installed into plasma-facing components (PFCs) in a Tokamak vacuum vessel are broken, water under high temperature and pressure in the cooling tubes will be discharged into a vacuum vessel (VV). Then, the discharged water will impinge on the hot surface of the PFCs and evaporate. As a result, the VV will be filled with vapor and the pressure in the VV will increase rapidly. This is called an Ingress-of-Coolant Event (ICE).

The physical phenomena under vacuum during the ICE were investigated by Takase et al. [1] using a preliminary ICE apparatus and led controlling factors on the pressure rise. Since the preliminary ICE apparatus did not simulate the actual shape of ITER (International Thermonuclear Experimental Reactor), however, it was difficult to estimate the ITER safety at the ICE from the results of the preliminary ICE experiments. Furthermore, the maximum pressure during the ICE is affected by the condensation in a suppression tank and the two-phase flow pressure drop at a divertor, whose effects strongly depend on the configuration of ITER components. Until now, however, there are no quantitative data about the maximum pressure available to the safety analysis codes of fusion reactors.

Based on these backgrounds an integrated ICE test facility was constructed referring to the ITER components to examine whether a concept of the ITER pressure suppression system is adequate. On

*Corresponding author: K. Takase, Tel.: +89 29 282 5351; Fax: +89 29 282 6439; E-mail: takase@popsvr.tokai.jaeri.go.jp.

1383-5416/01/02/$8.00

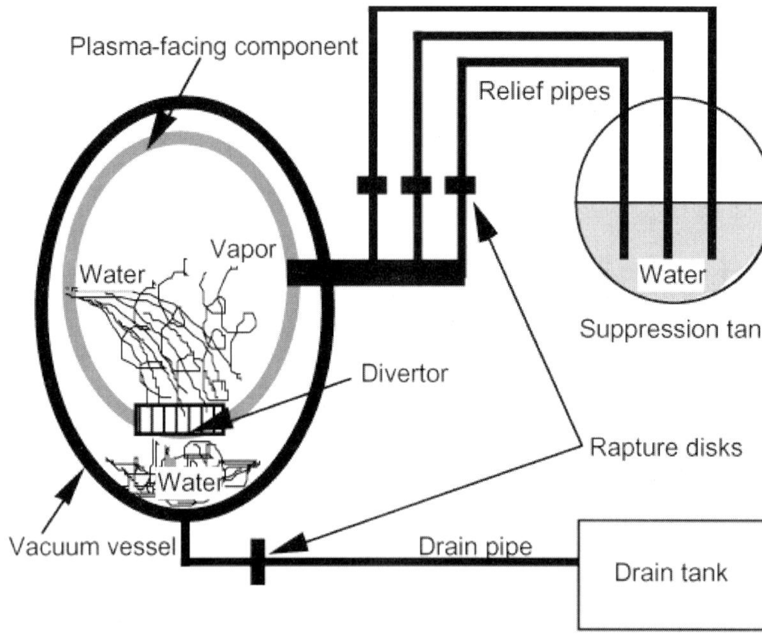

Fig. 1. A concept of the ITER vacuum vessel pressure suppression system.

the other hand, the numerical analyses with the modified TRAC-PF1 code [2] were performed and its prediction accuracy was verified using the quantitative and qualitative results of the integrated ICE experiments.

2. Pressure suppression system in ITER

Figure 1 shows a concept of the vacuum vessel pressure suppression system (VVPSS) in ITER [3]. The ITER VVPSS mainly consists of relief pipes, suppression tank (ST), drain tank (DT), drain pipe and rapture disks. Three relief pipes are installed to the side of the VV in ITER and connected to the ST. The DT is connected to the bottom of the VV through a drain pipe. The double rapture disks are settled to each relief pipe and also the drain pipe.

The water is injected from the cooling tubes into the plasma-facing component (PFC) and a great amount of vapor generates due to the flashing under vacuum and the boiling on the plasma-facing surfaces, and then, the pressure inside the PFC increases. Because of the pressurization the double rapture disks at each relief pipe are broken and then the vapor flows into the ST through the relief pipes. The ST initially holds water under low temperature and pressure (around 25°C and 2.3 kPa). Therefore, the vapor from the PFC is condensed in the ST, and consequently, the pressurization in the PFC is suppressed. The remained water inside the PFC flows through the divertor to the bottom of the VV and goes to the drain tank through the drain line after the double rapture disks at the drain pipe are broken. According to the ITER design, the design pressure of the rapture disk at the relief pipe is 150 kPa and that at the drain pipe is 110 kPa.

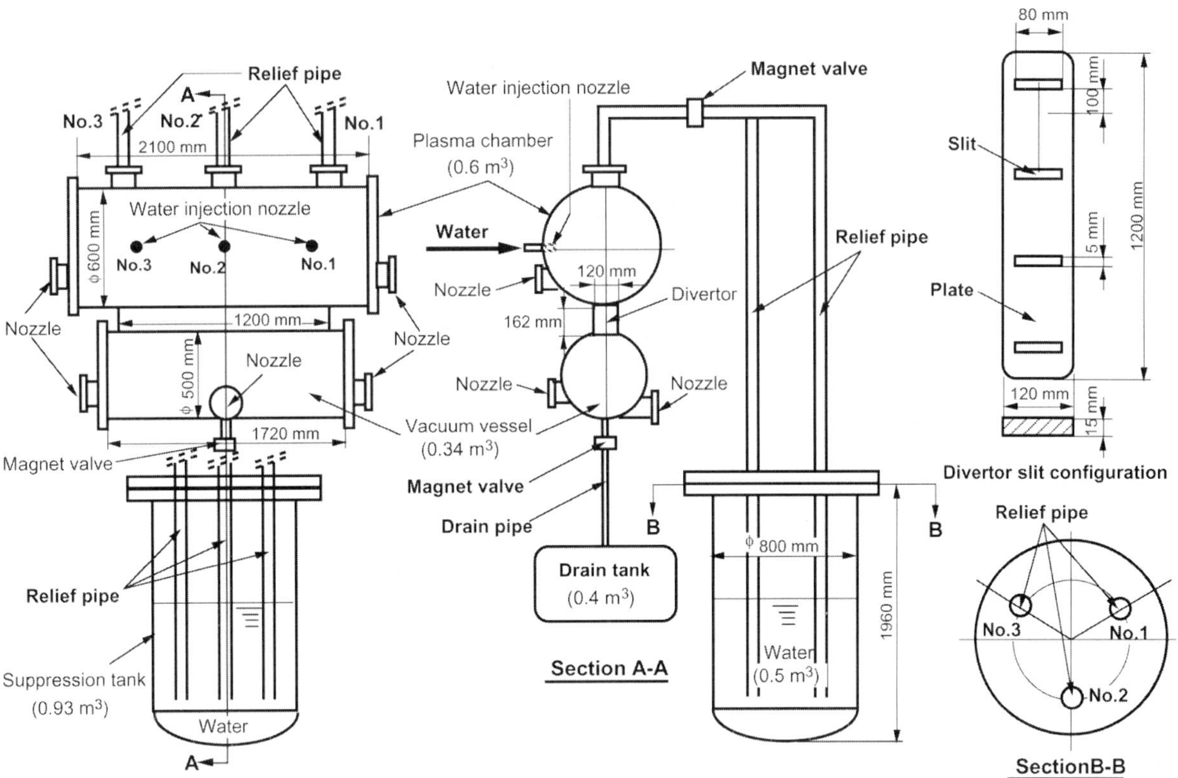

Fig. 2. A layout of structural components of the integrated ICE test facility.

3. Integrated ICE test facility

3.1. Scaling factor

The basic policy of the scaling factor is to functionally simulate the ICE phenomena, such as by scaling based on the volume without a modeling of the actual shape of the ITER machine. The proposed basic scaling laws between the test facility and ITER are: the volume size is 1/1600; and flow velocity, pressure and temperature are simulated by the scaling factor of 1/1, respectively. Most of structural components in the integrated ICE test facility were reduced with a scaling factor of 1/1600 for the size of ITER except for the relief pipe and drain pipe. Dimensions of the relief pipe such as the diameter and length were determined as the pressure difference in the relief pipe between PFC and ST is equal to that in ITER. Similarly, dimensions of the drain pipe were also determined. Here, the scaling factor of 1/1600 was mainly decided from the limitations of the electric capacity of compensation heaters around the test facility and heater rods in the boiler and of the height of the building where the test facility is constructed.

3.2. Structural components

Figure 2 shows a schematic drawing of an integrated ICE test facility. The integrated ICE test facility simulates the ITER structural components with a small-scale model of 1/1600 and mainly consists of a plasma chamber (PC), divertor, VV, relief pipes, ST, DT and drain pipe. In Fig. 2 the boiler simulates a

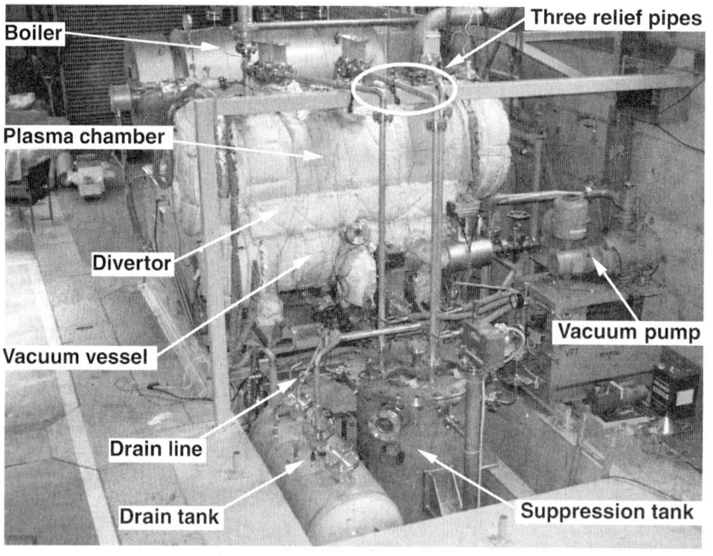

Fig. 3. Appearance of the integrated ICE test facility.

pressurizer at the primary cooling loop in ITER and the total cross-sectional area of three water injection nozzles corresponds to the supposed failure area in ITER. Figure 3 shows an appearance of the integrated ICE test facility.

The PC simulates the PFC volume of ITER and has a volume of $0.6\ \mathrm{m}^3$. The VV simulates a simplified VV in ITER and has a volume of $0.34\ \mathrm{m}^3$. The divertor simulates the simplified divertor cassette section in ITER and its dimensions are a width of 120 mm, height of 162 mm and length of 1200 mm. A plate with four slits is installed into the divertor to simulate the ITER evacuation slits. The slit dimensions are 5 mm in width, 80 mm in length, 15 mm in depth and 100 mm in pitch, and those values are determined based on the ITER design values. Compensation heaters are set to the outside of the PC, divertor and VV to control the outer temperature. The ST has a volume of $0.93\ \mathrm{m}^3$ and the initial water volume is less than $0.5\ \mathrm{m}^3$.

The ST is connected to the PC by three relief pipes with an inner diameter of 35 mm. A magnetic valve is installed to each relief pipe instead of the double rapture disks in case of ITER. The magnetic valve is opened when the PC pressure reaches 150 KPa after the injection of water as same as the ITER design value. The DT with a volume of $0.4\ \mathrm{m}^3$ is located below the VV and connected to the VV with the drain pipe with an inner diameter of 12 mm. A magnetic valve is also set up to the drain pipe instead of the double rapture disks in ITER.

Three sets of water injection nozzles were installed to simulate the multiple first wall pipe break event in ITER. A diameter of the water injection nozzle is 7.3 mm. Total cross-sectional area of three water injection nozzles corresponds to the supposed maximum failure area in ITER based on the scaling factor of 1/1600. Water in the boiler is pressurized by nitrogen gas up to maximum 4 MPa and heated by the heater rods up to maximum 250°C. The maximum water volume in the boiler is $0.6\ \mathrm{m}^3$. The water pressure in the boiler during the injection of water is controlled by nitrogen gas to keep at a constant value because the injected water flow rates from the water injection nozzles must not change with time.

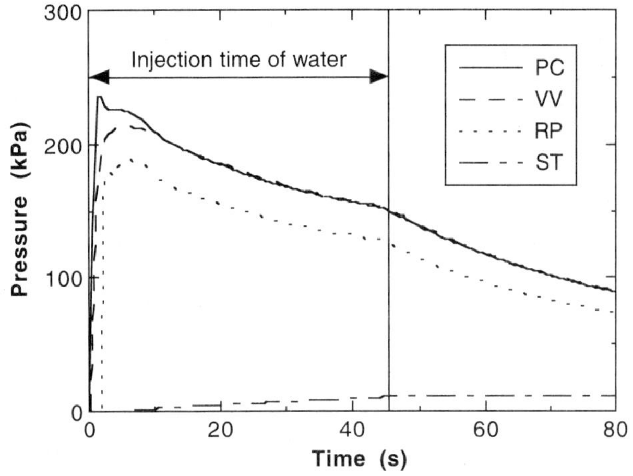

Fig. 4. An example of measured pressure transients.

3.3. Experimental conditions

Experimental conditions were: water temperature 125–150°C; water pressure 2–4 MPa; PC wall temperature 230°C; VV wall temperature 100–230°C; initial pressure inside the test facility less than 100 Pa; number of used water injection nozzles 3; injection time period of water 45 s; and number of relief pipes 1–3. The total amount of water (i.e., $0.26\ \mathrm{m}^3$) which is injected into the PC of the integrated ICE test facility by the combination of 3 water injection nozzles and 45 s injection time period corresponds to 1/1600 of the total injected water volume (i.e., $420\ \mathrm{m}^3$) in ITER during the ICE. At the design of ITER it is considered that 80% or more of total amount of water in ITER cooling loops is discharged into the PC until 45 s after the supposed maximum ICE phenomenon occurred. Then, experimental conditions on the total injected water volume and water injection nozzle size for the integrated ICE test facility were determined based on an assumption that the total amount of water in ITER cooling loops is injected into the PC during 45 s. Therefore, it is evident that the integrated ICE tests give the results of the safety side to ITER.

4. Results and discussion

Figure 4 shows the measured pressure transients after the ICE occurred. The solid, dashed, dotted and one-pointed solid lines represent the pressures in the PC, VV, relief pipe (RP) and ST, respectively. The experimental conditions are: water temperature 125°C; water pressure 2 MPa; PC wall temperature 230°C; VV wall temperature 100°C; and two relief pipes were used. The pressures in the PC and VV increases very quickly due to the flashing and the boiling of water on a heated wall just after the injection of water. The pressure rise in the relief pipe comes late in comparison with those in the PC and VV because of the opening of the magnetic valve. The pressure increase in the ST is quite small because of the condensation of vapor. The maximum pressures in the PC, VV, relief pipe and ST were 237 kPa at 2 s, 214 kPa at 5.5 s, 188 kPa at 5.5 s and 11 kPa at 45 s, respectively.

Figure 5 shows the effect of the wall and water temperatures on the pressure rise inside the PC. Experimental conditions are: water pressure 2 MPa; PC wall temperature 230°C; and three relief pipes were used. The solid line presents the measured pressure transient in the PC under the conditions of

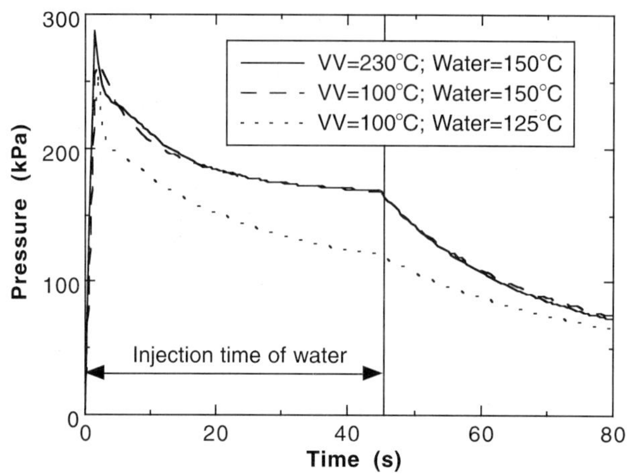

Fig. 5. Effect of wall and water temperatures on pressure rise in the PC.

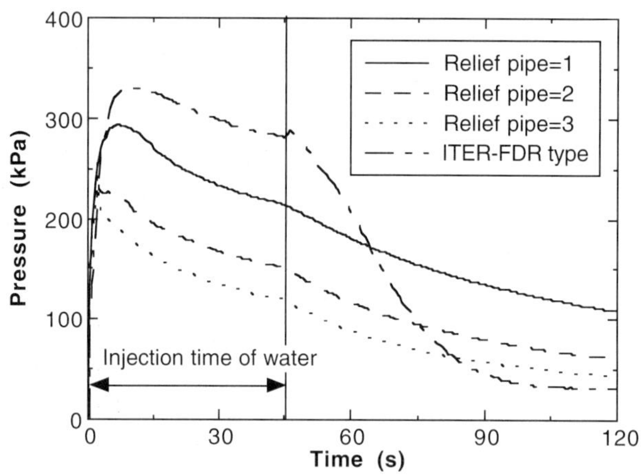

Fig. 6. Effect of cross-sectional area of the relief pipe on pressure rise in the PC.

the VV wall temperature 230°C and water temperature 150°C. Similarly, the dashed line represents that under the conditions of the VV wall temperature 100°C and water temperature 150°C and the dotted line shows that under the conditions of the VV wall temperature 100°C and water temperature 125°C. The solid and dashed lines show almost the same pressure transients. On the other hand, the dotted line shows a lower value than the solid and dashed lines. Therefore, it was found that the pressure rise in the PC under the ICE strongly depends on the water temperature in comparison with the VV wall temperature when the PC wall temperature is constant. About this matter the following reason is considered. Since the inside of the PC gets wet as soon as the water is injected into the PC, the PC wall is cooled down by the ingress water and then the PC wall temperature is almost equal to the water temperature. As the result the PC pressure is equivalent to the saturation pressure which is determined by the water temperature. Here, the saturation pressure decreases with decreasing the water temperature, and then, the PC is isolated from the VV by the divertor. From this reason it is concluded that the effect of the VV wall temperature to the PC pressure is quite small in comparison with the effect of the water temperature to that.

Figure 6 shows the effect of the cross-sectional area of the relief pipe on the pressure rise inside the

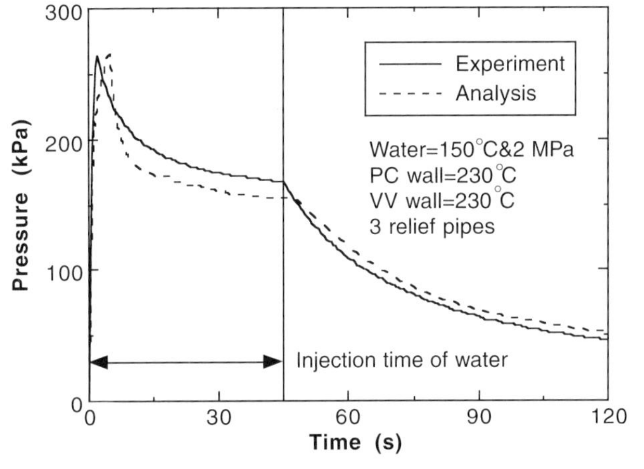

Fig. 7. Comparison of experimental and analytical results on pressure rise in the PC.

PC. Experimental conditions are: water temperature 150°C; water pressure 2 MPa; PC wall temperature 230°C; and VV wall temperature 230°C. The solid, dashed and dotted lines represent the measured pressure transients in the PC when the number of the used relief pipes is 1, 2 and 3, respectively. The pressure in the PC decreases as the number of the used relief pipes increases. That is, the pressure rise in the PC during the ICE depends on the cross-sectional area of the relief pipe. Because a large amount of vapor is sent to the ST and condenses immediately in case that the cross-sectional area of the relief pipe is large. The one-pointed solid line shows the measured pressure transient in the PC in cases that the relief pipe is located between the VV and ST and the DT is nothing such the ITER-FDR type [4]. In case of ITER-FDR type, since the mixture of hot water and vapor is sent to the ST, the water temperature inside the ST increases and the condensation efficiency of the ST is down. Therefore, the maximum pressure transient of the ITER-FDR type is larger than those of the other lines. It is evident from this results that the performance of ITER VVPSS is higher than that in ITER-FDR.

Figure 7 shows an example of the experimental and analytical pressure transients in the PC. Here, the solid line represents the experimental result and the dashed line shows the analytical results predicted by the modified TRAC-PF1 code. The TRAC is an advanced, best-estimate computer program to calculate the transient behavior of a pressurized water reactor (PWR). The original TRAC-PF1 code [5], which is one of a series of TRAC codes, uses a two-phase two-fluid model and was modified to predict the transient behavior of a fusion reactor [2]. The predicted maximum pressure (i.e., 265 kPa) agreed well with an error less than 1% with the measured maximum pressure (i.e., 263 kPa). The authors [6] have already confirmed that the predictions by the modified TRAC-PF1 code are in good agreement with the experimental results under the relief pipe condition of ITER-FDR. From a series of the numerical analyses, it was concluded that the modified TRAC-PF1 code can be applied suffici ently for the two-phase flow analysis under the relief pipe condition of the current ITER design.

Figure 8 shows the void fraction distributions inside the PC and VV predicted by three-dimensional computations with the modified TRAC-PF1 code. Here, (a) 0.5 s, (b) 2 s, (c) 4 s and (d) 45 s present the time from the start of the water injection. The gradation bar shows the void fraction rate: the blue and red mean 100% water and 100% vapor, respectively. Calculated conditions are the same as the experimental conditions of Fig. 4. The water injected from the three water nozzles into the PC hits on the opposite wall (Fig. 8(a)). The water is accumulated inside the PC with time (Fig. 8(b)–(d)). After the water injection into the PC, the accumulated water in the PC moves gradually to the VV through the divertor slits according to the pressure balance between the ST and DT.

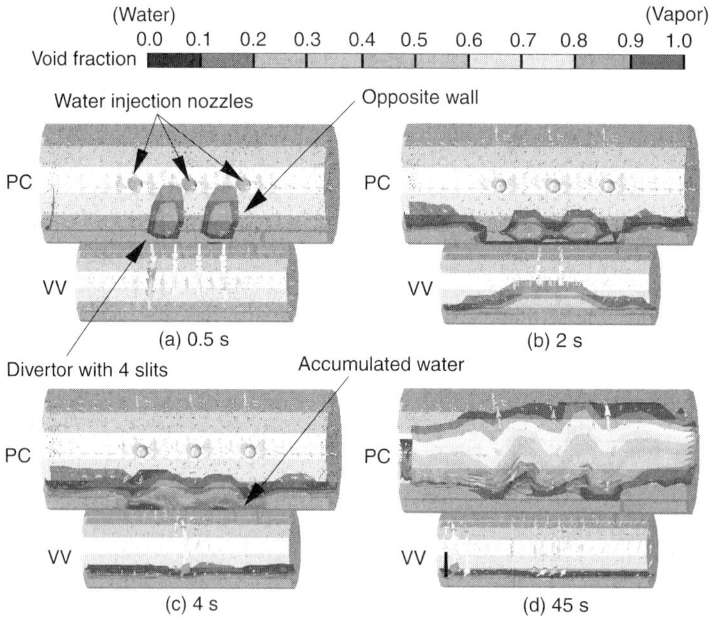

Fig. 8. Void fraction distributions in the PC and VV predicted by three-dimensional computations with the modified TRAC-PF1 code.

5. Concluding remarks

The present experimental and analytical results are summarized as follows:

1) The ITER VVPSS is very effective to reduce the pressurization during the ingress of water into vacuum;
2) The pressure rise characteristics strongly depend on the water temperature in comparison with the VV wall temperature when the PC wall temperature is constant;
3) The pressure decrease inside the PC caused by the condensation of vapor in the ST depends on the cross-sectional area of the relief pipe because the amount of vapor from the PC to the ST depends on that;
4) The modified TRAC-PF1 code can predict the maximum pressure in the PC during the water injection with sufficient accuracy to the integrated ICE experimental results; and,
5) The present three-dimensional numerical analysis with the modified TRAC-PF1 code is very useful to understand and clarify quantitatively the water-vapor two-phase flow behavior in the fusion reactor during the ICE.

References

[1] K. Takase et al., *Fusion Technology* **34** (1998), 640–644.
[2] K. Takase, *Fusion Technology* **39** (2001), 1050–1055.
[3] IAEA, ITER Documentation Series, No. 20, IAEA, Vienna, 2000.
[4] IAEA, IAEA, ITER Documentation Series, No. 12, IAEA, Vienna, 1997.
[5] D.R. Liles, NUREG/CR-3858 (LA-10157-MS), 1986.
[6] K. Takase et al., *Fusion Engineering and Design* **54** (2001), 593–603.

International Journal of Applied Electromagnetics and Mechanics 13 (2001/2002) 373–379
IOS Press

Modeling of liquid walls in APEX study

S. Smolentsev[a,*], M. Abdou[a], T. Kunugi[b], N. Morley[a], S. Satake[c] and A. Ying[a]
[a]*UCLA, 420 Westwood Plaza, 43-133 Engineering IV, Los Angeles, CA 90095-1597, USA*
[b]*Kyoto University, Yoshida, Sakyo, Kyoto, 606-8501, Japan*
[c]*Toyama University, Toyama, Toyama, 930-0887, Japan*

Abstract. Liquid wall concept has a significant place in the Advanced Power Extraction (APEX) study as a part of the US program on Nuclear Fusion. In the present paper, different approaches currently developed for modeling liquid wall magnetohydrodynamics and heat transfer in APEX have been presented. Turbulent flows of low-conductivity liquids, such as molten salts (Flibe), are analyzed using a two-equation turbulence model and Direct Numerical Simulation. The analysis of liquid metal flows under a strong reactor magnetic field is based on the magnetohydrodynamic equations for laminar flows.

1. Introduction

In fusion applications, liquid metals (LMs) are traditionally assumed to be the best working fluids. Due to their high electrical conductivity, LMs have strong magnetohydrodynamic (MHD) interaction, however turbulence is completely suppressed except of some particular situations. Along with LMs, molten salts (10^4 times poorer conductors) are being carefully studied as a practical candidate for fusion applications [1]. For example, a 2 cm thick flow of Flibe moving in the poloidal direction along the reactor First Wall (FW) is used in the APEX (Advanced Power Extraction) study, that is an exploration of innovative concepts for fusion chamber technology carried out in the US since 1998 [1]. Unlike LMs, such flows do not experience significant MHD forces and remain turbulent. However, under a reactor strong magnetic field, turbulence pulsation in low-conductivity fluids can be partially suppressed with an accompanying reduction in heat transfer. Because of the significant differences in the behavior of LMs and low-conductivity fluids, different approaches are needed. In the APEX study, "K-ε" two-equation turbulence model extended to MHD flows and Direct Numerical Simulation (DNS) are used for modeling of low-conductivity fluids. Laminar models based on the Navier-Stokes-Maxwell equations are applied to the analysis of LM flows. All approaches require the induced currents and the free surface location to be calculated simultaneously with other flow quantities. The paper describes briefly the approaches currently developed in the APEX study and gives some illustrations under conditions relevant to particular fusion designs. All details have not been shown here because of the paper limitations. More detailed analyses were presented in [1–5]. Unlike previous numerical studies, in which only relatively simple cases for laminar MHD flows in open channels were considered (developing flows in a one component constant magnetic field [6], fully developed flows in a two-component constant magnetic field [7]), the present ones introduce much more details, with more emphasize on MHD turbulence effects, and those related to multi-component and space-varying reactor magnetic fields.

*Corresponding author: S. Smolentsev, Tel.: +1 310 794 5366; Fax: +1 310 825 2599; E-mail: sergey@fusion.ucla.edu.

1383-5416/01/02/$8.00

2. Low-conductivity fluid modeling using two-equation model

"K-ε" turbulence model extended to MHD free surface flows has been used. The model is restricted to low magnetic Reynolds numbers. After applying Reynolds averaging to the Navier-Stokes-Maxwell equations with the conventional closure approximations, one can derive equations for the turbulent kinetic energy, K, and the dissipation rate, ε:

$$\frac{\partial K}{\partial t} + \langle \nu_j \rangle \frac{\partial K}{\partial x_j} = \underbrace{\nu_t \left(\frac{\partial \langle \nu_i \rangle}{\partial x_j} \right)^2}_{\text{Production}} + \underbrace{\frac{\partial}{\partial x_j}\left[\left(\nu + \frac{\nu_t}{\sigma_K} \right) \frac{\partial K}{\partial x_j} \right]}_{\text{Diffusion}} - \underbrace{\varepsilon - \varepsilon_{em}^K}_{Dissipation}; \tag{1}$$

$$\frac{\partial \varepsilon}{\partial t} + \langle \nu_j \rangle \frac{\partial \varepsilon}{\partial x_j} = C_1 \frac{\varepsilon}{K} \nu_t \left(\frac{\partial \langle \nu_i \rangle}{\partial x_j} \right)^2 + \frac{\partial}{\partial x_j}\left[\left(\nu + \frac{\nu_t}{\sigma_\varepsilon} \right) \frac{\partial \varepsilon}{\partial x_j} \right] - C_2 \frac{\varepsilon}{K} \varepsilon - \varepsilon_{em}^{\varepsilon}. \tag{2}$$

All terms and coefficients in (1–2) are standard except of ε_{em}, which stands for the Joule dissipation. If the liquid flows in the poloidal direction, the wall-normal and the toroidal (streamwise) fields will have the strongest effect on the flow. The closures for the electromagnetic terms for these two field orientations, have been obtained as follows [2]:

$$\varepsilon_{em}^K = 1.9 \exp\{-1.0N\} \frac{\sigma}{\rho} B_0^2 K; \quad \varepsilon_{em}^{\varepsilon} = 1.9 \exp\{-2.0N\} \frac{\sigma}{\rho} B_0^2 \varepsilon. \tag{3}$$

The Stuart number, N, is built through $2h_0$ (h_0 is the characteristic flow thickness); σ is the electrical conductivity; ρ is the density, and B_0 is the applied magnetic field. The boundary conditions on K and ε at the free surface extend the standard ones by taking into consideration the effect of the magnetic field on the dissipation length scale. When modeling turbulent heat transfer, two effects were incorporated. The first one is the turbulence reduction by a magnetic field, and the second one is the turbulence redistribution in the near-surface region. Both effects are taken into account through the extension of the Fourier law as follows:

$$\langle t' \nu' \rangle = -\frac{\nu_t}{Pr_t} \frac{\partial T}{\partial y}. \tag{4}$$

The turbulent viscosity, ν_t, is defined in a conventional way by the Kolmogorov-Prandtl relation, while the turbulent Prandtl number, P_{r_t}, was evaluated using experimental data for open channel flows [8] as follows:

$$Pr_t = 0.7 \times [1 + \exp\{37 \times (y/h - 0.89)\}]. \tag{5}$$

Using the turbulence model (1–5) along with the mean flow equations and boundary conditions, Flibe flows with a free surface exposed to a high-density heat flux along the outboard FW were calculated under conditions related to the ARIES RS reactor (a high-power-density tokamak with an advanced, reversed shear plasma containment) for flows with axial symmetry. The initial thickness of 2.3 cm with the initial velocity of 10 m/s used in the CLiFF design (a part of the APEX study, which stands for the Convective Liquid Flow First Wall) results in almost uniform flow. Thick flows experience significant contraction due to gravity (Fig. 1). The surface temperature rise in CLiFF calculated with a heat flux of 2 MW/m^2 is about 80 K at the end of the 8 m FW section (Fig. 2). As calculations showed, in liquid walls thinner than 2 cm, the MHD effects will be almost negligible. For thicker walls, MHD effects on both mean flow and turbulence become pronounced.

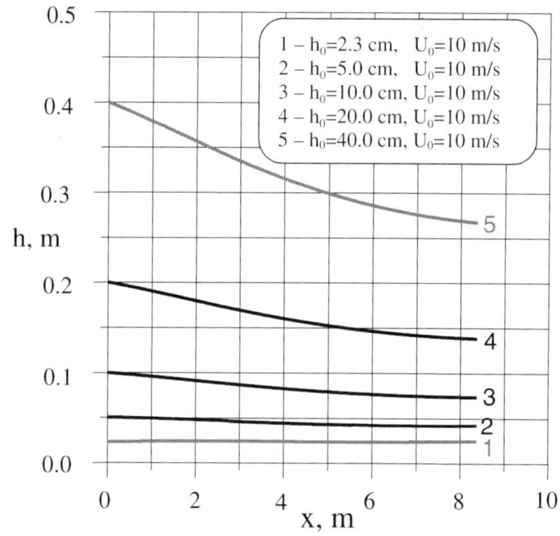

Fig. 1. Downstream variations of the thickness of the thick and thin Flibe liquid walls flowing over ARIES RS fusion reactor outboard FW under the influence of MHD forces and gravity. The liquid enters the chamber at the top and then flows in the poloidal direction to its bottom. Coordinate x measures the distance from the flow inlet along the FW in the main flow direction. Calculated with the present $K-\varepsilon$ MHD turbulence model.

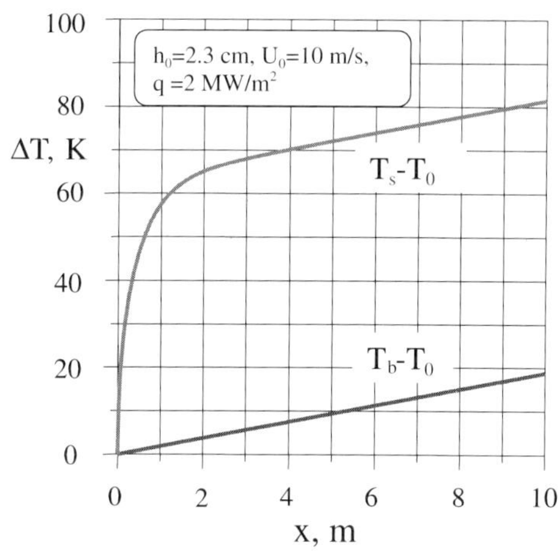

Fig. 2. Surface (T_s) and bulk (T_b) temperature rise in the thin Flibe liquid wall (ARIES RS, CliFF) exposed to a high density uniform surface heat flux. T_0 is the temperature in the flow inlet. As it can be seen, the surface temperature grows fast within the entry section and then changes linearly within the thermally fully developed segment. Calculated with the present $K-\varepsilon$ model.

3. Direct numerical simulation of open channel flows under a magnetic field

When developing the closures in the $K-\varepsilon$ model, DNS calculations were used. In the present study, because of the limitations on the DNS technique, the simulations were restricted to open channel MHD flows with a non-wavy flat free surface. The Reynolds number based on the bulk velocity and the flow height was 2300. The magnetic field was applied in the streamwise ($Ha = 20.0, 30.0$) or in the spanwise direction ($Ha = 5.0, 10.0$). In terms of the tokamak vacuum chamber, spanwise and streamwise stand for the toroidal and poloidal directions respectively. The number of computational grids used in this

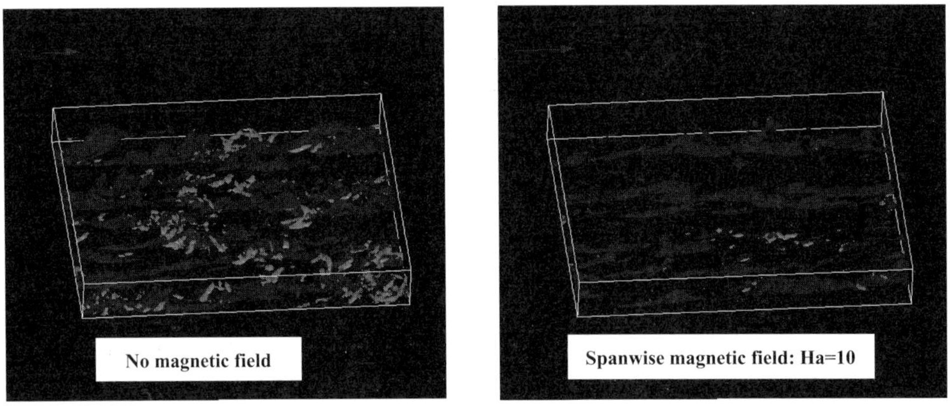

Fig. 3. Turbulence structures in open channel flow with and without a magnetic field: low-pressure regions, $p^+ < -3$, (grey); low-speed streaky strike structures, $u+ < -3$, (blue). DNS for $Re_t = 150$. The turbulence structures are significantly reduced in the presence of a magnetic field.

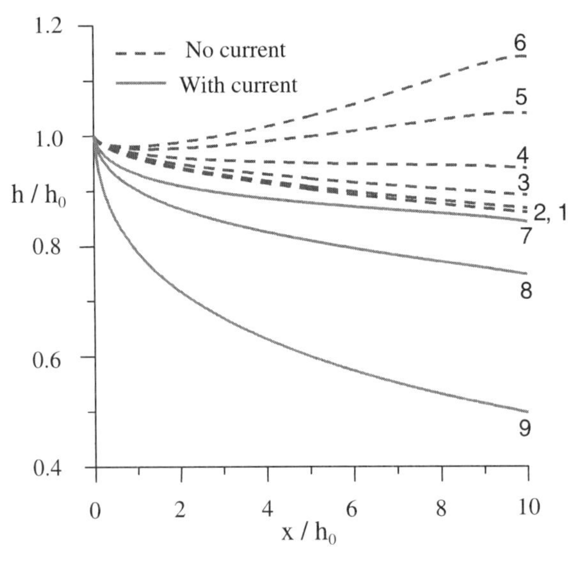

Fig. 4. Downstream flow thickness variations in open surface lithium flows with axial symmetry imposed to a linearly varying spanwise magnetic field with and without an applied electric current (j). Calculated with the present model (6-14). In the calculations, the flow length, L, is 20 cm, the initial flow thickness, h_0, is 0.02 m, the initial flow velocity, U_0, is 5 m/s. The liquid is propelled by applying a current. 1: $\Delta B_z/L = 0, j = 0$; 2: $\Delta B_z/L = 1.0$ T/m, $j = 0$; 3: $\Delta B_z/L = 2.0$ T/m, $j = 0$; 4: $\Delta B_z/L = 3.0$ T/m, $j = 0$; 5: $\Delta B_z/L = 4.0$ T/m, $j = 0$; 6: $\Delta B_z/L = 4.5$ T/m, $j = 0$; 7: $\Delta B_z/L = 4.5$ T/m, $j = 4.0$ kA/m^2; 8: $\Delta B_z/L = 4.5$ T/m, $j = 8.0$ kA/m^2; 9: $\Delta B_z/L = 4.5$ T/m, $j = 40.0$ kA/m^2.

study was $256 \times 128 \times 128$ (streamwise, wall-normal, and spanwise). The electromagnetic part of the governing equations included the equation for the electric potential and Ohm's law. The turbulence structures, the low-speed streaky structures and the low-pressure regions, are shown in Fig. 3. Both spanwise and streamwise magnetic fields lead to turbulence suppression as it is evidenced by a reduced number of the turbulence structures, however the effect is much stronger in a spanwise field, which directly affects the fluctuating velocity field through the Lorentz force. The turbulence suppression by a streamwise field occurs indirectly through changing the pressure field. A magnetic field results in reducing turbulence in the bulk and suppressing its generation near the wall. Some features showing transition to 2-D turbulence can be seen, such as enlargement of vortices (inverse energy cascade). However, stretching vortices along the magnetic field lines typical for developed 2-D MHD turbulence is not observed probably because of the low turbulence Reynolds number regime ($Re_t = 150$) realized in the present DNS calculations.

4. Modeling LM flows with axial symmetry

As examples of MHD flows with axial symmetry we refer to the lithium flow over the central column in the NSTX reactor[1] and outboard FW flow in the already mentioned ARIES RS reactor. When developing a mathematical model, the flow was assumed to be thin. It means the flow thickness is much smaller than the characteristic dimensions of the vacuum chamber and both main radii of curvature of the structural wall. In this approximation, only streamwise variations of the external magnetic field are important. Besides that the diffusion transport in the flow direction is negligible compared to that by convection. The second assumption is the inductionless approximation that implies the magnetic Reynolds number is much smaller than unity. Under this condition, the induced magnetic field, B', is much less than the external one, B^0, and can be neglected where they appear together. We also assume the external magnetic field does not vary in time. Third, in a strong "reactor-type" magnetic field, the LM flow will be laminarized, so that introduction of turbulence effects in the model is not needed. The governing equations were formulated in boundary fitted coordinates. In what follows "ξ", "η", and "ϕ" are used for the streamwise (poloidal), wall-normal, and spanwise (toroidal) coordinates respectively. Omitting details, the equations have been derived in terms of the velocity components and the induced magnetic field as follows:

$$U\frac{\partial U}{\partial \xi} + V\frac{\partial U}{\partial \eta} = -\frac{1}{\rho}\frac{\partial P}{\partial \xi} + \nu\frac{\partial^2 U}{\partial \eta^2} + g_\xi + \frac{1}{\rho}(j_\eta B^0_\phi - j_\phi B^0_\eta); \tag{6}$$

$$0 = -\frac{1}{\rho}\frac{\partial P}{\partial \eta} + g_\eta + \frac{1}{\rho}(j_\phi B^0_\xi - j_\xi B^0_\phi); \tag{7}$$

$$U\frac{\partial W}{\partial \xi} + V\frac{\partial W}{\partial \eta} = \nu\frac{\partial^2 W}{\partial \eta^2} + \frac{1}{\rho}(j_\xi B^0_\eta - j_\eta B^0_\xi); \tag{8}$$

$$\frac{1}{r}\frac{\partial U r}{\partial \xi} + \frac{\partial V}{\partial \eta} = 0; \tag{9}$$

$$\frac{1}{r}\frac{\partial j_\xi r}{\partial \xi} + \frac{\partial j_\eta}{\partial \eta} = 0; \tag{10}$$

$$j_\xi = -\frac{1}{\mu_0}\frac{\partial B'_\phi}{\partial \eta}; \tag{11}$$

[1]The National Spherical Torus Experiment (NSTX) is a fusion device constructed by the Princeton Plasma Physics Lab.

$$j_\eta = -\frac{1}{\mu_0}\frac{1}{r}\frac{\partial}{\partial \xi}B'_\phi r; \tag{12}$$

$$j_\phi = \sigma(UB^0_\eta - VB^0_\xi); \tag{13}$$

$$0 = \frac{1}{\sigma\mu_0}\left[\frac{1}{r}\frac{\partial}{\partial \xi}\left(r\frac{\partial B'_\phi}{\partial \xi}\right) + \frac{\partial^2 B'_\phi}{\partial \eta^2}\right] + \frac{\partial}{\partial \xi}(WB^0_\xi - UB^0_\phi) - \frac{\partial}{\partial \eta}(VB^0_\phi - WB^0_\eta). \tag{14}$$

Equations (6)–(8) are the projections of the momentum equation on the coordinate axes in the boundary layer approximation. Equation (9) is the mass conservation equation and Eq. (10) is the continuity equation for the electric current. Relations (11–13) are the expressions for the current components, and Eq. (14) is the induction equation. The notation used is standard. Symbols g_ξ and g_η stand for the acceleration due to gravity along with centrifugal or centripetal accelerations due to the motion of the liquid over the curved wall; $r(\xi)$ denotes the distance from the chamber axis to a given point in the liquid.

As the numerical analysis based on the finite-difference solution of (6–14) shows, the most significant MHD effects are caused by the wall-normal field and the spanwise space-varying magnetic field (both fields lead to an extra MHD drag and flow thickening). However in the presence of a gradient spanwise magnetic field, an excessive MHD drag can be overcome by applying a streamwise electric current that results in the "magnetic propulsion effect" [9]. Figure 4 illustrates the effectiveness of the magnetic propulsion by the example of a Lithium flow over a 20-cm length section. The flow becomes thicker as the spanwise magnetic field gradient grows because of the flow opposing effect caused by the induced electric currents. Applying an electric current creates a pressure gradient that accelerates the flow and finally makes it thinner.

5. Concluding remarks

The models and approaches described in the present paper can be applied to low-conductivity fluid turbulent MHD flows and laminar LM flows in channels with open surface. The application of these models to the liquid walls under conditions relevant to a fusion reactor reveals a variety of different MHD effects caused by a multi-component space-varying magnetic field. The detailed analysis of these effects is the essential part of the design work. An accurate mathematical model has been derived for open channel LM flows with axial symmetry. This model includes all three components of the applied magnetic field and all velocity components. The flows with no axial symmetry (for example, the axial symmetry will be broken if the flow along the FW is divided into sections by poloidal partitions) are essentially 3-D. In the presence of the partitions, the most crucial effect is that caused by a space-varying wall-normal magnetic field. As preliminary considerations show, changing the wall-normal magnetic field over the flow length gives rise to an axial potential difference within each section, which drives currents in the axial direction. These currents, while interacting with the magnetic field, produce forces driving the liquid in the toroidal direction from the core to the partitions that may lead to spilling the liquid over the partitions. This also results in a higher MHD drag than that in a uniform magnetic field. At present, a detailed 3-D numerical analysis of such flows under strong complex tokamak magnetic fields can be hardly performed. However, a number of simplified models has been derived. Many illustrations can be found on the official APEX site (www.fusion.ucla.edu/APEX). Future studies will concentrate on both the design options and further improvements of the models and computer codes.

Acknowledgment

This work was performed under US Department of Energy Contract DE-FG-03-86ER-52123.

References

[1] M. Abdou, The APEX Team, On the exploration of innovative concepts for fusion chamber technology, *Fusion Engineering and Design* **54** (2001), 181–247.

[2] S. Smolentsev et al., Development and adjustment of $K - \varepsilon$ turbulence model for MHD channel flows with large aspect ratio in a transverse magnetic field, *Proc. of the 4th Int. PAMIR Conference,* France, Sep 2000, pp. 21–26.

[3] A. Ying et al., Free surface heat transfer and innovative designs for thin and thick liquid walls, *Fusion Engineering and Design* **49–50** (2000), 397–406.

[4] S. Smolentsev et al., New multi-component MHD code and its application to APEX/NSTX, APEX-11 Project Meeting at ANL (www.fusion.ucla.edu/APEX), May 2000.

[5] A. Ying et al., MHD and heat transfer issues and characteristics for Li free surface flows under NSTX conditions, *Fusion Technology* **39** (2001), 739–745.

[6] I. Evtushenko, E. Kirillina, S. Smolentsev and A. Tananaev, Hydrodynamics and heat transfer of thin liquid-metal films in a magnetic field, in: *Progress in Astronautics and Aeronautics. Metallurgical Technologies, Energy Conversion, and Magnetohydrodynamic Flows,* (Vol. 148), H. Branover and Y. Unger, eds, 1993, 500–508.

[7] V. Kudrin, S. Smolentsev and A. Tananaev, Fully developed flow of a liquid metal thin film in an inclined magnetic field, *Magnetohydrodynamics* **29**(1) (1993), 66–70.

[8] H. Ueda et al., Eddy diffusivity near the free surface of open channel flow, *Int. J. Heat Mass Transfer* **20** (1977), 1127–1136.

[9] L. Zakharov et al., Magnetic propulsion of conducting fluid and the theory of controlled tokomak fusion reactor, Meeting on Liquid Lithium, Controlled Tokamak Fusion Reactors & MHD, Int. Sherwood Fusion/Plasma Conference, Atlanta GA, March 21, 1999.

International Journal of Applied Electromagnetics and Mechanics 13 (2001/2002) 381–385
IOS Press

Radial electric field control by electron injection in tohoku university heliac

Sumio Kitajima*, Shigeyuki Nakanishi and Hidetoshi Hashizume
Department of Quantum Science and Energy Engineering, Faculty of Engineering, Tohoku University, Sendai, Miyagi 980-8579, Japan

Abstract. In this paper, we present experimental results of improvement for plasma confinement by controlling electric field with electron injection [1]. In order to generate the negative radial field, we inserted a hot cathode into the plasma and then biased negative voltage. We observed the plasma parameter changes after the hot cathode biasing: the increase of line density, the fluctuation suppression, the negative radial electric field formation, the plasma rotation, and the increase of the energy confinement time. These results suggest that the improvement of plasma confinement by the formation of negative radial electric field is achieved.

1. Introduction

In Tokamak, the H mode is generally observed where the confinement is improved by formation of radial electric field. Moreover, the electrode biased H mode has been achieved on the Continuous Current Tokamak (CCT) [2] by applying the radial electric field biased with an electrode inserted into the plasma edge. In the ideal case, the H mode's scenario becomes like the followings;

1. the formation of the radial electric field
2. the formation of plasma flow or shear flow
3. the fluctuation suppression
4. the improvement for plasma confinement

In helical systems, the H mode like phenomenon has been also achieved in Wendelstein 7-AS [3]. However, this phenomenon is not common in the various helical systems.

Motivated by these backgrounds, we attempted the formation of a negative radial electric field by negative biasing and measured the fluctuation and plasma rotation to evaluate the improvement of energy confinement time by changing the input power from the hot cathode.

2. Experimental setup

The TU – Heliac is a small simple system [4], whose magnetic field is produced by three sets coils: thirty two toroidal field coils, a center conductor coil and two vertical field coils.

*Corresponding author: Sumio Kitajima, Tel.: +81 22 217 7927; Fax: +81 22 217 7900; E-mail: sumio.kitajima@qse.tohoku.ac.jp.

1383-5416/01/02/$8.00

Table 1
Plasma parameters before biasing

Major radius	48 cm
Average radius of the last closed flux surface (LCFS)	6.6 cm
Magnetic field on axis	0.3 T
Working Gas	He (1.2×10^{-2} Pa)
Electron density on axis	6×10^{11} cm^{-3}
Electron temperature on axis	~ 25 eV
Line density	9×10^{12}cm^{-2}
RF frequency	18.8 kHz
RF output	30 kW

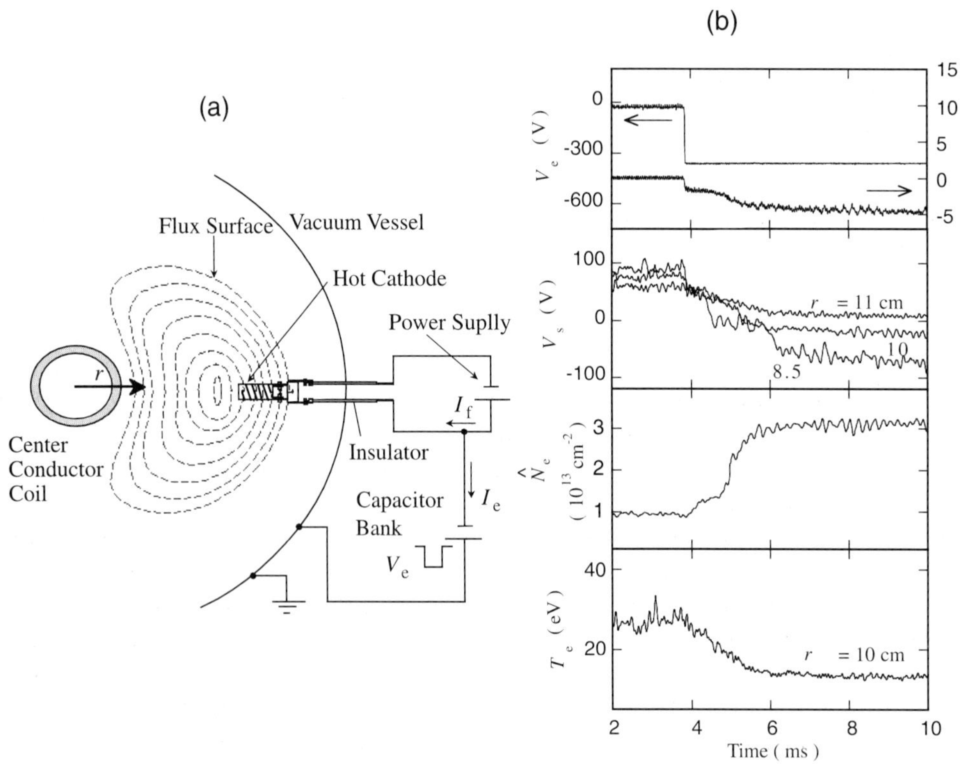

Fig. 1. Experimental arrangement of the hot cathode (a) and typical history of the plasma parameters (b).

The hot cathode experiments were performed by using plasma sustained only by low frequency ($f_{rf} < f_{ci}$, f_{ci}: ion cyclotron frequency) ohmic heating which is provided by poloidal coils located outside the vacuum vessel. The plasma parameters of the TU – Heliac before biasing are listed in Table 1.

The cathode is a cylindrical form made of LaB_6: 10 mm in diameter, 17 mm in length, and a parallel spiral geometry. We control the emission current I_e by changing both the current I_f supplied to heat the cathode and its biasing voltage V_e. The experimental arrangement of the hot cathode is shown in Fig. 1(a).

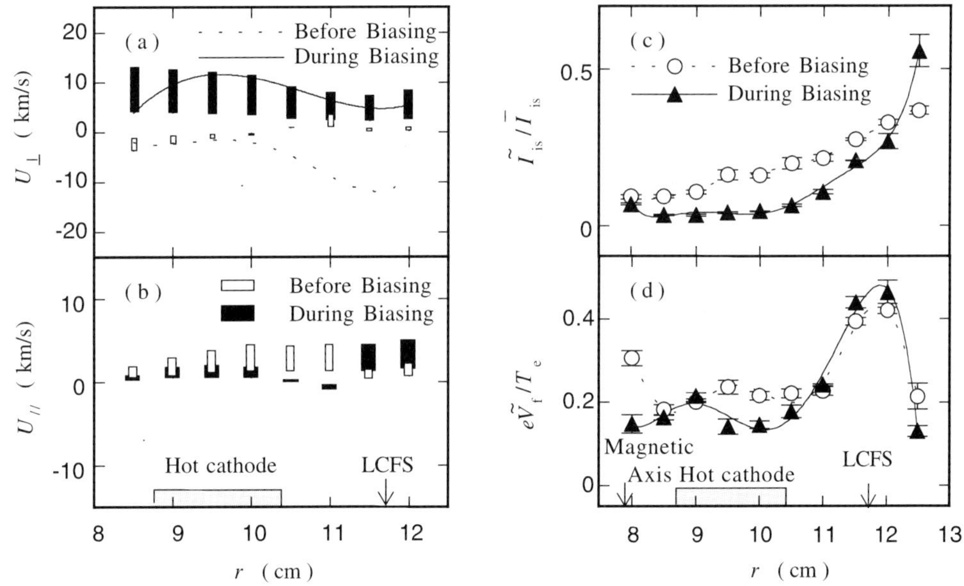

Fig. 2. Radial profiles of perpendicular flow velocity (a), parallel flow velocity (b), fluctuation of ion saturation current (c) and fluctuation of floating potential (d).

3. Experimental results

Typical history of the plasma parameters during the discharge is shown in Fig. 1(b). The parameter r used in the figure means the minor radius as shown in Fig. 1(a). The last closed flux surface (LCFS) is located at $r = 11.7$ cm. In Fig. 1(b), the typical five kinds of parameter are plotted with the elapsed time. V_e , I_e and V_s are the hot cathode bias voltage, its emission current, and space potential measured by a Langmuir probe, respectively. $\hat{N}_e$ means the line density along a vertical chord passing through the magnetic axis, which is measured by microwave interferometer. T_e denotes the electron temperature at $r = 10$ cm, measured by the Langmuir probe. The hot cathode is biased from $t = 3.75$ ms to $t = 10$ ms. Followed by biasing the hot cathode and flowing of the emission current, the space potential suddenly changes to negative value. Then the line density increases about three times while the electron temperature decreases because of cold electrons supplied with the hot cathode. The difference in the space potential V_s among three measured points denotes that the radial electric field was formed. The sudden decrease in the space potentials means that the formation of the radial electric field is initiated abruptly by the hot cathode biasing.

We measured the plasma flow velocity from the Mach probe using the Hudis and Lidsky model [5],

$$U = \sqrt{2}/4 \cdot \sqrt{T_e/T_i}\ \ln\left(I_{\text{up}}/I_{\text{down}}\right)\sqrt{T_e/m_i} \tag{1}$$

where we assume that $1 < T_e/T_i < 10$. I_{up} and I_{down} means the ion saturation currents in upstream and downstream sides, respectively. Figure 2 shows the radial profile of (a) the perpendicular flow velocity $U_\perp$ and (b) the parallel flow velocity $U_\parallel$ where open and closed rectangles correspond to the results before and during biasing, respectively. The plasma does not rotate in either the perpendicular or parallel directions before biasing. On the other hand, the perpendicular flow velocity is formed during biasing.

The perpendicular flow velocity $U_\perp$ is also given by the following kinetic equation,

$$U_\perp = E_r/B - (1/n_i Z_i e B)\partial p_i/\partial r \tag{2}$$

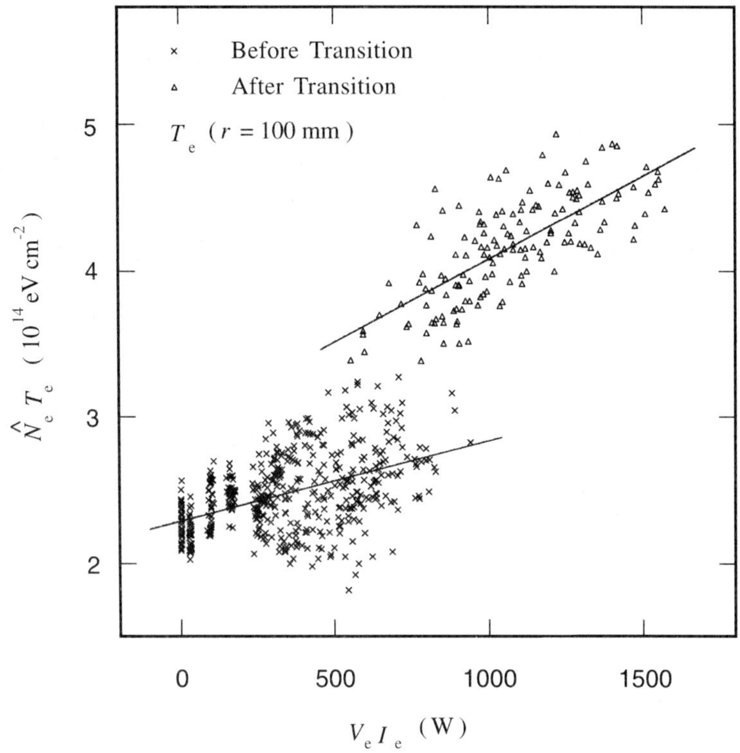

Fig. 3. The stored energy dependency on the input power.

where n_i, $Z_i e$ and p_i are ion density, charge of the ion and ion pressure, respectively. Using $T_i < T_e \approx$ 10 eV, $n_i = n_e \approx 5 \times 10^{16}$ m^3 and $\delta r = 40$ mm, the ion diamagnetic drift velocity becomes 0.7 km, which is negligibly small compared with the $\boldsymbol{E} \times \boldsymbol{B}$ drift velocity. Then, $U_\perp$ is calculated from the $\boldsymbol{E} \times \boldsymbol{B}$ drift velocity using the radial profile of space potential V_s, whose result is also shown in Fig. 2(a) by solid and broken lines. The profiles of $U_\perp$ evaluated by the two methods agree well. This result suggests the plasma flow does occur.

Figure 2 also shows radial profiles of (c) the normalized fluctuation level of the ion saturation current, (d) the normalized fluctuation level of the floating potential. When an electron temperature fluctuation is assumed negligible, the normalized fluctuation level of the ion saturation current is equal to that of electron density and the normalized fluctuation level of the floating potential is equal to that of plasma potential. From Figs 2(c) and (d), it can be seen that both fluctuations are suppressed by the biasing.

The stored energy dependency on input power from the hot cathode is shown in Fig. 3. The horizontal axis indicates the input power from the hot cathode. The vertical axis indicates the product of the line density and the electron temperature, which is proportional to the stored energy. The linear gradient after the formation of the negative radial electric field is about twice larger than that before the formation. Since the linear gradient is proportional to the energy confinement time, this result suggests the improvement of the confinement time occurs by the formation of the negative radial electric field.

4. Conclusions

– By biasing the hot cathode negatively, the negative electric field is formed and the line density increased.

- The formation of the perpendicular flow occurs followed by the fluctuation suppression of the ion saturation current and the floating potential.
- After the formation of the negative radial electric field, the energy confinement time becomes twice larger than that before the formation. This result suggests that the improvement of the confinement time by the formation of the negative radial electric field.

These results correspond with the **H** mode's theory and suggest the improvement for plasma confinement by controlling electric field.

References

[1] S. Kitajima et al, *Proc. 11th Int. Toki Conf. on Plasma Physics and Controlled Nuclear Fusion*, 2000, pp. 125.
[2] R.J. Taylor et al, *Phys. Rev. Lett.* **63** (1989), 2365.
[3] V. Erckmann et al, *Phys.Rev. Lett.* **70** (1993), 2085.
[4] S. Kitajima et al, *Jpn. J. Appl. Phys.* **30** (1991), 2606.
[5] M. Hudis et al, *J. Appl. Phys.* **41** (1970), 5011.

International Journal of Applied Electromagnetics and Mechanics 13 (2001/2002) 387–392
IOS Press

Efficiency properties of adaptive mesh generation

Lutz Jänicke* and Arnulf Kost
Brandenburgische Technische Univeristät Cottbus, Allgemeine Elektrotechnik und Numerische Feldberechnung, Universitätsplatz 3–4, D-03044 Cottbus, Germany
Tel.: +49 355 69 4154 (Kost); Fax: +49 355 69 4153

Abstract. One of the major advantages of adaptive mesh generation is the optimum fitting of the mesh to the problem, leading to good memory efficiency. By suitably choosing the refinement parameters, the adaptive mesh generation is also time efficient. As the best refinement parameters depend on the problem to be solved and can also change during the adaptive solution process, they must be chosen dynamically. The choice is based on the error distribution, that is available from the local error estimation. Therefore no additional computational cost arises due to the dynamic parameter setting.

1. Introduction

Adaptive mesh generation offers a good memory efficiency by fitting the mesh to the problem, so that the problem can be solved with a small number of unknowns. There is however a tradeoff with the computer time needed, that often leads to the opinion that adaptive mesh generation is a slow process due to the additional iterative loop involved.

During the adaptive mesh generation process, the refinement percentage decides about the efficiency of the computation. The smaller the percentage, the better the mesh is fitted to the problem, but the more adaptive iterations are required. It however turns out, that the adaptive iterations are often performed on a very small mesh and are therefore very fast. Even more, we found that for an adaptively generated mesh the convergence of the equation system is better. Utilizing these two effects, the adaptive mesh generation can be performed so fast, that even its time efficiency is better than that of an unadapted mesh.

2. Approximation error

In order to compare the efficiency of mesh generation strategies, the results obtained must be compared on an "objective" basis. We use the global error expressed in energy norm for comparison.

2.1. Error estimation

As the real solution is only available for special test cases, the error can only be estimated. We use the "local error problem" technique [1], that gives very reliable results for both the local and the global error and that can therefore be used instead of the real error.

*Corresponding author: Lutz Jänicke, E-mail: Lutz.Jaenicke@aet.TU-Cottbus.DE; Arnulf.Kost@aet.TU-Cottbus.DE.

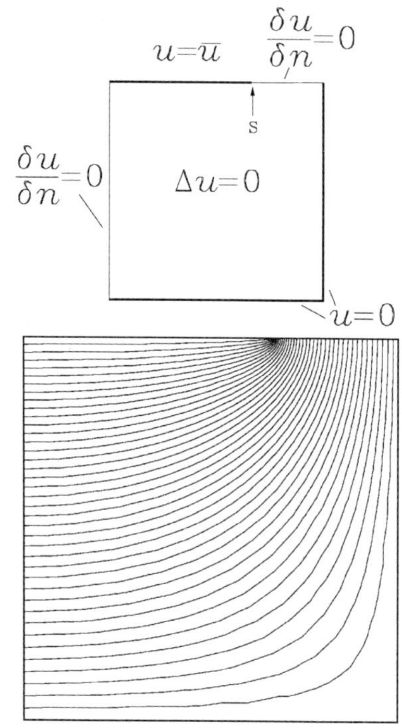

Fig. 1. Test problem: electrostatic 2d sample problem.

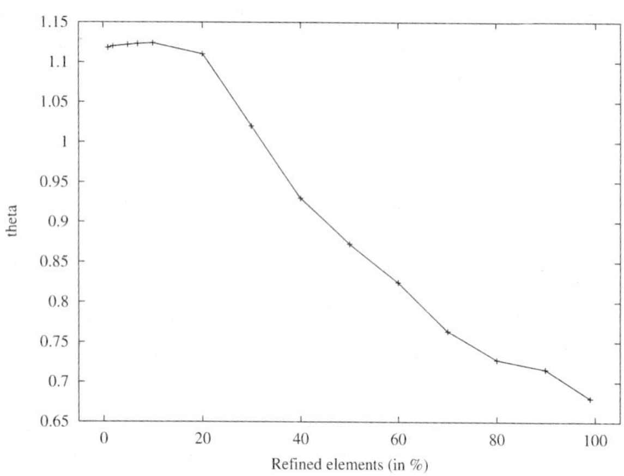

Fig. 2. Electrostatic example: convergence rate over refinement.

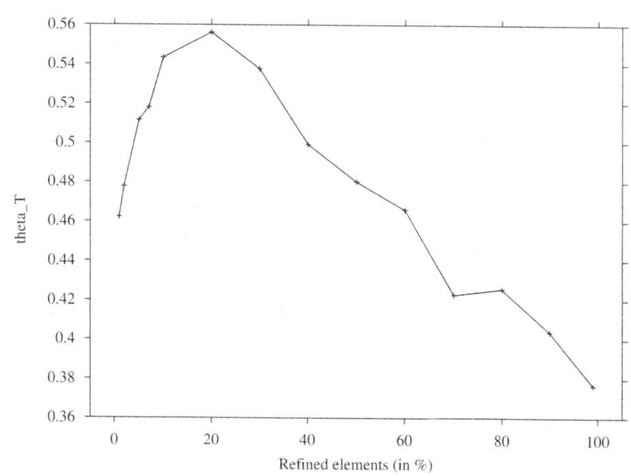

Fig. 3. Electrostatic example: time rate over refinement.

2.2. *Convergence behaviour*

While the usual convergence estimate for the finite element method $|||e||| = O(h^p)$ is derived from the differential equation, the presence of singularities in real life problems leads to the extended convergence estimate

$$|||e||| = O(h^{\vartheta}) \qquad |||e||| = O(\mathrm{NDF}^{-\vartheta/2}) \quad \text{(for 2D)} \qquad \text{with} \quad \vartheta = \min(p, \lambda) \tag{1}$$

showing that in the presence of singularities of the intensity λ the convergence is not so good as expected [2]. Consequently the use of adaptive mesh generation is intended to correct this loss.

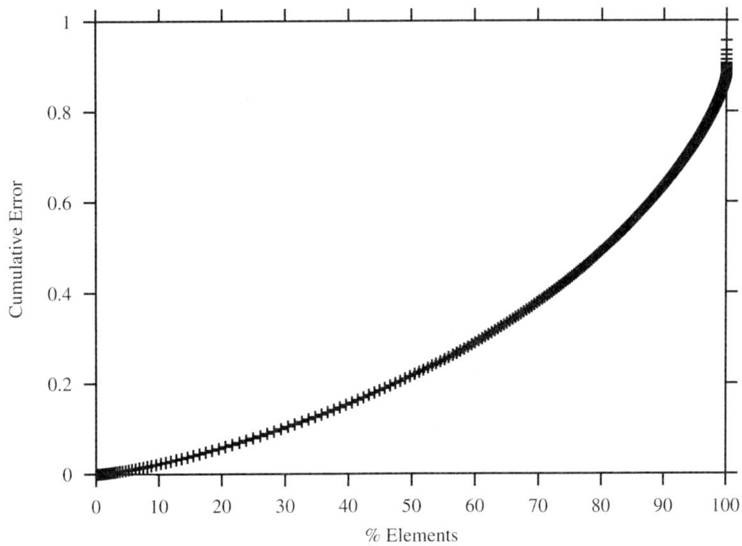

Fig. 4. Electrostatic example: cumulative error distribution for 20 % refinement.

3. Refinement parameters

By evaluating the convergence rate ϑ for different problems and several different refinement percentages we could show, that there is an optimum refinement percentage, below which the mesh fitting is not improved anymore. This gives an upper bound for the convergence parameter ϑ even if the percentage of refined elements is chosen smaller than this optimum value [3]. In the same investigation it also turned out, that the best time efficiency, expressed by the time convergence rate

$$|||e||| = O(t^{-\vartheta_{\mathrm{T}}}), \tag{2}$$

can be achieved by running the adaptive iteration at the same optimum refinement percentage. (It should be noted, that unlike Eq. (1) the time convergence rate Eq. (2) is not derived from any mathematical relationship. Since the adaptive mesh generation includes several steps following different time laws (refinement, mesh smoothing, solving the equation system, error estimation), Eq. (2) is normally not perfectly valid and should here be used as an estimate only.) Figures 2 and 3 show the convergence behaviour for an electrostatic example problem, see Fig. 1 [3].

This optimum refinement percentage is problem specific and is related to the error distribution in the problem [4]. When the refinement percentage is at the optimum point or below, the error distribution is smooth (but not perfectly homogeneous). If the refinement percentage is chosen too high, the error in the problem is dominated by a small number of elements with high error values.

3.1. Dynamic parameter choice

The results obtained in [3] have only been obtained by an extensive number of computations to find out the optimum value for each problem.

Based on these results a curve describing the relationship between the error distribution (described by the area under the cumulative error curve, see Figs 4 and 5) and the optimum refinement percentage was presented [4,5]. The area under the cumulative curve can easily be obtained at each refinement step and can be used to set the percentage for the next refinement step (see Fig. 6).

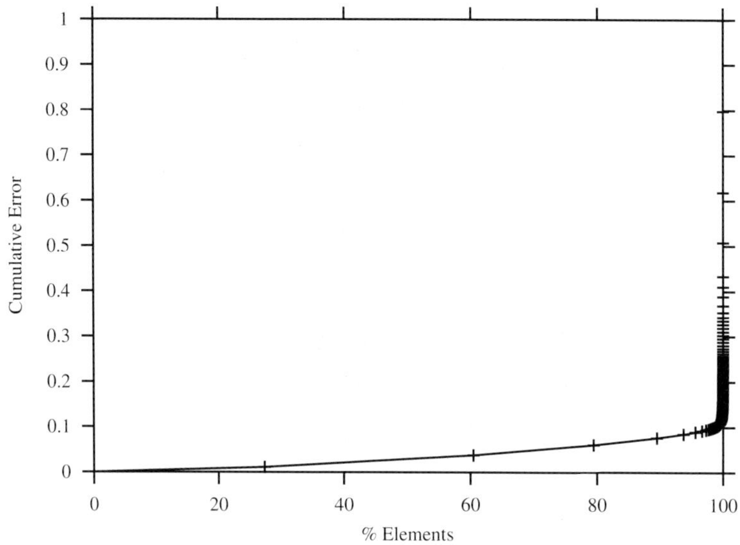

Fig. 5. Electrostatic example: cumulative error distribution for 50 % refinement.

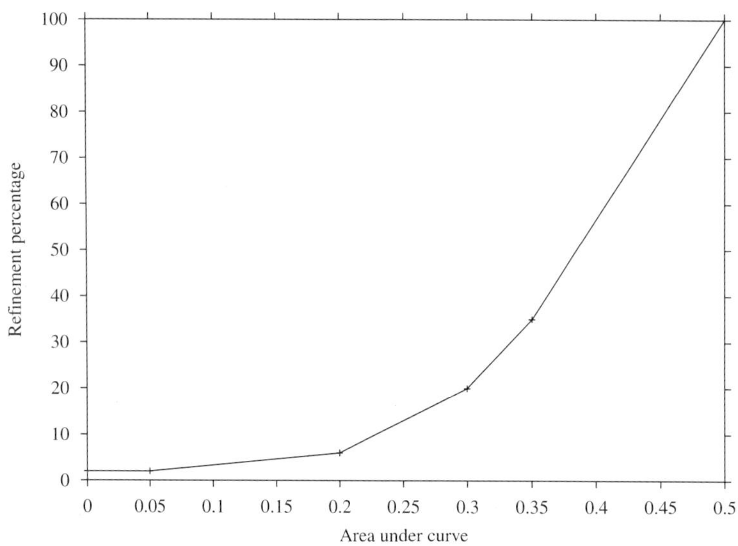

Fig. 6. Proposed curve with refinement percentage over area under cumulative error curve.

3.2. Resulting refinement procedure

This refinement procedure has been applied to a number of different problems, most of them being computations of electrical machines, hence containing a lot of sharp material corners. It turned out, that typically the refinement process starts up with a large number of steps with a very selective refinement percentage (indicating a very inhomogeneous error distribution). The error distribution then becomes more homogeneous and the percentage of elements refined in each step increases. Figure 8 shows the refinement characteristics for a switched reluctance motor [6].

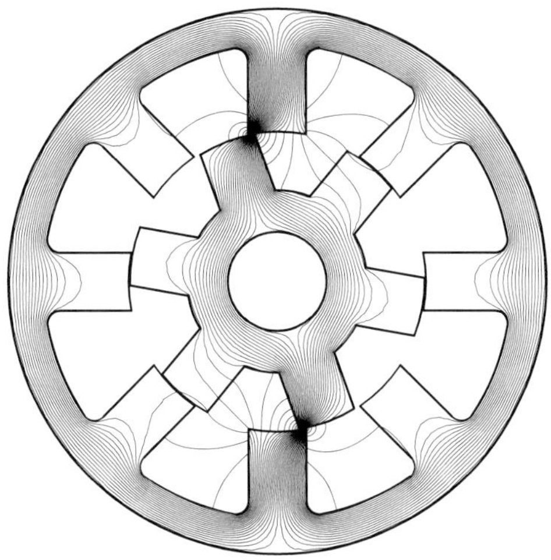

Fig. 7. Test problem: switched reluctance motor.

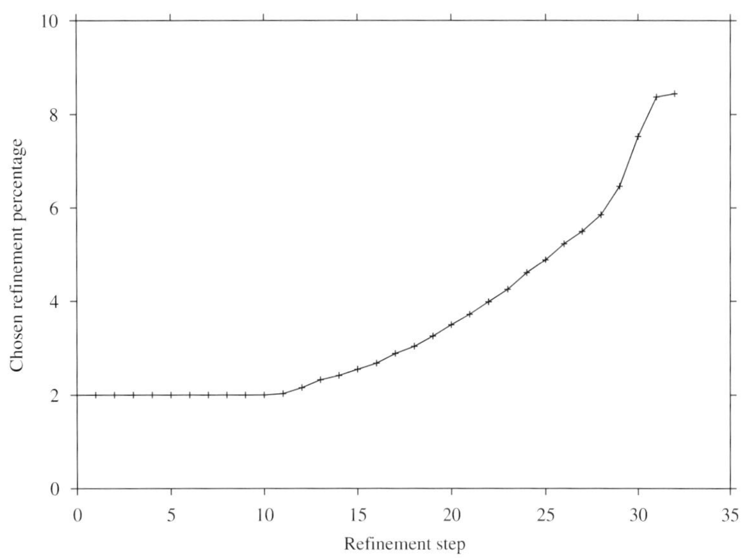

Fig. 8. Chosen refinement percentage during adaptive process for SR motor.

3.3. Evaluating the procedure

Due to the large number of iterations with small refinement percentage at the beginning of the adaptation, the overall number of iterations needed becomes high. Therefore, on the first glance, the procedure seems to be inefficient. A large number of iterations is however performed during the initial steps, when the number of unknowns is low, so that the computer time required for these steps is more or less negligible. Figures 8 and 9 show, that the first 10 iterations are performed on the minimum level of 2 % refinement. The cumulative computer time for these 10 iterations is around 10 seconds. (This example was intended to show the behaviour of the process up to a very high number of unknowns (400,000); an accurate solution of the field problem is already available with a much smaller amount of unknowns and iterations.)

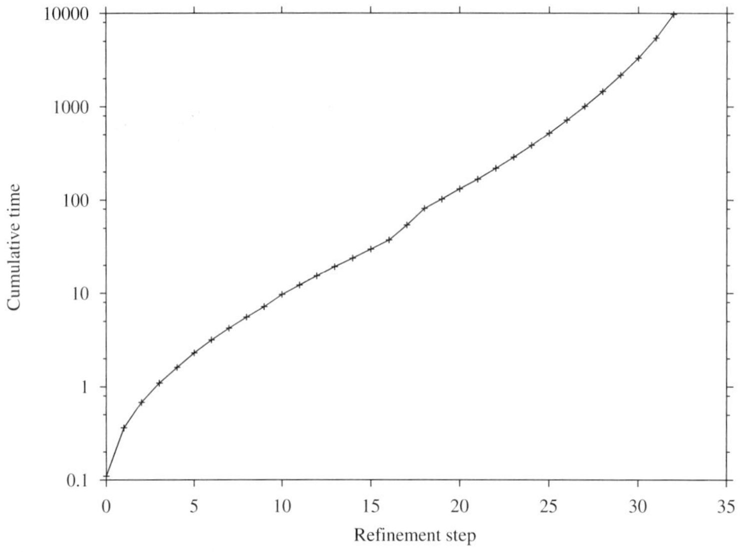

Fig. 9. Cumulative computer time during adaptive process for SR motor.

4. Conclusions

While it looks promising to choose a more aggressive refinement strategy (refining more elements in each step in order to reduce the total computer time), it turns out, that the dynamic strategy with high selectivity for inhomogeneous error distributions shows advantages with respect to the efficiency.

The adaptive steps with very selective refinement only occur at the beginning of the process, where iterations are cheap. The selective refinement also leads to an excellent fitting of the mesh to the problem and therefore to very low error values. The optimum procedure to find is hence not based on the number of iterations but on the relation between error (quality of the solution) and overall computation time.

As a side effect, the memory efficiency is also kept at its optimum value by using the dynamic refinement strategy due to the selective refinement.

References

[1] R.E. Bank and A. Weiser, Some a posteriori error estimators for elliptic partial differential equations, *Mathematics of Computation* **44**(170) (April 1985), 283–301.

[2] O.C. Zienkiewicz and J.Z. Zhu, Adaptivity and mesh generation, *International Journal for Numerical Methods in Engineering* **32** (1991), 783–810.

[3] L. Jänicke and A. Kost, Investigations on the efficiency of adaptive mesh generation, in: *IGTE 98, Graz, Austria*, 1998, pp. 197–202.

[4] L. Jänicke, A. Kost and J.P.A. Bastos, Local error distribution in adaptively generated meshes, *IEEE Transactions on Magnetics* **36**(4) (July 2000), 1627–1630.

[5] L. Jänicke, A. Kost and J.P.A. Bastos, Choosing refinement percentage in adaptive mesh generation, in: *Proceedings of CEFC Milwaukee,* USA, 2000.

[6] L. Jänicke and A. Kost, Efficiency properties of adaptive mesh generation, in: *Proceedings of ICEF Tianjin,* China, 2000, pp. 189–194.

International Journal of Applied Electromagnetics and Mechanics 13 (2001/2002) 393–398
IOS Press

Theoretical background of parallel electromagnetic field computations

Andrzej Jordan
Polish-Japanese Institute of Information Technology, ul. Koszykowa 86, 02-008 Warsaw, Poland
Tel.: +48 85 74 216 51 ext. 123; Fax: +48 85 74 216 57; E-mail: jordana@cksr.ac.bialystok.pl

Abstract. This paper presents a general concept of electromagnetic field parallel computations using super-computers and cluster system of workstations. In principle, all numerical algorithms of electromagnetic field computations (FEM, FDM, FEBM, FDTD) lead to the analysis of a large set of algebraic equations. The properties of parallel implementation of FDTD algorithm and the conjugate gradient method (CGM) were presented.

1. Introduction

Parallel computations of electromagnetic field are a very useful tool for the analysis of electromagnetic wave propagation. These computations let us achieve a significant speedup and lead to an increase of computational space. This increase also enables us to introduce homogeneous boundary conditions of the first kind, which follow from the nature of the physical phenomenon considered.

The analysis of a high frequency electromagnetic field is carried out using the finite-difference time-domain method (FDTD) [1,3,7]. In this case the Maxwell's system of vector equations is reduced into 6 scalar equations and next they are integrated in the time and space domain (Fig. 1).

A low frequency electromagnetic field is analyzed using FEM, FDM or FEBM (finite element-boundary integral method) [6,9,11]. After basic transformations of the Maxwell's equations we have to solve the following set of algebraic equations (Fig. 1).

$$\mathbf{K} \cdot \mathbf{x} = \mathbf{b}. \tag{1}$$

where $\mathbf{K}, \mathbf{x}, \mathbf{b}$ are a square matrix, an unknown vector and a known vector on the right-hand side, respectively. The Eq. (1) is joined with the steady state of wave propagation. The solution in the transient state is carried out in successive time steps.

2. Parallel algorithms of the electromagnetic field analysis

The development of parallel algorithms of the electromagnetic field analysis is a very complex problem. The distribution of the electromagnetic field is described by two conjugate Maxwell equations. The development of a parallel algorithm requires knowledge of physical, mathematical and numerical aspects of the problem considered. Physical aspects follow from alternate propagation of the electric and magnetic field and mutual connection of the electric field components E_x, E_y, E_z with the magnetic field components H_x, H_y, H_z, which is mathematically described by the operation of a rotation.

1383-5416/01/02/$8.00

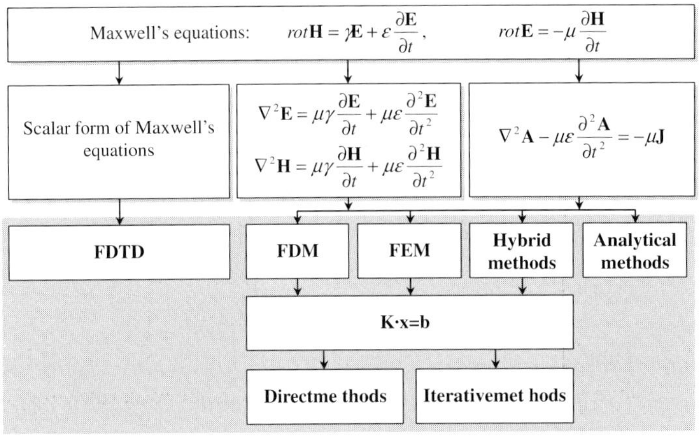

Fig. 1. Methods of electromagnetic wave computations.

Numerical problems arise from some fundamental numerical aspects. The round-off errors in numerical methods, the stability of an algorithm, and the selection of a method resulting from physical problems (FDTD or FEM) or the structure of the matrix **K**. Either a direct method or an iterative algorithm has to be implemented for the numerical solution of the Eq. (1).

In general, the parallel algorithms of the electromagnetic field analysis are developed on the basis of domain decomposition or a sequential algorithm itself [1]. Most often we deal with the domain decomposition, which leads to the partition of computational space on some number of subspaces. We also can meet with a decomposition of the matrix **K** controlled by the decomposition of vector x [9,11].

3. Parallel algorithm of the FDTD method

As it was mentioned above, the FDTD method is frequently applied to the analysis of the high frequency electromagnetic field. In this way the radar problems, high frequency antennas or the electromagnetic field influence on live organisms [3,7] are analyzed. By transforming the Maxwell's equations into a scalar form, 6 equations are obtained. They are transformed on the basis of K.S. Yee's cell and then differential schemes are transferred into a numerical form [8].

A decomposition of the computational area on three sub-regions assigned to the slave processors P_1, P_2 and P_3 is presented in Fig. 2. The source of an electromagnetic wave is placed at the point $P_S(x_s, y_s)$ and a non-ideal insulator body is located at the center of the middle subspace.

Presently, we shall consider the problem of a TM wave propagation in a two dimensional space. In this case, a distribution of the electromagnetic field components E_z, H_x and H_y has to be determined.

When the Maxwell's differential form equations are examined, it can be seen that the time derivative of the E field depends on the curl of the H field. This should be simplified to determine that the change in the E field (the time derivative) is dependent on the change in the H field across space (the curl). The new value of the E field is dependent on the old value of the E field and the difference in the old value of the H field on either side of the E field point in space.

In the parallel algorithm of the analysis, the processor P_1 (containing the wave source) sends the values of E_z from the edge B at even time steps. In this way the processor P_2 can compute the component H_x at the next time step. The results are sent back to the processor P_1 from the edge A. The above exchange of information is carried out on the edges between neighboring slave processors. It should be pointed

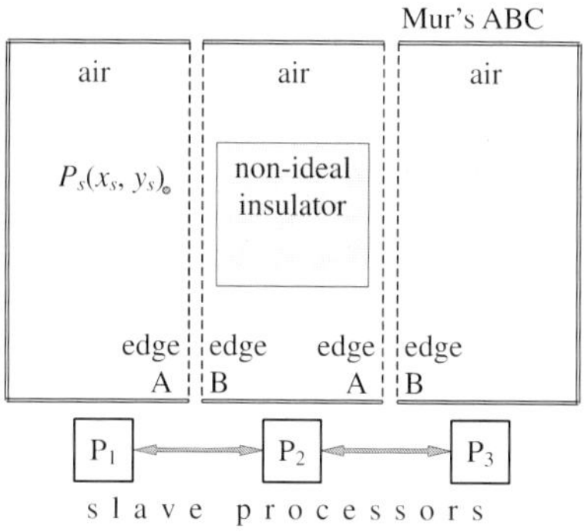

Fig. 2. The method of the domain decomposition into interdependent sub-regions.

out that the components of the electric field are computed at even steps in the time domain, while the magnetic field components at odd time steps respectively.

By analyzing the wave propagation in a three dimensional space, the exchange of temporary results between neighboring processors is carried out from the planes separating neighboring slave processors. Absorption boundary conditions (ABC) are assumed on the edges of the analyzed region [3,4]. The introduction of a large number of slave processors analyzing the interior fragments of the space leads to a proportional enlargement of the size of a computational region. Because the computational code of the processor P_2 can be repeated N-2 times (N – total number of slave processors) we can say that in this case the algorithm is scaleable.

In the parallel computations we attempt to achieve a maximal speedup of the computations

$$S = \frac{t_c(1)}{t_c(n)} \tag{2}$$

where $t_c(1)$ is the time of execution of the best version of the algorithm on a single processor, $t_c(n)$ is the time of computation of the parallel algorithm, n determines the number of the processors ($n = 1, 2, \ldots, N$).

A dependence of the speedup of computations of 3D model on the number of slave processors is presented in Fig. 3 [1,7].

4. Parallel algorithms for a large set of algebraic equations

The majority of numerical methods connected with the analysis of partial differential equations (Fig. 1) leads to the analysis of a system of algebraic equations. In the professional software, a solver is the main module being subject of parallelization. Basically, it is connected with the analysis of Eq. (1). In the case of large dimensions of the matrix $\mathbf{K}$ supercomputers should be used for the computations and parallel algorithms should be applied. Their structure is complicated and in most of the analyzed problems it follows from the sequential methods: direct and iterative.

Iterative methods are based on a proper selection of the initial vector and such transformation of the matrix $\mathbf{K} \rightarrow \mathbf{G}$. After this transformation the eigen-values of a new matrix fulfill the convergence condition of the method ($\rho(\mathbf{G}) < 1$). Among sequential iterative methods, the conjugate gradient method is frequently applied because of a fast convergence. It consists of four basic matrix operations (e.g. matrix multiplication), which can be subject to a parallelization in a simple way.

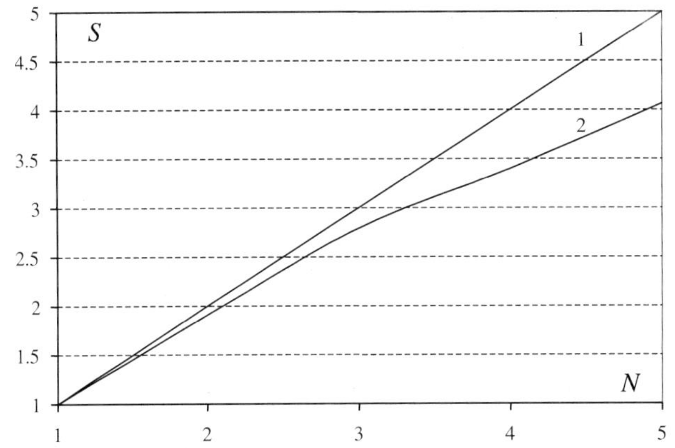

Fig. 3. Speedup of the 3D FDTD algorithm (cluster system of IBM computers): 1 – ideal case, 2 – obtained speedup.

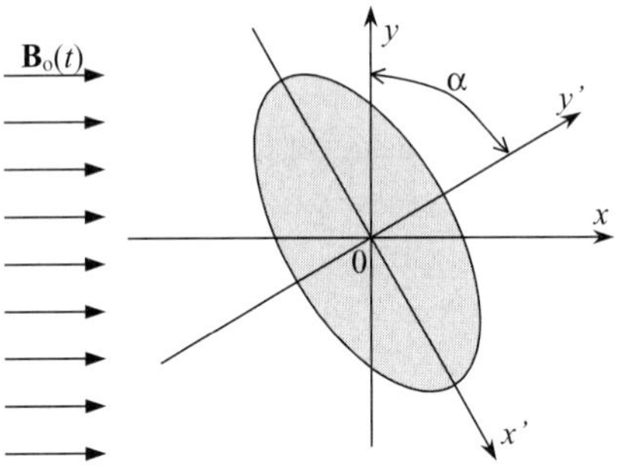

Fig. 4. The crossection of the analyzed model.

In order to discuss direct and iterative methods more precisely, the problem presented in [13] will be considered. In this case the eddy current losses in a ferromagnetic conductor of an elliptic cross section should be determined (Fig. 4). Introducing a vector potential A, the nonlinear equation of diffusion should be solved

$$\text{div}\,(\nu\ \text{grad}\ \mathbf{A}) = \sigma \frac{\partial \mathbf{A}}{\partial t}, \tag{3}$$

where ν is a converse of the magnetic permeability. For $\nu = \text{const}$ the Eq. (3) can be reduced to the form Eq. (1). Applying basic transformations connected with the finite element method, assuming conditions of the solution periodicity at a steady state and using the Crank-Nicholson scheme of integration in the time domain, we obtain the following system of algebraic equations

$$\left(\frac{1}{\Delta t}\mathbf{C} - \frac{1}{2}\mathbf{K}\right) \cdot \mathbf{A}_1 - \left(\frac{1}{\Delta t}\mathbf{C} + \frac{1}{2}\mathbf{K}\right) \cdot \mathbf{A}_{1+1} = -\mathbf{b}_{1+\frac{1}{2}} + \frac{1}{2}\mathbf{R} \cdot (\mathbf{A}_{1+1} + \mathbf{A}_1), \tag{4}$$

where $\mathbf{A}$ is the matrix of unknowns of vector potential ($\mathbf{A} = \text{x}$), $\mathbf{C}, \mathbf{R}$ – the additional square matrices [13], Δt – time step. In this case, Brouwer's fixed-point theorem [13] was applied to the solution of the nonlinear problem.

Substituting $\mathbf{K}^T\mathbf{K} = \mathbf{D}$ and $\mathbf{K}^T\mathbf{b} = \mathbf{h}$, the following iterative procedure is performed in a sequential version of the conjugate gradient method

$$\mathbf{r}_1 = \mathbf{h} - \mathbf{D} \cdot \mathbf{x}_1, \tag{5a}$$

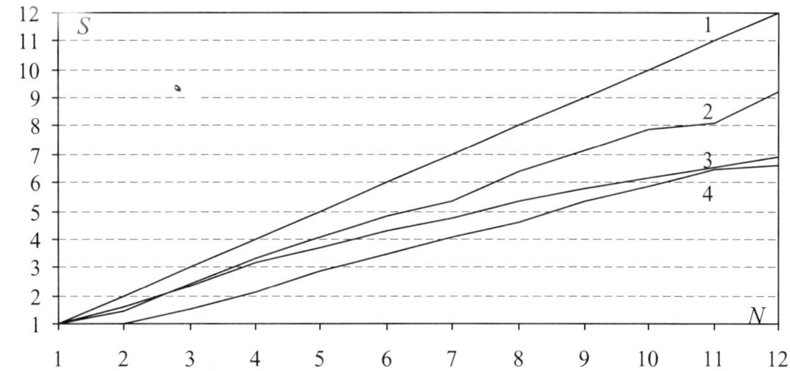

Fig. 5. A comparison of the implemented parallel algorithms (Hitachi SR 2201). 1 – ideal case, 2 – LU algorithm, dynamic loading of the processors, dim(**K**) = 1000, 3 – CGM algorithm, dim(**K**) = 1013, 4 – LU algorithm, static loading of the processors, dim(**K**) = 1000.

$$\mathbf{p}_1 = \mathbf{r}_1, \tag{5b}$$

where $\mathbf{r}_1$ is a residuum and $\mathbf{x}_1$ – the first approximation of the solution.

For the successive iterations ($q = 1, 2, 3, \ldots$) we compute

$$\begin{aligned} \alpha_q &= \frac{(\mathbf{r}_q, \mathbf{r}_q)}{(\mathbf{D}\mathbf{p}_q, \mathbf{p}_q)} \longrightarrow \mathbf{x}_{q+1} = \mathbf{x}_q + \alpha_q \mathbf{p}_q \longrightarrow \mathbf{r}_{q+1} = \mathbf{r}_q - \alpha_q \mathbf{D}\mathbf{p}_q \longrightarrow \\ \gamma_q &= \frac{(\mathbf{r}_{q+1}, \mathbf{r}_{q+1})}{(\mathbf{r}_q, \mathbf{r}_q)} \longrightarrow \mathbf{p}_{q+1} = \mathbf{r}_{q+1} + \gamma_q \mathbf{p}_q \end{aligned} \tag{6}$$

The iterative process is accomplished, when

$$\frac{\|\mathbf{r}_{q+1}\|}{\|\mathbf{h}\|} < \varepsilon \tag{7}$$

where $\|\cdot\|$ denotes the Euclidean norm of a vector.

Analyzing the presented algorithm, the following operations on matrices and vectors can be distinguished: the multiplication of matrices, the multiplication of a matrix by a vector, the scalar product of two vectors, the subtraction and summation of vectors. The above operations are subject to a parallelization. This process can be performed by the use of some professional programs (Hitachi library) or carried out in independent research [5]. In this work the presented results are based on the programs elaborated by the author, in which the parallel algorithms of the matrix and vector operations were applied on the basis of a simple decomposition of vector $\mathbf{x}$. Figure 5 presents a comparison of the CGM method (iterative method) with a direct solution (LU decomposition) of Eq. (1). It should be pointed out that in the LU decomposition the time of the solution of Eq. (1) is negligible with respect to the time of the LU decomposition ($q \times M$ simple operations in relation to M^2 operations of the LU decomposition, where M is the dimension of matrix **K**).

5. Conclusions

The parallel computations of the electromagnetic field propagation are used in numerical analysis of systems, which are characterized by a large number of Yee's cells (HF) or a large dimension of the

K matrix. In this way we obtain a significant increase of the speedup of computations and enormous possibilities to increase the computational region of three-dimensional problems.

The basic disadvantages of the parallel computations are: high cost of the systems of a distributed memory (MIMD) and complicated structure of the programs in which the communication libraries such as MPI [10] and PVM [12] are utilized. The high costs of hardware are currently compensated by using clusters of workstations and PC-class computers.

Another important problem is the optimization of parallel programs when the speedup of computations is the objective function. The example of such an optimization with the use of the VAMPIR program is presented in [2].

Acknowledgments

This project has been carried out within the framework of research grants at PJIIT No 5/2000-1 and the Polish Committee for Scientific Research No 8T10A 050 18.

References

[1] A. Jordan and B. Butrylo, Parallel computations electromagnetic wave propagation, PARELEC'1998, International Conference on Parallel Computing in Electrical Engineering, Bialystok, Poland, 1998, pp. 296–300.
[2] A. Smyk, P. Cieslak, M. Tudruj, A. Jordan and B. Butrylo, Optimization of FDTD Parallel Software Based on the Use of Vampir Monitor, SCI'2001, 5th World Multiconference on Systemics, Cybernetics and Informatics, Orlando, Florida, USA, 2001, pp. 343–348.
[3] A. Taflove and S.C. Hagness, *Computational electrodynamics, the finite-difference time-domain method,* Artech House Inc., Boston – New York, 2000.
[4] G. Mur, Absorbing boundary conditions for the finite-difference approximation of the time-domain electromagnetic field equations, *IEEE Trans. on Biomed. Eng.* **BME-34**(2) (1987), 148–157.
[5] Hitachi Fortran / C user's guide reference, www.pjwstk.waw.pl.
[6] J. Dongarra et al., ScaLAPACK user's guide, SIAM, Philadelphia, 1997.
[7] J. Forenc and A. Skorek, Analysis of high frequency electromagnetic wave propagation using parallel MIMD computer and cluster system, PARELEC'2000, International Conference on Parallel Computing in Electrical Engineering, Trois-Rivieres, Canada, 2000, pp. 176–180.
[8] K.S. Yee, Numerical solution of initial boundary value problems Maxwell's equations in izotropic media, *IEEE Transactions on Antennas and Propagation* **AP-14**(3) (1966), 302–307.
[9] L. Nicolas and C. Vollaire, A survey of computational electromagnetics on MIMD systems, PARELEC'1998, International Conference on Parallel Computing in Electrical Engineering, Bialystok, Poland, 1998, pp. 7–19.
[10] MPI Forum, http://www.mpi-forum.org.
[11] P. Tronczek, Application of LU decomposition for solving a large set of linear algebraic equations on supercomputer Hitachi SR 2201, MSc dissertation, Polish-Japanese Institute of Information Technology, Warsaw, 2000.
[12] PVM forum, http://www.epm.ornl.gov/pvm.
[13] W. Peterson, The fixed point technique in computing eddy currents within ferromagnetic conductors exposed to an alternating transverse magnetic field, *Archives Of Electrical Engineering* **XLVII**(1) (1998), 57–68.

International Journal of Applied Electromagnetics and Mechanics 13 (2001/2002) 399–404
IOS Press

Performance analysis of permanent magnet generators by adaptive mode

Bijan Sobhi-Najafabadi, Özdemir Göl* and Andrew Nafalski
University of South Australia, Division of Information Technology, Engineering and the Environment, School of Electrical and Information Engineering, Mawson Lakes, SA 5095, Australia

Abstract. Simple impedance models are routinely used in the design of permanent magnet generators. Unfortunately, this invariably results in poor prediction accuracy, particularly because saturation is not taken into account. This paper presents a method, which helps improve prediction accuracy by using field models as an adjunct to adjust the model parameters to the prevailing flux conditions in an adaptive mode.

1. Introduction

It is common practice to employ impedance models as the main tools in designing permanent magnet machines to a given specification. Although expeditious in terms of computational effort, such models suffer from a lack of detail in their formation. One major drawback is the assumption that the magnetic core is not subject to saturation. Consequently, such models are not capable of predicting the terminal characteristics under varying operating conditions with reasonable accuracy. One can expect that accounting for saturation would go a long way toward improving the accuracy of prediction.

2. Use of adaptive mode in permanent magnet generator design

In this paper, the term "adaptive mode" denotes the joint use of field models and "complexor diagrams" for the purpose of establishing the viability of a certain design topology to satisfy the design specification. Figure 1 depicts a simple per phase impedance model for a polyphase permanent magnet generator with surface mounted prismatic magnets on the rotor [1]. This model forms the basis of the complexor diagram, which combines the phasor representation of electrical quantities with the vector representation of magnetic quantities. The model parameters R_{ph}, X_ℓ and X_{as} in Fig. 1 represent the phase resistance, leakage reactance and the armature reaction reactance for the stator, respectively. All these parameters can be determined from design calculations [2]. In this case, both the generated EMF, $\dot{E}_g$, and X_{as} are depicted as being variable to represent the effect of core nonlinearity. The terminal voltage equation for each phase can be obtained on the basis of Fig. 1 as:

$$\dot{E}_{ag} = \dot{E}_g - j\dot{I}_{ph}X_{as} = \dot{V}_t + \dot{I}_{ph}R_{ph} + j\dot{I}_{ph}X_\ell \tag{1}$$

*Corresponding author: Özdemir Göl, Tel.: +61 8 8302 3285; Fax: +61 8 8302 3384; E-mail: o.gol@unisa.edu.au.

1383-5416/01/02/$8.00

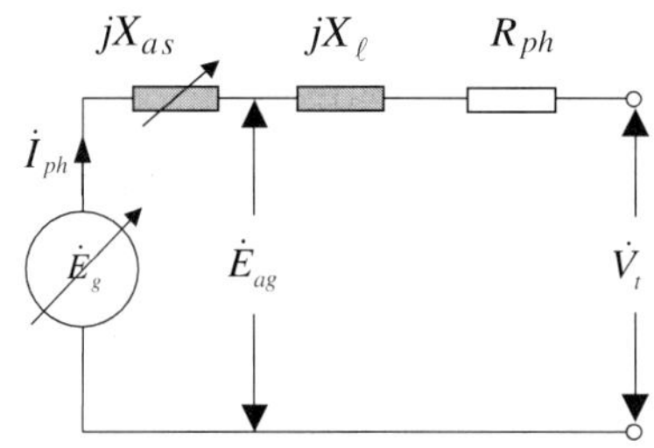

Fig. 1. Impedance model of a permanent magnet generator.

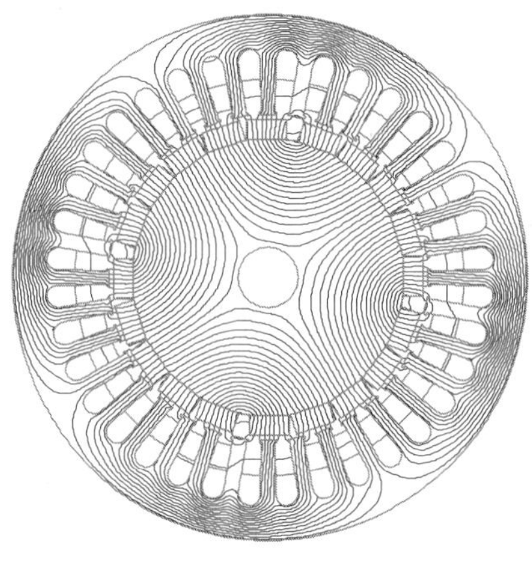

Fig. 2. Field distribution at rated load.

which can be represented as a phasor diagram. It needs to be noted that the term $\dot{E}_{ag}$ in Eq. (1) constitutes the so-called air gap EMF, the variation of which reflects the variation of the composite air gap flux with varying load as a result of the nonlinear nature of the ferromagnetic core. The knowledge of the air gap EMF, $\dot{E}_{ag}$, facilitates the prediction of steady-state performance characteristics by allowing the model parameters to be determined for varying operating conditions. The magnitude of the air gap flux, needed to quantify $\dot{E}_{ag}$, can be obtained by coupling the simple impedance model of Fig. 1 to a field model representing the flux distribution for a given load, as the one illustrated in Fig. 2.

3. Morphology of the adaptive mode

The adoption of the adaptive mode for performance prediction implies the sequence of actions described below:

Step 1: Identifying the direct and quadrature axes of the rotor reference frame

The orientation of the direct and quadrature axes can be established by an inspection of the actual rotor geometry – as in the case of surface mounted magnet structures – where the axis alignment is obvious. When this is not the case, as for embedded magnet topologies, the axis positions can be ascertained by performing a field analysis for the no-load operating condition of the generator as illustrated in Fig. 3 for a four-pole generator. Evidently, in this case, the permanent magnet excitation is the sole contributor to the available air gap field, since the armature windings carry no current.

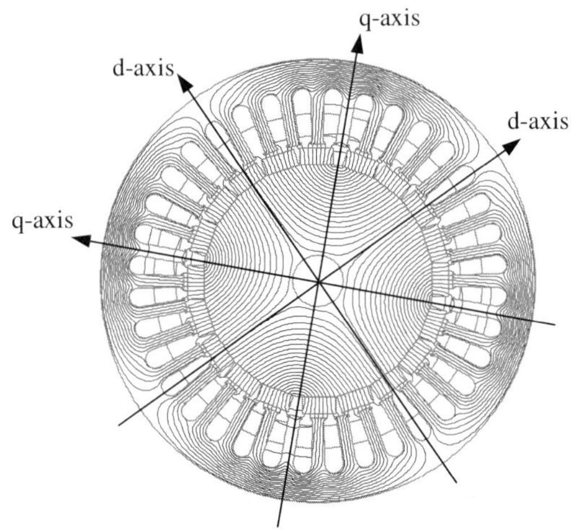

Fig. 3. Field distribution at no load, determining d–q axes.

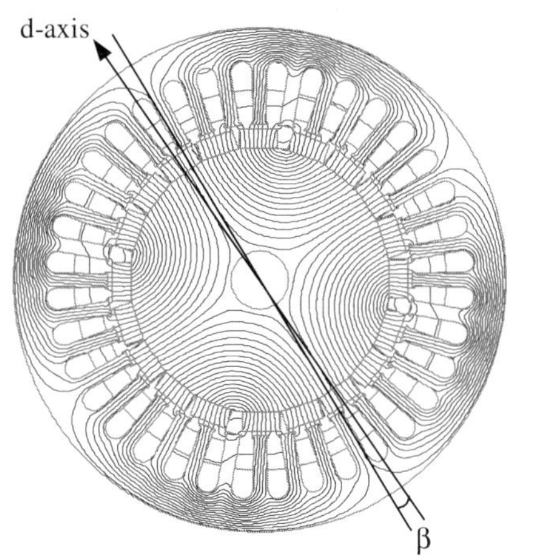

Fig. 4. Field distribution at rated load showing phase angle of air gap EMF relative to d-axis.

Step 2: Determining the air gap EMF

This step involves determining the air gap EMF, $\dot{E}_{ag}$, for a given load condition. The mid-air-gap field distribution within the machine space, corresponding to a given load current, is used as the medium to obtain a representative quantity for the air gap flux. This is done by performing a Fourier series analysis on the mid-air-gap radial flux density distribution [3–5]. The fundamental component of the air gap flux density, B_{ag}, is used to obtain $\dot{E}_{ag}$: the knowledge of B_{ag}, hence flux $\tilde{\phi}_{ag}$, enables the magnitude of $\dot{E}_{ag}$ and its phase angle relative to the direct or the quadrature axis to be determined, as alluded to in Figs 4 and 5. In this case it is arguably adequate to use the fundamental component only: the adoption of the complexor diagram as the main performance analysis tool predisposes the use of the fundamental components of the nonsinusoidal electromagnetic quantities.

Step 3: Determining phase angle of the phase current

The armature current, $\dot{I}_{ph}$, is responsible for setting up $\tilde{\mathcal{F}}_a$, the armature magnetotomotive force (MMF). One obvious way of finding information about $\tilde{\mathcal{F}}_a$ is to conduct a field analysis with the condition that the

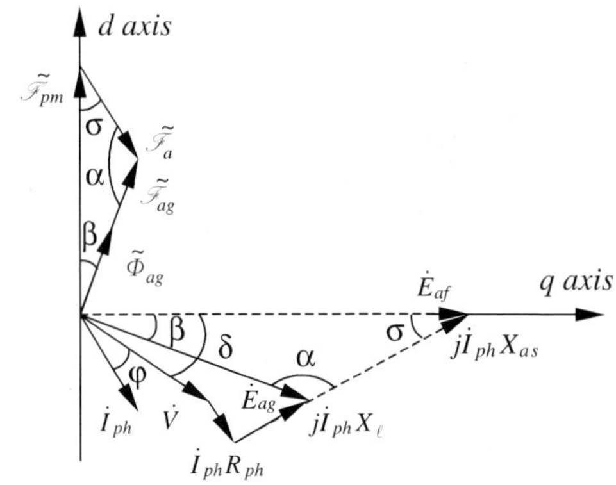

Fig. 5. Complexor diagram.

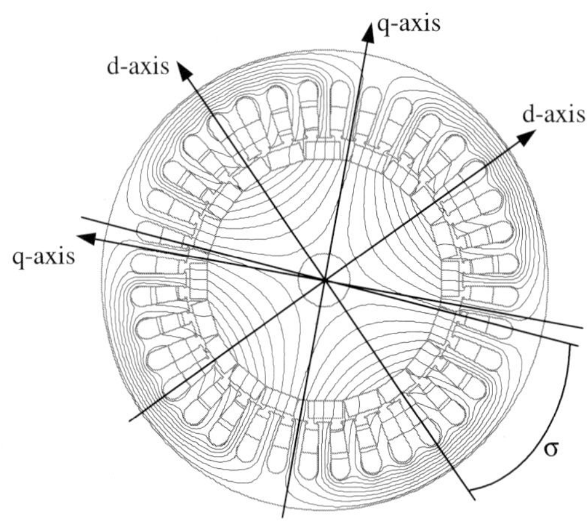

Fig. 6. Field distribution at rated load and deenergised permanent magnets.

armature winding is solely responsible for setting up the air gap field without any contribution from the rotor magnets. In computational terms this means passing a given current through the armature winding with the magnets "turned off" to obtain the field distribution. Again, a Fourier analysis is performed to find the mid-air-gap magnetic flux density from which the phase angle of $\tilde{\mathcal{F}}_a$, and hence that of i_{ph}, relative to either the q-axis or the d-axis can be found, as delineated in Fig. 6. It is worth to mention that it is not essential to complete the MMF vector triangle in the complexor diagram of Fig. 5 unless so desired, since knowing about the alignment of $\dot{I}_{ph}$ relative to the axes is sufficient to proceed with the rest of the computations. If it desired to complete the MMF vector triangle, this can be done further down the track after the voltage phasors have been established as the magnitude of $\tilde{\mathcal{F}}_{pm}$ can not be known independedly of the load condition, and hence of the corresponding terminal voltage. This is in sharp contrast with the case of a conventional synchronous generator with field windings, where the magnitude of the main field is controllable by means of the excitation current, which allows the corresponding field MMF to be uniquely determined by adjusting the field current. This, however, does not apply in the case of magnet excitation, where the magnet MMF is not controllable and varies depending on the so-called working point on the permanent magnet demagnetisation characteristic, dictated by the load condition.

Table 1
Comparison of predicted and measured results (NS: no saturation, ENS: % error with no saturation; S: saturation, ES: % error with saturation, VR: voltage regulation)

	Phase A					Phase B				
	Measured	Predicted				Measured	Predicted			
		NS	ENS	S	ES		NS	ENS	S	ES
No-load voltage, V	149.1	155	3.8	147.4	1.7	139.4	146.5	4.8	142.6	2.2
Full load voltage, V	121	127	4.7	119	1.7	114.1	120	4.9	112.1	1.8
Short circuit current, A	4	4	0.0	4	0.0	3.8	4	5.0	3.9	2.5
VR at full load, %	18.8	18	4.3	19.3	2.6	18.1	18	0.6	21.3	1.5

Step 4: Constructing the complexor diagram

On the basis of the foregoing steps, a complexor diagram can now be contructed so as to find the terminal voltage for the load condition under consideration. Determining the terminal behaviour – that is, the variation of terminal voltage with the load current drawn from the generator – is the object of performance analysis in this case.

In order to find $\dot{V}_t$, voltage drops across the relevant model components of the circuit model of Fig. 1 need to be obtained. Figure 5 alludes to the method of obtaining the terminal voltage $\dot{V}_t$ on the basis of phasor relationships.

A line coinciding with the phasor $j\dot{I}_{ph}X_\ell$, extended to intersect with the q-axis, enables the values of X_{as} and $\dot{E}_g$ to be found if desired. The dashed lines denote the nonlinear nature of both the $\dot{E}_g$ and the X_{as}. It must be emphasised that the knowledge of neither is significant in terms of the object of determining the terminal behaviour of the generator.

Applying the above described steps once does not necessarily yield the specified power factor, since there is no way of knowing from the outset what the angle between $\dot{V}_t$ and $\dot{I}_{ph}$ will be. To ensure that the phase angle between those two phasors corresponds to the specified power factor, the steps will have to be repeated; each time either changing the injected current [3] or the rotor position in the field model. Evidently the entire procedure has to be repeated if a new terminal condition with a different load current and power factor is to be analysed.

4. Verification

The performance analysis for a two-phase four-pole permanent magnet generator with surface mounted rare earth magnets in the rotor, the topology of which is implicit in Fig. 2, substantiates the validity of the solicited use of the adaptive mode. Figure 7 presents the predictions of the external characteristic: first based on the straight use of the model of Fig. 1 without saturation, then with saturation in adaptive mode. The measured results are superimposed to facilitate comparison. It is evident that the use of the adaptive mode, in the course of which the appropriate model parameters are manipulated so as to adapt the model to the varying terminal conditions, results in better prediction accuracy. Table 1, which gives the values for important points of operation, further emphasises that error margins are impressively reduced when the adaptive mode is employed.

5. Conclusion

Saturation affects the performance of a permanent magnet generator decisively and need to be taken into account during design for better prediction accuracy. It has been shown that it is possible to do that

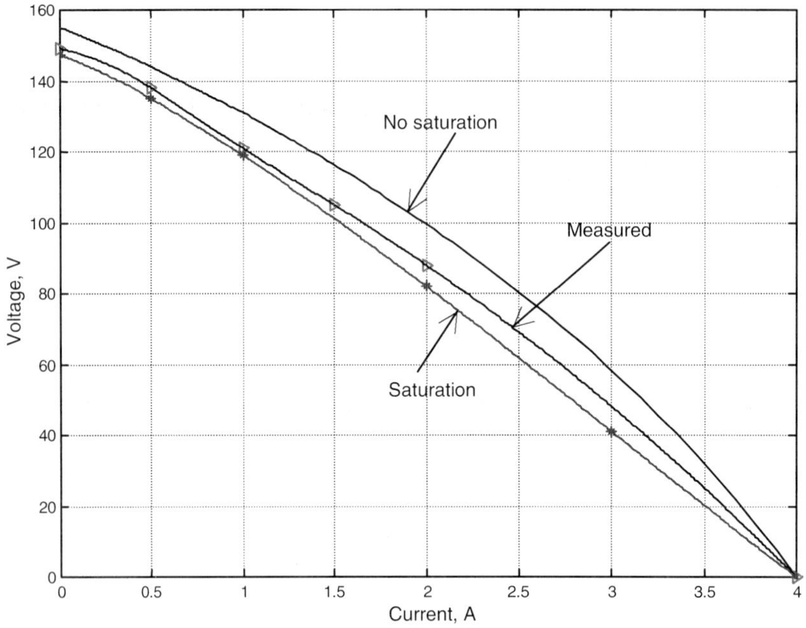

Fig. 7. External characteristic for Phase A.

even when a relatively simple impedance model is used for steady state performance analysis, provided the model is employed in an adaptive mode. The approach convincingly demonstrates that the prediction accuracy can be considerably improved whilst retaining model simplicity.

References

[1] Ö. Göl, B. Sobhi-Najafabadi and A.M. Parker, A Rare Earth Permanent Magnet Generator of Novel Design, *Proceedings of International Conference on Electrical Machines, ICEM92,* Manchester, UK, 1992, pp. 809–813.

[2] B. Sobhi-Najafabadi, An Enhanced Methodology for Permanent Magnet Generator Design, M.Eng. (Research) Thesis, University of South Australia, 1997.

[3] A.A. Arkadan, Computer-Aided Dynamic Performance Prediction of Permanent Magnet Generator Systems with Damping Circuits and Electronically Switched Loads, PhD Thesis, Clarkson University, USA, 1988.

[4] N.A. Demerdash and H.B. Hamilton, A simplified Approach to Determination of Saturated Synchronous Reactances of Large Tubogenerators Under Load, *IEEE Transactions on Power Apparatus and Systems* **PAS-95**(2) (1976), 560–569.

[5] N.A. Demerdash, T.W. Nehl, F.A. Fouad and A.A. Arkadan, Analysis of The Magnetic Field in Rotating Armature Electronically Commutated DC Machines By Finite Elements, *IEEE Transactions on Power Apparatus and Systems* **PAS-103**(7) (1984), 1829–1836.

International Journal of Applied Electromagnetics and Mechanics 13 (2001/2002) 405–410
IOS Press

Minimizing winding losses under trapezoidal current waveforms

M. Vasiliu*, A. Moraru and F. Hantila
Polytechnic University of Bucharest, Spl.Independentei 313, RO77206, Bucharest, Romania

Abstract. Time-periodic non-sinusoidal solutions of the magnetic diffusion equation are expressible either in the form of time-harmonic series or as space eigenfunction series. Based on the latter, eddy-current losses in a multilayer winding carrying a trapezoidal current are evaluated. Optimum layer thickness is obtained by minimizing the resistance factor of the multilayer winding. Formulae for square and symmetrical triangular waveforms are also included.

1. Introduction

It is well known that the ac resistance of multilayer transformer windings can be optimized such that eddy-current losses in the windings are minimized. This minimization concerns finding the optimum layer thickness. Optimum thickness of the layer or foil, for sinusoidal currents, is well documented. Recently, minimizing ac losses due to non-sinusoidal current waveforms, has gained a renewed interest in connection with transformers [1] and inductors [2] that are operated at high frequencies in order to reduce their size. Due to the linearity of the winding conducting materials, the classical method of harmonic analysis is applicable. The current waveform is decomposed into its Fourier components; time harmonic losses for each component are calculated, then summed up, to give the non-sinusoidal winding losses. Loss evaluation in the time-harmonic case is based on the one dimensional field model and the result is encapsulated in the Dowel's formula that is thus extensively used for both sinusoidal and non-sinusoidal waveforms [3,5,6]. Fourier time-series derived from this classical approach based on spectral decomposition of the non-sinusoidal waveforms is thus regarded as the natural approach to non-sinusoidal analysis. There is, however, an alternative approach, based on eigenmodes and eigenfunctions of the field configuration [7]. In this 1D case, the field problem reduces to a Sturm-Liouville boundary value problem with mixed boundary conditions. Field solution is expressible in term of trigonometric eigenfunctions. Losses can then be evaluated by time-averaging the instantaneous flux of the Poynting vector. The classical approach implies slow convergent series particularly for discontinuous waveforms. Our approach based on space eigenfunction series of the field problem, leads to series with accelerated convergence. In our approach piecewise continuous waveforms are rather preferable than avoidable.

*Corresponding author: Mihai Vasiliu, Tel.: +40 1 4111217; Fax: +40 1 4111217; E-mail: amv@elth.pub.ro.

1383-5416/01/02/$8.00

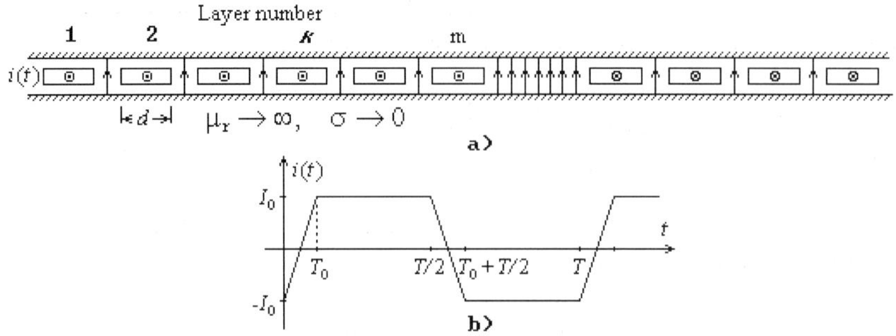

Fig. 1. a) Planar m-layer inductor; b) trapezoidal current waveform.

2. Time periodic non-sinusoidal magnetic field in a layer

To illustrate the method we consider a planar inductor sandwiched between two ferromagnetic layers [7]. Consider the k-layer, $k = 1, 2, \ldots, m$ of an m-layer winding of the inductor (Fig. 1(a)). We assume that all winding layers have the same size $d \times c$, carry the same time-periodic non-sinusoidal current $i_{\text{per}}(t)$ and are characterized by constant electrical conductivity σ and magnetic permeability μ. As in [8], if a layer consists of several conductors, it is regarded as a single foil conductor of reduced conductivity.

Under one-dimensional assumptions, magnetic diffusion in the k-layer is described by the equation

$$\mu\sigma \frac{\partial h}{\partial t} - \frac{\partial^2 h}{\partial x^2} = 0,\ 0 < x < d,\ t > 0 \tag{1}$$

the inhomogeneous boundary conditions

$$h(0,t) = \frac{1}{c}(k-1)i_{\text{per}}(t),\ h(d,t) = \frac{1}{c}k i_{\text{per}}(t) \tag{2}$$

and the condition of periodicity; $h(x,t) = h(x,t+T)$. Here d is thickness of the layer. The current waveform is trapezoidal (Fig. 1(b)). Due to the odd symmetry, $i_{\text{per}}(t) = -i_{\text{per}}(t-T/2)$, only restrictions of the time-periodic functions to one half-period, say the first one, $[0, T/2]$, have to be specified. For instance, the restriction of iper(t) to the first half period is given by

$$\frac{i(t)}{I_0} = \begin{cases} 2t/T - 1, 0 \leqslant t \leqslant T_0, \\ \quad 1 \qquad T_0 \leqslant t \leqslant T/2. \end{cases} \tag{3}$$

I_0 is the peak value of the trapezoidal current, $T_0 \in (0, T/2)$ is the duration of the linear variation between peak and valley values, as shown in Fig. 1(b). Due to current odd symmetry, the magnetic field satisfies:

$$h(x,t) = -h(x,t+T/2),\ 0 < x < d. \tag{4}$$

The solution of Eq. (1) subject to Eqs (2) and (3), can be expressed as a two-component function:

$$h(x,t) = h_1(x,t) + (k-1)h_2(x,t),\ 0 \leqslant x \leqslant d,\ k = 1, 2, \ldots, m. \tag{5}$$

The first component, h_1, corresponds to the magnetic diffusion into the first layer. The basic process in this layer is the skin effect. This component can be expressed as

$$\frac{h_1(x,t)}{H_0} = \begin{cases} \left(2\frac{t}{T_0} - 1\right)\frac{x}{d} + \frac{\tau}{T_0}\left[\frac{x}{3d}\left(\frac{x^2}{d^2} - 1\right) + M_1(x,t)\right], 0 \leqslant t \leqslant T_0, \\ \frac{x}{d} - \frac{\tau}{T_0}M_3(x, t - T_0), \qquad T_0 \leqslant t \leqslant T/2, \end{cases} \tag{6}$$

where $H_0 = I_0/c$. The second component, h_2, corresponds to the magnetic diffusion in the superior layers, by both skin and eddy-current effects. It is given by

$$\frac{h_2(x,t)}{H_0} = \begin{cases} 2\frac{t}{T_0} - 1 + \frac{\tau}{T_0}\left[\frac{x}{d}\left(\frac{x}{d} - 1\right) + M_2(x,t)\right], 0 \leqslant t \leqslant T_0, \\ 1 - \frac{\tau}{T_0}M_4(x, t - T_0), \qquad T_0 \leqslant t \leqslant T/2. \end{cases} \tag{7}$$

The M – functions in Eqs (6) and (7) have the following expressions:

$$M_1(x,t) = \frac{4}{\pi^2}\sum_{k=1}^{\infty}\frac{(-1)^{k+1}}{k^3}C_k^I \exp(-\lambda_k t)\sin\left(k\pi\frac{x}{d}\right), \tag{8}$$

$$M_2(x,t) = \frac{8}{\pi^2}\sum_{k=1}^{\infty}\frac{1}{(2k-1)^3}C_{2k-1}^I \exp(-\lambda_{2k-1} t)\sin\left((2k-1)\pi\frac{x}{d}\right), \tag{9}$$

and

$$M_3(x,t) = \frac{4}{\pi^2}\sum_{k=1}^{\infty}\frac{(-1)^{k+1}}{k^3}C_k^{II} \exp(-\lambda_k t)\sin\left(k\pi\frac{x}{d}\right), \tag{10}$$

$$M_4(x,t) = \frac{8}{\pi^2}\sum_{k=1}^{\infty}\frac{1}{(2k-1)^3}C_{2k-1}^{II} \exp(-\lambda_{2k-1} t)\sin\left((2k-1)\pi\frac{x}{d}\right), \tag{11}$$

where

$$C_k^I = \frac{1 + \exp(-\lambda_k(t/2 - T_0))}{1 + \exp(-\lambda_k T/2)}, \; C_k^{II} = \frac{1 - \exp(-\lambda_k T_0)}{1 + \exp(-\lambda_k T/2)}. \tag{12}$$

Here

$$\lambda_k = \frac{1}{\mu\sigma}\left(k\frac{\pi}{d}\right)^2 = \frac{k^2\pi^2}{\tau}, \; k = 1, 2, \ldots \tag{13}$$

are the eigenvalues of the field problem, of order k. The penetration time constant τ is given by $\tau = \mu\sigma d^2$. The C – coefficients in Eqs (9) and (11) are to be obtained from Eq. (12), for odd values of the index.

The validity of the solution Eqs (5)–(7) can be easily verified by means of the uniqueness theorem. Indeed, this solution satisfies the magnetic diffusion Eq. (1), boundary conditions Eq. (2) and the condition

$$h(x,t) = -h(x, T/2), \; 0 \leqslant x \leqslant d \tag{14}$$

The last condition is fulfilled because the C – coefficients satisfy the following relations:

$$C_k^I \exp(-\lambda_k T_0) + C_k^{II} = 1, \; C_k^I - C_k^{II}\exp(-\lambda_k(T/2 - T_0)) = 1. \tag{15}$$

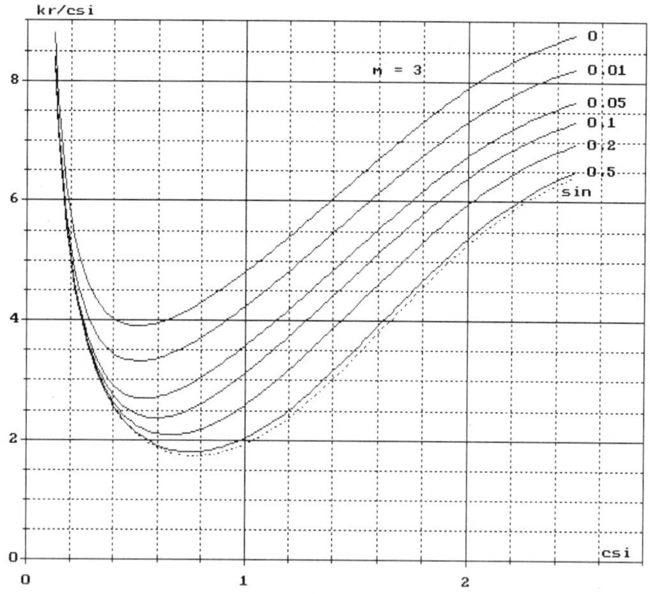

Fig. 2. Resistance increase function K_R/ξ versus relative thickness ξ.

3. AC resistance for trapezoidal waveforms

To simplify loss formula description we introduce several notations. The ratio rise time to the half-period of the symmetrical trapezoidal current is denoted by γ, i.e. $\gamma = 2\,T_0/T$; the skin depth at the fundamental frequency $f = 1/T$ is denoted by δ; and the ratio thickness of the layer to the skin depth is $\xi = d/\delta$. The penetration time constant into the layer is given by $\tau = \mu\sigma d^2$, and its ratio to the half-period is denoted by ν, thus $\nu = 2f\pi$. Clearly: $\xi^2 = \pi\upsilon/2$. The resistance factor, defined as the ratio the trapezoidal ac resistance to the dc resistance, of the winding, is denoted by K_R. Field solutions allow us the evaluation of the power influx by Poynting vector in the k-layer. By time averaging we obtain losses in the k-layer. Finally by averaging for all m layers, we obtain

$$K_R = K_1 + (m^2 - 1)K_2, \tag{16}$$

where

$$K_1 = 1 + \frac{1}{3-2\gamma}\frac{\upsilon^2}{\gamma}\frac{4}{15}\left[1 - \frac{2}{21}\frac{\upsilon}{\gamma}(1 - S(\upsilon) + S_1(\upsilon,\gamma))\right]. \tag{17}$$

$$K_2 = \frac{1}{3-2\gamma}\frac{\upsilon^2}{3\gamma}\left[1 - \frac{1}{10}\frac{\upsilon}{\gamma}(1 - T(\upsilon) + T_1(\upsilon,\gamma))\right]. \tag{18}$$

The resistance factor involves the sums

$$S(\upsilon) = \frac{1890}{\pi^6}\sum_{n=1}^{\infty}\frac{1}{n^6}\frac{1}{1+\exp(n^2\pi^2/\upsilon)} \tag{19}$$

$$S_1(\upsilon,\gamma) = \frac{945}{\pi^6}\sum_{n=1}^{\infty}\frac{1}{n^6}\frac{\exp((\gamma-1)n^2\pi^2/\upsilon) - \exp(-\gamma n^2\pi^2/\upsilon)}{1+\exp(n^2\pi^2/\upsilon)} \tag{20}$$

and the sums T and T_1 that occur in Eq. (3) are similar to S and S_1 in Eq. (2) but for factors 1920 and 960 instead of 1890 and 945, respectively; the summation being carried out for odd values of the index. Recall that the sums are functions of γ, the relative rise time of the trapezoidal waveform, and the relative penetration time into the layer ν. As in the sinusoidal case, in order to obtain the optimum layer thickness, it is required to minimize the function K_R/ξ versus ξ, the relative layer thickness. The rationale for using this function follows from the fact that losses increase with both dc resistance and resistance increase factor. Since the dc resistance decreases with the thickness of the layers, the function to be minimized implies the relative thickness at the denominator. Plot of K_R/ξ is shown in Fig. 1 for a three layers inductor and different rise times. The sums in K_R show a rapid convergence in the domain of interest, i.e. in the vicinity of the optimum relative thickness ξ.

4. Square wave and triangular waveforms

The resistance factor for square waveforms, which are encountered in full bridge converters and triangular waveforms, which occur in filter chokes, can be obtained from the trapezoidal waveforms. For square waveforms we get

$$K_{1|_{\gamma=0}} = 1 + \frac{2}{3}\upsilon[1 - S_0(\upsilon)],\ K_{2|_{\gamma=0}} = \frac{2}{3}\upsilon[1 - T_0(\upsilon)], \tag{21}$$

where

$$S_0(\upsilon) = \frac{12}{\pi^2}\sum_{n=1}^{\infty}\frac{1}{n^2}\frac{1}{1+\exp(n^2\pi^2/\upsilon)} \tag{22}$$

For triangular waveforms we have

$$K_{1|_{\gamma=1}} = 1 + \frac{4}{15}\upsilon^2\left[1 - \frac{4}{21}\upsilon(1 - S(\upsilon))\right],\ K_{2|_{\gamma=1}} = \frac{1}{3}\upsilon^2\left[1 - \frac{1}{5}\upsilon(1 - T(\upsilon))\right]. \tag{23}$$

5. Conclusions

The mode theory leads to a novel expression of the resistance factor for the m-layer winding carrying currents of trapezoidal, square or triangular waveforms. The new series are equivalent to those obtained by Fourier decomposition method; however, they show an improved convergence in the range of optimum layer thickness.

References

[1] W.G. Hurley, E. Gath and J.G. Breslin, Optimizing the AC resistance of multilayer transformer windings with arbitrary current waveforms, *IEEE Trans. Power Electronics* **15** (March 2000), 369–375.

[2] K.W.E. Cheng, Computation of the AC resistance of multistranded conductor inductors with multilayers for high frequency switching converters, *IEEE Trans. Magnetics* **36** (July 2000), 831–834.

[3] B.S. Ram, Loss and current distribution in foil windings of transformers, *IEE Proc.C, Generation, Transmission and Distribution* **145** (Nov. 1998), 709–716.
[4] N.H. Kutkut, A simple technique to evaluate winding losses including two-dimensional edge effects, *IEEE Trans. Power Electronics* **13** (Sep. 1998), 950–958.
[5] P.D. Evans and W.J.B. Heffernan, Transformer for multimegahertz power applications, *IEE Proc. Electric Power Applications* **142** (Nov. 1995), 379–389.
[6] A. Semlyen and F. de Leon, Eddy current add-on for frequency dependent representation of winding losses in transformer models used in computing electromagnetic transients, *IEE Proc. Generation, Transmission and Distribution* **141** (May 1994), 209–214.
[7] A. Moraru and M. Vasiliu, Field distribution in a planar winding carrying a non-sinusoidal current, *Rev. Roum. Sci. Techn. – Electrotechn. Energ.* **38** (Oct.–Dec. 1993), 519–528.
[8] L.L. Dowell, Effects of eddy-currents in transformer windings, *Proc. IEE* **113** (Aug. 1966), 8.

International Journal of Applied Electromagnetics and Mechanics 13 (2001/2002) 411–419
IOS Press

Magnetic loss measurement and calculation of a DC/DC converter

Hans Hauser*, Felix A. Himmelstoss and Peter H. Wurm
Vienna University of Technology, Faculty of Electrical Engineering and Information Technology, Gusshausstr. 27-29, A-1040 Vienna, Austria

Abstract. The use of a tapped inductor in the classical sepic converter leads to a high step-down ratio thus avoiding the extreme duty cycles of the active switch and the high peak currents in the active and passive switches. The function and some stress data of such a converter are summarized and the flux density and the current through the inductor during start-up are measured and simulated by the energetic model of ferromagnetic hysteresis and the magnetic losses of the inductors are calculated.

1. Introduction

The traditional converter topology of a sepic converter is extended by an additional inductor (in series with the capacitor C_1) which is magnetically coupled with the original inductor. Due to the magnetic coupling, both windings are on the same core and therefore no additional magnetical part is necessary. Figure 1 shows the principal circuit.

The converter is useful as step-down converter. In [1] boost converters with high step-up ratios are treated, in [1,3] tapped inductor converters in continuous inductor current mode with fixed switching frequency are proposed. In [4] the buck and the buck-boost converter with autotransformer are treated.

A combination of frequency control and duty cycle control is used to operate the converter at the border between continuous and discontinuous mode. The control law of the converter is still simple and there are no additional losses due to the reverse recovery of the free wheeling diode.

The loss mechanisms are ohmic losses in all parts, switching losses (during turn on and turn off) in the switching devices, and magnetic loss in the inductor. In this paper especially the magnetic losses are treated.

2. Circuit description

Subsequently the function of the buck-boost converter with autotransformer according to Fig. 1. is described and some important results are summarized.

Figure 2 shows the current through L_{22} when changing from the continuous to the discontinuous mode of operation. At time t_0 the active switch is turned on and the current rate of rise is determined by the

*Corresponding author: Hans Hauser, E-mail: hans.hauser@tuwien.ac.at.

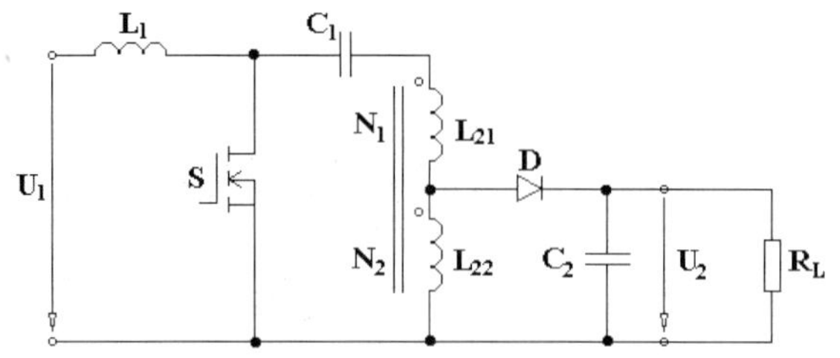

Fig. 1. Principal circuit of the tapped inductor sepic converter.

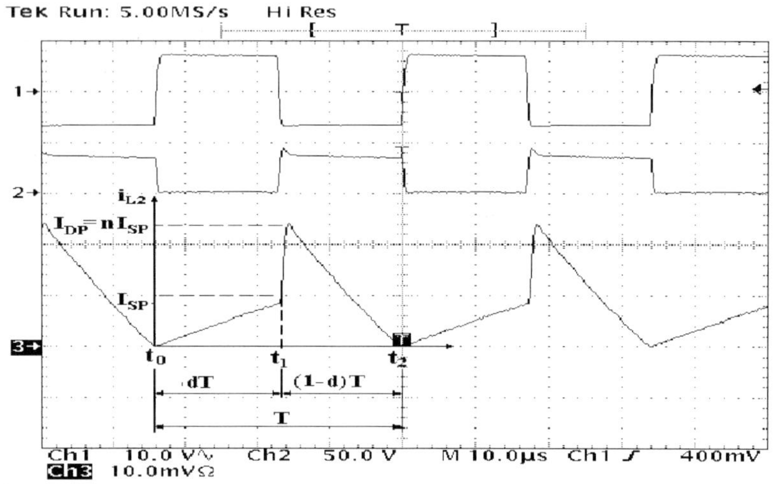

Fig. 2. Ch1: gate-source voltage, Ch2: drain-source voltage, Ch3: current through L_{22} (1A/div).

value of L_{gen} (the complete inductance of the tapped inductor) and the input voltage. At time t_1 the active switch is turned off and the current commutates to the diode branch. The magnetic flux in the core being steady, the value of the inductor is now determined by N_2 and therefore the current is jumping according to about

$$I_{DP} = \frac{N_1 + N_2}{N_2} \cdot I_{SP} \tag{1}$$

where I_{DP} stands for the diode peak current and I_{SP} for the switch peak current. The current decreases according to the output voltage and the inductor value L_{22}

$$L_{gen} = \left(\frac{N_1 + N_2}{N_2}\right)^2 \cdot L_{22} = n^2 \cdot L_{22}. \tag{2}$$

The energy stored in L_{gen} at time t_1 is completely transferred to the output at t_2 (if the converter is in the discontinuous mode of operation). For the continuous mode it is easy to calculate the voltage transfer ratio m according to

$$m = \frac{U_2}{U_1} = \frac{d}{(1-d)} \cdot \frac{1}{n} \tag{3}$$

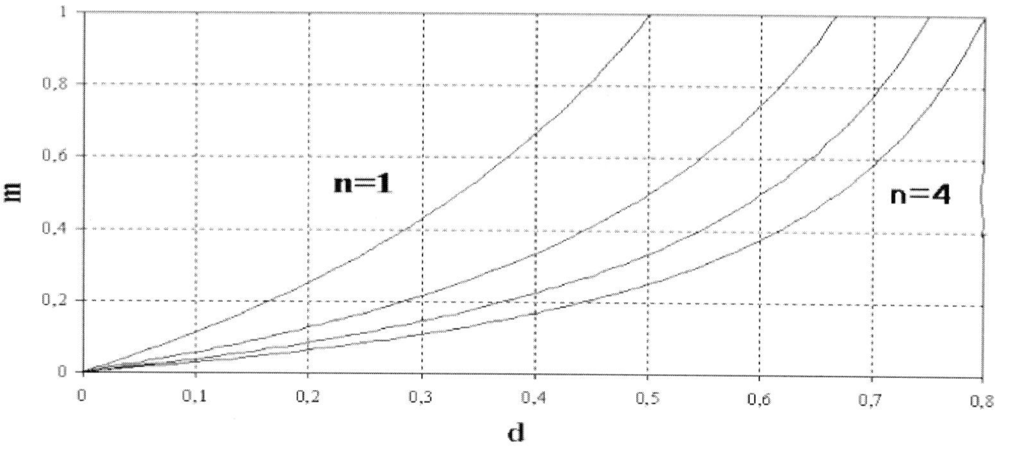

Fig. 3. Voltage transfer ratio m over the duty cycle d (step-down mode).

with the winding ratio

$$n = \frac{N_1 + N_2}{N_2}. \tag{4}$$

By setting $n = 1$ one gets the well-known transfer ratio of the classical sepic converter. The analysis of the continuous mode (including the border between continuous and discontinuous mode) is sufficient for dimensioning the converter even when used in the discontinuous mode under full load condition. In the discontinuous mode the converter is optimally utilized when working near the continuous mode. Figure 3 shows the dependence of the output to input voltage ratio m according to the duty cycle d for different winding ratios. One can see that for the same duty cycle much lower voltage transformation rates can be obtained. This special converter allows 50% duty ratio for a high step-down ratio leading to a much better utilization of the semiconductor devices.

It is also possible to use the circuit as a step-up converter. However, as can be seen in Eq. (3), high duty cycles of the active switch are necessary. Therefore, the sepic converter with autotransformer is especially useful for step-down mode and only small step-up ratios.

The voltage stress of the active switch U_S and the diode U_D normalized to the input voltage is given by

$$\frac{U_S}{U_1} = 1 + m \cdot n \tag{5}$$

and

$$\frac{U_D}{U_1} = -\left(\frac{1}{n} + m\right). \tag{6}$$

The equations show an increase of the voltage stress of the active switch and a decrease of the voltage stress of the diode when the winding ratio is increased. The average currents of the diode $\overline{I}_D$ and the switch $\overline{I}_S$ normalized to the load current are

$$\frac{\overline{I}_D}{I_{Load}} = 1 \tag{7}$$

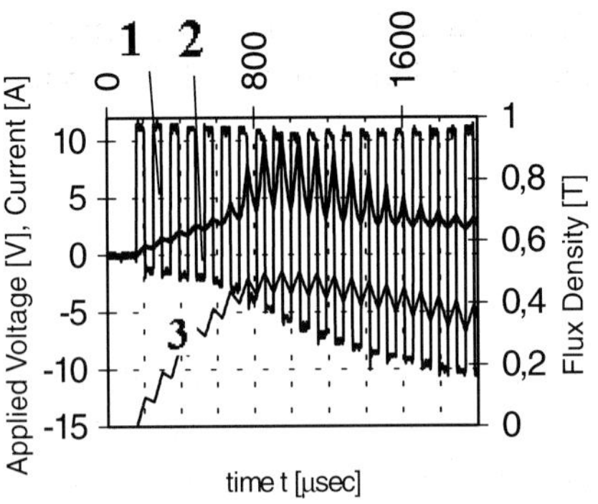

Fig. 4. Start-up, voltage (1) across and current (2) through the inductor (also shown in Fig. 6), calculated flux density (3).

and

$$\frac{\overline{I}_S}{I_{Load}} = \frac{m \cdot (1 + m \cdot n)}{n \cdot (1 + m)}. \tag{8}$$

For cooling design the root mean square values are necessary. For the switch the normalized rms current results in

$$\frac{I_{Srms}}{I_{Load}} = \sqrt{\frac{m \cdot (1 + m \cdot n)^3}{3 \cdot n^3 \cdot (1 + m)^2}} \tag{9}$$

and for the diode in

$$\frac{I_{Drms}}{I_{Load}} = \sqrt{\frac{1 + m \cdot n}{3}}. \tag{10}$$

Using a synchronous rectifier instead of the diode, the losses of the semiconductors can easily be estimated. Detection of zero crossing is also used for the control to operate the converter always at the border between continuous and discontinuous mode. Therefore, no additional acquisition electronic for the synchronous rectifier is necessary. With r_S and r_D as on-resistance of the switch and of the switch parallel to the diode respectively, normalizing with $I_{load}^2 \cdot r_s$ and using a factor k_{red} connecting the on-resistances of the two switches together ($r_D = r_s \cdot k_{red}$) one gets for the losses of the semiconductor devices

$$\frac{P_L}{r_S \cdot I_{load}^2} = I_{Srms}^2 + k_{red} \cdot I_{Drms}^2. \tag{11}$$

3. Measurement

Figure 4 shows a start-up of a simple sepic converter. The nearly rectangular trace shows the voltage across the output inductor, representing the derivation of the flux in the inductor; the lower trace starting at the left-hand zero depicts the current through the inductor. The third trace (scale on the right side) shows the flux density. Here no tapping is used and the converter starts without frequency control leading to a short period of continuous mode of operation and saturation. Due to saturation of the core the velocity of current rise (and fall) increases (decreases) significantly. The current was measured with a current probe and the flux density was calculated out of the voltage across to avoid the drift of the integrator output voltage.

4. Magnetic model

The energetic model of ferromagnetic hysteresis [6] calculates the magnetic state of ferromagnetic materials by minimizing the total energy function w_T, the sum of the energy of the magnetic field w_H and the material energy w_M. Considering isotropic magnetization only [7], the domains are assigned to two classes – either with a component parallel or antiparallel to the applied filed, respectively. The latter term is divided into the energy of demagnetizing fields w_D, and into contributions described by statistical domain behaviour: The reversible energy w_R and the irreversible energy w_I. The statistical domains are characterized by their volume fraction v_i and the direction $\vec{n}_i$ of their spontaneous magnetization M_s. The total magnetization of the ferromagnetic material is:

$$\vec{M} = M_s \sum_{i=1}^{n} v_i \vec{n}_i \tag{12}$$

with

$$\sum_{i=1}^{n} v_i = 1. \tag{13}$$

In general, there are n easy directions, mainly due to the magnetocrystalline and strain anisotropy for anisotropic materials, but only $n = 2$ easy directions for isotropic materials (the directions parallel and antiparallel to the effective field). Stable points of the magnetization curve are points with minimum total energy, i.e. a state of vanishing total variation

$$dw_T = \sum_{i=1}^{n} \left(\frac{\partial w_T}{\partial v_i} dv_i + \frac{\partial w_T}{\partial \vec{n}_i} d\vec{n}_i \right) = 0. \tag{14}$$

As there exist, per definition, no macroscopic easy directions in isotropic materials, the second term in the equation above is always negligible. Therefore, only the derivatives of the energy terms with respect to the volume fraction v_1 have to be considered [8].

The reversible energy represents the elastic interactions of the domain walls with grain boundaries, microscopic strains and internal stray fields or rotation of domain magnetization, for example (adaptive constants g and h).

The domain walls are moving reversibly (e.g. wall bowing) by the displacements s_i until an individual Barkhausen jump starting position s_j is reached. The probability density $p_j(s_j)$ is assumed to be decreasing with increasing s_j (adaptive constant: q). The function κ describes the influence of the total magnetic state on p_j at points of magnetization reversal [9]. κ depends on the covered displacements $s_i = 2|v_1 - v_0|$ (for 180°-walls) at a point of field reversal

$$\kappa = 2 - \kappa^{old} \exp\left[-2 \frac{q}{\kappa^{old}} |v_1 - v_0| \right]. \tag{15}$$

The losses of a single domain wall during an irreversible Barkhausen jump are assumed to increase with the displacement $s_i - s_j$ (wall friction: k in J/m^3). The calculations starts with the initial magnetization curve ($v_0 = 1/2, \kappa = 1$). At a point of field reversal the new κ is calculated by the equation above and v_0 is set to the actual value of v_1 at this point. In the following parametric form of the magnetization curve the vectorial properties of the fields have been eliminated and substituted by the signum-function:

$$M(v_1) = M_s(2v_1 - 1) \tag{16}$$

and

$$\begin{aligned} H(v_1) = &\operatorname{sign}(2v_1 - 1)h\left\{\left[2v_1^{v1}(1 - v_1)^{(1-v_1)}\right]^g - 1\right\} + N_d M_s(2v_1 - 1) \\ &+\operatorname{sign}(v_1 - v_0)\frac{k}{\mu_0 M_s}\left\{1 - \kappa \exp\left[-2\frac{q}{\kappa}|v_1 - v_0|\right]\right\} \end{aligned} \tag{17}$$

The identification of the model with measurements or data sheets can be done as following. The demagnetizing factor is approximately determined by the differential susceptibility χ_c at coercivity

$$N_d \approx 1/\chi_c|_{H=H_C} \tag{18}$$

At given M_s the parameters of the magnetization curve are then directly computed by the macroscopic values (approximation: $\exp[-q] \ll 1$ of coercivity H_c, technical saturation field $H_s(M_s)$, initial susceptibility χ_0, maximum magnetization M_m, and remanence M_r (squareness $S = M_r/M_s$). The model parameters are two constants

$$g \approx \frac{M_s}{M_m - M_r} \ln \frac{2H_c}{H_c - h\left([1+S)^{1+S}(1-S)^{1-S}]^{g/2} - 1\right) + N_d M_s} \tag{19}$$

and

$$h \approx \frac{H_s - H_c - N_d M_s}{2^g - 1} \tag{20}$$

for the reversible energy contribution, and two constants for the irreversible energy loss – the probability density constant q of pinning centers and the wall friction constant k

$$q \approx \frac{M_s}{H_c}\frac{1 + N_d\chi_0}{\chi_0} \tag{21}$$

and

$$k \approx \mu_0 M_s H_c. \tag{22}$$

Alternatively to one of the values H_s, M_r, H_c and χ_0, the energy loss density w represented by the area of the major hysteresis loop can be used for identification [10]

$$q = \frac{8\mu_0 M_s H_c}{4\mu_0 M_s H_c - w}. \tag{23}$$

Using the equations above, the parameters for the MnZn ferrite core material N67 (see Fig. 5) are

$$\mu_0.M_S = 500 \text{ mT},\ N_d = 1.01 \cdot 10^{-4},\ g = 13.3,\ h = 0.652 \text{ A/m},\ q = 8.35,\ k = 17.2 \text{ J/m}^3.$$

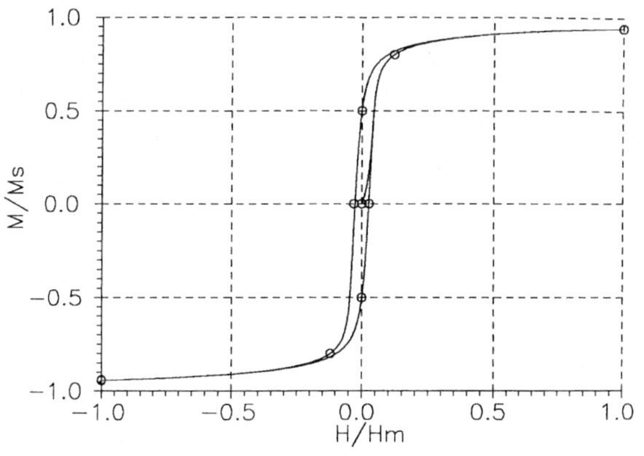

Fig. 5. Calculated (−) magnetization curve and measurements (o) of the material N67 (MnZn ferrite); maximum applied field Hm = 640 A/m.

The results agree within 3.5% with the measurements (see Fig. 5). $B(H)$ differs from $M(H)$ by $\mu_0.H$ which is negligible at weak fields. The ETD29 core volume is 5350 mm^3, corresponding to a core mass of 25.7 g. The effective area of the windings (two layers of 30 windings each) is 71 mm^2. Due to the large geometrical demagnetizing factor (1/70) of the core (with airgap), the $B(H)$ curve is almost free of hysteresis.

The applied hysteresis model is well suited for application in magnetic devices in SMPS because the input value is M (related to the voltage integral) and the output is H (related to the current). In contrast to other models, the identification is based on easily available data and the parameters have a physical meaning. They can be derived from the spontaneous magnetization M_s, magnetostriction constant λ, Young's modulus E, magnetocrystalline energy constant K_1, applied stress σ_a, and temperature T [11].

5. Simulation and calculation of the losses

The magnetic model was implemented in Microsoft Excel. The parameters of the magnetic core were taken from a magnetic databook [5]. For validation of the model the hysteresis curve without and with airgap was calculated and compared with the databook. This was the basis for the simulation model.

The voltage across the inductor is given by the switching state of the converter. This leads to a current through the inductor. The magnetization is used as variable in the magnetic calculation algorithm according Chapter 4. Therefore, the applied voltage has to be changed into a magnetization function which can be used as input for this algorithm leading to the magnetic field strength. The current through the inductor can be calculated out of the winding geometry and the magnetic field strength. The encircled area of the hysteresis curve is proportional to the loss energy and can be calculated easily.

Figure 6 compares the measured and the calculated current. In the linear part of the magnetizing curve a very good agreement is achieved. In the saturation region (which is not a normal operation state) the model shows a small displacement to the measured current waveform. During a normal start-up procedure the duty cycle will be reduced to avoid saturation of the core and the linked overcurrent. The model displacement in the saturation region is therefore of no practical importance. In Fig. 7 the summed core loss energy is depicted.

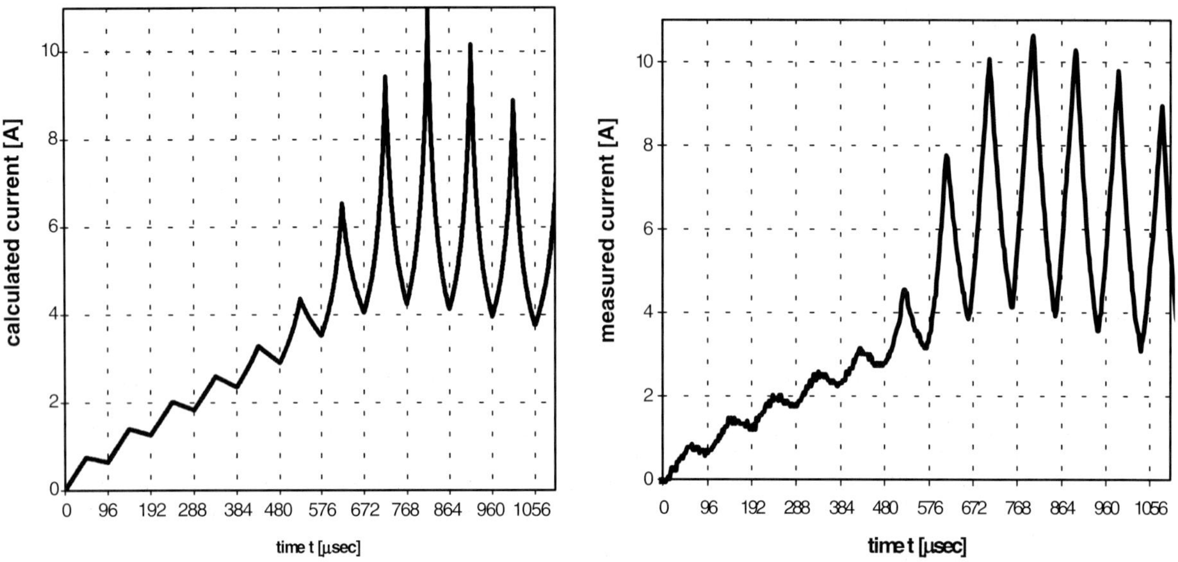

Fig. 6. Calculated (left) and measured (right) current.

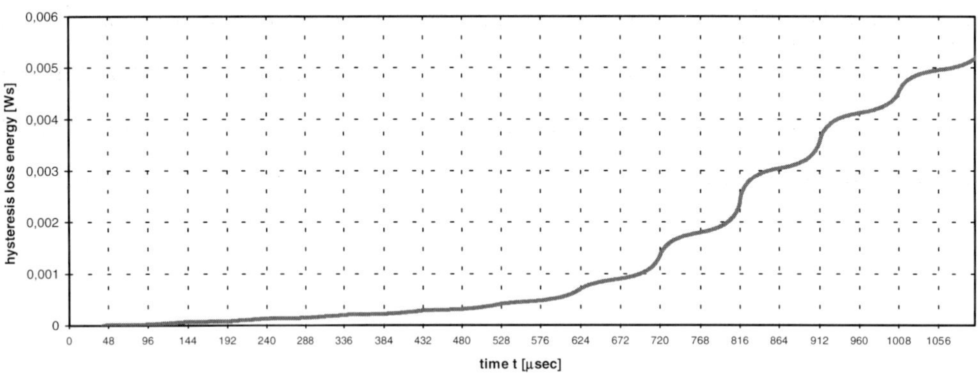

Fig. 7. Hysteresis loss energy over time.

6. Conclusions

Due to the simultaneous regulation of the switching frequency and the duty cycle in combination with the autotransformer applied to a sepic converter, a better utilization of the semiconductor devices, reduced losses and a very high step-down ratio of 1:5 can be achieved. In the steady state case the converter should work at the border between continuous and discontinuous inductor current mode. Reducing component stress of the semiconductor devices reduces their costs.

Depending on the semiconductors used and on improvements by zero voltage or zero current switching, respectively, the efficiency of the converter is between 85% and 90%. The parametric form of the energetic model allows a rapid and qualitatively accurate hysteresis loss prediction, which is about 0.7 W in our case. The ohmic losses are 1.5–2 times higher due to the proximity effect. The total loss of the inductor is finally about 3%–5% of the total loss of the converter.

Acknowledgement

The authors are very much indebted to the "Fonds zur Förderung der Wissenschaftlichen Forschung" (FWF) for the support under Grant no. P13894-TEC.

References

[1] R.D. Middlebrook, A Continuous Model for the Tapped-Inductor Boost Converter, *IEEE PESC*, 1975 Record, pp. 63.

[2] N. Mohan, T. Undeland and W. Robbins, *Power Electronics,* John Wiley & Sons, New York, 1995.

[3] J. Sebastian, J. Uceda, M.A. Perez, M. Rico and F. Aldana, A Very Simple Method to Obtain One Additional Fully Regulated Output in Zero-Current-Switched Quasiresonant Converters, *IEEE PESC*, 1990 Record, pp. 536–542.

[4] F. Himmelstoss, P. Wurm and H. Hauser, Magnetic Measurement and Modelling of a DC-DC Converter Start-Up, *Proceedings of the Conference Magnetic Measurement 2000,* MM2000, Prague, Czech Republic, September 11–12, 2000, pp. 9–11.

[5] Siemens Matsushita Components: Ferrite, B461–P6136.

[6] H. Hauser, Energetic Model of Ferromagnetic Hysteresis, *J. Appl. Phys.* **75** (1994), 2584–2597.

[7] P.L. Fulmek and H. Hauser, Statistical Hysteresis Model for Anisotropic and Isotropic Magnetic Materials, *J. Phys.* **IV 8** (1998), 335–338.

[8] H. Hauser and P.L. Fulmek, Magnetization Process and Statistical Domain Behaviour, *IOS Studies in Applied Electromagnetics and Mechanics* **11** (1998), 535–538.

[9] H. Hauser and P.L. Fulmek, Magnetization Reversal in an Energetic Hysteresis Model, *J. Magn. Magn. Mater* **160** (1996), 35–37.

[10] P. Andrei, L. Stoleriu and H. Hauser, Hysteresis Model Identification for Particulate Media, *J. Appl. Phys.* **87** (2000), 6555–6557.

[11] H. Hauser and R. Grössinger, Isotropic Hysteresis Modelling of Fe-Co-B Alloys, *J. Appl. Phys.* **85** (1999), 5133–5135.

International Journal of Applied Electromagnetics and Mechanics 13 (2001/2002) 421–426
IOS Press

Topology optimization of switched reluctance motor using mutual energy method

Jin-kyu Byun* and Song-yop Hahn
School of Electrical Engineering #051, Seoul National University, Kwanak P.O. Box 34, Seoul 151-742, Korea

Abstract. In this paper, the rotor part of the SRM (switched reluctance motor) is designed by topology optimization method. The objective function is the desired system energy profile considering the rotation angle of the rotor. The topology optimization method can design the optimal shape and topology of SRM rotor without any prior knowledge of the basic rotor layout. The density method is used to establish the relationship between the design variables and material properties, and mutual energy method is used to calculate the design sensitivity.

1. Introduction

Topology optimization is the optimal design method which uses the universal, gray-scale pixel-like representation of the device [1,2]. In the topology optimization, basic topology and layout of the device can be determined automatically during the design process, contrary to the conventional shape optimization where the basic layout of the device is determined before optimization and where only the boundary surfaces between materials are modified during the optimization. Topology optimization was researched actively in the field of structural engineering in the last decade. On the other hand, there has been relatively little work on the topology optimization in the electromagnetic system. In the mid-90's, Dyck et al. were the first to adopt the universal representation of the device for the optimal design of the magnetic systems [1]. After the papers proposed by Dyck, few works have been reported on this subject. However, in most of the papers, the applications have been limited to basic C-core design and the comprehensive formulations and applications in practical cases were not covered.

In this paper, the topology optimization method is applied to the design of the SRM (switched reluctance motor) considering the rotation angle of the rotor. The mutual energy method is used for design sensitivity analysis (DSA), and density method is used to parametrize the design space and establish the relationship between the design variables and material properties.

*Corresponding author: Jin-kyu Byun, Fax: +82 2 878 1452; E-mail: aiolia@dreamwiz.com.

1383-5416/01/02/$8.00

2. Density method

In the topology optimization method, the entire design region where the materials can exist is divided into small cells. These cells can coincide with the finite element method (FEM) mesh. The material properties are allowed to vary continuously in each cell. There are several techniques for associating design variables with these material properties.

The homogenization method considers microscopic composite materials and calculates the 'average' material property using asymptotic expansion. The homogenization method gives a systematic and reasonable way to interpret intermediate densities. However, the homogenization method requires computational efforts for calculating the material constants for each finite element. It also creates a large number of design variables for modeling the size and orientation of the microstructure.

The density method was first proposed in order to overcome these difficulties from the engineer's point of view. In the density method, the material density of each finite element is the design variable, and the relationship between the material density and the material property is defined by a simple mathematical equation.

To use the density method with the finite element method, the normalized density of material is assumed to be identical over one finite element. The normalized density within each finite element controls the electromagnetic properties of the material such as permeability, permittivity, etc. For example, the relationship between the permeability and the normalized density can be written as:

$$\mu = \mu_0(1 + (\mu_r - 1)p^n) \quad (0 \leqslant p \leqslant 1,\ 2 \leqslant n \leqslant 4) \tag{1}$$

where p is the normalized density, μ_r is the relative permeability of the real material, and n is the exponent which defines the relationship between material property and normalized density. The normalized density takes a value between 0 and 1.

When $n > 1$, the intermediate density is penalized and as a result the normalized density is forced to either 0 or 1. As n grows larger, the intermediate density is more heavily penalized, thus eliminating 'gray' materials more efficiently. But larger value of n can also limit the design space at the same time. Usually, the value of n can be chosen between 2 and 4 considering this trade-off [3]. The density method is widely used because it is easy to implement, has less design variables than homogenization method, and the intermediate density can be eliminated efficiently by adjusting the value of n.

3. Design sensitivity analysis: The mutual energy method

In the topology optimization, the number of design variables grows dramatically as the number of meshes increases to get reasonable resolution. Thus, quick and efficient calculation of the sensitivity information considering these characteristics is essential. The method using the mutual energy concept is presented in this section.

The two mutual energy terms in the magnetostatic system can be defined as follows:

$$L_1\left(\vec{A}_2\right) = \int_V \vec{A}_2 \cdot \vec{J}_1 dv, \quad L_2\left(\vec{A}_1\right) = \int_V \vec{A}_1 \cdot \vec{J}_2 dv \tag{2}$$

where $\vec{A}_1$ is vector potential by source current density $\vec{J}_1$, and $\vec{A}_2$ is associated with $\vec{J}_2$.

The total energy functional in the magnetostatic system can be written in terms of mutual energy as,

$$\chi = \frac{1}{2}a\left(\vec{A}_1, \vec{A}_2\right) - \frac{1}{2}L_1\left(\vec{A}_2\right) - \frac{1}{2}L_2\left(\vec{A}_1\right) \tag{3}$$

where $a\left(\vec{A}_1, \vec{A}_2\right) = \int_V \left(\nabla \times \vec{A}_1\right) \cdot \nu \left(\nabla \times \vec{A}_2\right) dv$ is an energy bilinear term and ν is the reluctivity of the material.

For the vector potential distribution $\vec{A}_1$ and $\vec{A}_2$ which satisfy Maxwell's equation, the following relations are obtained from the principle of virtual work,

$$a\left(\vec{A}_1, \vec{A}_2'\right) = L_1\left(\vec{A}_2'\right) \text{ for all } \vec{A}_2' \text{ in admissible space } U \tag{4}$$

$$a\left(\vec{A}_2, \vec{A}_1'\right) = L_2\left(\vec{A}_1'\right) \text{ for all } \vec{A}_1' \text{ in admissible space } U \tag{5}$$

so that the energy functional χ satisfies at equilibrium,

$$\chi = \frac{1}{2}a\left(\vec{A}_1, \vec{A}_2\right) - \frac{1}{2}L_1\left(\vec{A}_2\right) - \frac{1}{2}L_2\left(\vec{A}_1\right) = -\frac{1}{2}L_2\left(\vec{A}_1\right) \tag{6}$$

Now let p be the design variable, and consider χ as a function of $\vec{A}_1, \vec{A}_2$ and p. The first variation of χ at equilibrium yields,

$$\begin{aligned}\delta\chi\left(\vec{A}_1, \vec{A}_2, p\right) &= \frac{1}{2}a\left(\vec{A}_1, \frac{\partial \vec{A}_2}{\partial p}\delta p + \delta \vec{A}_2\right) - \frac{1}{2}L_1\left(\frac{\partial \vec{A}_2}{\partial p}\delta p + \delta \vec{A}_2\right) \\ &+ \frac{1}{2}a\left(\vec{A}_2, \frac{\partial \vec{A}_1}{\partial p}\delta p + \delta \vec{A}_1\right) - \frac{1}{2}L_2\left(\frac{\partial \vec{A}_1}{\partial p}\delta p + \delta \vec{A}_1\right) \\ &+ \frac{1}{2}\int_V \left(\nabla \times \vec{A}_1\right) \cdot \frac{\partial \nu}{\partial p}\left(\nabla \times \vec{A}_2\right) dv\delta p\end{aligned} \tag{7}$$

The first four terms in Eq. (7) can be cancelled by the principle of virtual work Eqs (4) and (5). Therefore, the sensitivity of χ with respect to p is given by,

$$\frac{\partial \chi}{\partial p} = \frac{1}{2}\int_V \left(\nabla \times \vec{A}_1\right) \cdot \frac{\partial \nu}{\partial p}\left(\nabla \times \vec{A}_2\right) dv \tag{8}$$

From Eqs (6) and (8), the sensitivity of mutual energy term $L_2\left(\vec{A}_1\right)$ as to the design variable p is obtained as follows,

$$\frac{\partial L_2\left(\vec{A}_1\right)}{\partial p} = -\int_V \left(\nabla \times \vec{A}_1\right) \cdot \frac{\partial \nu}{\partial p}\left(\nabla \times \vec{A}_2\right) dv \tag{9}$$

4. Application to the switched reluctance motor design

Figure 1 shows the simple schematic picture of the SRM with 6 stator poles and stator coils. The rotor region is made of magnetic material and marked by gray-colored central region in Fig. 1, and it is designed by the topology optimization method. Note that the initial rotor has no salient pole shape because the basic layout or topology of the design region is not needed in the topology optimization.

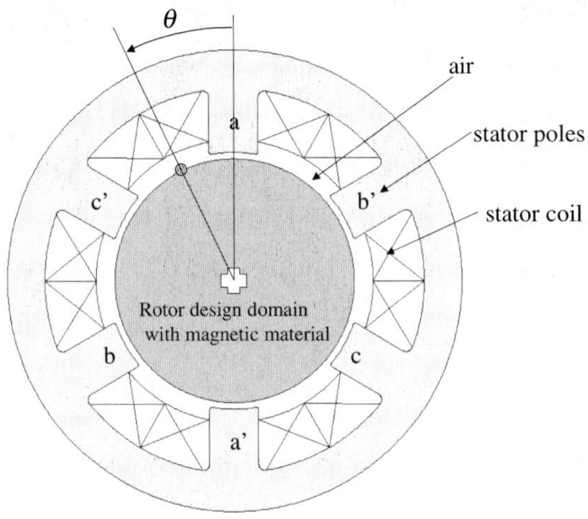

Fig. 1. The schematic picture of the initial SRM model.

In the design region, the finite elements take the shape of squares to regularize the design space. The normalized density of material is assumed to be identical over one square to further regularize the design space. The rotation angle of the rotor is 5 degrees, and the total number of rotations is 36 to represent a total rotation of 180 degrees. Because of the symmetry, the rotation of 180 degrees is enough for optimization. The desired system energy profile is given by a cosine function of the rotor rotation angle θ when the current is applied at phase a and a' in Fig. 1.

The objective function F in this problem can be defined as:

$$F = \frac{1}{2}\sum_{i=1}^{nr}[W_i - W_{i0}]^2 \tag{10}$$

where nr is the number of rotation angles at which the system energy is calculated, W_i is the calculated system energy at ith rotation angle, and W_{i0} is the desired system energy at ith rotation angle according to the energy profile.

The sensitivity of this objective function can be written as:

$$\frac{dF}{dp_j} = \sum_{i=1}^{nr}[W_i - W_{i0}]\frac{dW_i}{dp_j} \quad (j = 1, 2, \dots, np) \tag{11}$$

where np is the number of the design variables and p_j is the jth design variable. Using the chain rule, dW_i/dp_j term in Eq. (11) can be represented as:

$$\frac{dW_i}{dp_j} = \frac{\partial W_i}{\partial p_j} + \frac{\partial W_i}{\partial \vec{A}_i}\frac{\partial \vec{A}_i}{\partial p_j} \tag{12}$$

where $\vec{A}_i$ is the magnetic vector potential at ith rotor location. From the extremum condition for energy functional, $\partial W_i/\partial \vec{A}_i$ term in Eq. (12) is zero for any potential distribution satisfying Maxwell's equations, so that Eq. (12) becomes:

$$\frac{dW_i}{dp_j} = \frac{\partial W_i}{\partial p_j} \tag{13}$$

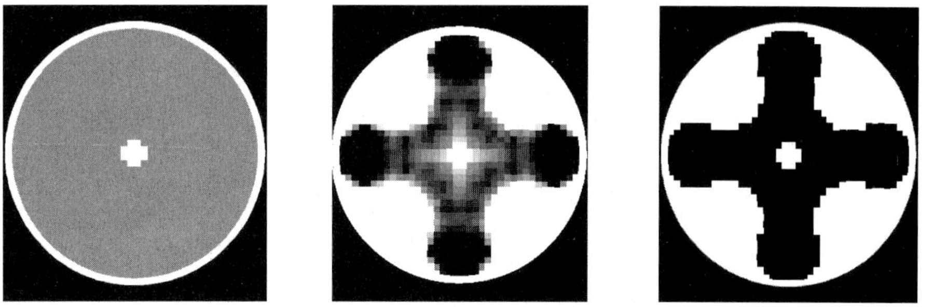

(a) 0 iterations (b) 5 iterations (c) 50 iterations

Fig. 2. The material distribution of SRM rotor.

This is the sensitivity of the self-energy of the system with respect to the design variable. Following the formulations in Section 3 and noting that W_i is the self-energy instead of the mutual energy, Eq. (13) can be rewritten as:

$$\frac{dW_i}{dp_j} = \frac{1}{2}\int_V \left(\nabla \times \vec{A}_i\right) \cdot \frac{\partial \nu}{\partial p_j} \left(\nabla \times \vec{A}_i\right) dv \tag{14}$$

The $\partial \nu / \partial p_j$ term in Eq. (14) is given from density method as:

$$\frac{\partial \nu}{\partial p_j} = -n\mu_0(\mu_r - 1)p_j^{n-1} / \{\mu_0(1 + (\mu_r - 1)p_j^n)\}^2 \tag{15}$$

Substituting Eqs (14) and (15) into Eq. (11), the sensitivity of the objective function can be readily obtained.

The calculated sensitivity is used in the optimization algorithm to search the optimum design variables. The steepest descent method is adopted as an optimization algorithm because there is no constraint on the volume of the magnetic material used for rotor in this case. The change in the design variable vector $\Delta[p]$ is defined as $\Delta[p] = l\hat{m}$ where the normalized search direction $\hat{m}$ is given by,

$$\hat{m} = \left.\frac{-dF/d[p]}{\|dF/d[p]\|}\right|_i \tag{16}$$

and the step size l is searched by linearizing the system energy W_i and given as:

$$\begin{aligned} l_{op} &= -\left[\sum_{i=1}^{nr} k_i(W_i - W_{i0})\right] \Big/ \left[\sum_{i=1}^{nr} k_i^2\right], \\ k_i &= \frac{dW_i}{d[p]} \cdot \hat{m} = -\left\|\frac{dF}{d[p]}\right\|^{-1} \frac{dW_i}{d[p]} \cdot \sum_{j=1}^{nr} [W_j - W_{j0}] \frac{dW_j}{d[p]} \end{aligned} \tag{17}$$

The gray-scale pictures of the rotor at the various iteration numbers are shown in Fig. 2. It can be seen from the results that the topology optimization method can 'find' the optimal shape and topology of the salient-pole rotor without any prior knowledge of the basic rotor layout. However, the shape of the final device is somewhat coarse because of the limited number of the design variables in the topology optimization. The system energy as the function of rotation angle is shown in Fig. 3. It is clear that the energy function converges to the desired cosine function from initial flat line.

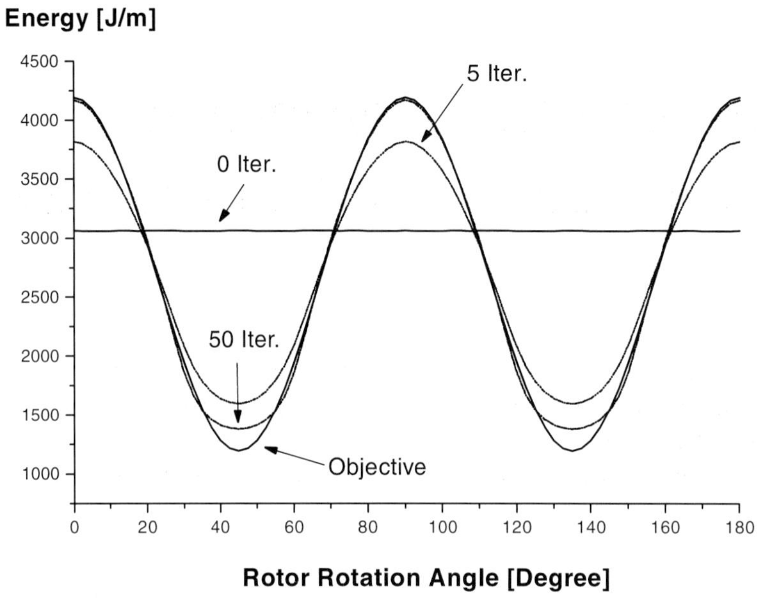

Fig. 3. The system energy variation with iteration number.

5. Conclusion

The topology optimization was applied to the optimal design of SRM rotor in this paper. The optimal topology and shape of the salient-pole rotor could be found according to the system energy profile without prior knowledge of the basic rotor layout. However, because of the limited resolution, the topology optimization should be used to find the basic topology and layout of the device as a pre-processor module of the optimum design tool. And the conventional shape design method can be used based on the result of the topology optimization to determine the detailed shape and size of the device.

References

[1] D.N. Dyck and D.A. Lowther, Automated design of magnetic devices by optimizing material distribution, *IEEE Trans. Magn.* **32**(3) (May 1996), 1188–1193.

[2] J. Yoo, N. Kikuchi and J.L. Volakis, Structural optimization in magnetic devices by the homogenization design method, *IEEE Trans. Magn.* **36**(3) (May 2000), 574–580.

[3] R.J. Yang, Multidiscipline topology optimization, *Computer & Structures* **63**(6) (1997), 1205–1212.

International Journal of Applied Electromagnetics and Mechanics 13 (2001/2002) 427–430
IOS Press

Modeling acoustic resonance in high-pressure discharge lamp arc chambers

Attila Vágvölgyi[a,*], Ágoston Böröczki[b], Szabolcs Gyimóthy[a] and Imre Sebestyén[a]
[a]*Department of Electromagnetic Theory, Budapest University of Technology and Economics H-1521 Budapest, Egry J. u. 18, Hungary*
[b]*Automotive Engineering Department, GE Lighting Tungsram Rt., H-1340 Budapest, Váci út 77, Hungary*

Abstract. A numerical method was developed for modeling acoustic standing waves emerging in high-pressure discharge lamps. The applied model takes the inhomogeneous distribution of propagation velocity into account, which is caused by the temperature profile. Arc physics was not considered. Software was implemented based on the finite element method. In addition to the sole computation of waves in some practical arrangements, also the validity of the applied model was investigated.

1. Introduction

The headlamps of some contemporary car designs contain so called high intensity discharge (HID) lamps [1]. The physics of the arc burning here in the high-pressure gas is very complicated. The arc is excited by nearly symmetric square-waves to input nearly constant power. Nevertheless the polarity switches cause some pressure fluctuations – i.e. acoustic waves – in the gas. The waves are reflected back from the quartz glass wall of the arc chamber, and can result in standing waves.

This acoustic resonance can cause fast changes in the electrical and photometrical parameters of the arc; sometimes it leads to strong instabilities or even arc extinction. On the other hand, certain resonant modes have beneficial effects, as claimed by several patents: among others they can help in intermixing ingredients in the gas, they can straighten the arc against convection, or enforce it into a desired shape.

The discharge process is kept in its operating point by a control electronics. A sophisticated controller enhances the useful modes while prevents the others. However, controller design must be preceded by the analysis of possible frequencies and waveforms in the bulb.

Numerous research works were carried on to build accurate but efficient models of the phenomena [2, 3]. This is definitely not a simple eigenvalue-problem, since there is a strong and mutual coupling of arc, thermal profile and waveforms (eigenmodes) as explained in the following. First, the arc does not constrain waveforms directly – e.g. like a boundary condition – but rather through the temperature profile. The core temperature of the arc is approx. 4000 K while that of the wall is only about 1500 K, which implies an inhomogeneous wave-propagation velocity distribution throughout the bulb, thereby influencing mode shapes. On the other side, waveforms are primarily responsible for arc bending (besides

*Corresponding author: Attila Vágvölgyi, Tel.: +36 1 463 2915; Fax: +36 1 463 3189; E-mail: wigy@null.net.

1383-5416/01/02/$8.00

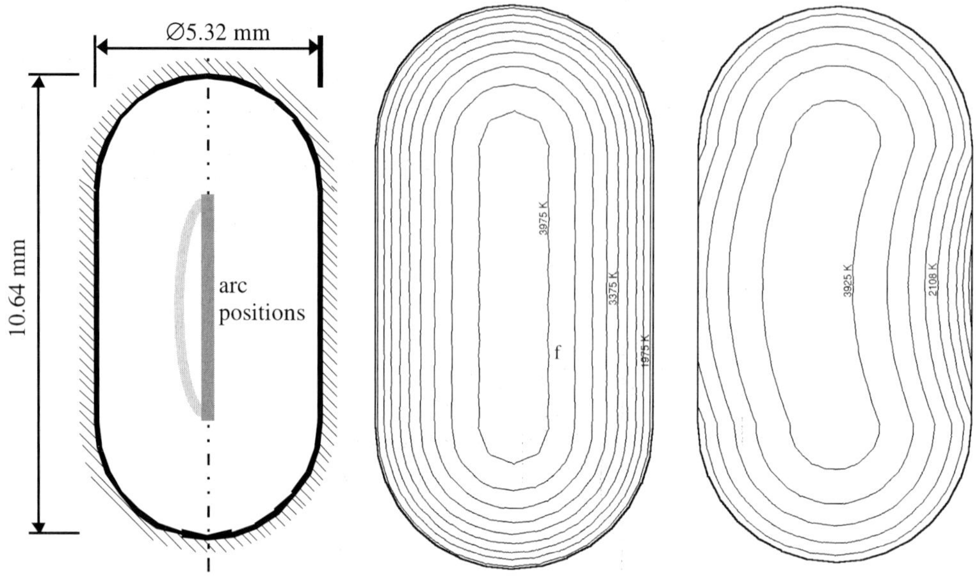

Fig. 1. Geometry of the arc chamber (left) and temperature profiles for cases "b" and "c".

the effect of convection), i.e. for the thermal profile. The significance of each effect still has not been clarified soothingly.

The aim of our work was to predict acoustic resonance frequencies and corresponding mode shapes as accurately as possible but without modeling the arc itself. In addition to this we investigated whether (or to what extent) we must use more complex models including arc physics.

2. Model and solving method

Neglecting shear stresses in the gas we can formulate the scalar Helmholz-equation for the pressure φ as follows:

$$\Delta\varphi(x,y,z) = -(2\pi f \cdot v(x,y,z))^2 \cdot \varphi(x,y,z);\ \Delta = \frac{\partial^2}{\partial x^2} + \frac{\partial^2}{\partial y^2} + \frac{\partial^2}{\partial z^2} \tag{1}$$

Note that sound-velocity v is a function of the location. Resonance occurs on frequencies, where the equation becomes singular. We can calculate the resonant frequencies from the eigenvalues of the equation; the forms of standing waves originate from the eigenfunctions.

The main problem is that if a given temperature profile is provided for the calculation, then only part of the calculated eigenmodes might produce such an arc shape, that matches the assumed profile. The selection of the "false" modes would need the modeling of the arc itself, which is out of the scope of our simple model. So we must posses some preliminary knowledge (from experiments in general) about how arc bending depends on the waveform.

The generic geometry of the bulb requires some numerical method; we have chosen the finite element method using linear tetrahedral elements with nodal variables. We transformed Eq. (1) into a linear system of equations by using the Galerkin-method.

We use self-implemented software on Linux platform for 3D mesh generation and for assembling the sparse global matrices of the generalized eigenvalue problem. The actual geometry and temperature

Table 1
Comparison of frequencies (given in MHz)

Index	001	002	003	004	005	010	011	012	013	020	021	022	100	101
case a	0.116	0.225	0.323	0.410	0.489	0.238	0.279	0.351	0.434	0.390	0.424	0.484	0.492	0.540
case b	0.116	0.225	0.323	0.413	0.512	0.243	0.288	0.363	0.450	0.408	0.447	0.510	0.452	0.479
case c	0.116	0.223	0.325	0.417	0.517	0.250	0.287	0.363	0.458	0.413	0.439	0.507	0.458	0.481

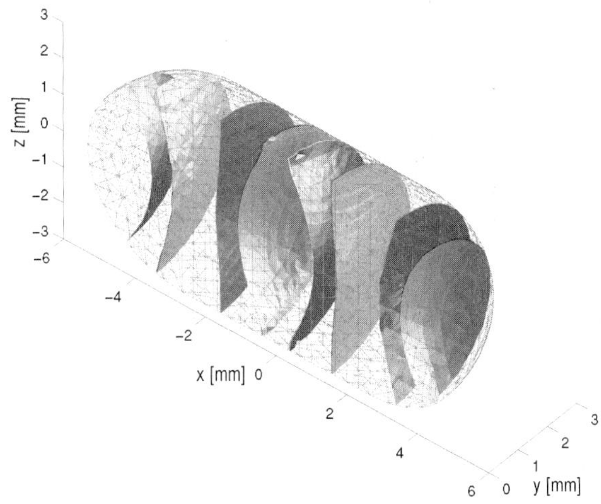

Fig. 2. Shape of mode #004, case “b”.

profile are part of the input of the software. Eigenvalues and eigenvectors are computed by Matlab. The resulting eigenmodes are loaded back onto the original mesh in order to preprocess them for the 3D plot utility of Matlab.

3. Results

Simple bulb model was created, which resembles to a real one. In order to be more realistic, material data, temperature values and radial dimension of the arrangement were taken from [2]. Figure 1 (left) shows the geometry of the bulb. We must use 3D model because in a 2D model only a small subset of modes and frequencies would appear.

Three fundamental cases were investigated and compared to each other. In the first one (case “a”) homogeneous distribution of the average temperature was supposed. In case “b” the temperature profile belongs to an arc in central position, while case “c” takes also a slight arc bending into account (see Fig. 1, middle and right).

Table 1 summarizes the calculated frequencies for the first 14 harmonics. We use here the “cylindrical” terminology for identifying and indexing the modes, i.e. we distinguish radial, azimuthal and longitudinal components, as far as it is imaginable. Note that in case “c” there must be some “false” modes mentioned above, but we could not make a selection. The highest relative difference between cases “a” and “b” is 12.7%; between “b” and “c” is 2.6%; and between “a” and “c” is 12.4%. Although frequencies do not show significant difference in some modes (e.g. mode #004, cases “b” and “c”), the shapes of the corresponding eigenmodes do (see Figs 2–3, explanation follows)!

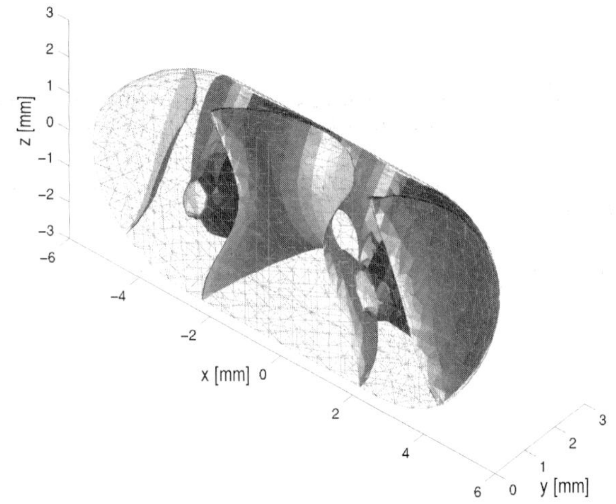

Fig. 3. Shape of mode #004, case "c".

The eigenfunctions were normalized in the range $[-1, +1]$. Each figure shows two sets of surfaces cut out from the eigenmode at a specified positive and negative level (± 0.2), together with the enclosing wire-frame of the half-bulb. We can say that these surfaces separate regions with high pressure-oscillations from regions with fast convections. Note that it is the gradient of the pressure that is proportional to the local displacement of the material in stationary case. The experimental validation of model predictions is underway.

4. Conclusion

The comparison of mode shapes – rather than only frequencies, as the usual method – underlines the significance of the effect of temperature profile in the investigated phenomenon. However, with our method one must – to some extent – know the waveform in advance (the one he is just about to compute) in order to set the estimated inhomogeneity in velocities. This paradox situation can only be overcome by using a more sophisticated model, which covers the coupling of acoustic phenomenon and arc physics.

Acknowledgments

This work was sponsored primarily by the GE Lighting Tungsram Rt. Hungary through the "Aschner Lipót" fellowship, and supported by the Hungarian Scientific Research Fund through grants F-030570 and T-035264.

References

[1] W. Elenbaas, *Light sources*, Macmillan, 1972.
[2] H. L. Witting, Acoustic resonances in cylindrical high-pressure arc discharges, *Journal of Applied Physics* **49**(5) (May 1978), 2680–2683.
[3] J.M. Davenport and R.J. Petti, Acoustic resonance phenomena in low wattage metal halide lamps, *Journal of the Illuminating Engineering Society* (April 1985), 633–642.

International Journal of Applied Electromagnetics and Mechanics 13 (2001/2002) 431–435
IOS Press

Application of a rational quartic triangular Bézier patch to field calculation by a boundary subdivision method

Shoji Hamada* and Tadasu Takuma
Department of Electrical Engineering, Kyoto University, Yoshida-Honmachi, Sakyo-ku, Kyoto, 606-8501, Japan

Abstract. A rational quartic triangular Bézier patch enables us to numerically represent an octant of a sphere with neither shape errors nor degeneration of nodes, when it adopts the control points and weights derived by Farin. This patch allows us to avoid undesirable influences caused by the misrepresentation of a spherical shape, if the objects to be treated involve an octant of a sphere in the field analysis by a boundary subdivision method. We have carried out a static field calculation for a spherical dielectric placed in a homogeneous applied field with a surface charge method by introducing a simple method that constructs arbitrary mesh patterns on this patch.

1. Introduction

A boundary subdivision method is widely used for field analysis, when the objects to be treated consist of solid regions with uniform and linear mediums. In particular, it is suited to the problems with an infinite or a semi-infinite domain, which are frequently encountered in practical calculations. The performance of this method depends on that of the numerical representations of the boundary surface shapes, the distribution of unknowns on the surfaces, and the boundary condition of the Laplace equation. A sphere or its octant frequently appears in basic and practical problems. The representation of a sphere is apparently quite easy if we consider only the accuracy of the shape. However, it is not easy when we also consider the representation of unknowns on it, because the singularity caused by the degeneration of nodes, such as the North and South poles in the polar coordinate or on a revolution of NURBS (Non-Uniform Rational B-Splines), disturbs the natural representation of them. A rational quartic triangular Bézier patch enables us to represent the octant of a sphere without such difficulties [1]. In this paper, we show a simple method that permits us to design arbitrary mesh patterns on this octant patch, and we demonstrate a static field calculation result with a thus generated mesh pattern.

*Corresponding author: Shoji Hamada, Tel.: +81 75 753 5289; Fax: +81 75 751 1576; E-mail: shamada@kuee.kyoto-u.ac.jp.

1383-5416/01/02/$8.00

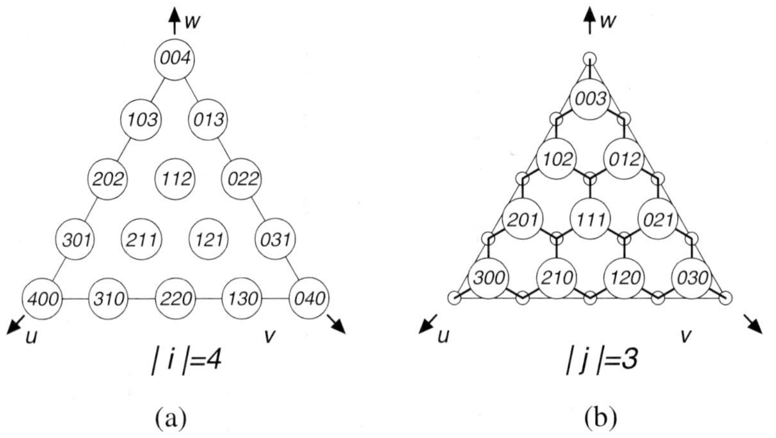

Fig. 1. Integer set i and j in the area coordinate (u, v, w).

2. Representation of an octant of a sphere

A rational quartic triangular Bézier patch is determined by fifteen pairs of control points $\mathbf{b_i}$ and weights $\omega_\mathbf{i}$ [1]. Here, $i = (\mathrm{i}, \mathrm{j}, \mathrm{k})$ means an integer set of i, j, k among the all (fifteen) combinations that satisfy both $|\mathbf{i}| = \mathrm{i} + \mathrm{j} + \mathrm{k} = 4$ and $4 \geqslant \mathrm{i}, \mathrm{j}, \mathrm{k} \geqslant 0$. A point on this patch $\mathbf{b}^4(\mathbf{u})$ is expressed by the following equation with a triangular area coordinate $\mathbf{u} = (\mathrm{u}, \mathrm{v}, \mathrm{w})$.

$$\mathbf{b}^4(\mathbf{u}) = \left(\sum_{|\mathbf{i}|=4} \omega_\mathbf{i} \mathbf{b_i} B_\mathbf{i}^4(\mathbf{u}) \right) \Big/ \left(\sum_{|\mathbf{i}|=4} \omega_\mathbf{i} B_\mathbf{i}^4(\mathbf{u}) \right) \tag{1}$$

where, $B_\mathbf{i}^4(\mathbf{u}) = (4!/(\mathrm{i}!\mathrm{j}!\mathrm{k}!))\mathrm{u}^\mathrm{i}\mathrm{v}^\mathrm{j}\mathrm{w}^\mathrm{k}$ stands for the quartic Bernstein polynomials. The correspondence between the integer sets $\mathbf{i}$ and the locations in $\mathbf{u}$ indicated by them are illustrated in Fig. 1(a), where the number in a circle means a list of i, j, and k. The control points $\mathbf{b_i}$ and weights $\omega_\mathbf{i}$ at these locations for the octant of a sphere are given in [1,2]. The directional derivative $D_\mathbf{d}\mathbf{b}^4(\mathbf{u})$ is expressed by the following equations.

$$D_\mathbf{d}\mathbf{b}^4(\mathbf{u}) = 4\left(\sum_{|\mathbf{j}|=3} (\omega\mathbf{b})_\mathbf{j}(\mathbf{d}) B_\mathbf{j}^3(\mathbf{u}) - \mathbf{b}^4(\mathbf{u}) \sum_{|\mathbf{j}|=3} (\omega_\mathbf{j}(\mathbf{d}) B_\mathbf{j}^3(\mathbf{u}) \right) \Big/ \left(\sum_{|\mathbf{i}|=4} \omega_\mathbf{i} B_\mathbf{i}^4(\mathbf{u}) \right)$$

$$(\omega\mathbf{b})_\mathbf{j}(\mathbf{d}) = \mathrm{d}(\omega_\mathbf{j+e1}\mathbf{b}_\mathbf{j+e1}) + \mathrm{e}(\omega_\mathbf{j+e2}\mathbf{b}_\mathbf{j+e2}) + \mathrm{f}(\omega_\mathbf{j+e3}\mathbf{b}_\mathbf{j+e3}) \tag{2}$$

$$\omega_\mathbf{j}(\mathbf{d}) = \mathrm{d}\omega_\mathbf{j+e1} + \mathrm{e}\omega_\mathbf{j+e2} + \mathrm{f}\omega_\mathbf{j+e3}$$

where, $\mathbf{j}$ means another integer set of i, j, k among the all (ten) combinations that satisfy both $|\mathbf{j}| = \mathrm{i} + \mathrm{j} + \mathrm{k} = 3$ and $3 \geqslant \mathrm{i}, \mathrm{j}, \mathrm{k} \geqslant 0$. Three integer sets $\mathbf{e1} = (1, 0, 0)$, $\mathbf{e2} = (0, 1, 0)$ and $\mathbf{e3} = (0, 0, 1)$ are used to relate $\mathbf{i}$ to $\mathbf{j}$. The influence of $\mathbf{i}$ on $\mathbf{j}$ is illustrated in Fig. 1(b), where the branches from $\boldsymbol{j}$ to three $\boldsymbol{i}$s correspond to the operation (addition) of $\mathbf{e1}, \mathbf{e2}$, and $\mathbf{e3}$ to $\mathbf{j}$. A direction in the area coordinate of the domain triangle is indicated by a real number set $\mathbf{d} = (\mathrm{d}, \mathrm{e}, \mathrm{f})$ that satisfies $\mathrm{d} + \mathrm{e} + \mathrm{f} = 0.0$. For example, a directional derivative with $\mathbf{d} = (1.0, 0.0, -1.0)$ corresponds to a partial derivative of $\boldsymbol{b}$ with respect to u along an isoparametric curve satisfying v=constant. The Jacobian determinant of the

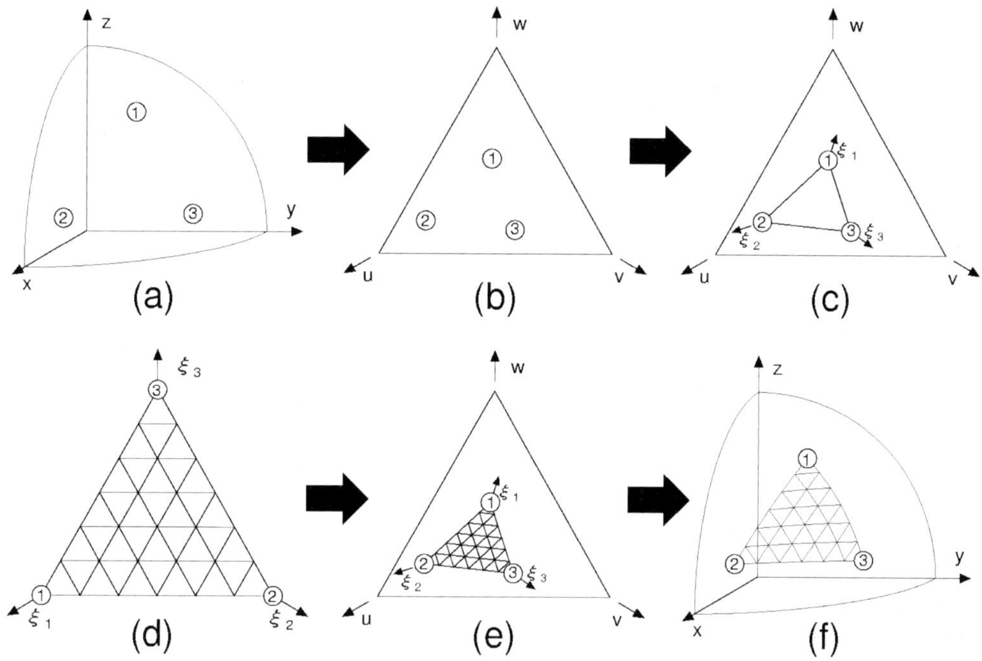

Fig. 2. Definition of shape representation function on the octant of a sphere.

mapping from $\mathbf{u}$ to $\mathbf{b}$ is obtained by two directional derivatives, for example, with $\mathbf{d} = (1.0, 0.0, -1.0)$ and $\mathbf{d} = (0.0, 1.0, -1.0)$. The Jacobian always has non-zero value, because the mapping has no degeneration.

3. Subdivision of an octant of a sphere

In order to represent arbitrary distribution of unknown variables on the octant patch, it is advantageous to subdivide the patch into relatively small surface patches, because the collection of piecewise representation with lower-order function makes a more flexible representation than that by single higher-order representation. We assume that the locations of all node-points of triangular sub-patches on the octant have already been designed in an arbitrary way without hole or gap on the whole surface. The procedure required here is to represent a triangular sub-patch that consists of three node-points (x_1, y_1, z_1), (x_2, y_2, z_2), and (x_3, y_3, z_3) on the octant in the $\mathbf{x} = (x, y, z)$ coordinate, as shown in Figs 2(a) and (f). First, we search for the corresponding node-points (u_1, v_1, w_1), (u_2, v_2, w_2), and (u_3, v_3, w_3) in the $\mathbf{u} = (u, v, w)$ coordinate, as shown in Fig. 2(b). This procedure requires numerical iterative calculation based on the following equations.

$$(\mathbf{x} - \mathbf{b}^4(\mathbf{u})) \cdot \left.\frac{\partial \mathbf{b}^4(\mathbf{u})}{\partial u}\right|_{v=\text{const.}} = 0, \quad (\mathbf{x} - \mathbf{b}^4(\mathbf{u})) \cdot \left.\frac{\partial \mathbf{b}^4(\mathbf{u})}{\partial v}\right|_{u=\text{const.}} = 0, \tag{3}$$

From thus obtained three points in the (u, v, w) coordinate, we define the three sides of a triangle as a linear connection of the nodes, and also define an area coordinate (ξ_1, ξ_2, ξ_3) for the linear sub-triangle, as shown in Fig. 2(c). In this step, the three points could be on a straight line in the (u, v, w) coordinate,

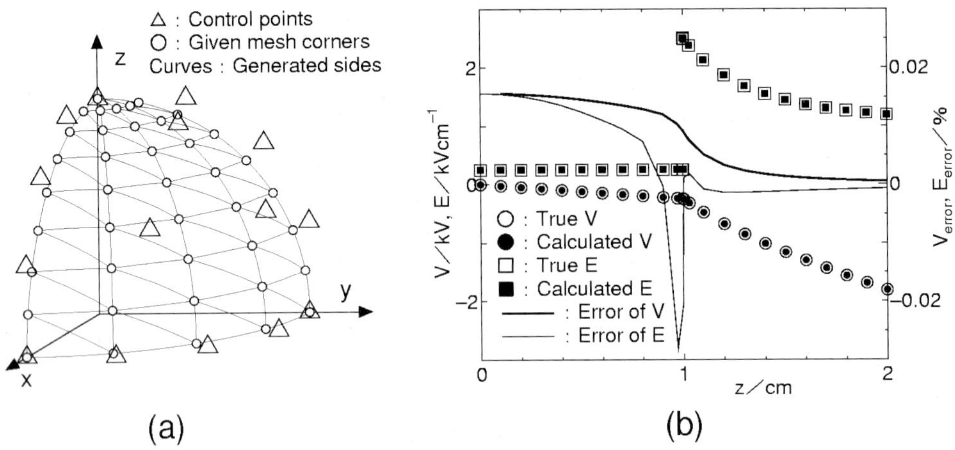

Fig. 3. Calculated potential and field on the z-axis.

and this procedure could fail in such cases. But we think it a so rare case that we could ignore such situations.

A linear mapping from (ξ_1, ξ_2, ξ_3) to $(\mathrm{u}, \mathrm{v}, \mathrm{w})$, as shown in Figs 2(d) and (e), is defined by a matrix equation, and a real number set $(\delta, \varepsilon, \zeta)$ that represents a direction in the (ξ_1, ξ_2, ξ_3) coordinate, which satisfies $\delta + \varepsilon + \zeta = 0.0$, is also mapped into $(\mathrm{d}, \mathrm{e}, \mathrm{f})$ in the $(\mathrm{u}, \mathrm{v}, \mathrm{w})$ coordinate by the following equations.

$$\begin{pmatrix} \mathrm{u} \\ \mathrm{v} \\ \mathrm{w} \end{pmatrix} = \begin{pmatrix} \mathrm{u}_1 & \mathrm{u}_2 & \mathrm{u}_3 \\ \mathrm{v}_1 & \mathrm{v}_2 & \mathrm{v}_3 \\ \mathrm{w}_1 & \mathrm{w}_2 & \mathrm{w}_3 \end{pmatrix} \begin{pmatrix} \xi_1 \\ \xi_2 \\ \xi_3 \end{pmatrix}, \quad \begin{pmatrix} \mathrm{d} \\ \mathrm{e} \\ \mathrm{f} \end{pmatrix} = \begin{pmatrix} \mathrm{u}_1 & \mathrm{u}_2 & \mathrm{u}_3 \\ \mathrm{v}_1 & \mathrm{v}_2 & \mathrm{v}_3 \\ \mathrm{w}_1 & \mathrm{w}_2 & \mathrm{w}_3 \end{pmatrix} \begin{pmatrix} \delta \\ \varepsilon \\ \zeta \end{pmatrix} \tag{4}$$

The mapping procedure from $(\mathrm{u}, \mathrm{v}, \mathrm{w})$ into $(\mathrm{x}, \mathrm{y}, \mathrm{z})$, as shown in Figs 2 (e) and (f), has been already given in the previous section, so that a shape representation function from the (ξ_1, ξ_2, ξ_3) coordinate to a subdivided triangle on the spherical surface is defined.

4. Calculated result

As a benchmark test, we have carried out an electric field calculation concerning a dielectric sphere in vacuum in the presence of a uniform applied field parallel to the z-axis. This example is frequently used for testing the accuracy of numerical field calculation methods, because the analytical solution is available for checking the calculated results. We use a surface charge method (SCM) for this calculation, which is known as one of boundary subdivision methods. SCM simulates the charge density distribution on the boundary surfaces. If the unknown distribution of the charge is determined numerically by satisfying the boundary equations on the surface, the electric field strength and potential are calculated by integrating the fractional effects of piecewise charge elements. The calculation of the magnetic field is possible based on the same principle.

The radius, the relative permittivity of the sphere, and the applied field strength are 1 cm, 10, and 1 kVcm^{-1}, respectively. We placed the corners of subdivided triangles on the octant so as to cover the whole surface according to the longitudinal and latitudinal division of the earth, as shown in Fig. 3(a).

The curved sides of the subdivided patches obtained by the proposed procedure are also illustrated in Fig. 3(a). The whole sphere is represented by eight octants, and is subdivided into 480 pieces of sub-triangles. The unknown charge density distribution for SCM is represented by the quadratic function with six degrees of freedom, which is defined on every subdivided triangular patch on the octants. The boundary condition of the field between two dielectrics (the dielectric sphere and the vacuum) is the continuity of the normal component of dielectric flux density through the boundary surface. This condition is formulated by the point matching method, that is, N equations are expressed at N boundary points for solving N unknown variables required for the quadratic representation of the charge density distribution.

Figure 3(b) shows the calculated electric field and potential on the z-axis. Theoretical values and relative errors are also plotted in the figure. The center of the sphere is placed at $z = 0.0$, and the North Pole of the sphere is placed at $z = 1.0$. The maximum error of the field is -0.0279%, which is observed in the vicinity of the North Pole. This error is caused mainly by the inaccuracy of the quadratic representation of the charge, which corresponds to the limitation of the quadratic representation concerning this test case.

5. Conclusion

A rational quartic triangular Bézier patch is useful for the shape representation if the considering objects include a sphere or its octants. A field calculation has been carried out with a surface charge method, by introducing a simple method that permits us to design arbitrary mesh patterns on this patch. This technique is also applicable to a boundary element method (BEM).

References

[1] G. Farin, *Curves and surfaces for computer aided geometric design,* (fourth edition), ACADEMIC PRESS, 1997.

[2] G. Farin, B. Piper and A.J. Worsey, The octant of a sphere as a non-degenerate triangular Bézier patch, *Computer Aided Geometric Design* **4** (1987), 329–332.

International Journal of Applied Electromagnetics and Mechanics 13 (2001/2002) 437–440
IOS Press

Active magnetic shielding with magneto-impedance sensor

Yasuo Okazaki*, Shunji Yanase and Noriko Sugimoto
Department of Electrical & Electronics Engineering, Gifu University, 1-1 Yanagido, Gifu, 501-1193, Japan

Abstract. Active shielding effect was examined in a negative feedback circuit system consisting of a magneto-impedance effective sensor, an amplifier and a canceling coil to compensate external magnetic field noise. The frequency dependent active shielding effectiveness depends on the phase difference between the input and output sensor signals in a loop, whch were less than 90 degree up to 20,000 Hz. An excellent characteristic of active shielding effectiveness of 48 dB was obtained for the external magnetic field noise from dc field to the frequency of 2,000 Hz.

1. Introduction

Active magnetic shielding is a technology to shield an open space using canceling coils like Helmholtz coils that diminish external magnetic field noise by generating opposite field to the external field. It has advantages over conventional magnetic shielding in respects of lightweight skeleton device and contemporary setting. The active shielding needs only Helmholtz coils to compensate magnetic field by electric current so that low cost of equipment and installation would be expected, while magnetic shielding conventionally performed by magnetic materials with high permeability are generally expensive and heavy. The active magnetic shielding has been applied to SEM or TEM electron microscope shielding, NMR or MASS shielding and magnetic room shielding for SQUID being combined with the conventional shielding [1]. It would be a key technology in active shielding to control current to a canceling coil. Kelha used a fluxgate magnetic sensor and PID controller and obtained about 20 dB improvement in shielding effectiveness at 1 Hz [2]. In the current active shielding technology, analog signal of external field detected by a magnetic sensor is once converted to digital signal by A/D converter. The equivalent current to the opposite external field is generated through digital signal processor and D/A converter [1]. This system requires time and cost in calculating and converting signals so that shielding frequency is limited up to less than 10 Hz of external field. Matsumoto proposed a method to feedback SQUID current directly to a canceling coil without digitized treatment and processing [3]. Active shielding effectiveness of 60 dB was obtained at 1 Hz but the shielding factor decreased rapidly at higher frequency.

Here a very simple active shielding system is presented using a negative feedback circuit with a magneto-impedance effect sensor to cancel magnetic field noise from dc to higher frequency with no digitized processing.

*Corresponding author: Yasuo Okazaki, Tel.: +81 58 293 2696; Fax: +81 58 293 2696; E-mail: okazaki@cc.gifu-u.ac.jp.

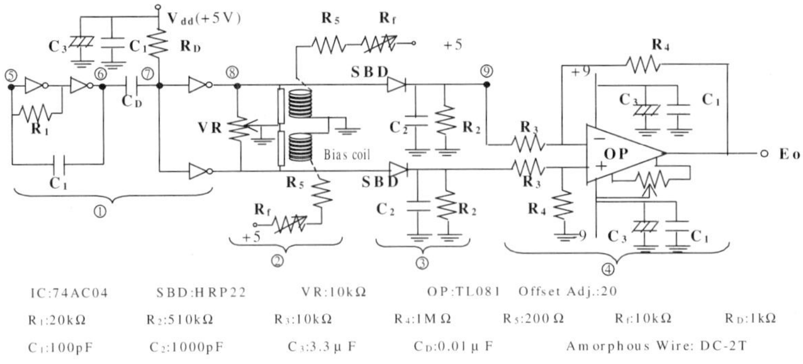

Fig. 1. Magneto-impedance effect sensor with two heads.

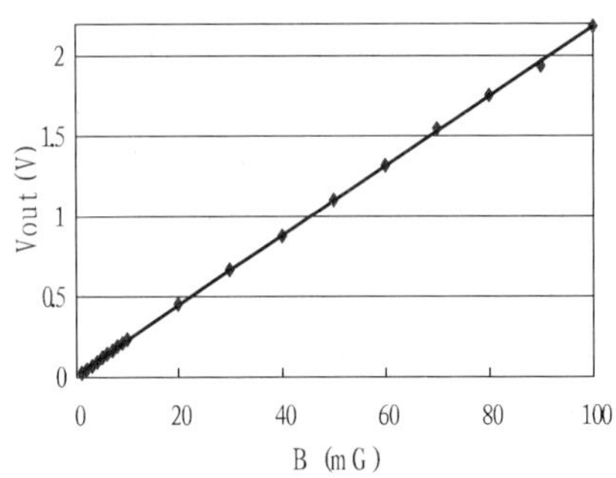

Fig. 2. External field vs. output voltage.

2. Experimental

A basic system for active shielding consists of a magnetic sensor, an amplifier and a canceling coil. A magnetic sensor applying twin heads differential typed magneto-impedance effect [4] has been produced by a couple of CoFeSiB amorphous drawn wires of 30 μm diameter and 3 mm long by Unitika Co. as shown in Fig. 1. The amorphous wires were connected consecutively with a variable resistor between them. Biased DC field was applied to each wire by 30 turn coils, which was controlled to obtain ultimate sensitivity as a device (Fig. 1②). The oscillating frequency from a CMOS multi-vibrator in Fig. 1① was 230 kHz. For a detection circuit Schottky barrier diodes for high frequency were used as shown in Fig. 1③. The cut-off frequency was practically 16 kHz. The output characteristics of the sensor are shown in Fig. 2 for the external field of 60 Hz. Output voltage showed good linear relationship against external ac magnetic field from 0 to 100 mG (10 μT) at 60 Hz. The frequency dependent output voltage was constant from 0 to 10,000 Hz as shown in Fig. 3.

Active shielding measurement was carried out by setting the magneto-impedance effect sensor in the center of canceling coils consisting of a pair of coils with 10 turns respectively. The coils were 450 mm square and 250 mm apart each other like a Helmholtz coil. The field distribution below 3% was obtained between the two coils at the center of 200 mm wide. In order to measure active shielding effectiveness, the canceling coil and the magnetic sensor were set in the center of a large square type Helmholtz coil of 1800 mm long and 980 mm apart. This square type Helmholtz coil has been proved to have parallel constant magnetic field in 500 mm wide cubic space and generated alternating magnetic field, Bex, from 0 to 100 mG (10 μT). Output signals from the sensor were amplified so as to generate equivalent current to the opposite external field.

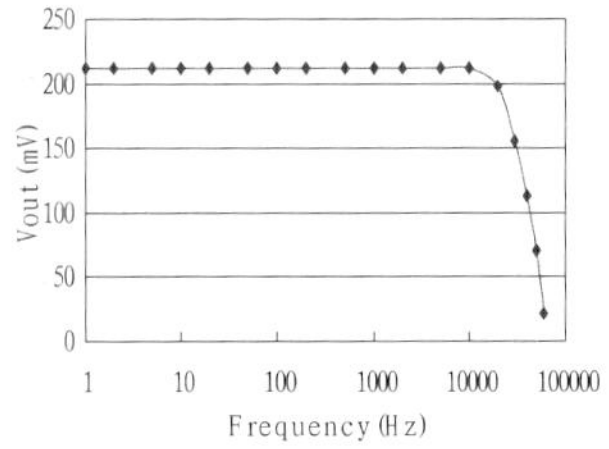

Fig. 3. Frequency vs. output voltage.

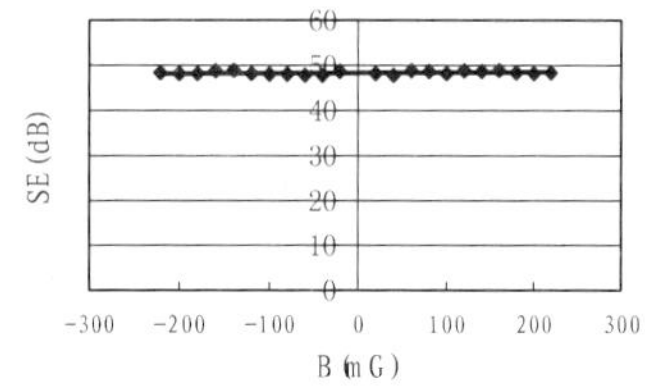

Fig. 4. Shielding effectiveness for dc field.

3. Results and discussion

Sinusoidal and non-sinusoidal magnetic field as external magnetic fields, Bex, were applied to measure active shielding effectiveness. The sinusoidal magnetic fields generated by the square type Helmholtz coil were from 1 mG to 80 mG at the frequency of 1–10,000 Hz. For example, the external of $Bex = 60$ mG at 50 Hz was shielded to $Bi = 0.24$ mG. For non-sinusoidal field, practical external fields from electric machines like motors were applied. As an example, $Bex = 12.5$ mG from a universal motor was applied and shielded to $Bi = 0.05$ mG. The shielding effectiveness, SE, calculated from SE (dB) $= 20$ log (Bex/Bi) was to be 48 dB for the both examples. Figures 4 and 5 summarize the results of the shielding effectiveness of dc (static) and ac (alternating) external magnetic field, $Bex = 0$ to 220 mG (22 μT) for dc and 0–80 mG (8 μT) at 50 Hz, respectively. Shielding effectiveness of 48 dB was obtained for dc magnetic field $Bex = -220$ to $+220$ mG and ac magnetic field $Bex = 30$ to 80 mG. At smaller ac field below $Bex = 30$ mG shielding effectiveness decreased slightly to 43 dB, which would be affected by residual noise in the system. The frequency dependent shielding effectiveness is shown in Fig. 6 for $Bex = 50$ mG.

The excellent shielding effectiveness at high frequency made possible to shield non-sinusoidal field. The decrease at higher frequency is due to the phase difference between input and output signals in the system. Figure 7 shows the frequency dependent phase difference. No phase difference was seen up to 10 kHz and that below 90-degree was observed till 20 kHz.

Active shielding effectiveness, SE, in the system can be defined by the factors of the canceling coil, the magneto-impedance effect sensor and the amplifier. These factors are conversion coefficients of input and output signals for each element. Here, let K, S and A be factors for the canceling coil, the magneto-impedance sensor and the amplifier, respectively, as shown in Fig. 8. The shielded field, Bi, and SE can be calculated in the negative feedback loop as follows:

$$Bi = V/S \tag{1}$$

$$Bex = Bi + VAK = Bi + ASKBi \tag{2}$$

$$Bi/Bex = 1/(1 + AKS) \tag{3}$$

$$SE = 20\log(Bex/Bi) = 20\log(1 + AKS) \tag{4}$$

V is output signal of magneto-impedance sensor.

As easily seen from Eq. (3), AKS should be large enough in order to shield Bex by the system. Since the factors in the experiments were $A = 180$–220 (V/V), $S = 24$ (V/G), $K = 0.045$ (G/V), and

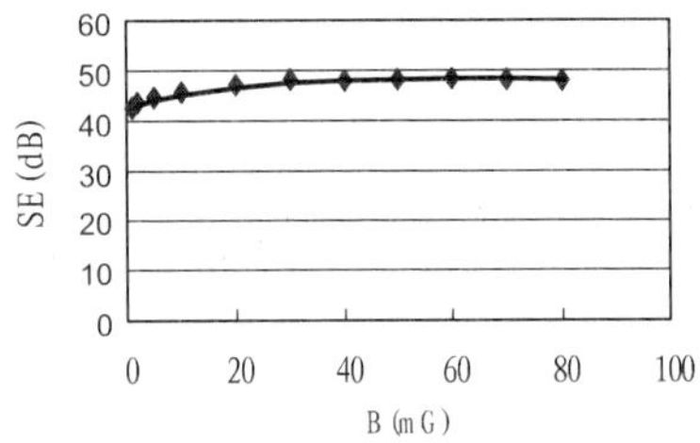

Fig. 5. Shielding effectiveness for ac field.

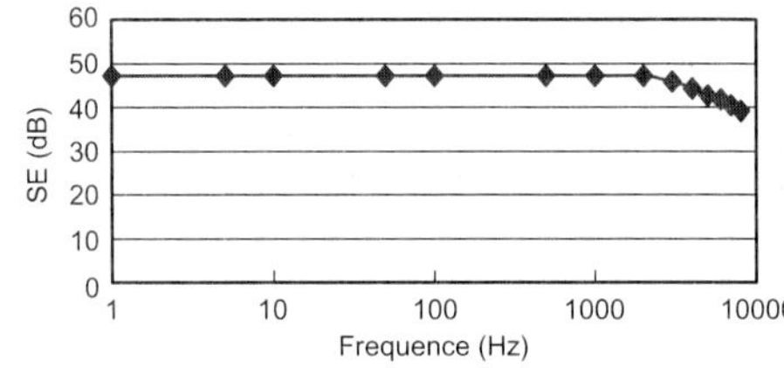

Fig. 6. Shielding effectiveness vs. frequency ($Bex = 50$ mG).

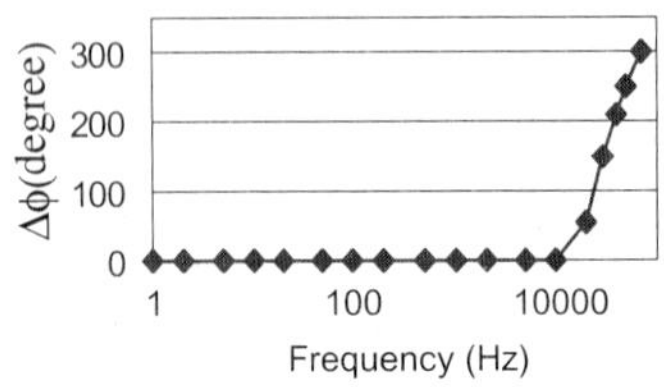

Fig. 7. Frequency dependent phase difference between input and output signals.

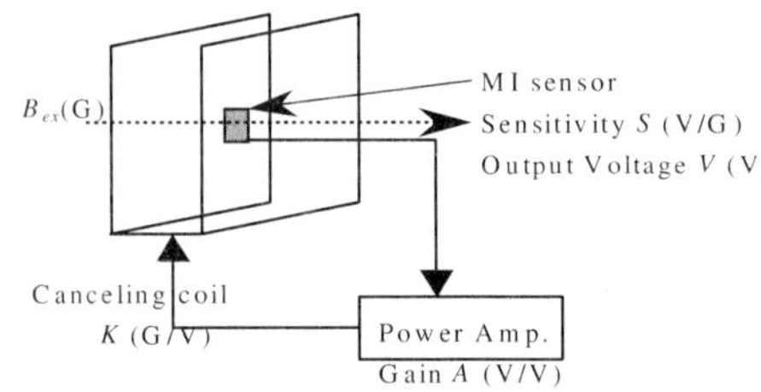

Fig. 8. Active shielding effectiveness measurement.

$AKS = 195$–238, the shielding effectiveness, SE is to be 46–48 dB. The SE value agrees quite well the experimental results shown in Figs 4 and 5.

We also measured active shielding effectiveness in the same system by substituting a flux gate magnetic sensor for the magneto-impedance effect sensor, as a comparison. The frequency dependent shielding effectiveness was 15 dB for 1 Hz and rapidly decreased till 5 Hz. The phase difference reached 90 degree at 8 Hz, while for the magneto-impedance effect sensor the phase difference was 90 degree at 20,000 Hz as shown in Fig. 7.

Here a one-dimensional canceling coil was applied but this system will be easily applicable to a three-dimensional canceling coil [4]. In this experiment the magneto-impedance sensor was placed in the center of the canceling coil. In practice the center would be occupied with an object to be shielded like an electron microscope. In that case, a magnetic sensor may be moved to off-center position according to field distribution of a canceling coil.

4. Conclusion

Active shielding was examined in a negative feedback system with a magneto-impedance effect sensor. The shielding effectiveness of 48 dB was obtained from static to 2,000 Hz magnetic fields. The excellent shielding effectiveness is clarified to come from the characteristic of frequency dependent phase difference of magneto-impedance sensor.

References

[1] Description of active field compensation MK3-MRS, VACUUMSHMELZE GMBH(1997).

[2] V. Kelha, J. Pukki, R. Peltonen, A. Penttinen, R. Ilmoniem and J. Heino, Design. Construction, and Performance of Large-Volume Magnetic shield, *IEEE Trans. Magn.* **18**(1) (January 1982), 260–269.

[3] K. Matsumoto, K. Fujioka and Y. Kuraoka, SQUID Based Active Shield, *MAG-92-84, IEE Jpn.* (1992), 77–89.

[4] T. Kanno and K. Mohri, MI magnetic Sensor Using a C-MOS IC and Amorphous Wire, *J. Magn. Soc. Jpn.* **21**(4-2) (1997), 645–648.

International Journal of Applied Electromagnetics and Mechanics 13 (2001/2002) 441–446
IOS Press

Magnetic shielding on low-frequency

Tamás Barbarics* and Péter Kis
Department of Electromagnetic Theory,, Budapest University of Technology and Economics, H-1521, Budapest, Hungary

Abstract. This paper deals with the examination of low frequency magnetic shielding. In the designe of magnetic shielding, plate can play a significant role in emphasizing the importance of shielding. The magnetic field was examined including the shielding sheets with various magnetic properties. The field of the linear magnetic sheet was calculated first, and the non-linear properties of shielding plate were taken into account with a non-linear $H - M$ characteristic in the second case.

1. Introduction

Invisible environmental pollution is generated by the power current bus-bars. These bus-bars are usually located in the basement of buildings. The sensitive electronical devices can be disturbed on the ground floor because of the magnetic field. This undesired effect of magnetic and electric field can be reduced with the appropriately designed shielding plate. The efficiency of shielding depends on the shape of the shielding plate as well.

Aim of the calculation is to determine the magnetic field at a given height above the shielding sheet, as shown in Fig. 1. The model contains two conductors and a magnetic shielding sheet. The excitation in the conductors is selected as a 50 Hz sinusoidal alternating current. The calculations are completed for $I = 10$ A effective value current in the loop.

The shielding of bus-bars is realized with a plate above the conductors. The current of the bus-bars induced eddy currents in the shielding plate, which has opposite direction to the original field reducing the magnetic field above the plate. The applied method is the global variational method, based on the minimalization process of a suitable chosen energy related functional.

The problem can be divided into two parts. First a non-ferromagnetic shielding plate is examined as a linear problem. Then the ferromagnetic property is taken into consideration, with a nonlinear model [3].

2. Governing equations

The magnetic vector potential is introduced for the determination of the minimum value of the energy related functional, where $\boldsymbol{A}$ is the magnetic vector potential.

$$\boldsymbol{B}(x,y,t) = \nabla \times \boldsymbol{A}(x,y,t), \tag{1}$$

*Corresponding author: Tamás Barbarics, Tel.: +36 1 463 3190; Fax: +36 1 463 3189; E-mail: barbarics@evtsz.bme.hu.

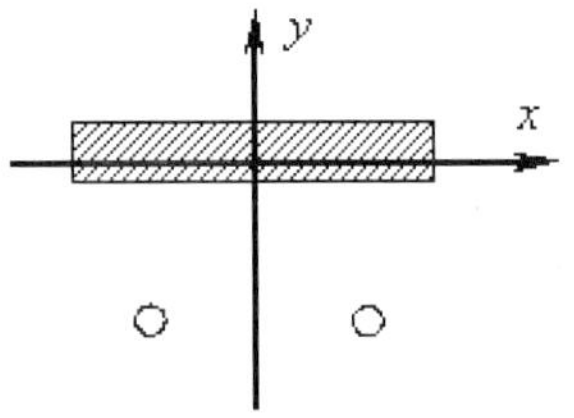

Fig. 1. The cross-section of the examined region.

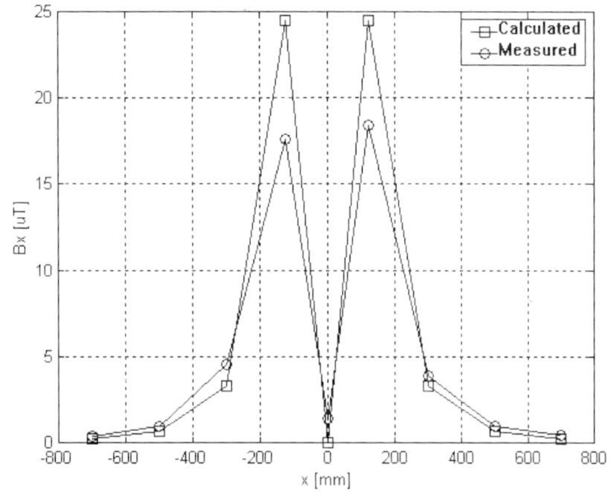

Fig. 2. The magnetic induction in the x direction at 75 mm above the bus-bars.

$$W(\boldsymbol{A}) = \frac{1}{2}\int_{\Omega}[\nabla \times \boldsymbol{A} \cdot \nabla \times \boldsymbol{A}]d\Omega - \int_{\Omega_s}\left[\mu_0 \boldsymbol{M} \cdot \nabla \times \boldsymbol{A} - \mu_0 \sigma \boldsymbol{A}\dot{\boldsymbol{A}}\right] d\Omega. \tag{2}$$

The extreme value of the energy related functional according to the Ritz method can be approximated by a space dependent entire function set with time dependent unknown parameters

$$\boldsymbol{A}(x,y,t) = \boldsymbol{A}_s(x,y,t) + \boldsymbol{A}_m(x,y,t) = \boldsymbol{A}_s(x,y,t) + \sum_{k=1}^{n} \mathbf{F}_k(x,y)a_k(t), \tag{3}$$

where $\boldsymbol{A}_S$ denotes the source field from the current excitation and the $\boldsymbol{A}_m$ is the unknown component, which modified the field, $\mathbf{F}_k(x,y)$ are the elements of the basis vector constructed by the help of the $\boldsymbol{R}$-functions, and $a_k(t)$ are the elements of the unknonws parameter, which have to determined for each time step during the solution.

To determine the magnetic field in the investigated region, the solution of the ordinary differential equation with nonlinear excitation for the unknown parameters $\mathbf{a}(t)$ can be formed as

$$\mathbf{K}_1\mathbf{a} + \mu_0\sigma\mathbf{K}_2\dot{\mathbf{a}} = \mu_0\mathbf{G}_1 - \mu_0\sigma\mathbf{G}_2, \tag{4}$$

where the elements of the matrixes can be constructed as:

$$\begin{aligned}\mathbf{K}_1[k,q] &= \int \nabla \times F_k \cdot \nabla \times F_q d\Omega,\ \mathbf{K}_2[k,q] = \int F_k \cdot F_q d\Omega \\ \mathbf{G}_1[k] &= \int \nabla \times F_k \cdot M d\Omega,\ \mathbf{G}_2[k] = \int F_k \cdot \dot{\mathbf{A}}_s d\Omega\end{aligned} \tag{5}$$

In the first case, when the linear shielding is examined, then $\mathbf{G}_1$ is selected to be zero. The solution of the nonlinear ordinary differential equation is evaluated by Δt_n time steps, for every time moment $t_{n+1} = t_n + \Delta t_n$. To represent the time derivatives an iteration scheme can be introduced

$$\mathbf{a}_{n+1} = \mathbf{a}_n + \Delta t_n\left(\theta\dot{\mathbf{a}}_{n+1}(1-\theta)\dot{\mathbf{a}}_n\right). \tag{6}$$

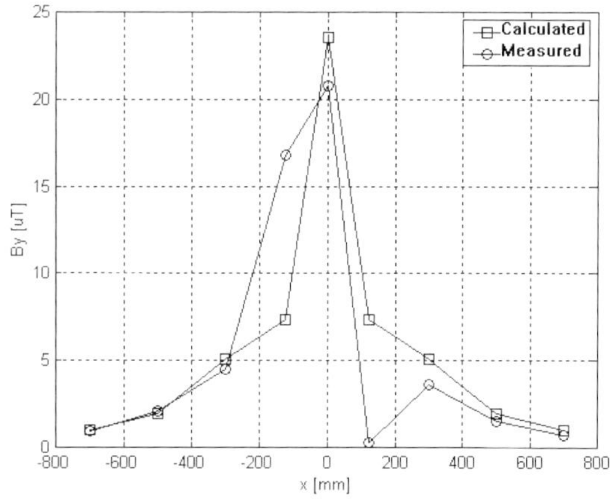

Fig. 3. The magnetic induction in the y direction at 75 mm above the bus-bars.

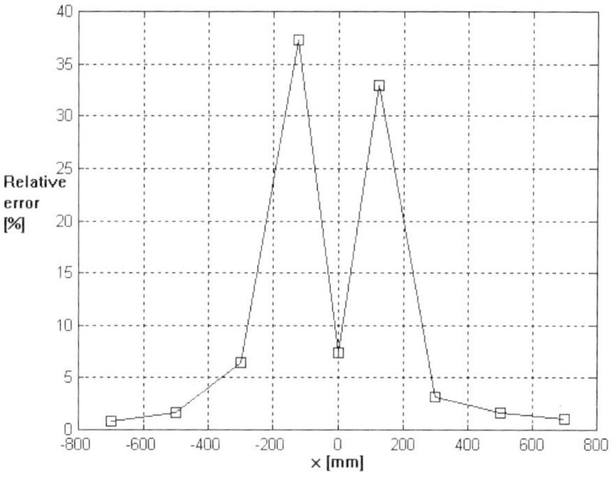

Fig. 4. The relative error in the x direction at 75 mm above the bus-bars.

The above iteration algorithm for marching in time can be applied for the calculation $\mathbf{a}_n$ parameters. The value of θ was selected to be 0.5, so the Crank-Nicolson scheme was applied

$$(\theta\Delta t\mathbf{K}_1 + \mu_0\sigma\mathbf{K}_2)\mathbf{a}_{n+1} = (1-\theta)\Delta t\mathbf{G}_n + \theta\Delta t\mathbf{G}_{n+1}(\mu_0\sigma\mathbf{K}_2 - (1-\theta)\Delta t\mathbf{K}_1)\mathbf{a}_n. \quad (7)$$

The nonlinear system of the above equation can be solved by recomputing the magnetization in each iteration step from previously approximated point of the nonlinear characteristics [1,2].

3. Numerical results

First the arrangement is examined without the magnetic shielding, and in this case the numerical result can be compared with measured ones. The results can be seen in Fig. 2 for the magnetic induction in the

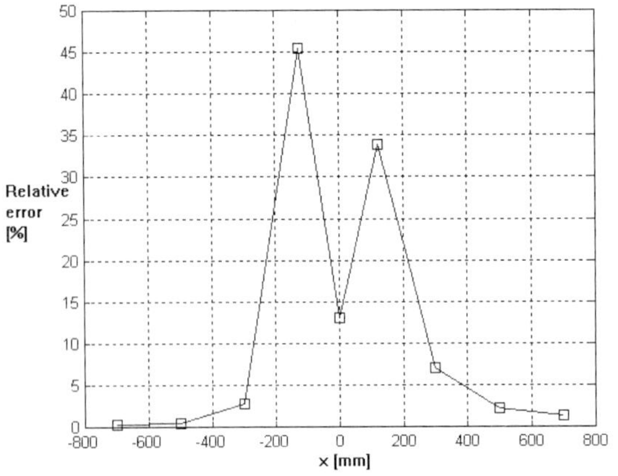

Fig. 5. The relative error in the x direction at 75 mm above the bus-bars.

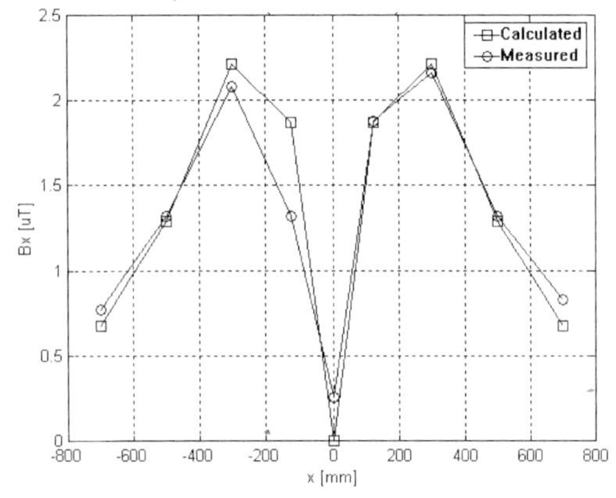

Fig. 6. The magnetic induction in the x direction at 355 mm above the bus-bars.

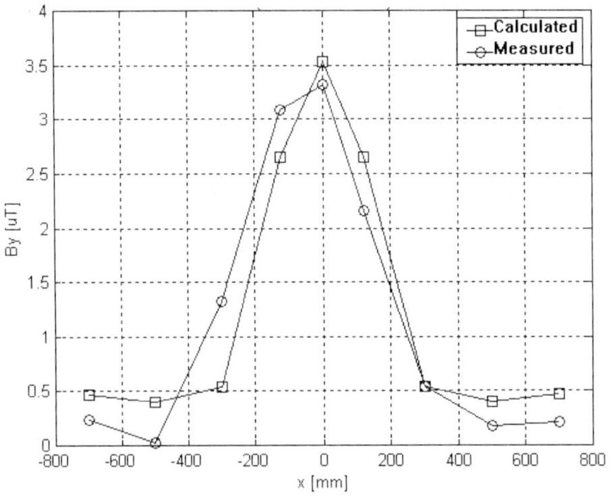

Fig. 7. The magnetic induction in the y direction at 355 mm above the bus-bars.

x direction and in Fig. 3 for the y direction at 75 mm above the plate, while the relativ error in percent between the results in Figs 4 and 5 respectively.

As it can be seen from the figures the shape of the functions are the same, the errors accure only above the bus-bars. The arrangement is a symmetrical one, and thus the numerical results are symmetrical as well, however the measured values are not symmetrical, and this makes the difference between the numerical and measured results. The same effect can be followed in Figs 6 and 7, where the magnetic induction is plotted in x and y direction respectively at 355 mm above the bus-bars.

The result of the determination of non-ferromagnetic shielding are plotted in Figs 8 and 9. The figures show the effect of the eddy current. The effective value and the time function of the magnetic flux density can be seen in the figures. There are only x and y component because the direction of the excitation current is z.

The transient effect can be well observed in Fig. 9, where the sinusoidal function is deformed by the

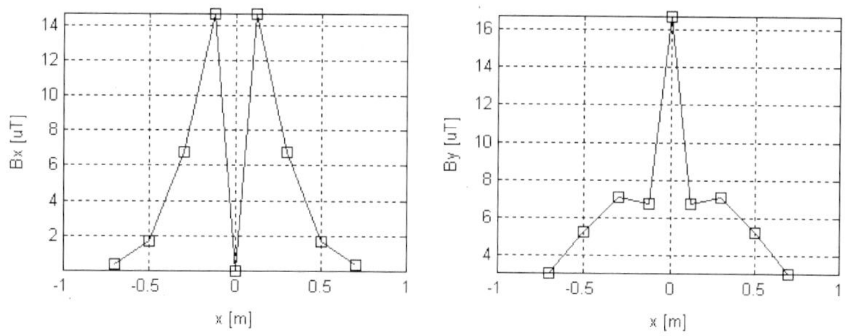

Fig. 8. The effective value of B_x and B_y at 75 mm above the shielding plate.

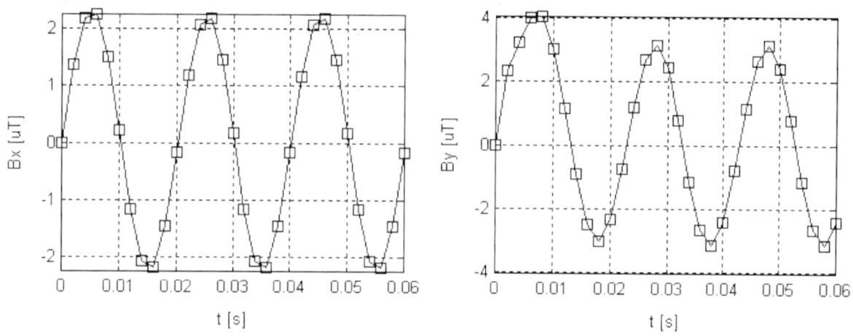

Fig. 9. The time function of B_x and B_y above the sheet at 355 mm.

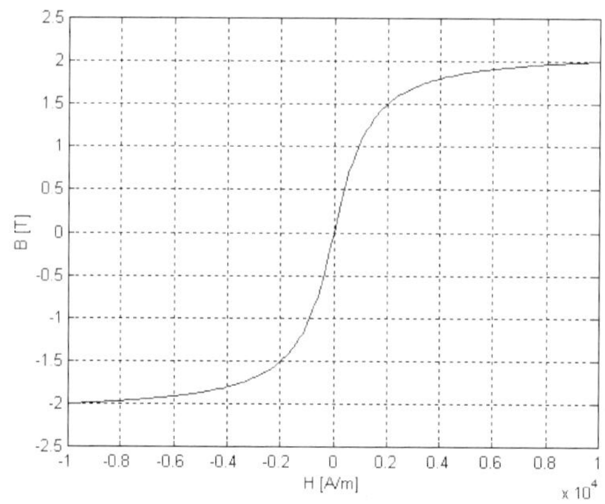

Fig. 10. The non-linear material characteristics.

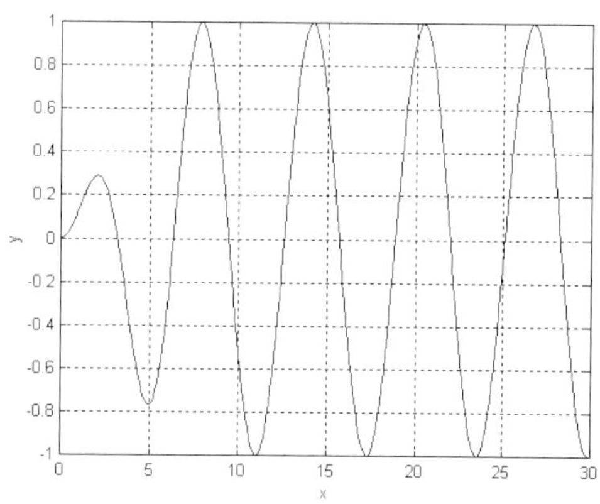

Fig. 11. The applied excitation.

eddy currents.

In the second case the magnetic shielding sheet is taken into account as a nonlinear one. The considered nonlinear property of the material is plotted in Fig. 10, and the time function of the excitation can be seen in Fig. 11. The effective value of the magnetic induction in x and y direction with the non-linear shielding can be checked in Fig. 12, and the transients in the time function in Fig. 13. With the knowledge of the source component of the vector potential, approximation in the field analysis can be evaluated with respect to the material parameters.

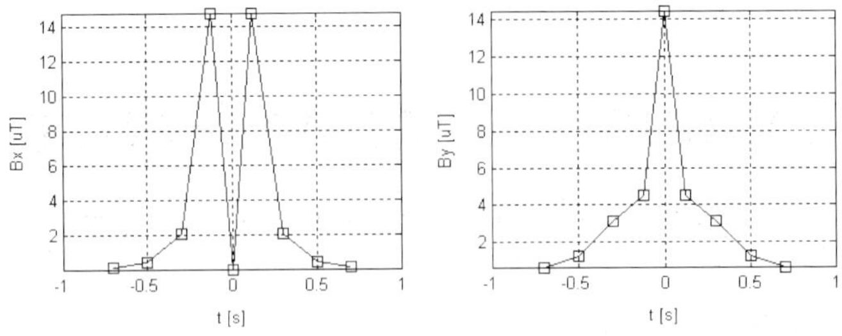

Fig. 12. The effective value of the B_x and B_y with the nonlinear magnetic sheet.

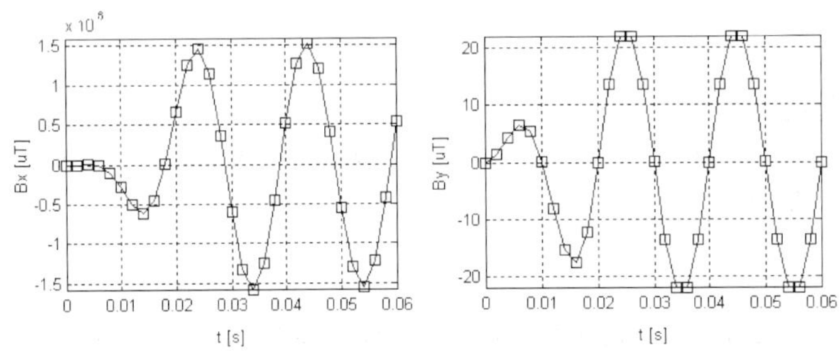

Fig. 13. The time function of the B_x and B_y with the nonlinear magnetic sheet.

4. Conclusion

The examination of low frequency magnetic shielding above bus-bars was introduced. The magnetic field was examined for a 2 dimensional model, including the shielding sheets with various magnetic properties. The magnetic field of the arrangement was examined and compared with measured data. The effects of the linear and non-linear shielding were determined, taking into account the non-linear $H - M$ characteristic of the material.

Acknowledgements

The author would like to thank for the financial support to the OTKA Foundation.

References

[1] A. Ivanyi, Electromagnetic field calculation for Magnetic Shielding with ferromagnetic material, Budapest, 1998.

[2] T. Barbarics, A. Kost, D. Lederer and P. Kis, Electromagnetic field calculation for Magnetic Shielding with ferromagnetic material, *IEEE Trans. on Magn.* **36**(4) (2000), 986–989.

[3] W.M. Frix, G.G. Karady and K. Maracas, Reduction of power frequency magnetic field by the use of nonferrous metal shields, Proceedings of the 27th Annual north American power Symposium 2–3, 1995, pp. 98–105.

International Journal of Applied Electromagnetics and Mechanics 13 (2001/2002) 447–450
IOS Press

Finite element analysis of deflection yoke using a novel technique for analyzing current density distribution

Chang-Hwan Im* and Hyun-Kyo Jung
School of Electrical Engineering (#040), Seoul National University, San 56-1, Shillim-dong, Kwanak-gu, 151-742, Seoul, Korea

Abstract. A recently developed method for the accurate calculation of current density distribution is firstly applied to a practical case with the 3-D finite element analysis. The deflection yoke (DY) for the CRT system is adopted for the simulation. The simulated results are verified by comparing with an experimental data.

Keywords: Finite element method, deflection yoke, current density distribution analysis

1. Introduction

When analyzing electric devices using the 3-D FEA with magnetic vector potential formulation, the coil parts are usually modeled as solid forms. Therefore, to calculate the magnetic field, the current density distribution of the coil should be analyzed. For a simple shaped coil as cylindrical or planar type, the processes are usually very easy. However, when analyzing the electric devices with very complex coil system such as deflection yoke (DY), some special methods should be used. When conventional current vector potential (T) method in [2] is applied to the curved shape coil, the calculated current flows more inside the coil than outside the coil as shown in Fig. 1(b). For avoiding this problem, coil is divided into several sections and current density is calculated separately for each section [3]. However, because the criterion for the division of the sections is not determined in the method, the method can be applied to only simple shaped coils. Recently, a novel method for resolving the above problem is proposed by the authors [1]. However the method has not been applied to a practical model especially with the finite element method (FEM). In this paper, the proposed method is applied to the finite element analysis of the deflection yoke. From the comparison with the experimental result, the accuracy of the method will be verified.

*Corresponding author: Chang-Hwan Im, Tel.: +82 2 880 7262; Fax: +82 2 878 1452; E-mail: ichich2@snu.ac.kr.

1383-5416/01/02/$8.00

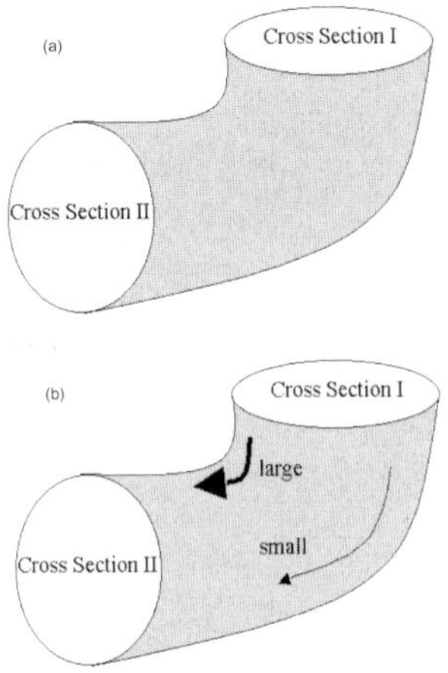

Fig. 1. Example for current density analysis.

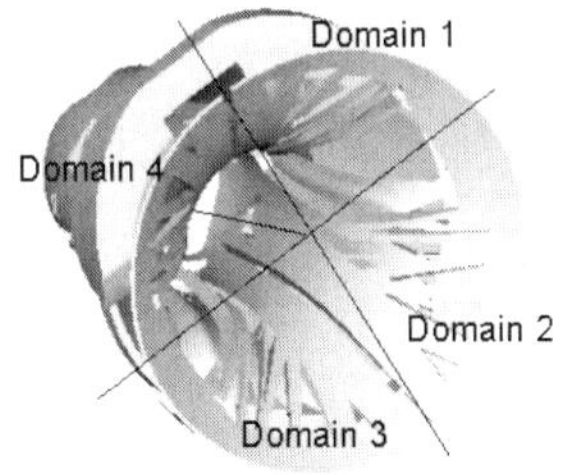

Fig. 2. Analysis model.

2. Technique for current density distribution analysis [1]

In case that the cross-sectional areas perpendicular to the current directions are uniform, the magnitudes of current densities at every point have constant values as well. Therefore, the current density distribution can be easily found just by multiplying the constant magnitude of the current density by the directional vectors of electric field intensity (E) in the following processes. However, for the case of DY, whose coil is highly pressed during the manufacturing processes with maintaining its thickness, the cross-sectional areas perpendicular to the current direction becomes non-uniform [5]. In such a case, a special technique should be applied. The procedure for the technique is as follows [1]:

2.1. Electric field analysis

Constant potentials are imposed as a boundary condition to the cross Sections I and II in the Fig. 1(a). By the electrostatic analysis using the finite element method, electric field intensity vector $\boldsymbol{E}$ can be evaluated for every element. The unit directional vectors of $\boldsymbol{E}$ are found by dividing $\boldsymbol{E}$ with its own magnitude.

2.2. Calculation of discrete current densities

Between two potentials imposed to the cross Sections I and II, n potentials are determined ($\phi_i, i = 1, 2, \ldots, n$) and equipotential surfaces for each potential are found. Then the areas of each equipotential surface, $S_i (i = 1, 2, \ldots, n)$ are calculated. For each equipotential surface, magnitudes of current density J_i are calculated by

$$J_i = \frac{N \cdot i}{S_i} \tag{1}$$

2.3. Calculation of current density for each element

Average potential values are evaluated for all elements. If the average potential of an element (e), $\phi^{(e)}$, is between ϕ_i and ϕ_{i+1}, the magnitude of current density $J^{(e)}$ is calculated by

$$J^{(e)} = \frac{\phi - \phi_i}{\phi_{i+1} - \phi_i} J_{i+1} + \frac{\phi_{i+1} - \phi}{\phi_{i+1} - \phi_i} J_i. \tag{2}$$

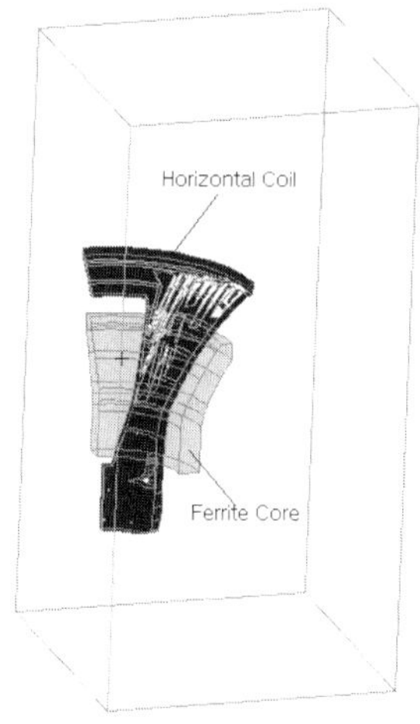

Fig. 3. Structures of modeled deflection yoke (DY).

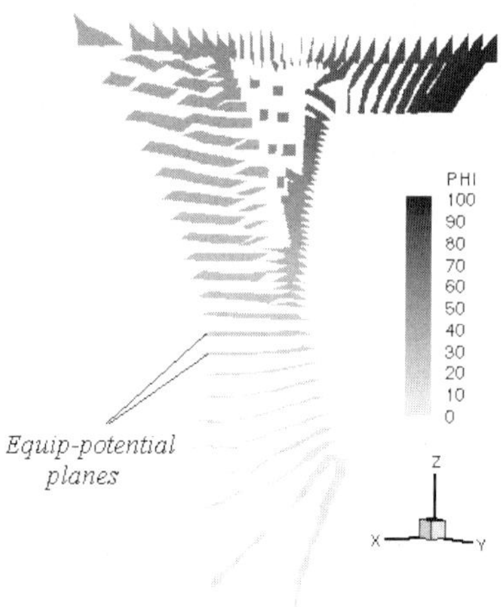

Fig. 4. Equipotential planes for calculating current density distribution.

Fig. 5. Manufactured deflection yoke for experiments.

The current density vectors of all elements can be found by multiplying the unit directional vectors of $\boldsymbol{E}$ and the magnitudes of current densities calculated by Eq. (2).

3. Simulation and result

The analysis model is a deflection yoke (DY) that is illustrated in Fig. 2. To make an accurate modeling data, line element model is generated by a very accurate winding modeler and it is transformed as a solid model [5]. In the complex geometry of the deflection yoke, only two parts are considered – horizontal coils and ferrite core. As seen in Fig. 2, the domains $1 \sim 4$ are geometrically symmetric with respect to two orthogonal planes, but the analysis region cannot be reduced to a half of the whole model even if the periodic boundary condition is applied. To consider the full symmetric property of DY, a recent technique proposed by the authors is used [4]. By considering the geometrical symmetry of the DY, only

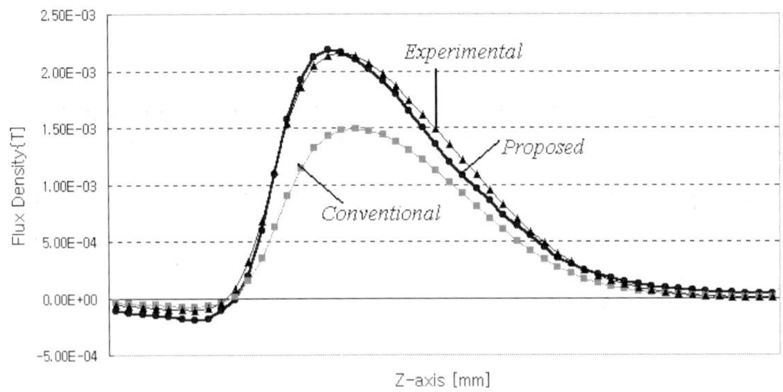

Fig. 6. y-component of flux density along z-axis.

a quarter model can be analyzed. Figure 3 shows a quarter of modeled geometry for the analysis. The generated meshes for the quarter model are mirrored with respect to two orthogonal planes, and make the whole system matrix equation for the DDM [4].

The current density of the deflection coil is calculated by using the conventional current vector potential (T) formulation in [2] and the novel method presented in this paper. Figure 4 shows the equipotential plane for applying the proposed method. In the analysis, total numbers of elements, nodes and edges are 41524, 7083, and 79278, respectively. The ferrite core is assumed to be linear as the magnetic field is not strong enough to saturate it. Figure 5 shows the manufactured deflection yoke model for the experiment, which consists of ferrite core and horizontal coil. Figure 6 shows the y component of flux density along z-axis. It can be found that the result by the proposed method is well agreed with the experimental data.

4. Conclusion

In this paper, a novel technique for analyzing the current density distribution of solidly modeled filamentary coil is applied to the deflection yoke. By the simulation, it is verified that the unbalance and miscalculation of current density in the solidly modeled coil can be resolved. By comparing the analysis results with experimental ones, accuracy and validity of the method are verified.

References

[1] Chang-Hwan Im, Hong-Kyu Kim and Hyun-Kyo Jung, Novel Technique for Current Density Distribution Analysis of Solidly Modeled Coil, *Proceedings of IEEE 13th Conference on the Computational of Electromagnetic Fields (Compumag 2001)*, Evian, 2001, pp. II–82.

[2] Y. Kawase, H. Mori and S. Ito, 3-D finite element analysis of electrodynamic repulsion forces in stationary electric contacts taking into account asymmetric shape (Invited), *IEEE Trans. Magn.* **33** (1997), 1994–1999.

[3] Y. Kawase, T. Mori, K. Hirata and Y. Mitsutake, Performance analysis of electromagnetic ID transmission system using 3-D finite element method, *IEEE Trans. Magn.* **34** (1998) 3206–3209.

[4] Chang-Hwan Im, Hong-Kyu Kim and Hyun-Kyo Jung, Efficient technique for 3-D edge element method considering geometrical symmetry, *Biennial IEEE Conference on Electromagnetic Field Computation – CEFC-2000,* Milwaukee, USA, June 4–7, 2000, pp. 189 (will be published in IEEE Trans. Magn. In September, 2001).

[5] Chang-Hwan Im, Hong-Kyu Kim and Hyun-Kyo Jung, Optimization of Coil Shape in Deflection Yoke Considering Practical Coil Winding Processes, *Proceedings of IEEE 13th Conference on the Computational of Electromagnetic Fields (Compumag 2001)*, Evian, 2001, pp. IV-143.

International Journal of Applied Electromagnetics and Mechanics 13 (2001/2002) 451–456
IOS Press

Performance characteristics improvement of a PM disk motor by using soft magnetic composite material

Goga Cvetkovski[a], Lidija Petkovska[a], Milan Cundev[a] and Sinclair Gair[b]
[a]*Sts. Cyril & Methodius University, Faculty of Electrical Engineering, P.O. Box 574, 1000 Skopje, Macedonia*
Tel.: +389 2 399 143; Fax: +389 2 364 262; E-mail: gogacvet@yahoo.co.uk
[b]*Napier University, School of Engineering, 219 Colinton Road, Edinburgh EH14 1DJ, UK*

Abstract. The paper presents a comparison of the magnetic field and magnetic characteristics of a PM disk motor with laminated magnetic stator open slot core, laminated magnetic stator closed slot core with soft magnetic composite and soft magnetic composite stator core. The data for this analysis is provided by a 2D FEM calculation of the magnetic field. After the proper modelling of the motor a calculation of the magnetic field for no load and current load is performed, for the three models. The results of the magnetic field distribution, air gap flux density distribution and the values of the magnetic and electric parameters for the three models are also presented.

1. Introduction

The object applied in the research is a brushless three phase synchronous permanent magnet disk motor fed by an inverter and battery or fuel cell. The motor is a double sided axial field motor with two stators with 36 slots and a centred rotor with 8 skewed neodium-iron-boron permanent magnets. The stator cores of the permanent magnet disk motor (PMDM) have been made of wound strip cores with open stator slots. In this type of motor the slots are not easily placed in a laminated core since the laminations need to be spirally wound and the slots become progressively farther apart as the laminations occupy a greater diameter. For the PMDM prototype the slots were cut into the previously wound stator cores, which as a procedure brings a big risk of damaging the cores. The reduction of the torque pulsation due to the stator slot openings is very important due to the fact that this type of motor is designed for a direct wheel drive in electric vehicles. It is evident that the research should be focused on reducing the torque pulsation and reducing the complexity of the stator core production.

2. Soft magnetic composite materials application

Powder metallurgy techniques have been in use for nearly three decades and their main application is as cores in the high frequency excitation range of magnetic circuit devices. In this, their relatively high permeability and low loss values at high frequencies, have provided significant cost and performance advantages over other materials, especially their isotropic characteristics along all three axes. Although

Fig. 1. Side view of the PMDM.

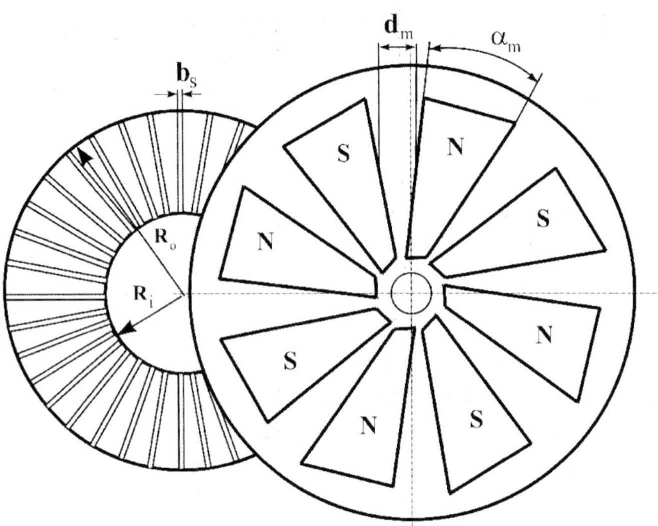

Fig. 2. Stator-rotor presentation.

the soft magnetic composite materials have low permeability and high iron loss at low frequencies they became very attractive for use as stator slot closure and for magnetic core parts or for manufacturing the stator cores [4] of permanent magnet motors in general, in the recent years. The low value of the soft magnetic composite (SMC) material permeability ($\leqslant 800$) in permanent magnet motor application does not affect the overall performance of the motor as it does in induction and reluctance motor applications. Other very important positive properties of the SMC material are the isotropic properties of the material and the rather good iron losses at high frequencies produced by the inverter performance. All of the above mentioned magnetic disadvantages can be overcame by the most important property of the SMC material which is the ability to be easily shaped in complex forms which can be after that surface finished

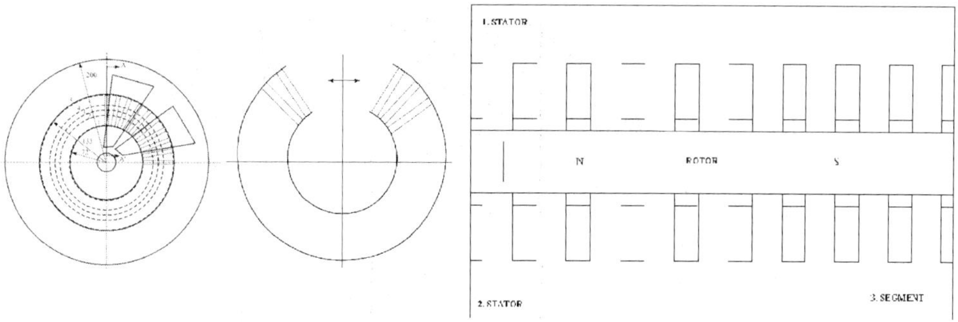

Fig. 3. Radial division of the motor into 5 segments and transformation into linear presentation.

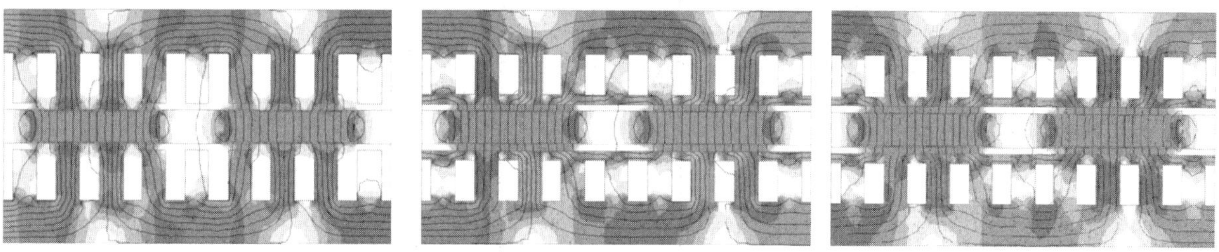

Fig. 4. Magnetic field distribution in the middle segment for the three models.

very smoothly. This property makes the SMC material very attractive for stator slots closure, that can be easily shaped or even for producing a complex shaped stator, as it is the disk stator of the permanent magnet disk motor. The implementation of SMC materials as stator slots closure for the laminated stators seems to be the first step towards improving the performance characteristics of the permanent magnet disk motor. In Fig. 1 a side view of the prototype PMDM, which is analysed, is presented, while in Fig. 2 one stator and one rotor are presented. In this paper three models of the PM disk motor are going to be analys ed. The first model is the disk motor with laminated magnetic stator open slot core, the second model is the laminated magnetic stator closed slot core with SMC material and the third model is the SMC material stator closed slot core.

3. Structural modelling of the PMDM for FEM analysis

In order to be able to analyse the three models of the PMDM with different stators accurately, a calculation of the magnetic field has to be performed for all three models. The quasi-3D method which is adopted for this analysis consists of a 2D FEM calculation of the magnetic field in a three dimensional radial domain of the axial field motor. For this purpose, a notional radial cut through the two stators and one rotor of the disk motor is performed and then opened out into linear form [3], as shown in Fig. 3. With this linear quasi three-dimensional model of the disk motor, which is divided into five segments, it is possible to model the skewing of the magnets and also to simulate the vertical displacement and rotation of the rotor [1].

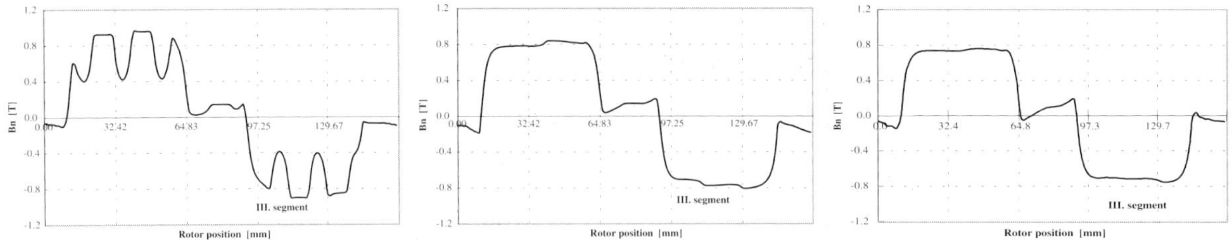

Fig. 5. Air gap flux density distribution in the middle segment for the three models.

Table 1
Electric and magnetic parameters of the three PMDM models

	Laminated open slot stators	Laminated closed slot stators	SMC stators with closed slots
Φ_g[Wb]	0.0021905	0.0022974	0.0020829
$B_{g\ avg}$ [T]	0.44334	0.47108	0.43016
$X_s[\Omega]$	3.71	4.5	3.12
P_{Fe} [W]	7.87704	8.6576	90.613

4. Magnetic field calculation and distribution

After the motor has been properly modelled, a 2D FEM magnetic field calculation of the three models of motors for each segment separately at no load and at rated current load ($I_{\text{ph}} = 8.723$ A) is performed. The distribution of the magnetic field and the air gap flux density for the motor with laminated open slot stators, laminated closed slot stators and the motor made of soft magnetic composite material stators for the middle segment at rated current is presented in Figs 4 and 5, respectively.

The distributions of the magnetic field and the air gap flux density for the middle segment show evident improvement of the closed slot stator model in relation to the open slot stator model and the soft magnetic composite stator core model. This has been accomplished only by inserting the soft magnetic composite material in the slots as a result of which the distribution of the flux density in the air gap is improved and the torque pulsation due to the slots openings is minimised. This improvement can be of great importance knowing that this type of motor is built for EV drive where the smoothness of the ride is of great importance. The advantage of the soft magnetic composite stators model is in the much easier manufacturing procedure in relation to the laminated iron strip stators.

5. Performance characteristics and parameters calculation

In order to be able to perform a proper comparison analysis of the three models a calculation of the magnetic and electric parameters of the three motor models is performed using the FEM data. The values of the air gap flux, air gap flux density, synchronous reactance and iron loss calculated from the FEM data, for comparison analysis, are presented in Table 1. Although the difference of the specific iron losses of the iron strip ($<$ 5 W/kg@50 Hz) and the SMC material ($<$ 10 W/kg@50 Hz) is quite large, in practice their overall iron losses difference is much smaller due to the inverter (harmonics) and manufacturing (punching, not perfect insulation, mechanical stress) influence on the laminated stators. The air gap flux, air gap flux density and synchronous reactance value change is a result of the magnetic circuit configuration change.

The torque-speed characteristics for the three models, based on the equivalent circuit model [2] and FEM data, are presented in Fig. 6, where the torque of the laminated open slots stators, laminated closed

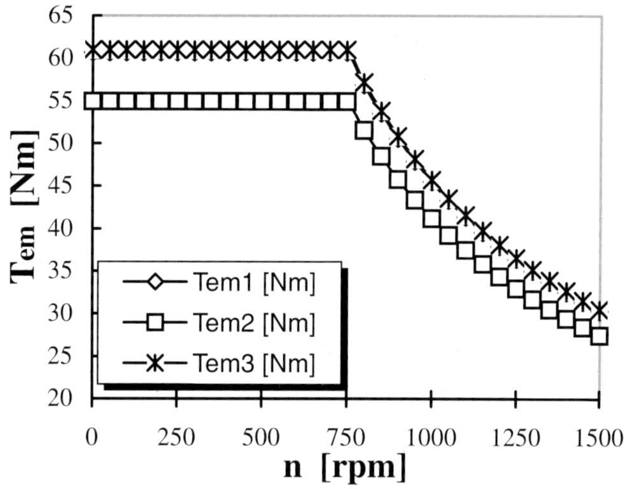

Fig. 6. Torque-speed characteristics comparison.

slots stators and SMC material closed slots stators are labelled as $T_{\mathrm{em}1}$, $T_{\mathrm{em}2}$ and T_{rmem3}, respectively. The torque is calculated by applying Eq. (1):

$$T_{\mathrm{em}} = \frac{m_{\mathrm{ph}} \cdot U_{\mathrm{ph}} \cdot E_{\mathrm{ph}}}{\omega_{\mathrm{m}} \cdot X_s} \cdot \sin\delta \qquad (1)$$

where:

m_{ph} – number of phases
U_{ph} – phase supply voltage (circuit model)
E_{ph} – induced voltage (circuit model)
X_{s} – synchronous reactance (FEM model)
ω_{m} – rotor speed
δ – load angle (circuit model)

6. Conclusion

In this paper a comparison of the magnetic field of a PM disk motor with soft magnetic composite and laminated magnetic stator core with opened and closed slots is presented. The data for this analysis is provided by a 2D FEM calculation of the magnetic field. For this purpose, in order to cover the whole 3D domain of the motor, a special modelling of the motor is performed. A calculation of the magnetic field for current load is performed, for the three models. The results of the magnetic field distribution and air gap flux density distribution for the three models are also presented. The values of the magnetic and electric parameters for the three models, based on the values from the FEM analysis, are calculated and presented. The results from this paper show that the SMC material is quite attractive for PMDM stator application and has a great manufacturing and performance characteristics advantage over the PMDM laminated stators.

References

[1] G. Cvetkovski, L. Petkovska, M. Cundev and S. Gair, Quasi 3D FEM in Function of an Optimisation Analysis of a PM Disk Motor, *Proceedings of ICEM2000* **4**(4) (2000), 1871–1875.
[2] S.B. Dewan, G.R. Slemon and A. Straughen, *Power Semiconductor Drives*, John Wiley & Sons, New York, 1984.
[3] S. Gair, A. Canova, J.F. Eastham and T. Betzer, *A New 2D FEM Analysis of a Disc Machine with Offset Rotor*, Proceedings of the 1996 Int. Conference on Power Electronics, Drives and Energy Systems, New Delhi, 1996, pp. 617–621.
[4] C. Mecrow, A.G. Jack and S.A. Evans, *Permanent Magnet Machine with Soft Magnetic Composite Stators*, Proceedings of ICEM'98, Istanbul, 1998, pp. 346–351.

International Journal of Applied Electromagnetics and Mechanics 13 (2001/2002) 457–463
IOS Press

Author Index Volume 13 (2001/2002)

Set Price 3 Volumes £605 Shankar

Vol 1 only £220